David D. Boehr (Ed.)

Viral Replication Complexes: Structures, Functions, Applications and Inhibitors

MDPI

This book is a reprint of the Special Issue that appeared in the online, open access journal, *Viruses* (ISSN 1999-4915) in 2015 (available at: http://www.mdpi.com/journal/viruses/special_issues/replication-complexes).

Guest Editor
David D. Boehr
The Pennsylvania State University
USA

Editorial Office
MDPI AG
Klybeckstrasse 64
Basel, Switzerland

Publisher
Shu-Kun Lin

Managing Editor
Delphine Guérin

1. Edition 2016

MDPI • Basel • Beijing • Wuhan • Barcelona

ISBN 978-3-03842-167-2 (Hbk)
ISBN 978-3-03842-168-9 (PDF)

Table of Contents

List of Contributors VII

About the Guest Editor XIII

Preface to "Viral Replication Complexes: Structures, Functions,
Applications and Inhibitors" XV

Chapter 1: Molecular Perspectives on Virus Replication Proteins

Cristina Ferrer-Orta, Diego Ferrero and Núria Verdaguer
RNA-Dependent RNA Polymerases of Picornaviruses: From the Structure to
Regulatory Mechanisms
Reprinted from: *Viruses* **2015**, 7(8), 4438–4460
http://www.mdpi.com/1999-4915/7/8/2829 3

Ester Sesmero and Ian F. Thorpe
Using the Hepatitis C Virus RNA-Dependent RNA Polymerase as a Model to
Understand Viral Polymerase Structure, Function and Dynamics
Reprinted from: *Viruses* **2015**, 7(7), 3974–3994
http://www.mdpi.com/1999-4915/7/7/2808 29

**Xinran Liu, Derek M. Musser, Cheri A. Lee, Xiaorong Yang, Jamie J. Arnold,
Craig E. Cameron and David D. Boehr**
Nucleobase but not Sugar Fidelity is Maintained in the Sabin I RNA-Dependent
RNA Polymerase
Reprinted from: *Viruses* **2015**, 7(10), 5571–5586
http://www.mdpi.com/1999-4915/7/10/2894 52

**Ibrahim M. Moustafa, David W. Gohara, Akira Uchida, Neela Yennawar and
Craig E. Cameron**
Conformational Ensemble of the Poliovirus 3CD Precursor Observed by MD
Simulations and Confirmed by SAXS: A Strategy to Expand the Viral Proteome?
Reprinted from: *Viruses* **2015**, 7(11), 5962–5986
http://www.mdpi.com/1999-4915/7/11/2919 74

Chapter 2: Coordinated Interactions of Viral Proteins and Nucleic Acids

Erin Noble, Michelle M. Spiering and Stephen J. Benkovic
Coordinated DNA Replication by the Bacteriophage T4 Replisome
Reprinted from: *Viruses* **2015**, 7(6), 3186–3200
http://www.mdpi.com/1999-4915/7/6/2766 111

Jason W. Rausch and Stuart F. J. Le Grice
HIV Rev Assembly on the Rev Response Element (RRE): A Structural Perspective
Reprinted from: *Viruses* **2015**, 7(6), 3053–3075
http://www.mdpi.com/1999-4915/7/6/2760 127

Bárbara Rojas-Araya, Théophile Ohlmann and Ricardo Soto-Rifo
Translational Control of the HIV Unspliced Genomic RNA
Reprinted from: *Viruses* **2015**, 7(8), 4326–4351
http://www.mdpi.com/1999-4915/7/8/2822 153

Valerie J. Klema, Radhakrishnan Padmanabhan and Kyung H. Choi
Flaviviral Replication Complex: Coordination between RNA Synthesis and 5'-RNA Capping
Reprinted from: *Viruses* **2015**, 7(8), 4640–4656
http://www.mdpi.com/1999-4915/7/8/2837 180

Alexandre Carpentier, Pierre-Yves Barez, Malik Hamaidia, Hélène Gazon, Alix de Brogniez, Srikanth Perike, Nicolas Gillet and Luc Willems
Modes of Human T Cell Leukemia Virus Type 1 Transmission, Replication and Persistence
Reprinted from: *Viruses* **2015**, 7(7), 3603–3624
http://www.mdpi.com/1999-4915/7/7/2793 198

Chapter 3: Interplay Between Virus and Host in Virus Replication

Santiago Guerrero, Julien Batisse, Camille Libre, Serena Bernacchi,
Roland Marquet and Jean-Christophe Paillart
HIV-1 Replication and the Cellular Eukaryotic Translation Apparatus
Reprinted from: *Viruses* **2015**, *7*(1), 199–218
http://www.mdpi.com/1999-4915/7/1/199 223

Luc Willems and Nicolas Albert Gillet
APOBEC3 Interference during Replication of Viral Genomes
Reprinted from: *Viruses* **2015**, *7*(6), 2999–3018
http://www.mdpi.com/1999-4915/7/6/2757 244

Jun Xiao, Jiang Deng, Liping Lv, Qiong Kang, Ping Ma, Fan Yan, Xin Song,
Bo Gao, Yanyu Zhang and Jinbo Xu
Hydrogen Peroxide Induce Human Cytomegalovirus Replication through the
Activation of p38-MAPK Signaling Pathway
Reprinted from: *Viruses* **2015**, *7*(6), 2816–2833
http://www.mdpi.com/1999-4915/7/6/2748 266

Meng Zhu, Hao Duan, Meng Gao, Hao Zhang and Yihong Peng
Both ERK1 and ERK2 Are Required for Enterovirus 71 (EV71) Efficient Replication
Reprinted from: *Viruses* **2015**, *7*(3), 1344–1356
http://www.mdpi.com/1999-4915/7/3/1344 285

Chapter 4: Virus Replication Organelles

Colleen R. Reid, Adriana M. Airo and Tom C. Hobman
The Virus-Host Interplay: Biogenesis of +RNA Replication Complexes
Reprinted from: *Viruses* **2015**, *7*(8), 4385–4413
http://www.mdpi.com/1999-4915/7/8/2825 301

Evan D. Rossignol, Jie E. Yang and Esther Bullitt
The Role of Electron Microscopy in Studying the Continuum of Changes in
Membranous Structures during Poliovirus Infection
Reprinted from: *Viruses* **2015**, *7*(10), 5305–5318
http://www.mdpi.com/1999-4915/7/10/2874 332

Yu-Fu Hung, Melanie Schwarten, Silke Hoffmann, Dieter Willbold, Ella H.
Sklan and Bernd W. Koenig
Amino Terminal Region of Dengue Virus NS4A Cytosolic Domain Binds to
Highly Curved Liposomes
Reprinted from: *Viruses* **2015**, *7*(7), 4119–4130
http://www.mdpi.com/1999-4915/7/7/2812 351

Chapter 5: Antiviral Drugs

Lei Kang, Jiaqian Pan, Jiaofen Wu, Jiali Hu, Qian Sun and Jing Tang
Anti-HBV Drugs: Progress, Unmet Needs, and New Hope
Reprinted from: *Viruses* **2015**, *7*(9), 4960–4977
http://www.mdpi.com/1999-4915/7/9/2854 367

Lonneke van der Linden, Katja C. Wolthers and Frank J.M. van Kuppeveld
Replication and Inhibitors of Enteroviruses and Parechoviruses
Reprinted from: *Viruses* **2015**, *7*(8), 4529–4562
http://www.mdpi.com/1999-4915/7/8/2832 391

List of Contributors

Adriana M. Airo Department of Medical Microbiology and Immunology, University of Alberta, Edmonton, AB T6G 2E1, Canada.

Jamie J. Arnold Department of Biochemistry and Molecular Biology, The Pennsylvania State University, University Park, PA 16802, USA.

Pierre-Yves Barez Molecular and Cellular Epigenetics (GIGA) and Molecular Biology (Gembloux Agro-Bio Tech), University of Liège (ULg), 4000 Liège, Belgium.

Julien Batisse Architecture et Réactivité de l'ARN, CNRS, Université de Strasbourg, Institut de Biologie Moléculaire et Cellulaire, 15 rue René Descartes, 67084 Strasbourg cedex, France.

Stephen J. Benkovic Pennsylvania State University, Department of Chemistry, 414 Wartik Laboratory, University Park, PA 16802, USA.

Serena Bernacchi Architecture et Réactivité de l'ARN, CNRS, Université de Strasbourg, Institut de Biologie Moléculaire et Cellulaire, 15 rue René Descartes, 67084 Strasbourg cedex, France.

David D. Boehr Department of Chemistry, The Pennsylvania State University, University Park, PA 16802, USA.

Esther Bullitt Department of Physiology & Biophysics, Boston University School of Medicine, 700 Albany Street, W302, Boston, MA 02118-2526, USA.

Craig E. Cameron Department of Biochemistry and Molecular Biology, The Pennsylvania State University, University Park, PA 16802, USA.

Alexandre Carpentier Molecular and Cellular Epigenetics (GIGA) and Molecular Biology (Gembloux Agro-Bio Tech), University of Liège (ULg), 4000 Liège, Belgium.

Kyung H. Choi Department of Biochemistry and Molecular Biology, Sealy Center for Structural Biology and Molecular Biophysics, University of Texas Medical Branch at Galveston, Galveston, TX 77555-0647, USA.

Alix de Brogniez Molecular and Cellular Epigenetics (GIGA) and Molecular Biology (Gembloux Agro-Bio Tech), University of Liège (ULg), 4000 Liège, Belgium.

Jiang Deng Beijing Key Laboratory of Blood Safety and Supply Technologies, Beijing 100850, China; Beijing Institute of Transfusion Medicine, 27 (9) Taiping Road, Beijing 100850, China.

Hao Duan Department of Microbiology, School of Basic Medical Sciences, Peking University Health Science Center, 38 Xueyuan Road, Beijing 100191, China.

Diego Ferrero Molecular Biology Institute of Barcelona (CSIC), Barcelona Science Park (PCB), Baldiri i Reixac 10, Barcelona E-08028, Spain.

Cristina Ferrer-Orta Molecular Biology Institute of Barcelona (CSIC), Barcelona Science Park (PCB), Baldiri i Reixac 10, Barcelona E-08028, Spain.

Bo Gao Beijing Key Laboratory of Blood Safety and Supply Technologies, Beijing 100850, China; Beijing Institute of Transfusion Medicine, 27 (9) Taiping Road, Beijing 100850, China.

Meng Gao Department of Microbiology, School of Basic Medical Sciences, Peking University Health Science Center, 38 Xueyuan Road, Beijing 100191, China.

Hélène Gazon Molecular and Cellular Epigenetics (GIGA) and Molecular Biology (Gembloux Agro-Bio Tech), University of Liège (ULg), 4000 Liège, Belgium.

Nicolas Gillet Molecular and Cellular Biology, Gembloux Agro-Bio Tech, University of Liège (ULg), 13 avenue Maréchal Juin, Gembloux 5030, Belgium; Molecular and Cellular Epigenetics (GIGA) and Molecular Biology (Gembloux Agro-Bio Tech), University of Liège (ULg), 4000 Liège, Belgium.

David W. Gohara Department of Biochemistry and Molecular Biology, St Louis University School of Medicine, 1100 South Grand Ave, St Louis, MO 63104, USA.

Stuart F. J. Le Grice Reverse Transcriptase Biochemistry Section, Basic Research Program, Frederick National Laboratory for Cancer Research, Frederick, MD 21702, USA.

Santiago Guerrero Architecture et Réactivité de l'ARN, CNRS, Université de Strasbourg, Institut de Biologie Moléculaire et Cellulaire, 15 rue René Descartes, 67084 Strasbourg cedex, France; Present Address: Centre for Genomic Regulation, Dr. Aiguader, 88, PRBB Building, 08003 Barcelona, Spain.

Malik Hamaidia Molecular and Cellular Epigenetics (GIGA) and Molecular Biology (Gembloux Agro-Bio Tech), University of Liège (ULg), 4000 Liège, Belgium.

Tom C. Hobman Department of Cell Biology; Department of Medical Microbiology and Immunology, University of Alberta, Edmonton, AB T6G 2E1, Canada.

Silke Hoffmann Institute of Complex Systems, Structural Biochemistry (ICS-6), Forschungszentrum Jülich, 52425 Jülich, Germany.

Jiali Hu Department of Pharmacy, The Third Staff Hospital of Baogang Group, 15 Qingnian Road, Baotou 014010, China.

Yu-Fu Hung Institute of Complex Systems, Structural Biochemistry (ICS-6), Forschungszentrum Jülich,52425 Jülich, Germany; Institut für Physikalische Biologie, Heinrich-Heine-Universität Düsseldorf, Universitätsstraße 1,40255 Düsseldorf, Germany.

Lei Kang Department of Clinical Pharmacy, Shanghai First People's Hospital, Shanghai Jiao Tong University, 650 New Songjiang Road, Songjiang District, Shanghai 201620, China.

Qiong Kang Beijing Key Laboratory of Blood Safety and Supply Technologies, Beijing 100850, China; Beijing Institute of Transfusion Medicine, 27 (9) Taiping Road, Beijing 100850, China.

Valerie J. Klema Department of Biochemistry and Molecular Biology, Sealy Center for Structural Biology and Molecular Biophysics, University of Texas Medical Branch at Galveston, Galveston, TX 77555-0647, USA.

Bernd W. Koenig Institute of Complex Systems, Structural Biochemistry (ICS-6), Forschungszentrum Jülich,52425 Jülich, Germany; Institut für Physikalische Biologie, Heinrich-Heine-Universität Düsseldorf, Universitätsstraße 1,40255 Düsseldorf, Germany.

Cheri A. Lee Department of Biochemistry and Molecular Biology, The Pennsylvania State University, University Park, PA 16802, USA.

Camille Libre Architecture et Réactivité de l'ARN, CNRS, Université de Strasbourg, Institut de Biologie Moléculaire et Cellulaire, 15 rue René Descartes, 67084 Strasbourg cedex, France.

Lonneke van der Linden Laboratory of Clinical Virology, Department of Medical Microbiology, Academic Medical Center, University of Amsterdam, Meibergdreef 15, Amsterdam 1105 AZ, The Netherlands.

Xinran Liu Department of Chemistry, The Pennsylvania State University, University Park, PA 16802, USA.

Liping Lv Beijing Key Laboratory of Blood Safety and Supply Technologies, Beijing 100850, China; Beijing Institute of Transfusion Medicine, 27 (9) Taiping Road, Beijing 100850, China.

Ping Ma Beijing Key Laboratory of Blood Safety and Supply Technologies, Beijing 100850, China; Beijing Institute of Transfusion Medicine, 27 (9) Taiping Road, Beijing 100850, China.

Roland Marquet Architecture et Réactivité de l'ARN, CNRS, Université de Strasbourg, Institut de Biologie Moléculaire et Cellulaire, 15 rue René Descartes, 67084 Strasbourg cedex, France.

Ibrahim M. Moustafa Department of Biochemistry and Molecular Biology, The Pennsylvania State University, University Park, PA 16802, USA.

Derek M. Musser Department of Chemistry, The Pennsylvania State University, University Park, PA 16802, USA.

Erin Noble Pennsylvania State University, Department of Chemistry, 414 Wartik Laboratory, University Park, PA 16802, USA.

Théophile Ohlmann CIRI, International Center for Infectiology Research, Université de Lyon, Lyon 69007, France; Inserm, U1111, Lyon 69007, France; Ecole Normale Supérieure de Lyon, Lyon 69007, France; Université Lyon 1, Centre International de Recherche en Infectiologie, Lyon 69007, France; CNRS, UMR5308, Lyon 69007, France.

Radhakrishnan Padmanabhan Department of Microbiology and Immunology, Georgetown University School of Medicine, Washington, DC 20057, USA.

Jean-Christophe Paillart Architecture et Réactivité de l'ARN, CNRS, Université de Strasbourg, Institut de Biologie Moléculaire et Cellulaire, 15 rue René Descartes, 67084 Strasbourg cedex, France.

Jiaqian Pan Department of Clinical Pharmacy, Shanghai First People's Hospital, Shanghai Jiao Tong University, 650 New Songjiang Road, Songjiang District, Shanghai 201620, China.

Yihong Peng Department of Microbiology, School of Basic Medical Sciences, Peking University Health Science Center, 38 Xueyuan Road, Beijing 100191, China.

Srikanth Perike Molecular and Cellular Epigenetics (GIGA) and Molecular Biology (Gembloux Agro-Bio Tech), University of Liège (ULg), 4000 Liège, Belgium.

Jason W. Rausch Reverse Transcriptase Biochemistry Section, Basic Research Program, Frederick National Laboratory for Cancer Research, Frederick, MD 21702, USA.

Colleen R. Reid Department of Medical Microbiology and Immunology, University of Alberta, Edmonton, AB T6G 2E1, Canada.

Bárbara Rojas-Araya Molecular and Cellular Virology Laboratory, Program of Virology, Institute of Biomedical Sciences, Faculty of Medicine, University of Chile, Independencia 834100, Santiago, Chile.

Evan D. Rossignol Department of Physiology & Biophysics, Boston University School of Medicine, 700 Albany Street, W302, Boston, MA 02118-2526, USA.

Melanie Schwarten Institute of Complex Systems, Structural Biochemistry (ICS-6), Forschungszentrum Jülich, 52425 Jülich, Germany.

Ester Sesmero Department of Chemistry and Biochemistry, University of Maryland Baltimore County, 1000 Hilltop Circle, Baltimore, MD 21250, USA.

Ella H. Sklan Department Clinical Microbiology and Immunology, Sackler School of Medicine,Tel Aviv University, Tel Aviv 69978, Israel.

Xin Song Beijing Key Laboratory of Blood Safety and Supply Technologies, Beijing 100850, China; Beijing Institute of Transfusion Medicine, 27 (9) Taiping Road, Beijing 100850, China.

Ricardo Soto-Rifo Molecular and Cellular Virology Laboratory, Program of Virology, Institute of Biomedical Sciences, Faculty of Medicine, University of Chile, Independencia 834100, Santiago, Chile.

Michelle M. Spiering Pennsylvania State University, Department of Chemistry, 414 Wartik Laboratory, University Park, PA 16802, USA.

Qian Sun Department of Clinical Pharmacy, Shanghai First People's Hospital, Shanghai Jiao Tong University, 650 New Songjiang Road, Songjiang District, Shanghai 201620, China.

Jing Tang Department of Clinical Pharmacy, Shanghai First People's Hospital, Shanghai Jiao Tong University, 650 New Songjiang Road, Songjiang District, Shanghai 201620, China.

Ian F. Thorpe Department of Chemistry and Biochemistry, University of Maryland Baltimore County, 1000 Hilltop Circle, Baltimore, MD 21250, USA.

Akira Uchida Department of Biochemistry and Molecular Biology, The Pennsylvania State University, University Park, PA 16802, USA.

Frank J.M. van Kuppeveld Virology Division, Department of Infectious Diseases and Immunology, Faculty of Veterinary Medicine, Utrecht University, Yalelaan 1, Utrecht 3584 CL, The Netherlands.

Núria Verdaguer Molecular Biology Institute of Barcelona (CSIC), Barcelona Science Park (PCB), Baldiri i Reixac 10, Barcelona E-08028, Spain.

Dieter Willbold Institute of Complex Systems, Structural Biochemistry (ICS-6), Forschungszentrum Jülich, 52425 Jülich, Germany; Institut für Physikalische Biologie, Heinrich-Heine-Universität Düsseldorf, Universitätsstraße 1,40255 Düsseldorf, Germany.

Luc Willems Molecular and Cellular Biology, Gembloux Agro-Bio Tech, University of Liège (ULg), 13 avenue Maréchal Juin, Gembloux 5030, Belgium; Molecular and Cellular Epigenetics (GIGA) and Molecular Biology (Gembloux Agro-Bio Tech), University of Liège (ULg), 4000 Liège, Belgium.

Katja C. Wolthers Laboratory of Clinical Virology, Department of Medical Microbiology, Academic Medical Center, University of Amsterdam, Meibergdreef 15, Amsterdam 1105 AZ, The Netherlands.

Jiaofen Wu Department of Pharmacy, Ningbo Medical Treatment Center Lihuili Hospital, 57 Xingning Road, Ningbo 315040, China.

Jun Xiao Beijing Key Laboratory of Blood Safety and Supply Technologies, Beijing 100850, China; Beijing Institute of Transfusion Medicine, 27 (9) Taiping Road, Beijing 100850, China.

Jinbo Xu Beijing Key Laboratory of Blood Safety and Supply Technologies, Beijing 100850, China; Beijing Institute of Transfusion Medicine, 27 (9) Taiping Road, Beijing 100850, China.

Fan Yan Beijing Key Laboratory of Blood Safety and Supply Technologies, Beijing 100850, China; Beijing Institute of Transfusion Medicine, 27 (9) Taiping Road, Beijing 100850, China.

Jie E. Yang Department of Physiology & Biophysics, Boston University School of Medicine, 700 Albany Street, W302, Boston, MA 02118-2526, USA.

Xiaorong Yang Department of Chemistry, The Pennsylvania State University, University Park, PA 16802, USA.

Neela Yennawar Huck Institutes of life sciences, The Pennsylvania State University, University Park, PA 16802, USA.

Hao Zhang Department of Microbiology, School of Basic Medical Sciences, Peking University Health Science Center, 38 Xueyuan Road, Beijing 100191, China.

Yanyu Zhang Beijing Key Laboratory of Blood Safety and Supply Technologies, Beijing 100850, China; Beijing Institute of Transfusion Medicine, 27 (9) Taiping Road, Beijing 100850, China.

Meng Zhu Department of Microbiology, School of Basic Medical Sciences, Peking University Health Science Center, 38 Xueyuan Road, Beijing 100191, China.

About the Guest Editor

David D. Boehr, PhD, is currently an Associate Professor in the Department of Chemistry at the Pennsylvania State University. His PhD research focused on the study of antibiotic resistance enzymes, under the guidance of Dr. Gerard Wright at McMaster University in Canada. He also conducted postdoctoral research on the biophysical characterization of enzymes using solution-state Nuclear Magnetic Resonance with Dr. Peter Wright at The Scripps Research Institute. David's current research has focused on understanding protein structure, dynamics, regulation and binding interactions of viral proteins, particularly the RNA-dependent RNA polymerase. Other research interests include understanding amino acid interaction networks and the control of structural dynamics of enzymes towards enzyme engineering applications.

Preface to "Viral Replication Complexes: Structures, Functions, Applications and Inhibitors"

Viruses are obligate intracellular parasites that need to co-opt a living cell's machinery for replication. At the heart of the viral replication machinery are the nucleic acid polymerases, which are responsible for efficiently copying the viral genome. This process must often be coordinated with other viral processes, including protein translation and viral packaging. The polymerases and other components of the replication machinery may serve as potential anti-viral targets.

Chapter 1 of this book is devoted to the basic structure -function relationships of viral nucleic acid polymerases and related proteins. Chapter 2 then discusses how interactions between the viral polymerase and viral nucleic acids and other proteins help to coordinate nucleic acid synthesis and other important viral processes. Chapter 3 builds on this discussion by including the roles of host proteins in these processes. Many viruses also co-opt host lipid membranes to provide more conducive environments for RNA synthesis and virus replication, which is discussed in Chapter 4. Finally, many of the processes discussed in the preceding chapters are potential antiviral drug targets, which is discussed in Chapter 5.

David D. Boehr
Guest Editor

Chapter 1:
Molecular Perspectives on
Virus Replication Proteins

RNA-Dependent RNA Polymerases of Picornaviruses: From the Structure to Regulatory Mechanisms

Cristina Ferrer-Orta, Diego Ferrero and Núria Verdaguer

Abstract: RNA viruses typically encode their own RNA-dependent RNA polymerase (RdRP) to ensure genome replication within the infected cells. RdRP function is critical not only for the virus life cycle but also for its adaptive potential. The combination of low fidelity of replication and the absence of proofreading and excision activities within the RdRPs result in high mutation frequencies that allow these viruses a rapid adaptation to changing environments. In this review, we summarize the current knowledge about structural and functional aspects on RdRP catalytic complexes, focused mainly in the *Picornaviridae* family. The structural data currently available from these viruses provided high-resolution snapshots for a range of conformational states associated to RNA template-primer binding, rNTP recognition, catalysis and chain translocation. As these enzymes are major targets for the development of antiviral compounds, such structural information is essential for the design of new therapies.

Reprinted from *Viruses*. Cite as: Ferrer-Orta, C.; Ferrero, D.; Verdaguer, N. RNA-Dependent RNA Polymerases of Picornaviruses: From the Structure to Regulatory Mechanisms. *Viruses* **2015**, *7*, 4438–4460.

1. Introduction

RNA dependent RNA polymerases (RdRPs) are the catalytic components of the RNA replication and transcription machineries and the central players in the life cycle of RNA viruses. RdRPs belong to the superfamily of template-directed nucleic acid polymerases, including DNA-dependent DNA polymerases (DdDP), DNA-dependent RNA polymerases and Reverse Transcriptases (RT). All these enzymes share a cupped right hand structure, including fingers, palms and thumb domains, and catalyze phosphodiester bond formation through a conserved two-metal ion mechanism [1]. A structural feature unique to RdRPs is the "closed-hand" conformation, in opposition to the "open-hand" found in other polynucleotide polymerases. This "close-hand" conformation is accomplished by interconnecting the finger and thumb domains through the N-terminal portion of the protein and several loops protruding from fingers, named the fingertips that completely encircle the active site of the enzyme [2,3]. In the prototypic RdRPs the closed "right hand" architecture encircles seven motifs (A to G) conserved in sequence and structure (Figure 1), playing critical roles in substrate recognition and catalysis. Three well-defined channels have been identified in the RdRP structures, serving as: the entry path for template (template channel) and for nucleoside triphosphates (NTP channel) and the exit path for the dsRNA product (central channel) (Figure 1B).

The *Picornaviridae* family is one of the largest virus families known, including many important human and animal pathogens. Picornaviruses are non-enveloped RNA viruses possessing a single-stranded RNA genome (7–8 kb) of positive polarity, with a small peptide (VPg; from 19 to 26 amino acids long) linked to its 5′-end. Their genomes have a long highly structured 5′ nontranslated region (NTR), a single large open reading frame (ORF) and a short 3′ NTR, terminated with a poly(A) tail. The ORF is translated in the cytoplasm of the host cell into a polyprotein, which is proteolytically processed by viral proteases to release the structural proteins (VP1-4), needed to assemble virus capsids and the nonstructural proteins (2A-2B-2C-3A-3B-3C^{pro}-3D^{pol} and in some genera L) as well as some stable precursors necessary for virus replication in host cells [4]. The picornavirus genome is replicated via a negative-sense RNA intermediate by the viral RdRP, named 3D^{pol}. This enzyme uses VPg (the product of 3B) as a primer to initiate the replication process. The structure and function of 3D^{pol} has been studied extensively in the past decades and, to date, the 3D^{pol} crystal structures have been reported for six different members of the enterovirus genus [poliovirus (PV), coxsackievirus B3 (CVB3), enterovirus 71 (EV71) and the human rhinoviruses HRV1B, HRV14, and HRV16], for the aphthovirus FMDV and for the cardiovirus EMCV, either isolated or bound to different substrates [5–15]. These structures provided insights into both initiation of RNA synthesis and the replication elongation processes. Furthermore, mutational analyses in PV and FMDV also have demonstrated that some substitutions in residues located far from the active site, in particular at the polymerase N terminus, have significant effects on catalysis and fidelity. All of these observations suggest that nucleotide binding and incorporation are modulated by a long-distance network of interactions [5,16–22].

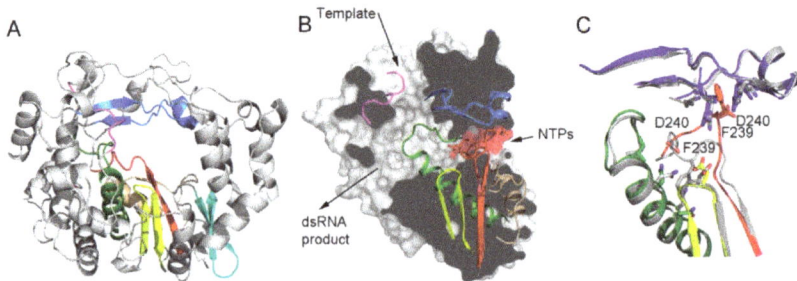

Figure 1. *Cont.*

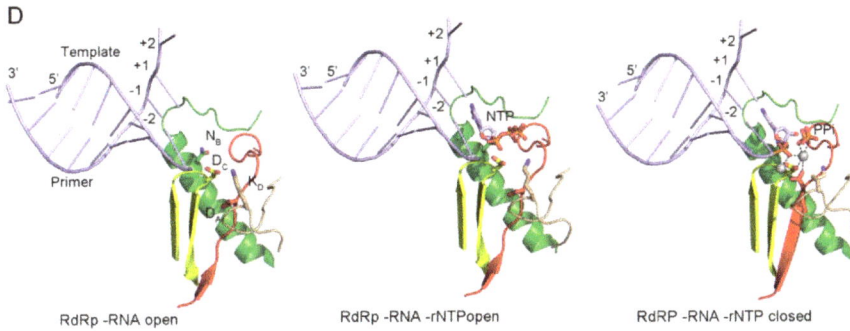

D

Template
+2
+1
-1
-2
3' 5'
Primer
N_B
D_C
K_D

RdRp -RNA open

+2
+1
-1
-2
3' 5'
NTP

RdRp -RNA -rNTPopen

+2
+1
-1
-2
5'
3'
PPi

RdRP -RNA -rNTP closed

Figure 1. Overall structure of a viral RdRP. (**A**) Ribbon representation of a typical picornaviral RdRP (model from the cardiovirus EMCV 3D^{pol}, PDB id. 4NZ0). The seven conserved motifs are indicated in different colours: motif A, red; motif B, green; motif C, yellow; motif D, sand; motif E, cyan; motif F, blue; motif G, pink; (**B**) Lateral view of a surface representation of the enzyme (grey) that has been cut to expose the three channels that are the entry and exit sites of the different substrates and reaction products. The structural elements that support motifs A–G are also shown as ribbons. This panel also shows the organization of the palm sub-domain with motif A shown in two alternative conformations: the standard conformation (PDB id. 4NZ0) found in the apo-form of most crystallized 3D^{pol} proteins and the altered conformation found int the tetragonal crystal form of the EMCV enzyme (PDB id. 4NYZ). The alterations affect mainly Asp240, the amino acid in charge of incoming ribonucleotide triphosphate (rNTP) selection, and the neighboring Phe239 that move ~10 Å away from its position in the enzyme catalytic cavity directed towards the entrance of the nucleotide channel, approaching to motif F; (**C**) Close up of the structural superimposition of the two alternative conformations of the EMCV motif A; (**D**) The PV replication-elongation complexes. Sequential structures illustrating the movement of the different palm residues from a binary PV 3D^{pol}-RNA open complex (left) to an open 3D^{pol}-RNA-rNTP ternary complex (middle) where the incoming rNTP is positioned in the active site for catalysis and, a closed ternary complex (right) after nucleotide incorporation and pyrophosphate (PPi) release. The residues D$_A$ (involved in rNTP selection through an interaction with the 2′ hydroxyl group), D$_C$ (the catalytic aspartate of motif C), K$_D$ (the general acid residue of motif D that can coordinate the export of the PPi group) and N$_B$ (a conserved Asn of motif B, interacting with D$_A$) have been highlighted as sticks. The different structures correspond to the 3D^{pol}-RNA (PDB id. 3OL6), 3D^{pol}-RNA-CTP open complex (PDB id. 3OLB) and 3D^{pol}-RNA-CTP closed complex (PDB id. 3OL7) structures of PV elongation complexes, respectively [7].

2. VPg Binding to 3D^{pol} and Initiation of RNA Synthesis

Correct initiation of RNA synthesis is essential for the integrity of the viral genome. There are two main mechanisms by which viral replication can be initiated: primer-independent or *de novo*, and primer-dependent initiation, reviewed in [23]. Briefly, in the *de novo* synthesis, one initiation nucleotide provides the 3'-hydroxyl for the addition of the next nucleotide whereas the primer dependent initiation requires the use of either an oligonucleotide or a protein primer as provider of the hydroxyl nucleophile. It is remarkable that the RdRPs of viruses that initiate replication using *de novo* mechanisms (*i.e.*, members of the *Flaviviridae* family) share a number of unique features which ensure efficient and accurate initiation, including a larger thumb subdomain containing structural elements that fill most of the active site cavity, providing a support platform for the primer nucleotides (reviewed in [24,25]). These protrusions also serve as a physical barrier preventing chain elongation. Therefore, it is necessary that the initiation platform can move away from the active site after stabilizing the initiation complex, allowing the transition from initiation to elongation [26–28]. By contrast, the members of the *Picornaviridae* and *Caliciviridae* families use exclusively the protein-primed mechanism of initiation. The RNA polymerases of these viruses use VPg as primer for both minus and plus strand RNA synthesis. These enzymes display a more accessible active site cavity, enabling them to accommodate the primer protein for RNA synthesis [13,29].

The very first step in protein-primed initiation in picornavirus is the uridylylation of a strictly conserved tyrosine residue of VPg [30]. In this process, the viral polymerase 3D catalyzes the binding of two uridine monophosphate (UMP) molecules to the hydroxyl group of this tyrosine using as template a cis-replicating element (cre) that is located at different positions of the RNA genome, in the different picornaviridae genera (see [31] for an extensive review). The nucleotidylylation reaction can, however, also occur in a template-independent manner in other viruses, for example in caliciviruses [32].

The picornaviral proteins VPg, 3D^{pol} and 3C^{pro}, alone or in the 3CD precursor form, together with the viral RNA *cre* elements comprise the so-called "VPg uridylylation complex" responsible for VPg uridylylation *in vivo*. Despite extensive structural and biochemical studies, there are several different models for the interactions established between VPg and 3D^{pol} or 3CD in the uridylylation complex and the precise mechanism of uridylylation remains uncertain [31].

Biochemical and structural studies performed for different members of the family: PV [33], HRV16 [12], FMDV [13], CVB3 [8] and EV71 [34] revealed three distinct VPg binding sites on 3D^{pol} (Figure 2). Strikingly, whereas most picornaviruses express only a single VPg protein, FMDV possesses three similar but not identical copies of VPg: VPg1, VPg2 and VPg3 [35], all of which are found linked to viral RNA [36]. Although not all the copies are needed to maintain infectivity [37,38], there

6

are no reports of naturally occurring FMDV strains with fewer than three copies of 3B, suggesting that there is a strong selective pressure towards maintaining this redundancy [39,40].

The structure of two complexes between FMDV 3D^{Pol} and VPg1: 3D^{Pol}-VPg1 and 3D^{Pol}-VPg1-UMP revealed a number of residues in the active site cleft of the polymerase involved in VPg binding and in the uridylylation reaction. Functional assays performed with 3D^{Pol} and VPg mutants with substitutions in residues involved in interactions, according to the structural data, showed important effects in uridylylation [13]. The position of VPg in complex with the FMDV 3D^{Pol} is remarkably similar to the position of the primer and RNA duplex product found in the complex with the same enzyme [14,15]. Most of the amino acids of 3D^{Pol} seen in contact with the RNA primer and duplex product are also involved in interactions with VPg. In fact, the structure shows how the VPg protein accesses the active site cavity from the front of the molecule through the large RNA binding cleft mimicking, at least in part, the RNA molecule. The N-terminal position of VPg projects into the active site where the hydroxyl moiety of the residue Tyr3 is in good proximity to the catalytic aspartates of motifs A and C (Figure 2). In this position, Tyr3 essentially mimics the 3' OH of the primer strand during the RNA elongation. Conserved residues in the fingers, palm and thumb domains of the polymerase were identified as being responsible for stabilizing VPg in its binding cavity. In the 3D^{Pol}-VPg1-UMP complex, the hydroxyl group of Tyr3 side chain was found covalently attached to the α-phosphate moiety of the uridine-monophosphate (UMP) molecule [13]. The positively charged residues of motif F also participate in the uridylylation process, stabilizing Tyr3 and UMP in a proper conformation for the reaction [13] (Figure 2B). Two divalent cations, together with the catalytic aspartic acid residues of motifs A and C, participate in VPg uridylylation. All the observed structural features suggest a conservation of the catalytic mechanism described for all polymerases [1]. Mutational analyses at the conserved FMDV 3D^{Pol} residues that strongly interact with VPg in the crystal structures show a drastic defect in VPg uridylylation [13]. This "front-loading" model for VPg binding, compatible with a *cis* mechanism of VPg uridylylation was further supported by the crystal structures of HRV16 3D^{Pol} [12] and of the PV 3CD precursor [20]. In the latter structure, the extensive crystal packing contacts found between symmetry-related 3CD molecules and the proximity of the N-terminal domain of 3C to the VPg binding site, in the way that VPg was positioned in the FMDV 3D-VPg complex, suggests a possible role of the contacting interfaces in forming and regulating the VPg uridylylation complex during the initiation of viral replication [20].

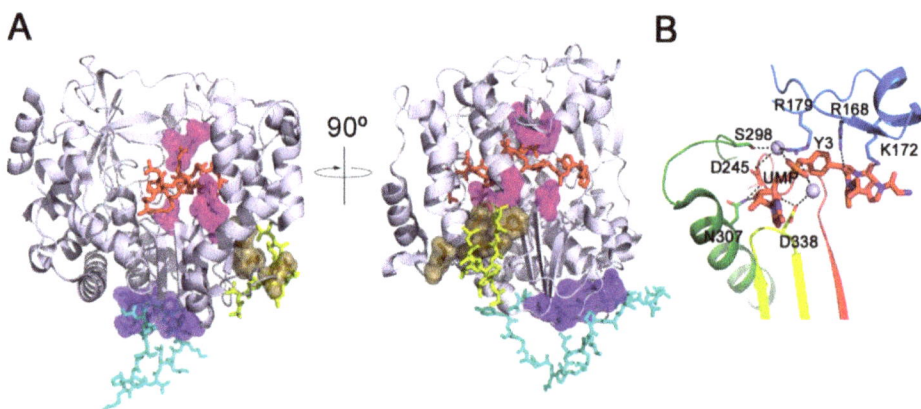

Figure 2. (**A**) Comparison of identified VPg binding sites in picornavirus 3DPols. Because all reported structures of picornavirus 3DPols share high structural similarities, we used the structure of the FMDV 3DPol (PDB id. 2F8E, [2]) as a representative model in this figure and colored it with a light-blue cartoon. The bound VPgs with FMDV [13], CVB3 (PDB id. 3CDW, [8]) and EV71 (PDB id. IKA4, [34]) are shown as red, yellow and cyan sticks, respectively. The residues for VPg (or 3AB) binding in PV (F377, R379, E382 and V391) [33], FMDV (E166, R179, D338, D387 and R388) [13] and EV71 (T313, F314, I317, L319, D320, Y335 and P337) [34] 3DPol are represented as a surface and colored as sand, magenta and blue, respectively, in the cartoon representation; (**B**) Details of the interactions described in the active site of the FMDV 3DPol during the uridylylation reaction. The VPg residues and UMP covalently linked are shown in red sticks, the divalent cations are shown as light-blue spheres and the amino acids involved in uridylylation reaction are shown as sticks. The motifs A, B, C and F are colored in red, green, yellow and blue, respectively.

After nucleotidylylation of VPg, some structural rearrangements of the 3DPol will follow, marking the transition from initiation to the elongation phase of RNA synthesis. There is experimental evidence supporting possible structural differences in 3DPol when involved in the priming reaction *vs.* elongation of RNA. *i.e.*, the nucleoside analog 5-Fluorouridine triphosphate (FUTP) is a potent inhibitor of VPg uridylylation but not of RNA elongation [19]. Furthermore, for the poliovirus, it has been suggested, based on the structure of the FMDV 3DPol-VPg complex that a conserved Asn in the polymerase motif B (Asn297) (equivalent to Asn307 in FMDV) interacts with the 3'-OH of the incoming nucleotide in the uridylylation complex, but with the 2'-OH in the elongation complex [41]. The interaction of Asn307 with the incoming rNTP 2' OH during RNA elongation has been confirmed in a number of picornavirus elongation complexes [6,7,15].

A second binding site for VPg was found in the structure of the CVB3 polymerase [8]. The VPg fragment solved, corresponding to the C-terminal half of the peptide, was bound at the base of the thumb sub-domain in an orientation that did not allow its uridylylation by its own carrier 3D^{pol} (Figure 2A). This VPg binding, partially agreed with previous data reported for PV, showing that a number of amino acids located in motif E of poliovirus (PV) were required for VPg or their precursor, 3AB, binding and affecting VPg uridylylation [33]. In light of these results, the authors proposed that VPg bound at this position was either uridylylated by another 3D molecule or that it played a stabilizing role within the uridylylation complex [8].

Finally, a third VPg binding site was discovered in the structure of the EV71 3D^{pol}-VPg complex. In this complex, VPg is anchored at the bottom of the palm domain of the polymerase, showing a V-shape conformation that crosses from the front side of the catalytic site to the back side of the enzyme (Figure 2A). Similarly to that occurring in the previously studied viruses, the mutational analyses of the interacting residues evidenced a reduced binding of VPg to the EV71 3D^{pol} affecting uridylylation [34,42]. Additional experiments performed by the same authors, by mixing the VPg-binding-defective mutants with catalytic defective mutant of the EV71 polymerase, demonstrated *trans* complementation of VPg uridylylation *in vitro*. However, the structure of the EV71 3D^{pol}-VPg complex showed that the VPg Tyr3 is buried at the base of the polymerase palm indicating that a conformational change should occur to expose the side chain of Tyr3 for uridylylation.

Taking into account the important sequence homology between the picornaviral VPg sequences [13,43] and the high similarities existing in the 3D^{pol} structures, it is tempting to speculate that the three different VPg binding sites observed in the different crystal structures might reflect distinct binding positions of VPg to both 3D^{pol} or its precursor 3CD at different stages of the virus replication initiation process. As discussed above, the FMDV genome codes for three VPg molecules, all of them are present in naturally occurring viruses [39,40]. A global picture of the assembly of the multicomponent complexes involved in replication initiation in Picornaviruses and its regulation would require the structural and functional analyses of higher order complexes formed by the polymerase 3D^{pol}, involving different proteins or protein precursors (VPg, 3AB, 3C^{pro}, 3CD) and RNA templates. Such structures would shed new light on the molecular events underlying the initiation of RNA genome replication in these viruses, and should provide crucial information for the design of new antiviral strategies.

3. Structural Elements Regulating Replication Elongation in RdRPs

3.1. Subtle Conformational Changes Associated with Nucleotide Selection and Active Site Closure for Catalysis

The replication elongation process can be roughly divided in three steps, including nucleotide selection, phosphodiester bond formation and translocation to the next nucleotide for the subsequent round of nucleotide addition. Structural biology has been crucial to elucidate the structural changes associated with each phase of catalysis for a wide number of polymerases [25,29,44,45]. Extensive biochemical and structural studies in the A- and B-families of open-hand nucleic acid polymerases indicate that the movement of an α-helix of the fingers sub-domain would control each step of the nucleotide-addition cycle and facilitates translocation along the template after catalysis [46,47]. In contrast in the "closed-hand" RdRPs, the presence of the fingertips encircling the catalytic site constrains the fingers' movement relative to the thumb, avoiding the swinging movement of the fingers that is associated with active site closure in open-hand polymerases.

The structures of a large number of RdRP-RNA-rNTP replication-elongation complexes determined, for different members of both, the *Picornaviridae* and the *Caliciviridae* families, have provided important insights into the structural changes associated to each catalytic step [3,6,7,25,29,48–51]. These structures indicate that RdRPs use subtle rearrangements within the palm domain to fully structure the active site for catalysis upon correct rNTP binding. Briefly, in a first state, the RdRP-RNA complex in the absence of an incoming rNTP, shows an open conformation of the polymerase active site characterized by a partially formed three-stranded β-sheet of the palm domain motifs A and C (Figure 1B,C). A second feature characterizing this state is the presence of a fully prepositioned templating nucleotide (t+1), sitting above the active site and stacked on the upstream duplex and ready for the binding of the incoming rNTP (Figure 1C). In a second state, an incoming rNTP reaches the active site and establishes base-pairing interactions with the template base (t+1), but catalysis has not taken place because the catalytic site is still in open conformation. The third state occurs after binding of the correct nucleotide to the active site. This binding induces the realignment of β-strands in the palm subdomain that includes the structural motifs A and C, resulting in the repositioning of the motif A aspartate to allow interactions with both metal ions required for catalysis [6,7,48,49]. It is important to remark that the active site closure in RdRPs is triggered by the correct nucleotide binding, suggesting that nucleotide selection in RdRPs is a simple process in which base-pairing interactions control the initial rNTP binding geometry and the resulting positioning of the ribose hydroxyls becomes the major checkpoint for proper incoming nucleotide selection. In particular, two residues within motif B (Ser and

10

Asn) and, a second Asp residue at the C-terminus of motif A, strictly conserved among picorna- and caliciviruses form the ribose binding pocket (Figure 1C). The interactions between these amino acids and the ribose hydroxyl groups of the incoming rNTP would stabilize the subtle restructuring of the palm domain that results in the formation of a functional active site. An incorrect nucleotide can bind, but its ribose hydroxyls will not be correctly positioned for active site closure, in consequence, the incorporation efficiency will be reduced.

In addition, growing amounts of data indicate that the conformational changes in motif D determine both efficiency and fidelity of nucleotide addition [52,53]. Biochemical studies of nucleotidyl transfer reactions catalyzed by RdRPs, RTs and single subunit DNA polymerases made the unexpected observation that two protons, not just one, are transferred during the reaction and that the second proton derives from a basic amino acid of the polymerase (termed general acid) and is transferred to the PPi leaving group [52]. PPi protonation is not essential but contributes from 50-fold to 1000-fold the rate of nucleotide addition. Additional data from mutagenesis and kinetics of nucleotide incorporation showed that the general acid was a lysine located in the conserved motif D of RdRPs and RTs [53]. Solution NMR studies were used to analyze the changes that occurred during nucleotide addition. A methionine within motif D, located in the vicinity of the conserved lysine, was found to be a very informative probe for the positioning of the motif. Authors have found that the constitution of the catalytically competent elongation complex (RdRP-RNA-NTP) required the formation of a hydrogen bond between the β-phosphate of incoming rNTP and motif D lysine [54]. The protonation state of this lysine was also observed to be critical to achieve the closed conformation of the active site. Moreover, the ability of motif D to reach the catalytically competent conformation seems to be hindered by the binding of an incorrect nucleotide and this ability continues to be affected after nucleotide misincorporation. Indeed, the NMR data correlates the conformational dynamics of motif D to the efficiency and fidelity of nucleotide incorporation [54].

A number of structures of enterovirus elongation complexes have been trapped in a post-catalysis state in which the newly incorporated nucleotide and the pyrophosphate product are still in the active site, and the active site opens its conformation [6,7]. This state will be followed by the translocation of the polymerase by one base pair to position the next templating nucleotide in the active site for the next round of nucleotide addition. A number of crystal structures have also been solved in a post-translocation state [6,7,15]; however, no structures of translocation intermediates are currently available for RdRPs and the precise mechanism is not yet known. Recent data also suggest that a conserved lysine residue within motif D can coordinate the export of the PPi group from the active site once catalysis has taken place [16], thereby triggering the end of the reaction cycle and allowing enzyme translocation. Very recently, structural and functional data in enteroviruses indicate that steric clashes between the motif-B loop and the template RNA would also promote translocation [51] (see below).

Viral RdRPs are considered to be low fidelity enzymes, generating mutations that allow the rapid adaptation of these viruses to different tissue types and host cells. Based on X-ray data of CVB3 and PV catalytic complexes, the laboratories of Peersen and Vignuzzi engineered different point mutations in these viral polymerases and studied their effects on *in vitro* nucleotide discrimination as well as virus growth and genome replication fidelity. Data obtained revealed that the palm mutations produced the greatest effects on *in vitro* nucleotide discrimination and that these effects appeared strongly correlated with elongation rates and *in vivo* mutation frequencies, with faster polymerases having lower fidelity. These findings suggested that picornaviral polymerases have retained a unique palm domain-based active-site closure as a mechanism for the evolutionary fine-tuning replication fidelity and provide a pathway for developing live attenuated virus vaccines based on engineering the polymerase to reduce virus fitness [55,56].

Finally, recent structural data on calicivirus RdRPs have provided evidence of new conformational changes occurring during catalysis. Structural comparison of the human Norovirus (NV) RdRP determined in multiple crystal forms, in the presence and absence of divalent metal cations, nucleoside triphosphates, inhibitors and primer-template duplex RNAs, revealed that in addition to the active site closure, the NV RdRP exhibits two additional key changes: a rotation of the central helix in the thumb domain by 22°, resulting in the formation of a binding pocket for the primer RNA strand and the displacement of the C-terminal tail region away from the central active-site groove, which also allows the rotation of the thumb helix [57].

3.2. Conformational Plasticity of the Motif B Loop Regulates RdRP Activity

The central role of the motif B loop in template binding, incoming nucleotide recognition and correct positioning of the sugar in the ribose-binding pocket was evidenced in the first structures of the FMDV catalytic complexes [15,29]. This loop, connecting the base of the middle finger to the α-helix of motif B, is able to adopt different conformations when it binds to different template and incoming nucleotides, being one the most flexible elements of the active site of RdRPs in picornaviruses, as well as in other viral families (reviewed in [50]). In fact, structural comparisons evidenced large movements of the B-loop, ranging from a conformation in which the loop is packed against the fingers domain leaving the active site cavity fully accessible for template entry, to a configuration where the loop protrudes towards the catalytic cavity and clashes with the template RNA (Figure 3). The key residue of this flexible region is a strictly conserved glycine, which acts as a hinge for the movement. The critical role of the B-loop dynamics was previously anticipated by site-directed mutagenesis in the picornavirus EMCV. Substitutions of the hinge glycine in $3D^{pol}$ essentially abolished RNA synthesis *in vitro* [58]. Furthermore, additional interactions established between the B-loop and the RNA phosphodiester backbone of the upstream duplex, between the −1 and −2 nucleotides, prompted researchers to hypothesize a function of the loop in modulating polymerase activity through effects on translocation [6,7,15].

Extensive structural and functional work in PV, using several polymerase mutants, harboring substitutions within the B-loop sequence Ser288-Gly289-Cys290, evidenced a major role of these residues in the $3D^{pol}$ catalytic cycle [51]. The work concluded that the B-loop is able to adopt mainly three major conformations, termed *in/up*, *in/down* and *out/down* and that each alternative conformation is important for the correct NTP binding and for the post-catalysis translocation step. The terms *in/out* refer to the loop conformations, packed against the fingers (in), or protruding into the catalytic cavity (out) (Figure 3A). The designation is based on whether the residue Cys290 is buried "in" a hydrophobic pocket directly behind the loop or is "out" of the pocket and exposed to solvent [59]. Moreover, the Sholders and Peersen work highlighted the role of the PV Ser288. The side chain of this residue may also adopt two alternative conformations: pointing "up" toward the ring finger and away from the active site, or "down" pointing toward the active site. These authors propose a sequential model for the structural changes occurred during the PV RdRP catalytic cycle where initially, in the apo-form structure of PV $3D^{pol}$, the B-loop is in the *in/up* conformation, allowing rNTP entry. Equivalent conformations of the B-loop were observed in the apo-forms of $3D^{pol}$ in FMDV [14], Rhinovirus [11] and Coxsackievirus [8,9]. On nucleotide binding, the Ser288 flips down toward the active site (*in/down* conformation), establishing a hydrogen bond with the Aspartic acid residue of motif A, involved in the selection of the incoming ribonucleotide (Asp_A). It is also important to remark that in the previous step, Asp_A was hydrogen bonded to a conserved Asn from the motif B (Asn_B) (Figure 1C,D). Following these changes, new rearrangements occur, including the realignment of the palm motif A, required for catalysis. After the phosphodiester bond formation, the Sholders and Peersen model proposes a movement of the B-loop from "in/down" to "out/down" configuration, resulting in a steric clash between the B-loop and the backbone of the RNA template strand that would facilitate translocation along the RNA and prevent backtracking after translocation. In addition, the finding that the PV $3D^{pol}$ G289A mutant was able to catalyze single nucleotide addition but was defective for processive elongation provided more evidence of the crucial role of this glycine that confers the flexibility required for the loop movements.

13

A

B loop

GDD

B

Template

C290
G289
S288

NTP

Primer

C

FMDV	288	KRI......TVEGGMPSGCSATSIINTILNNIYVLYAL....RRHY	323
EV71	279	KTY......CVLGGMPSGCSGTSIFNSMINNIIIRALL....IKTF	314
PV	278	KTY......CVKGGMPSGCSGTSIFNSMINNLIIRTLL....LKTY	313
HRV	277	TYY......EVEGGVPSGCSGTSIFNSMINNIIIRTLV....LDAY	312
CBV	279	KHY......FVRGGMPSGCSGTSIFNSMINNIIIRTLM....LKVY	314
NV	291	TI.......SINEGLPSCVPCTSQWNSIAHWLLTLCALS...EVTN	326
IBDV	475	LQI......K.TYGQGSGNAATFINNHLLSTLV...LDQ...WNLM	507
HCV	273	NC.......GYRRCRASGVLTTSCGNTLTCYLK.ASAA....CRAA	306
WNV	591	TVMD...VISREDQRGSGQVVTYALNTFTNLAVQLVR.M...MEGE	629
Reo	672	DFT......HMTTTFPSGSTATSTEHTANNSTMMETFLTV.WGPEH	710
φ6	377	TLLGDPSNPDLEVGLSSGQGATDLMGTLLMSITYLVMQL...DHTA	419
Qβ	1007	SVV......TYEKISSMGNGYTFELESLIFASLA..RSV...CEIL	1041

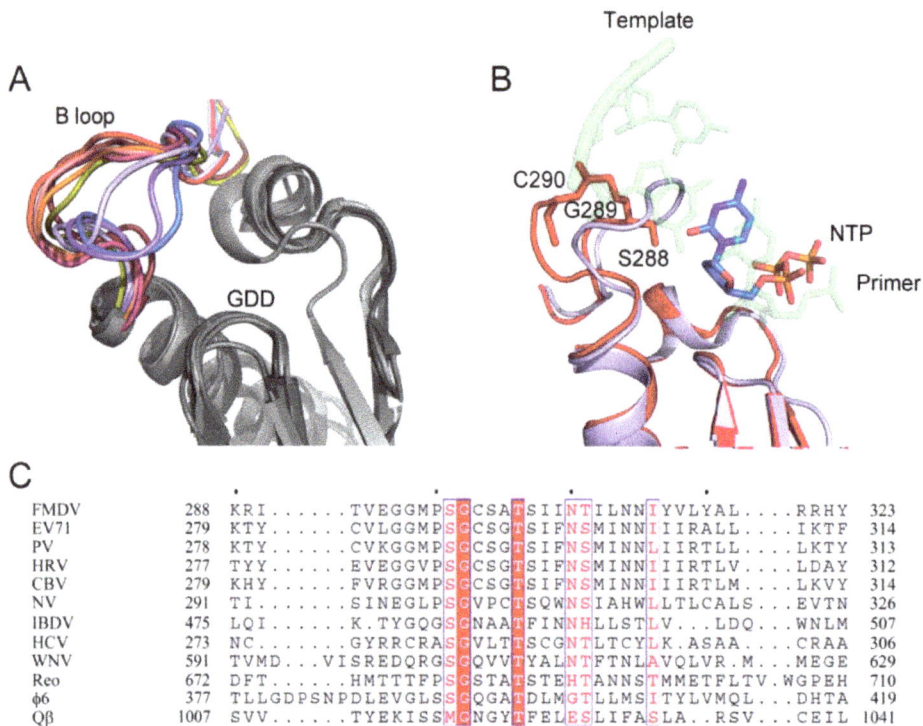

Figure 3. The conformational changes in the B-loop of RdRPs. (**A**) Superposition of the different conformations described for the B-loop. Motifs A, B and C are represented as ribbons and colored in gray tones. The B-loop is shown in different colors for each observed conformation, from red (up) to blue (down): NV NS7, Mg^{2+} bound (PDB id. 1SH3, chain A) chocolate; PV apo-form (PDB id, 1RA6) red; FMDV-RNA complex (PDB id, 1WNE) magenta; PV C290V mutant (PDB id. 4NLP) light-orange; IBDV VP1 + VP3 C-ter peptide (PDB id. 2R70) orange; NV NS7, Mg^{2+} bound (PDB id. 1SH3, chain B) yellow; PV C290F mutant (PDB id. 4NLQ) light-blue; IBDV VP1 apoform (PDB id. 2PUS) slate; FMDV K18E mutant (PDB id. 4WYL) blue; (**B**) Superimposition of the up conformation of PV apo-form (PDB id. 1RA6) red and the down conformation of PV C290F mutant (PDB id. 4NLQ) slate with the RNA template-primer and an incoming rNTP molecule are represented as sticks in semi-transparent representation; (**C**) Sequence alignment of the B-loop region of all the RdRPs from dsRNA and +ssRNA.

The high sequence and structural conservation of the B-loop among viral polymerases (Figure 3C) strongly suggest that its conformational dynamics would be a common feature of the RNA-dependent RNA polymerases from positive-strand RNA viruses.

3.3. Unusual Conformation of Motif A Captured in the Structure of the Cardiovirus EMCV 3D^{pol}

The crystal structure of the EMCV 3D^{pol} in its unbound state has been recently solved in two different crystal forms [60]. As expected, the overall architecture of the enzyme was similar to that of the known RdRPs of other members of the *Picornaviridae* family. However, structural comparisons revealed a large reorganization of the active-site cavity in one of the crystal forms. The rearrangement affects mainly the C-terminal loop of motif A, containing the aspartic acid residue involved in incoming rNTP selection (Asp240 in EMCV) (Figure 1B). The heart of this conformational change is that the Asp240 neighbor residue, Phe239, made a drastic movement whereby it is popped out of a hydrophobic pocket in the palm domain to participate in an intriguing set of cation-π interactions in the fingers domain at the edge of the NTP entry tunnel [60]. Another important feature of this altered active site conformation is that the active site of the enzyme was captured in a closed-like state, with the β-sheet supporting motif A totally formed and the catalytic Asp235 positioned in front of the motif C Asp333 (Figure 1B). This active-site conformation has never been observed before in the absence of RNA and a correctly base-paired rNTP. In addition, the N-terminal Gly1 residue was moved out of its binding site, anchored in the fingers domain, toward a totally exposed orientation in the polymerase surface. This is extremely intriguing because like most of the picornaviral RdRPs, the EMCV enzyme is only active when cleaved from the polyprotein to generate an N-terminus with a Gly1 residue [61]. Those observations prompt to hypothesize that this EMCV 3D^{pol} crystal form might represent the structure of the inactive form of the enzyme that would be present in the precursor protein, where Gly1 cannot be buried because is not a terminal residue. However, this hypothesis seems to be in conflict with the structural data currently available for the poliovirus precursor 3CD that also showed Gly1 exposed as part of the flexible linker joining 3C and 3D^{pol} but with the active site in the standard open conformation found in the pre-catalytic complexes [20]. The possible role of the altered conformation of the motif A loop, in particular, of the positioning of the rNTP binding residue Asp240 at the edge of the rNTP entry tunnel, is another mystery to decipher in order to gain insight on the regulation of this enzyme activity.

3.4. Conformational Flexibility in the Template Channel

The template channel also exhibits substantial flexibility as visualized by comparing the X-ray structures of different replication elongation complexes [6,7,14,15,19,29,62], as well as predicted by molecular dynamics simulations in a number of picornaviral polymerases [22]. This flexibility appears directly correlated with the role of this channel in driving the template nucleotides toward the catalytic cavity. Of particular importance is the flexible nature of a region included at the 3D^{pol} N-terminus (residues 16 to 20; FMDV numbering), lining the channel that appears to be interacting with the RNA near the single-strand/double-strand junction (Figure 4). The conformational changes occurring in this region would assist the movement of the template nucleotides at the +2 and +3 positions. Interestingly, structural comparisons of the wild type FMDV 3D^{pol} catalytic complexes showed that the basic side chain of Arg17 is involved in different interactions with the template nucleotide t+2 in all complexes analyzed [14,15]. In these complexes, the t+2 nucleotide points towards the active site cavity, stacked with the t+1 nucleotide that is located in the opening of the central cavity, in close contact with the motif B loop (Figure 4B). The equivalent residue of Arg17 in enteroviral polymerases is Pro20 (PV numbering). Structural comparisons of distinct enteroviruses elongation complexes show that Pro20 and its surrounding residues form a conserved pocket where the t+2 nucleotide binds [6,7] (Figure 4). This pocket found in the enterovirus 3D^{pol} seems to be a preformed structure that is also present in the unbound enzymes. In contrast, the FMDV wild type enzyme lacks a preformed pocket in the template channel and, as mentioned above, the t+2 nucleotide is oriented towards the active site cavity (Figure 4), constituting an important structural difference between the enterovirus and FMDV catalytic complexes. Surprisingly, the comparative structural analyses of FMDV 3D^{pol} mutants presenting alterations in RNA binding affinity and incoming nucleotide incorporation, including a remarkable increase or decrease in the incorporation of the nucleoside analog ribavirin, showed important movements in this polymerase region that result in the formation of distinct pockets where the t+2 nucleotide binds [19,62] (Figure 4).

Besides facilitating specific contacts with the RNA template, the dynamic nature of residues lining the template channel should permit the access of the t+1 nucleotide into the 3D catalytic site. The base of the template channel is built mainly by residues of the motif B loop. These B-loop residues are involved in interactions with the t+1 nucleotide in the active site, as well as with the incoming rNTP [15]. Putting together all data is tentative to speculate that the rearrangements in the template channel and the B-loop occur in a concerted manner and that these concerted changes serve to regulate both RNA replication processivity and fidelity.

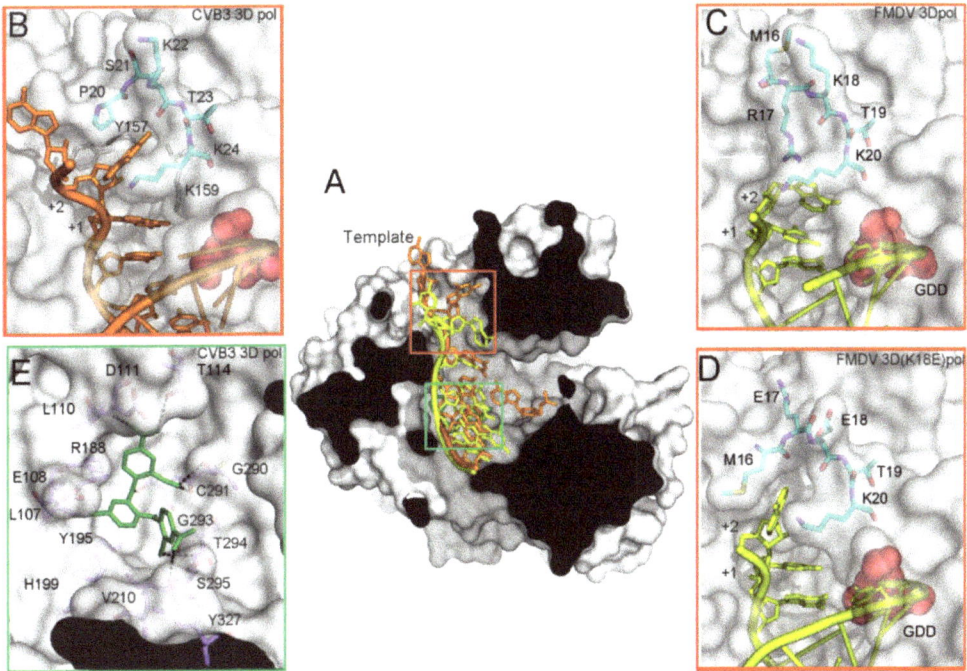

Figure 4. Structure and interactions in the template channel of a picornavirus 3D^{pol}. (**A**) The structure of the CVB3 3D^{pol} (PDB id. 4K4Y) has been used as a model, the molecular surface of the polymerase is shown in grey with the acidic residues of the active site in red and the RNA depicted as a cartoon in orange and the FMDV RNA is superimposed in yellow. The non-nucleoside analogue inhibitor is also superimposed in green; (**B**) Structure and interactions in the template channel at the entrance of the active site of CVB3 3D^{pol} (PDB id. 4K4Y), the N-terminal residues 20–24 depicted as sticks in cyan and the RNA in orange and others residues involved in the binding RNA are represented as grey sticks; (**C**) The wild type FMDV 3D^{pol}-RNA complex (PDB id. 1WNE); and (**D**) the FMDV 3D^{pol} (K18E)-RNA complex (PDB id. 4WZM); (**E**) Interaction network between GPC-N114 and its binding pocket of CVB3 3D^{pol} represented by surfaces (PDB id. 4Y2A). The polymerase residues in direct contact with the inhibitor are shown with sticks in atom type color with carbon in slate and explicitly labeled. Hydrogen bonds are depicted as dashed lines.

4. Polymerase Oligomerization

Proteins can oligomerize through reversible associations mediated by electrostatic and hydrophobic interactions, hydrogen bonds or by covalent stabilization by disulfide bonds. RdRPs, the enzymes that exclusively belong to the RNA virus world are not the exception. In recent years, the X-ray and

Cryo-electron microscopy (cryo-EM) analyses revealed the quaternary structures of a large number of RdRP oligomers, defining the critical residues that lead these associations. Otherwise, complementary biochemical analyses allowed deciphering of the functional roles in most of these arrangements.

RdRP-RdRP interactions to form dimers or higher order oligomers have been predominantly reported for (+) ssRNA viruses, including several Picornavirus [63–68], Flavivirus [68,69] and Calicivirus [67,70] enzymes, as well as, in RdRPs of plants [71] and Insect viruses [72]. The homo-interaction of RdRPs was also described in replicases of (−) ssRNA viruses such as influenza A virus [73], and in more distant dsRNA viruses like infectious pancreatic necrosis virus [74]. RdRP oligomerization has been predominantly observed *in vitro* during crystallization, probably produced by the high protein concentration, as well as by other environmental changes like pH or ionic strength. However, intracellular accumulation of oligomeric polymerases was also observed during viral infection of different RNA viruses including PV [64], Sendai virus [75], Rift Valley Fever virus [76] and norovirus [70], among others. These observations suggest that RdRP oligomerization can also be a natural event as a sort of post translational modification. The specificity for the dimerization/multimerization involves distinct surfaces depending on the enzyme, *i.e.*, HCV RdRP dimerization has been proposed to be mediated by the thumb domain [69], whereas, in PV, the polymerase fingers appear to be crucially involved [66]. Contacts between these domains during oligomer formation may cause small conformational changes that are transferred to the active site as an allosteric regulation or could even modify the accessibility of the substrate channels.

The first oligomerization state of an RdRP was described for PV and the nature of the molecular contacts at two different polymerase interfaces, termed I and II, were postulated from the first crystal structure of PV 3D^{pol} [63,64]. The Interface I derived from interactions between the front of the thumb subdomain of one molecule and the back of the palm subdomain of the neighbour molecule in the crystal (Figure 5) whereas interface II involved two N-terminal regions of the polymerase that appeared disordered in this structure [63]. Later on, Lyle *et al.*, using cryo-EM demonstrated that the purified PV 3D^{pol} was able to organize two-dimensional lattices and tubular arrangements formed by polymerase fibres [65] and, recently, the structure of these assemblies has been characterized at the pseudo-atomic level [77,78]. The planar lattices, forming a ribbon-like structure, consist of linear arrays of dimeric RdRPs supported by strong interactions through the interface-I as defined in the PV 3D^{pol} crystal structure [77]. The tubular structure is also formed via interface-I but is also assisted by a second set of interactions placed in interface-II, involving interactions between the fingertips of one molecule and the palm of its contacting neighbour. The fitting of the 3D^{pol} coordinates into the cryo-EM reconstructions showed that interface I connects adjacent dimers by head-to-tail contacts [78] (Figure 5). The relevance of a number of interface II residues in lattice formation was further confirmed by mutagenic analysis. In particular, mutations at residues Tyr32 and Ser438 involved interface II contacts both in planar and tubular array results in a disruption of PV 3D^{pol} lattice formation [77]. Furthermore, several lines of evidence suggested that the PV polymerase can change the conformation upon forming oligomers and, in the tubular assemblies, the porous nature of the polymerase lattice is likely to allow the participation of other viral and cellular proteins [78].

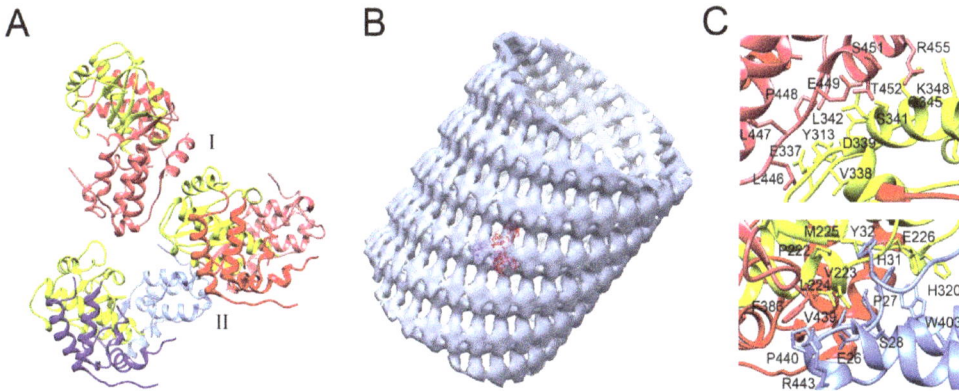

Figure 5. Oligomerization of the PV 3D^{pol}. (**A**) Polymerase-polymerase interactions mediated by interfaces I and II, explicitly marked (PDB id. 1RDR) in yellow palm subdomain, in red and blue fingers subdomain and in light colors thumb subdomain; (**B**) Volume map of the reconstructed 3D^{pol} tubes with the crystallographic model positioned inside. The volume map was reproduced from [78] (EM code emd2270); (**C**) Close up of the interface I (upper panel) and interfase II (bottom panel) in the oligomeric tubular array of PV 3D^{pol} according to [78].

All examples of high-order RdRP assemblies point out that oligomerization may be an advantageous feature providing a functional control, such as allosteric regulation in addition to increasing the stability against degradation and denaturation. The functioning of RdRPs in oligomeric arrays also has an additional advantage, to concentrate the reaction substrates in a physical place and dispose the active sites to use them iteratively.

5. Implications for Antiviral Drug Discovery

RdRP synthesize RNA using an RNA template. This biochemical activity, almost exclusive of RNA virures offers the opportunity to identify very selective inhibitors of this viral enzyme. Antiviral drugs targeting the RdRPs may either directly inhibit polymerase activity or essential interactions with the RNA template, or the RdRP-RdRP contacts promoting oligomerization, or interactions with other regulatory proteins. The detailed structural and mechanistic understanding of the conformational changes occurring during catalysis is essential not only for understanding of viral replication at the molecular level but also for the design of novel inhibitors capable of trapping the enzyme in specific conformational states. The Flaviviruses, Hepatitis C virus, Dengue virus and West Nile virus, as well as the calicivirus NV are clear examples of how much effort has been directed towards developing drugs that inhibit viral replication [79–82]. In general Direct-Acting Antivirals (DAAs), inhibiting RNA replication can be classified into two groups on the basis of their chemical structure and mechanisms of action: Nucleoside Analog (NA) inhibitors and Non-Nucleoside Inhibitors (NNIs). NAs target the active site of the polymerase and need to be converted by the host cell machinery to the corresponding nucleotides, which can either induce premature termination of RNA synthesis [83,84] or be incorporated by the viral polymerase into the nascent RNA, causing accumulation of mutations and contributing to virus extinction through lethal mutagenesis [85].

Conversely, NNIs bind mainly to allosteric pockets of the target polymerase causing alterations in the enzyme dynamics. They might either stabilize an inactive conformation or trap the enzyme in a functional conformation but impeding either the transition between initiation and elongation or the processivity of polymerase elongation [86,87].

Allosteric inhibitors directed against protein-RNA or protein-protein interactions involving viral polymerases are less explored as antiviral drugs. However, effective antiviral molecules that seem to inhibit interactions of the viral polymerase within the replicative complex have been for pestiviruses [88,89] and, a new compound against Dengue virus have been identified that appears to block the RdRP activity through binding to the RNA template channel [20].

In a very recent study, we have identified a novel non-nucleoside inhibitor of 3D^{pol}, the compound GPC-N114 (2,2′-[(4-chloro-1,2-phenylene)bis(oxy)]bis(5-nitro-benzonitrile), with broad-spectrum antiviral activity against both enteroviruses and cardioviruses [90]. The X-ray analysis of CVB3 3D^{pol}-GPC-N144 co-crystals revealed that the binding site of the compound was located at the junction of the palm and the fingers domains, partially overlapping with the binding site of the templating nucleotide (Figure 4E). The polymerase-inhibitor interactions involved different residues of the conserved motifs G, F, B and A, most found in direct contact with the RNA templates in all picornaviral 3D^{pol}-RNA complexes determined so far. Structural comparisons between unbound and GPC-N114 bound CVB3 3D^{pol} revealed that the polymerase did not undergo any major conformational change upon binding of the compound.

Surprisingly, GPC-N114-resistant enterovirus variants could not be obtained, but two EMCV resistance mutations (Met300Val and Ile303Val, in the motif B-loop) were readily selected in the presence of suboptimal concentration of GPC-N114 [90]. The reason for the inability of enteroviruses to develop resistance against GPC-N114 remains to be established. A possible explanation is that mutations that would confer resistance to GPC-N114 also impair binding of the template-primer, thereby preventing replication. Although the exact reason remains to be determined, the structural data suggest that, in contrast to most allosteric binding sites, the GPC-N114-binding cavity in enterovirus 3D^{pol} lacks the conformational plasticity required to develop resistance. In contrast, EMCV 3D^{pol} appears to be sufficiently plastic to allow for compound-resistance substitutions. As expected, structural comparisons showed high similarity between the CVB3 and the EMCV enzymes, but a major difference existed in the main interactions established between these polymerases with the inhibitor that could possibly underlie the differences observed in the emergence of resistant mutations. A key interaction of the CVB3 3D^{pol}-GPC-N114 was mediated by Tyr195 (Figure 4E). In contrast, the equivalent residue in EMCV 3D^{pol} is Ala, resulting in a weaker interaction with the compound. This weaker interaction together with an increment of the flexibility in the compound-binding area, induced by the B loop resistance mutations, might result in a decrease of GPC-N114 binding to the EMCV enzyme.

In summary, the identification of this novel drug-binding pocket in the picornaviral 3D^{pol} might serve as a starting point for the design of new antiviral compounds targeting the template-binding channel.

6. Conclusions

RNA-dependent RNA polymerases (RdRPs) play central roles in both transcription and viral genome replication. In picornaviruses, these functions are catalyzed by the virally encoded RdRP, termed 3D^{pol}. 3D^{pol} also catalyzes the covalent binding of two UMP molecules to a tyrosine on the small protein VPg. Uridylylated VPg then serves as a protein primer for the initiation of RNA synthesis.

The ever growing availability of structures of picornaviral catalytic complexes provided an increasingly accurate picture of the functional steps and regulation events underliying viral RNA genome replication. Data currently available provides high-resolution pictures for a range of conformational states associated to template and primer recognition, VPg uridylylation, rNTP recognition and binding, catalysis and chain translocation. Such structural information is providing new insights into the fidelity of RNA replication, and for the design of antiviral compounds.

Protein primed mechanism of replication initiation mediated by VPg appears to be a process that involves more than one VPg binding site in 3D^{pol} possibly at different stages of the virus replication initiation process. Although the number of

3D^{pol}-VPg structures available show individual snapshots of the process, to obtain a global picture of the assembly, regulation and dynamics of complete replication initiation complexes, requires further analyses of high order assemblies formed not only by the polymerase and VPg but also involving different viral and host proteins, protein precursors and RNA templates. Such structures should provide a more detailed view of the molecular events underlying the initiation of picornavirus genome replication.

The structures of a large number of picornavirus replication-elongation complexes captured subtle conformational changes associated with nucleotide selection and active site closure. Among these movements, the motif B-loop assists in the positioning of the template nucleotide in the active site. Binding of the correct nucleotide then induces the β-strands realignment in the palm subdomain and repositioning of the motif A aspartate for catalysis. Steric clashes between the motif B-loop and the template RNA would finally promote translocation.

Acknowledgments: Núria Verdaguer acknowledges funding from the Spanish Ministry of Economy and Competitiveness (BIO2011-24333).

Conflicts of Interest: The authors declare no conflicts of interest.

References

1. Steitz, T.A. A mechanism for all polymerases. *Nature* **1998**, *391*, 231–232.
2. Ferrer-Orta, C.; Arias, A.; Escarmis, C.; Verdaguer, N. A comparison of viral RNA-dependent RNA polymerases. *Curr. Opin. Struct. Biol.* **2006**, *16*, 27–34.
3. Ng, K.K.; Arnold, J.J.; Cameron, C.E. Structure-function relationships among RNA-dependent RNA polymerases. *Curr. Top. Microbiol. Immunol.* **2008**, *320*, 137–156.
4. Wimmer, E.; Paul, A.V. The making of a picornavirus genome. In *The Picornavirus*; Ehrenfeld, E., Domingo, E., Ross, R.P., Eds.; ASM Press: Washington, DC, USA, 2010; pp. 33–55.
5. Thompson, A.A.; Peersen, O.B. Structural basis for proteolysis-dependent activation of the poliovirus RNA-dependent RNA polymerase. *EMBO J.* **2004**, *23*, 3462–3471.
6. Gong, P.; Kortus, M.G.; Nix, J.C.; Davis, R.E.; Peersen, O.B. Structures of coxsackievirus, rhinovirus, and poliovirus polymerase elongation complexes solved by engineering RNA mediated crystal contacts. *PLoS ONE* **2013**, *8*, e60272.
7. Gong, P.; Peersen, O.B. Structural basis for active site closure by the poliovirus RNA-dependent RNA polymerase. *Proc. Natl. Acad. Sci. USA* **2010**, *107*, 22505–22510.
8. Gruez, A.; Selisko, B.; Roberts, M.; Bricogne, G.; Bussetta, C.; Jabafi, I.; Coutard, B.; de Palma, A.M.; Neyts, J.; Canard, B. The crystal structure of coxsackievirus B3 RNA-dependent RNA polymerase in complex with its protein primer VPg confirms the existence of a second VPg binding site on *Picornaviridae* polymerases. *J. Virol.* **2008**, *82*, 9577–9590.

9. Campagnola, G.; Weygandt, M.; Scoggin, K.; Peersen, O. Crystal structure of coxsackievirus B3 3D^{Pol} highlights the functional importance of residue 5 in picornavirus polymerases. *J. Virol.* **2008**, *82*, 9458–9464.

10. Wu, Y.; Lou, Z.; Miao, Y.; Yu, Y.; Dong, H.; Peng, W.; Bartlam, M.; Li, X.; Rao, Z. Structures of EV71 RNA-dependent RNA polymerase in complex with substrate and analogue provide a drug target against the hand-foot-and-mouth disease pandemic in China. *Protein Cell* **2010**, *1*, 491–500.

11. Love, R.A.; Maegley, K.A.; Yu, X.; Ferre, R.A.; Lingardo, L.K.; Diehl, W.; Parge, H.E.; Dragovich, P.S.; Fuhrman, S.A. The crystal structure of the RNA-dependent RNA polymerase from human rhinovirus: A dual function target for common cold antiviral therapy. *Structure* **2004**, *12*, 1533–1544.

12. Appleby, T.C.; Luecke, H.; Shim, J.H.; Wu, J.Z.; Cheney, I.W.; Zhong, W.; Vogeley, L.; Hong, Z.; Yao, N. Crystal structure of complete rhinovirus RNA polymerase suggests front loading of protein primer. *J. Virol.* **2005**, *79*, 277–288.

13. Ferrer-Orta, C.; Arias, A.; Agudo, R.; Perez-Luque, R.; Escarmis, C.; Domingo, E.; Verdaguer, N. The structure of a protein primer-polymerase complex in the initiation of genome replication. *EMBO J.* **2006**, *25*, 880–888.

14. Ferrer-Orta, C.; Arias, A.; Perez-Luque, R.; Escarmis, C.; Domingo, E.; Verdaguer, N. Structure of foot-and-mouth disease virus RNA-dependent RNA polymerase and its complex with a template-primer RNA. *J. Biol. Chem.* **2004**, *279*, 47212–47221.

15. Ferrer-Orta, C.; Arias, A.; Perez-Luque, R.; Escarmis, C.; Domingo, E.; Verdaguer, N. Sequential structures provide insights into the fidelity of RNA replication. *Proc. Natl. Acad. Sci. USA* **2007**, *104*, 9463–9468.

16. Shen, H.; Sun, H.; Li, G. What is the role of motif D in the nucleotide incorporation catalyzed by the RNA-dependent RNA polymerase from poliovirus? *PLoS Comput. Biol.* **2012**, *8*, e1002851.

17. Pfeiffer, J.K.; Kirkegaard, K. A single mutation in poliovirus RNA-dependent RNA polymerase confers resistance to mutagenic nucleotide analogs via increased fidelity. *Proc. Natl. Acad. Sci. USA* **2003**, *100*, 7289–7294.

18. Arnold, J.J.; Vignuzzi, M.; Stone, J.K.; Andino, R.; Cameron, C.E. Remote site control of an active site fidelity checkpoint in a viral RNA-dependent RNA polymerase. *J. Biol. Chem.* **2005**, *280*, 25706–25716.

19. Agudo, R.; Ferrer-Orta, C.; Arias, A.; de la Higuera, I.; Perales, C.; Perez-Luque, R.; Verdaguer, N.; Domingo, E. A multi-step process of viral adaptation to a mutagenic nucleoside analogue by modulation of transition types leads to extinction-escape. *PLoS Pathog.* **2010**, *6*, e1001072.

20. Marcotte, L.L.; Wass, A.B.; Gohara, D.W.; Pathak, H.B.; Arnold, J.J.; Filman, D.J.; Cameron, C.E.; Hogle, J.M. Crystal structure of poliovirus 3CD protein: Virally encoded protease and precursor to the RNA-dependent RNA polymerase. *J. Virol.* **2007**, *81*, 3583–3596.

21. Ferrer-Orta, C.; Sierra, M.; Agudo, R.; de la Higuera, I.; Arias, A.; Perez-Luque, R.; Escarmis, C.; Domingo, E.; Verdaguer, N. Structure of foot-and-mouth disease virus mutant polymerases with reduced sensitivity to ribavirin. *J. Virol.* **2010**, *84*, 6188–6199.

22. Moustafa, I.M.; Shen, H.; Morton, B.; Colina, C.M.; Cameron, C.E. Molecular dynamics simulations of viral RNA polymerases link conserved and correlated motions of functional elements to fidelity. *J. Mol. Biol.* **2011**, *410*, 159–181.

23. Van Dijk, A.A.; Makeyev, E.V.; Bamford, D.H. Initiation of viral RNA-dependent RNA polymerization. *J. Gen. Virol.* **2004**, *85*, 1077–1093.

24. Choi, K.H.; Rossmann, M.G. RNA-dependent RNA polymerases from *Flaviviridae. Curr. Opin. Struct. Biol.* **2009**, *19*, 746–751.

25. Lescar, J.; Canard, B. RNA-dependent RNA polymerases from flaviviruses and *Picornaviridae. Curr. Opin. Struct. Biol.* **2009**, *19*, 759–767.

26. Butcher, S.J.; Grimes, J.M.; Makeyev, E.V.; Bamford, D.H.; Stuart, D.I. A mechanism for initiating RNA-dependent RNA polymerization. *Nature* **2001**, *410*, 235–240.

27. Mosley, R.T.; Edwards, T.E.; Murakami, E.; Lam, A.M.; Grice, R.L.; Du, J.; Sofia, M.J.; Furman, P.A.; Otto, M.J. Structure of hepatitis C virus polymerase in complex with primer-template RNA. *J. Virol.* **2012**, *86*, 6503–6511.

28. Appleby, T.C.; Perry, J.K.; Murakami, E.; Barauskas, O.; Feng, J.; Cho, A.; Fox, D., 3rd; Wetmore, D.R.; McGrath, M.E.; Ray, A.S.; *et al.* Viral replication. Structural basis for RNA replication by the hepatitis C virus polymerase. *Science* **2015**, *347*, 771–775.

29. Ferrer-Orta, C.; Agudo, R.; Domingo, E.; Verdaguer, N. Structural insights into replication initiation and elongation processes by the FMDV RNA-dependent RNA polymerase. *Curr. Opin. Struct. Biol.* **2009**, *19*, 752–758.

30. Paul, A.V.; van Boom, J.H.; Filippov, D.; Wimmer, E. Protein-primed RNA synthesis by purified poliovirus RNA polymerase. *Nature* **1998**, *393*, 280–284.

31. Paul, A.V.; Wimmer, E. Initiation of protein-primed picornavirus RNA synthesis. *Virus Res.* **2015**, *206*, 12–26.

32. Goodfellow, I. The genome-linked protein VPg of vertebrate viruses—A multifaceted protein. *Curr. Opin. Virol.* **2011**, *1*, 355–362.

33. Lyle, J.M.; Clewell, A.; Richmond, K.; Richards, O.C.; Hope, D.A.; Schultz, S.C.; Kirkegaard, K. Similar structural basis for membrane localization and protein priming by an RNA-dependent RNA polymerase. *J. Biol. Chem.* **2002**, *277*, 16324–16331.

34. Chen, C.; Wang, Y.; Shan, C.; Sun, Y.; Xu, P.; Zhou, H.; Yang, C.; Shi, P.Y.; Rao, Z.; Zhang, B.; *et al.* Crystal structure of enterovirus 71 RNA-dependent RNA polymerase complexed with its protein primer VPg: Implication for a trans mechanism of VPg uridylylation. *J. Virol.* **2013**, *87*, 5755–5768.

35. Forss, S.; Schaller, H. A tandem repeat gene in a picornavirus. *Nucleic Acids Res.* **1982**, *10*, 6441–6450.

36. King, A.M.; Sangar, D.V.; Harris, T.J.; Brown, F. Heterogeneity of the genome-linked protein of foot-and-mouth disease virus. *J. Virol.* **1980**, *34*, 627–634.

37. Falk, M.M.; Sobrino, F.; Beck, E. VPg gene amplification correlates with infective particle formation in foot-and-mouth disease virus. *J. Virol.* **1992**, *66*, 2251–2260.

38. Pacheco, J.M.; Henry, T.M.; O'Donnell, V.K.; Gregory, J.B.; Mason, P.W. Role of nonstructural proteins 3A and 3B in host range and pathogenicity of foot-and-mouth disease virus. *J. Virol.* **2003**, *77*, 13017–13027.

39. Carrillo, C.; Lu, Z.; Borca, M.V.; Vagnozzi, A.; Kutish, G.F.; Rock, D.L. Genetic and phenotypic variation of foot-and-mouth disease virus during serial passages in a natural host. *J. Virol.* **2007**, *81*, 11341–11351.

40. MacKenzie, J.S.; Slade, W.R.; Lake, J.; Priston, R.A.; Bisby, J.; Laing, S.; Newman, J. Temperature-sensitive mutants of foot-and-mouth disease virus: The isolation of mutants and observations on their properties and genetic recombination. *J. Gen. Virol.* **1975**, *27*, 61–70.

41. Korneeva, V.S.; Cameron, C.E. Structure-function relationships of the viral RNA-dependent RNA polymerase: Fidelity, replication speed, and initiation mechanism determined by a residue in the ribose-binding pocket. *J. Biol. Chem.* **2007**, *282*, 16135–16145.

42. Sun, Y.; Wang, Y.; Shan, C.; Chen, C.; Xu, P.; Song, M.; Zhou, H.; Yang, C.; Xu, W.; Shi, P.Y.; *et al.* Enterovirus 71 VPg uridylation uses a two-molecular mechanism of 3D polymerase. *J. Virol.* **2012**, *86*, 13662–13671.

43. Sun, Y.; Guo, Y.; Lou, Z. Formation and working mechanism of the picornavirus VPg uridylylation complex. *Curr. Opin. Virol.* **2014**, *9*, 24–30.

44. Berdis, A.J. Mechanisms of DNA polymerases. *Chem. Rev.* **2009**, *109*, 2862–2879.

45. Steitz, T.A. Visualizing polynucleotide polymerase machines at work. *EMBO J.* **2006**, *25*, 3458–3468.

46. Kornberg, R.D. The molecular basis of eukaryotic transcription. *Proc. Natl. Acad. Sci. USA* **2007**, *104*, 12955–12961.

47. Steitz, T.A. The structural changes of T7 RNA polymerase from transcription initiation to elongation. *Curr. Opin. Struct. Biol.* **2009**, *19*, 683–690.

48. Zamyatkin, D.F.; Parra, F.; Alonso, J.M.; Harki, D.A.; Peterson, B.R.; Grochulski, P.; Ng, K.K. Structural insights into mechanisms of catalysis and inhibition in Norwalk virus polymerase. *J. Biol. Chem.* **2008**, *283*, 7705–7712.

49. Zamyatkin, D.F.; Parra, F.; Machin, A.; Grochulski, P.; Ng, K.K. Binding of 2′-amino-2′-deoxycytidine-5′-triphosphate to norovirus polymerase induces rearrangement of the active site. *J. Mol. Biol.* **2009**, *390*, 10–16.

50. Garriga, D.; Ferrer-Orta, C.; Querol-Audi, J.; Oliva, B.; Verdaguer, N. Role of motif B loop in allosteric regulation of RNA-dependent RNA polymerization activity. *J. Mol. Biol.* **2013**, *425*, 2279–2287.

51. Sholders, A.J.; Peersen, O.B. Distinct conformations of a putative translocation element in poliovirus polymerase. *J. Mol. Biol.* **2014**, *426*, 1407–1419.

52. Castro, C.; Smidansky, E.; Maksimchuk, K.R.; Arnold, J.J.; Korneeva, V.S.; Gotte, M.; Konigsberg, W.; Cameron, C.E. Two proton transfers in the transition state for nucleotidyl transfer catalyzed by RNA- and DNA-dependent RNA and DNA polymerases. *Proc. Natl. Acad. Sci. USA* **2007**, *104*, 4267–4272.

53. Castro, C.; Smidansky, E.D.; Arnold, J.J.; Maksimchuk, K.R.; Moustafa, I.; Uchida, A.; Gotte, M.; Konigsberg, W.; Cameron, C.E. Nucleic acid polymerases use a general acid for nucleotidyl transfer. *Nat. Struct. Mol. Biol.* **2009**, *16*, 212–218.

54. Yang, X.; Smidansky, E.D.; Maksimchuk, K.R.; Lum, D.; Welch, J.L.; Arnold, J.J.; Cameron, C.E.; Boehr, D.D. Motif D of viral RNA-dependent RNA polymerases determines efficiency and fidelity of nucleotide addition. *Structure* **2012**, *20*, 1519–1527.

55. Gnädig, N.F.; Beaucourt, S.; Campagnola, G.; Bordería, A.V.; Sanz-Ramos, M.; Gong, P.; Blanc, H.; Peersen, O.B.; Vignuzzi, M. Coxsackievirus B3 mutator strains are attenuated *in vivo*. *Proc. Natl. Acad. Sci. USA* **2012**, *109*, E2294–E2303.

56. Campagnola, G.; McDonald, S.; Beaucourt, S.; Vignuzzi, M.; Peersen, O.B. Structure-function relationships underlying the replication fidelity of viral RNA-dependent RNA polymerases. *J. Virol.* **2015**, *89*, 275–286.

57. Zamyatkin, D.; Rao, C.; Hoffarth, E.; Jurca, G.; Rho, H.; Parra, F.; Grochulski, P.; Ng, K.K. Structure of a backtracked state reveals conformational changes similar to the state following nucleotide incorporation in human norovirus polymerase. *Acta Crystallogr. D Biol. Crystallogr.* **2014**, *70*, 3099–3109.

58. Sankar, S.; Porter, A.G. Point mutations which drastically affect the polymerization activity of encephalomyocarditis virus RNA-dependent RNA polymerase correspond to the active site of *Escherichia coli* DNA polymerase I. *J. Biol. Chem.* **1992**, *267*, 10168–10176.

59. Boehr, D.D. The ins and outs of viral RNA polymerase translocation. *J. Mol. Biol.* **2014**, *426*, 1373–1376.

60. Vives-Adrian, L.; Lujan, C.; Oliva, B.; van der Linden, L.; Selisko, B.; Coutard, B.; Canard, B.; van Kuppeveld, F.J.; Ferrer-Orta, C.; Verdaguer, N. The crystal structure of a cardiovirus RNA-dependent RNA polymerase reveals an unusual conformation of the polymerase active site. *J. Virol.* **2014**, *88*, 5595–5607.

61. Hall, D.J.; Palmenberg, A.C. Cleavage site mutations in the encephalomyocarditis virus P3 region lethally abrogate the normal processing cascade. *J. Virol.* **1996**, *70*, 5954–5961.

62. Ferrer-Orta, C.; de la Higuera, I.; Caridi, F.; Sanchez-Aparicio, M.T.; Moreno, E.; Perales, C.; Singh, K.; Sarafianos, S.G.; Sobrino, F.; Domingo, E.; *et al.* Multifunctionality of a picornavirus polymerase domain: Nuclear localization signal and nucleotide recognition. *J. Virol.* **2015**, *89*, 6848–6859.

63. Hansen, J.L.; Long, A.M.; Schultz, S.C. Structure of the RNA-dependent RNA polymerase of poliovirus. *Structure* **1997**, *5*, 1109–1122.

64. Hobson, S.D.; Rosenblum, E.S.; Richards, O.C.; Richmond, K.; Kirkegaard, K.; Schultz, S.C. Oligomeric structures of poliovirus polymerase are important for function. *EMBO J.* **2001**, *20*, 1153–1163.

65. Lyle, J.M.; Bullitt, E.; Bienz, K.; Kirkegaard, K. Visualization and functional analysis of RNA-dependent RNA polymerase lattices. *Science* **2002**, *296*, 2218–2222.

66. Spagnolo, J.F.; Rossignol, E.; Bullitt, E.; Kirkegaard, K. Enzymatic and nonenzymatic functions of viral RNA-dependent RNA polymerases within oligomeric arrays. *RNA* **2010**, *16*, 382–393.

67. Kaiser, W.J.; Chaudhry, Y.; Sosnovtsev, S.V.; Goodfellow, I.G. Analysis of protein-protein interactions in the feline calicivirus replication complex. *J. Gen. Virol.* **2006**, *87*, 363–368.
68. Luo, G.; Hamatake, R.K.; Mathis, D.M.; Racela, J.; Rigat, K.L.; Lemm, J.; Colonno, R.J. *De novo* initiation of RNA synthesis by the RNA-dependent RNA polymerase (NS5B) of hepatitis C virus. *J. Virol.* **2000**, *74*, 851–863.
69. Chinnaswamy, S.; Murali, A.; Li, P.; Fujisaki, K.; Kao, C.C. Regulation of *de novo*-initiated RNA synthesis in hepatitis C virus RNA-dependent RNA polymerase by intermolecular interactions. *J. Virol.* **2010**, *84*, 5923–5935.
70. Hogbom, M.; Jager, K.; Robel, I.; Unge, T.; Rohayem, J. The active form of the norovirus RNA-dependent RNA polymerase is a homodimer with cooperative activity. *J. Gen. Virol.* **2009**, *90*, 281–291.
71. Cevik, B. The RNA-dependent RNA polymerase of *Citrus tristeza* virus forms oligomers. *Virology* **2013**, *447*, 121–130.
72. Ferrero, D.; Buxaderas, M.; Rodriguez, J.F. The structure of the RNA-dependent RNA polymerase of a Permutotetravirus suggests a link between primer-dependent and primer-independent polymerases. *PLoS Pathog.* **2015**. Submitted for publication.
73. Chang, S.; Sun, D.; Liang, H.; Wang, J.; Li, J.; Guo, L.; Wang, X.; Guan, C.; Boruah, B.M.; Yuan, L.; *et al.* Cryo-EM structure of influenza virus RNA polymerase complex at 4.3 Å resolution. *Mol. Cell* **2015**, *57*, 925–935.
74. Graham, S.C.; Sarin, L.P.; Bahar, M.W.; Myers, R.A.; Stuart, D.I.; Bamford, D.H.; Grimes, J.M. The N-terminus of the RNA polymerase from infectious pancreatic necrosis virus is the determinant of genome attachment. *PLoS Pathog.* **2011**, *7*, e1002085.
75. Smallwood, S.; Hovel, T.; Neubert, W.J.; Moyer, S.A. Different substitutions at conserved amino acids in domains II and III in the Sendai L RNA polymerase protein inactivate viral RNA synthesis. *Virology* **2002**, *304*, 135–145.
76. Zamoto-Niikura, A.; Terasaki, K.; Ikegami, T.; Peters, C.J.; Makino, S. Rift valley fever virus L protein forms a biologically active oligomer. *J. Virol.* **2009**, *83*, 12779–12789.
77. Tellez, A.B.; Wang, J.; Tanner, E.J.; Spagnolo, J.F.; Kirkegaard, K.; Bullitt, E. Interstitial contacts in an RNA-dependent RNA polymerase lattice. *J. Mol. Biol.* **2011**, *412*, 737–750.
78. Wang, J.; Lyle, J.M.; Bullitt, E. Surface for catalysis by poliovirus RNA-dependent RNA polymerase. *J. Mol. Biol.* **2013**, *425*, 2529–2540.
79. Powdrill, M.H.; Bernatchez, J.A.; Gotte, M. Inhibitors of the hepatitis C virus RNA-dependent RNA polymerase NS5B. *Viruses* **2010**, *2*, 2169–2195.
80. Gentile, I.; Buonomo, A.R.; Zappulo, E.; Coppola, N.; Borgia, G. GS-9669: A novel non-nucleoside inhibitor of viral polymerase for the treatment of hepatitis C virus infection. *Expert Rev. Anti-Infect. Ther.* **2014**, *12*, 1179–1186.
81. Caillet-Saguy, C.; Lim, S.P.; Shi, P.Y.; Lescar, J.; Bressanelli, S. Polymerases of hepatitis C viruses and flaviviruses: Structural and mechanistic insights and drug development. *Antivir. Res.* **2014**, *105*, 8–16.
82. Eltahla, A.A.; Lim, K.L.; Eden, J.S.; Kelly, A.G.; Mackenzie, J.M.; White, P.A. Nonnucleoside inhibitors of norovirus RNA polymerase: Scaffolds for rational drug design. *Antimicrob. Agents Chemother.* **2014**, *58*, 3115–3123.

83. Carfi, M.; Gennari, A.; Malerba, I.; Corsini, E.; Pallardy, M.; Pieters, R.; van Loveren, H.; Vohr, H.W.; Hartung, T.; Gribaldo, L. *In vitro* tests to evaluate immunotoxicity: A preliminary study. *Toxicology* **2007**, *229*, 11–22.

84. De Clercq, E.; Neyts, J. Antiviral agents acting as DNA or RNA chain terminators. *Handb. Exp. Pharmacol.* **2009**, *189*, 53–84.

85. Domingo, E. Virus entry into error catastrophe as a new antiviral strategy. *Virus Res.* **2005**, *107*, 115–228.

86. De Francesco, R.; Tomei, L.; Altamura, S.; Summa, V.; Migliaccio, G. Approaching a new era for hepatitis C virus therapy: Inhibitors of the NS3–4A serine protease and the NS5B RNA-dependent RNA polymerase. *Antivir. Res.* **2003**, *58*, 1–16.

87. Biswal, B.K.; Wang, M.; Cherney, M.M.; Chan, L.; Yannopoulos, C.G.; Bilimoria, D.; Bedard, J.; James, M.N. Non-nucleoside inhibitors binding to hepatitis C virus NS5B polymerase reveal a novel mechanism of inhibition. *J. Mol. Biol.* **2006**, *361*, 33–45.

88. Paeshuyse, J.; Leyssen, P.; Mabery, E.; Boddeker, N.; Vrancken, R.; Froeyen, M.; Ansari, I.H.; Dutartre, H.; Rozenski, J.; Gil, L.H.; *et al.* A novel, highly selective inhibitor of pestivirus replication that targets the viral RNA-dependent RNA polymerase. *J. Virol.* **2006**, *80*, 149–160.

89. Paeshuyse, J.; Chezal, J.M.; Froeyen, M.; Leyssen, P.; Dutartre, H.; Vrancken, R.; Canard, B.; Letellier, C.; Li, T.; Mittendorfer, H.; *et al.* The imidazopyrrolopyridine analogue AG110 is a novel, highly selective inhibitor of pestiviruses that targets the viral RNA-dependent RNA polymerase at a hot spot for inhibition of viral replication. *J. Virol.* **2007**, *81*, 11046–11053.

90. Van der Linden, L.; Vives-Adrián, L.; Selisko, B.; Ferrer-Orta, C.; Liu, X.; Lanke, K.; Ulferts, R.; de Palma, A.M.; Tanchis, F.; Goris, N.; *et al.* The RNA template channel of the RNA-dependent RNA polymerase as a target for development of antiviral therapy of multiple genera within a virus family. *PLoS Pathog.* **2015**, *23*, e1004733.

Using the Hepatitis C Virus RNA-Dependent RNA Polymerase as a Model to Understand Viral Polymerase Structure, Function and Dynamics

Ester Sesmero and Ian F. Thorpe

Abstract: Viral polymerases replicate and transcribe the genomes of several viruses of global health concern such as Hepatitis C virus (HCV), human immunodeficiency virus (HIV) and Ebola virus. For this reason they are key targets for therapies to treat viral infections. Although there is little sequence similarity across the different types of viral polymerases, all of them present a right-hand shape and certain structural motifs that are highly conserved. These features allow their functional properties to be compared, with the goal of broadly applying the knowledge acquired from studying specific viral polymerases to other viral polymerases about which less is known. Here we review the structural and functional properties of the HCV RNA-dependent RNA polymerase (NS5B) in order to understand the fundamental processes underlying the replication of viral genomes. We discuss recent insights into the process by which RNA replication occurs in NS5B as well as the role that conformational changes play in this process.

Reprinted from *Viruses*. Cite as: Sesmero, E.; Thorpe, I.F. Using the Hepatitis C Virus RNA-Dependent RNA Polymerase as a Model to Understand Viral Polymerase Structure, Function and Dynamics. *Viruses* **2015**, *7*, 3974–3994.

1. Introduction

Polymerases are crucial in the viral life cycle. They have an essential role in replicating and transcribing the viral genome and as a result are key targets for therapies to treat viral infection. A virus may not need to encode its own polymerase depending on where it spends most of its life cycle. Some small DNA viruses that spend all their time in the cell nucleus can make use of the host cell's polymerases. However, viruses that remain in the cytoplasm do need to encode their own [1].

For viruses that require their own polymerase, most of these enzymes display detectable activity *in vitro* without accessory factors. This is primarily because the sizes of genomes that can be packaged in the viral capsid are limited [1,2]. In addition, some polymerases perform other functions related to viral genome transcription and replication. Examples include the RNA-dependent RNA polymerases from the Flavivirus genus of the Flaviviridae family, retrovirus reverse transcriptases and some viral DNA-dependent polymerases. Flavivirus polymerases have a methyltransferase domain that catalyzes methylations of a 5′-RNA cap [3]. The retrovirus reverse transcriptase has an additional ribonuclease H domain that catalyzes degradation of the RNA strand in the RNA-DNA hybrid during genome replication [4]. Some viral DNA-dependent polymerases have a nuclease domain with proof-reading activity to correct nucleotides incorrectly incorporated during genome synthesis [5].

With regard to copying the viral genome, distinct replication mechanisms are used by different types of viral polymerases. A number of functions must be orchestrated depending on the specific virus in question [1]:

(1) Recognition of the nucleic acid binding site
(2) Coordination of the chemical steps of nucleic acid synthesis
(3) Conformational rearrangement to allow for processive elongation
(3) Termination of replication at the end of the genome

Viral polymerases are often classified into four main categories based on the nature of the genetic material of the virus as follows: RNA-dependent RNA polymerases (RdRps), RNA-dependent DNA polymerases (RdDps), DNA-dependent RNA polymerases (DdRps), and DNA-dependent DNA polymerases (DdDps) [1]. DdDps and DdRps are used for the replication and transcription, respectively, of DNA for both viruses and eukaryotic cells. In contrast, RdDps and RdRps are mainly used by viruses since the host cell does not require reverse transcription or RNA replication. RdDps are employed by retroviruses such as the human immunodeficiency virus (HIV). RdRps are employed by viruses such as Hepatitis C virus (HCV), poliovirus (PV), human rhinovirus (HRV), foot-and-mouth-disease virus (FMDV) and coxsackie viruses (CV) among others. We will primarily focus on RdRps in this review since they are crucial in the replication process of viruses that are important global pathogens.

There are seven classes of viruses according to the Baltimore classification [6] based on the genome type and method of mRNA synthesis. These are associated with the four classes of polymerases specified in the previous paragraph as shown in Table 1.

Table 1. Baltimore classification of viruses compared with the classification of viral polymerases based on their targeted genetic material.

Genetic Material	Baltimore Classification	Polymerase Classes	Examples
DNA	ssDNA viruses dsDNA viruses	DNA dependent DNA polymerases DNA dependent RNA polymerases	Human parvovirus B19 Bacteriophage φ29
RNA	(+) ssRNA viruses (−) ssRNA viruses dsRNA	RNA dependent RNA polymerases	HCV, PV, West Nile virus Influenza Bacteriophage φ6
RNA/ DNA	ssRNA-rt viruses dsDNA-rt viruses	RNA dependent DNA polymerases	Retrovirus Hepatitis B

30

2. General Structural Features of Viral Polymerases

The structure of all polymerases resembles a cupped right hand and is divided into three domains referred to as the palm, fingers and thumb (see Figure 1a) [1,7]. This nomenclature is based on an analogy to the structure of the Klenow fragment of DNA polymerase [8]. The palm domain is the most highly conserved domain across different polymerases and is the location of the active site. In contrast, the thumb domain is the most variable. Fingers and thumb domains vary significantly in both size and secondary structure depending on the specific requirements for replication in a given virus (*i.e.*, replicating single- or double-stranded RNA/DNA genomes). The fingers and thumb domains of different polymerases have similar positions with respect to the palm, which contains the active site in which catalytic addition of nucleotides occurs. Changes in the relative positions of the fingers and thumb domains are associated with conformational changes of the polymerase at different stages of replication [7]. Three well-defined channels have been identified on the polymerase, serving as the entry path for template and NTPs (*i.e.*, the template and NTP channels) and exit path for double stranded RNA (dsRNA) product (*i.e.*, the duplex channel) [9,10] (see Figure 1b,c,e).

In the active site, the correct NTP to be added to the daughter strand is selected by Watson-Crick base-pairing with the template base. The selectivity for ribose (rNTP) *vs.* deoxyribose NTPs (dNTP) is regulated by the interaction of the polymerase with the 2′-OH of the NTP. In general, DNA polymerases that incorporate dNTP in the growing daughter strand have a large side chain that prevents binding of an rNTP with a 2′-OH. However, RNA polymerases utilize amino acids with a small side chain and form H-bonds with the 2′-OH of the rNTP. The polymerase active site often binds the correct NTP with 10–1000-fold higher affinity than incorrect NTPs [11]. While viral polymerases often have domains in addition to the fingers, palm and thumb that carry out functions related to other aspects of viral genome transcription and replication (see Introduction), this is not the case for the HCV polymerase.

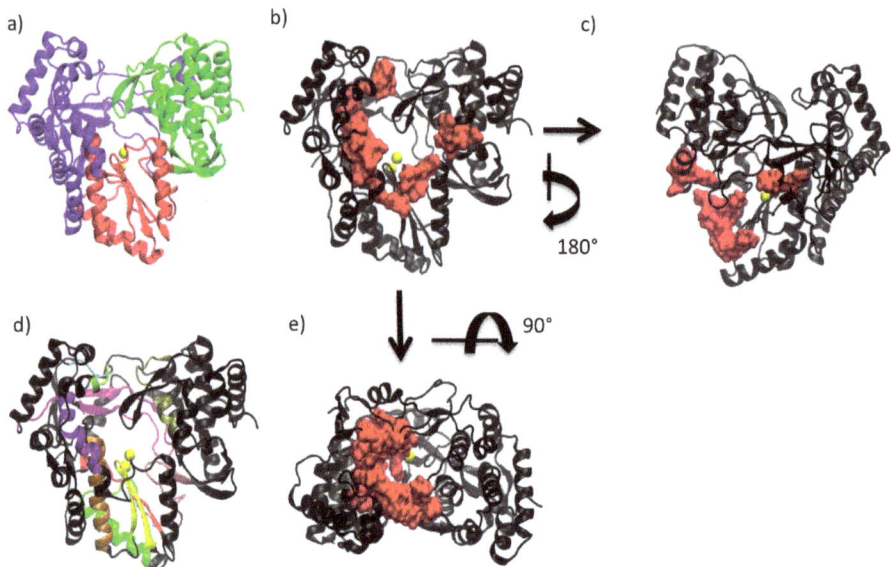

Figure 1. (**a**) Right-hand structure of HCV polymerase (NS5B). Palm, fingers and thumb domains are shown in red, blue and green respectively; (**b**) Duplex channel in NS5B (front of the enzyme); (**c**) NTP channel in NS5B (back of the enzyme); (**d**) Motifs and functional regions of NS5B. Motif A in red, B in orange, C in yellow, D in bright green, E in pink, F in purple and G in cyan. Functional regions: I in light green, II in violet and III in tan; (**e**) Template channel (top view of the enzyme).

3. Conserved Structural Motifs of Viral Polymerases

There are several structural motifs (designated A through G, see Figure 1d) that display varying levels of conservation among the different viral polymerases. Some motifs have been shown to be conserved across all viral polymerases (motifs A to E) while others (motifs F and G) have only been shown to be conserved for the RdRps. High levels of conservation despite the low sequence similarity among polymerases suggests that these motifs have functions that are vital for the action of these enzymes [1,7,9,12,13].

Motifs A and C have been closely studied because they are located in the active site. Motif C includes the GDD amino acid sequence that is the hallmark of RdRps. These conserved residues are bound to the metal ions (Mg^{2+} or Mn^{2+}) necessary for catalysis. Motif B contains a consensus sequence of SGxxxT and is located at the junction of the fingers and palm domains [7]. Motif F binds to incoming NTPs and RNA and is situated near the entrance of the RNA template channel. The sequence of this motif is not conserved in *de novo* initiating RdRps such as that present in HCV [14]. These polymerase sequence motifs have also been used to identify new polymerase genes in newly sequenced virus genomes [1]. Further details about the roles of each motif are shown in Table 2. Other regions have also been shown to have fundamental importance in RdRp function and have been named "functional regions". See Table 3 for a list of the residues included in these regions and the functional roles of each.

Table 2. Characteristic sequence motifs in polymerases from prototypical viruses: hepatitis C virus (HCV), poliovirus (PV) and foot-and-mouth-disease virus (FMDV) [7,9,12,15,16].

Conserved Elements		Role	Location	Residues		
				HCV	PV	FMDV
Motifs	A	Coordinates Magnesium and selects type of nucleic acid (RNA vs. DNA)	Palm	216–227	229–240	236–247
	B	Determines nucleotide choice (rNTP or dNTP)	Palm	287–306	293–312	303–322
	C	Coordinates Magnesium	Palm	312–325	322–335	332–345
	D	Helps accommodate active site NTPs	Palm	332–353	338–362	348–373
	E	Maintains rigidity of secondary structure that is required for relative positioning of thumb and palm domains	Palm	354–372	363–380	374–392
	F	Binds incoming NTPs and RNA	Fingers	132–162	153–178	158–183
	G	Binds primer and template	Fingers	95–99	113–120	114–121
Functional regions	I	Binds template	Fingers	91–94	107–112	108–113
	II	Binds template	Fingers	168–183	184–200	189–205
	III	Binds nascent RNA duplex	Thumb	401–414	405–420	416–430

Table 3. RdRps virus families and species [1,17].

	Virus family	Representative Species
(+) ssRNA	Picornaviradae	Poliovirus (PV) Human rhinovirus (HRV) Foot-and-mouth-disease virus (FMDV) Coxsackie viruses (CV) Hepatitis A virus (HAV)
	Caliciviridae	Rabbit hemorrhagic disease virus (RHDV) Norwalk virus (NV) Sapporo virus
	Togaviridae	Sindbis virus
	Flaviviridae	West Nile virus (WNV) Yellow fever virus Dengue virus (DENV) Japanese encephalitis disease virus (JEV) Hepatitis C virus (HCV) Bovine viral diarrhea virus (BVDV)
(−) ssRNA	Orthomyxoviridae Paramyxoviridae Bunyaviridae Rhabdoviridae Filoviridae Bornaviridae	Influenza virus Measles and mumps viruses Hantavirus Rabies virus Ebola and Marburg virus Borna disease virus
dsRNA	Cystoviridae Reoviridae Birnaviridae	Bacteriophage φ6 Reovirus Fish infectious pancreatic necrosis virus (IPNV) Infectious bursal disease virus (IBDV)

4. Structural Features of RdRps

RdRps replicate the genomic material in RNA viruses. Many of these viruses are significant public health concerns including HCV, Dengue virus, Japanese encephalitis and yellow fever. For this reason RdRps are key targets for new drugs and it is crucial to understand the mechanisms by which they replicate viral genomes. The fact that there are no mammalian homologs of RdRps [18,19] makes them an optimal drug target because potential therapeutics would tend to selectively affect the viral polymerases without interfering with the function of host polymerases.

Within RdRp encoding viruses there are ssRNA viruses (both + and − sense) and dsRNA viruses (see Table 3). Genome replication in (+) ssRNA viruses takes place in a membrane-bound replication complex [9,20,21]. (+) RNA serves as mRNA and can be translated immediately after entering the cell [1]. Thus, unlike the (−) RNA viruses, (+) RNA viruses do not need to package an RdRp within the virion [22].

The first X-ray structure of an RdRp was generated for Poliovirus (PV) polymerase in 1997 [23]. X-ray structures are currently available from seven families of RdRps. These include (+) RNA viruses: Picornaviridae (PV, HRV, FMDV, CV and HAV), Caliciviridae (RHDV, NV and Sapporo virus) and Flaviviridae (HCV and BVDV) as well as (−) RNA viruses: Orthomyxoviridae (Influenza virus) and dsRNA viruses: Cystoviridae (Bacteriophage φ6), Reoviridae (Reovirus and Rotavirus) and Birnaviridae (IBDV). A table listing each NS5B structure currently available in the PDB is included as supporting information. The PDB IDs, a description of each structure and their resolution is provided. Similar information for other viral polymerases is presented in Tables 1 and 2 of Subissi *et al.* [14].

A characteristic trait of RdRps is the extensive interaction between fingers and palm domains [24]. RdRps have an extension of the fingers domain called the fingertips that connects the fingers and thumb domains to form a fully enclosed active site. The fingertips also contribute to the formation of well-defined template and NTP channels in the front and back of the polymerase, respectively.

RdRps were originally thought to be found uniquely in viruses. However, in 1971 the first eukaryotic RdRp was found in Chinese Cabbage [25]. Later on cellular RdRps were also found in plants, fungi and nematodes [26–28]. Cellular RdRps play important roles in both transcriptional and post-transcriptional gene silencing [29]. Although viral and cellular RdRps show little sequence homology, both share the "right hand" shape containing palm, thumb and fingers domains. The palm domain of cellular RdRps is particularly well-conserved and contains four motifs maintained in all polymerases. These facts make it likely that the cellular RdRps share some of the basic mechanistic principles of viral RdRps and that knowledge obtained for viral RdRps may be transferable to cellular RdRps [13].

The Flaviviridae family has been widely studied because many members of this family cause diseases in humans. Within this family there are three genera: Flaviviruses, Hepaciviruses and Pestiviruses (see Table 4). HCV is part of the Hepaciviruses genus and is an important pathogen for which no vaccine is currently available. In explaining recent insights regarding the mechanism by which the HCV RdRp (gene product NS5B) replicates the viral genome, we will make comparisons with other members of the Flaviviridae family. However, we note that some differences may exist, particularly if the other family members are part of a different genus.

Table 4. Genera and species of the *Flaviviridae* family.

Virus Family	Genus	Species
Flaviviridae	*Flaviviridae*	West Nile virus Yellow fever virus Dengue virus Japanese encephalitis disease virus
	Hepaciviruses	Hepatitis C virus (HCV)
	Pestiviruses	Bovine viral diarrhea virus (BVDV)

5. Catalytic Mechanism and Polymerase Reaction Steps

All known polymerases synthesize nucleic acid in the $5'$ to $3'$ direction [9]. Thus, replication in positive-stranded RNA viruses occurs via a negative-stranded intermediate. The polymerase reaction has three stages: initiation, elongation and termination. For this cycle to take place the polymerase needs to have binding sites for: (a) the template strand; (b) the primer strand or initiating NTP (P-site) and (c) incoming NTP (N-site). The $3'$-nucleotide defining the site of initiation is designated "n". Residues at the "n" and "$n + 1$" positions of the template define the P-site and N-site.

At the initiation stage, the formation of the first phosphodiester bond is key for polymerization of the nucleotides to begin. To form this phosphodiester bond a hydroxyl group corresponding to a nucleotide $3'$-OH is needed. Depending on how this $3'$-OH is supplied two mechanisms are differentiated: primer dependent in the case that a primer provides the required hydroxyl group, or primer independent (also called *de novo*) if this hydroxyl group is provided by the first NTP [2]. The variety of mechanisms reflect the adaptation of the viruses to the host cell [1]. The size of the thumb domain seems to define whether a polymerase uses the primer-dependent or *de novo* mechanism. Most viruses in Picornaviridae and Caliciviridae families utilize a primer-dependent mechanism, but exceptions are found, such as noroviruses in the Caliciviridae family, that synthesize the (−) strand *de novo* [30]. In general these enzyme have a small thumb domain that provides a wider template channel to accommodate both template and primer. For this mechanism different primers such as polypeptides, capped mRNAs or oligonucleotides may be used. In contrast, the Flaviviridae family that employs the *de novo* mechanism has a large thumb domain and narrower template channel suited to accommodate only the ssRNA and NTP [1,14]. However, we note that under certain conditions *de novo* polymerases can be induced to become primer dependent [31].

When the *de novo* mechanism is used initiation takes place exactly at the 3'-terminus of the template RNA, so the initiating NTP (the first NTP of the growing strand) is dictated by the template. Both HCV and BVDV from the Flaviviridae family have been observed to require high concentrations of GTP for the initiation of RNA synthesis regardless of the RNA template nucleotide [32,33] which led to the suggestion that GTP may be needed for structural support of the initiating NTP. Harrus *et al.* [34] also suggested that GTP may act as the "initiation platform" and D'Abramo *et al.* [35] pointed out that this GTP may stabilize the interaction between the 3'-end of the template and the priming nucleotide. This stabilizing GTP binds inside the template channel, 6 Å from the catalytic site. It is not incorporated into the nascent RNA strand and is thought to be released from the active site during the elongation stage [9]. We note that another GTP molecule has been reported to bind at the rear of the thumb domain near the fingertips in NS5B. This GTP has been suggested to play a role in activating *de novo* initiation or in allosterically regulating the conformational changes needed for replication [36].

Because base-pairing alone is insufficient to stabilize the dinucleotide product in the "P-site", specialized structural elements are employed [13]. Besides the stabilizing GTP there is also a polymerase structural initiation platform, the so-called β-flap (residues 443–454). This β-flap likely supports the stabilizing GTP but would need to move out of the way in the elongation phase to allow the dsRNA product to exit [1,9,34]. Other researchers have suggested the C-terminal linker (residues 531 to 570) also plays a regulatory role in the initiation stage of replication by acting as a buttress during the initiation stage and moving out of the template channel in a similar way as the β-flap in order to allow egress of the double stranded RNA [14,37] (see Figure 2).

An advantage of the primer-dependent mechanism is that a stable elongation complex is formed more easily. There is limited abortive cycling, if any, and no requirement for large conformational rearrangements [13]. In contrast, for the *de novo* mechanism the first dinucleotide is not sufficiently stable and an initiation platform is needed to provide additional stabilization. This reduced stability sometimes results in abortive cycling for the *de novo* mechanism. However, an advantage of the *de novo* mechanism is that no additional enzymes are needed to generate the primer [38].

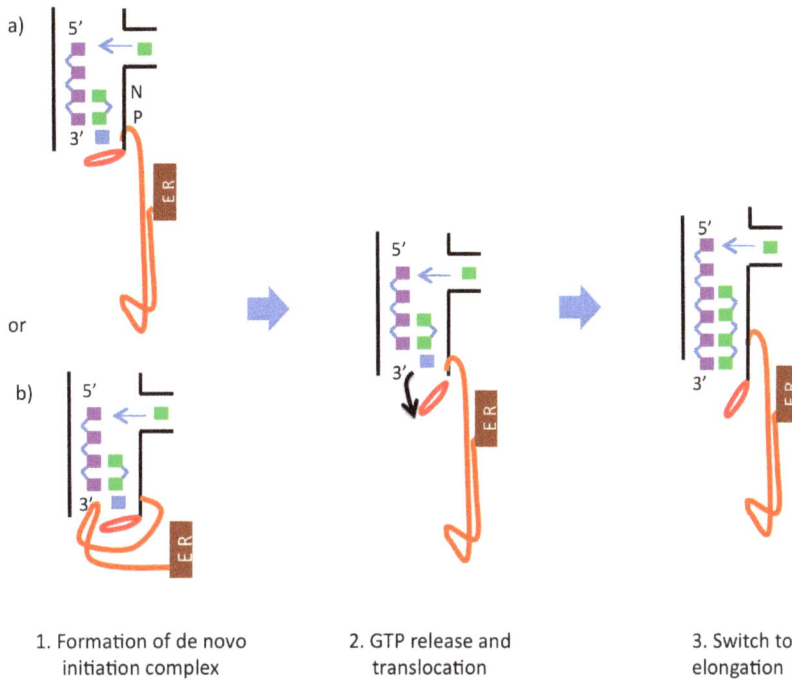

| | 1. Formation of de novo initiation complex | 2. GTP release and translocation | 3. Switch to elongation |

Figure 2. Schematic describing de novo initiation in Hepaciviruses and Pestiviruses. Note that Flaviruses do not anchor their C-terminus in the Endoplasmatic Reticulum (ER). This figure was generated by incorporating the descriptions provided by both Appleby *et al.* [35] and Choi [1]. The linker and C-terminal anchor are shown in orange as one contiguous element. The β-flap is colored red (as in Figure 3), the template strand in purple, the growing strand in green, the stabilizing GTP in blue and the Endoplasmic Reticulum (ER) in brown. The "N" and "P" indicate where the N-site (nucleotide-site) and P-site (priming-site) are. These correspond to the positions of the growing strand that bind to residues "*n*" and "*n* + 1" of the template strand respectively.

After the template and primer or initiating NTP are bound to the enzyme, the steps required for single-nucleotide addition are [1]:

(1) incorporation of the incoming NTP into the growing daughter strand by formation of the phosphodiester bond
(2) release of pyrophosphate
(3) translocation along the template.

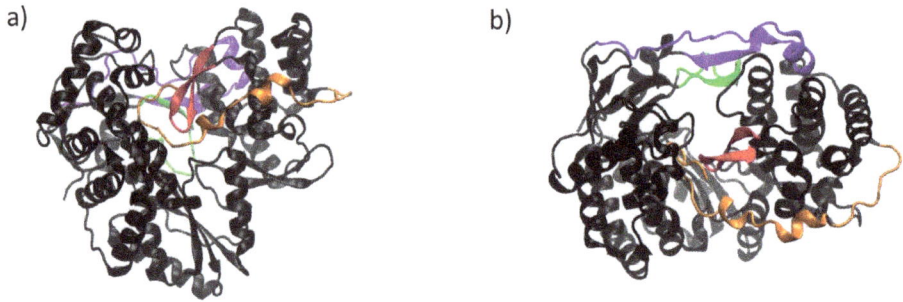

Figure 3. NS5B structure with characteristic elements highlighted. (**a**) Front view and (**b**) top view. The linker is shown in orange, the β-flap in red. The fingertips are shown in blue (the delta 1 loop) and green (the delta 2 loop).

These three steps are repeated cyclically during elongation until the full RNA strand is replicated.

In order to facilitate nucleotide addition, all polymerases have two metal ions (Mg^{2+} or Mn^{2+}) in the active site bound to two conserved aspartic acid residues. These metal ions have been shown to be essential for catalysis via the so-called "two metal ions" mechanism. This mechanism was proposed by Steiz in 1998 [39] and is as follows: the incoming NTP binds to metal ion B that orients the NTP in the active site and that may contribute to charge neutralization during catalysis. Metal ion B coordinates to the β- and γ-phosphate groups of the incoming NTP as well as the aspartic acid residue in motif A. Once the nucleotide is in place, the second divalent cation (Metal ion A) coordinates to the initiating NTP, lowering the pKa of the 3′-OH and facilitating nucleophilic attack on the α-phosphate. This then leads to formation of the phosphodiester bond and the release of pyrophosphate (PPi). Metal ion A coordinates to the α-phosphate group of the incoming NTP, the 3′-OH of the priming NTP and the aspartic acid residue in motif C (see Figure 4). Both metal ions stabilize the charge and geometry of the phosphorane pentavalent transition state during the nucleotidyl transfer reaction [1,13,38].

The switch to elongation requires a major conformational change in the polymerase structure. Both the β-flap and the linker need to be displaced and an opening of the enzymatic core occurs. This open conformation may be one of the factors that enables a higher processivity in the elongation stage compared to the initiation stage (for more detailed information about this change in conformation see Section 5 below).

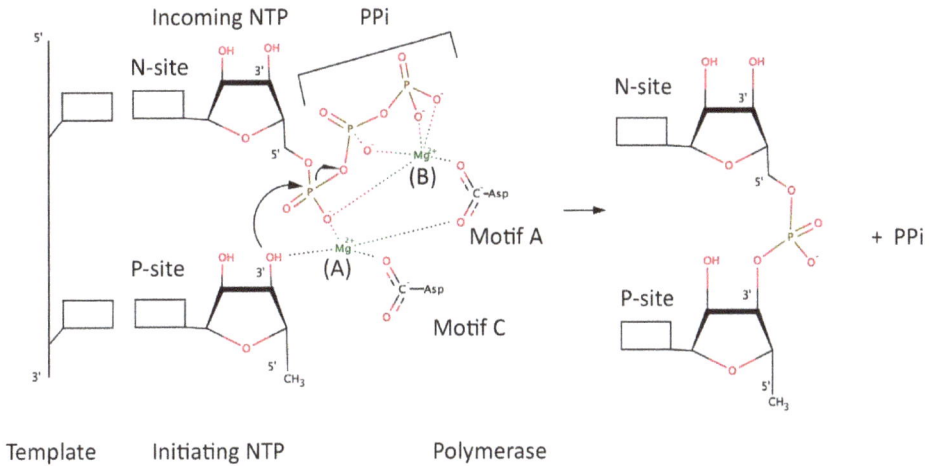

Figure 4. Two metals ions mechanism in RdRps. The squares represent the bases that are part of the nucleotides. This figure is inspired by a similar figure from Choi *et al.* [1].

Little is known about the termination of RNA synthesis. It has been suggested that the polymerase may simply fall off the end of the template once the complementary strand has been synthesized [14,40]. It is important to note that RNA synthesis by NS5B is error-prone due to the lack of proofreading activity of the RdRp enzymes. The mutation rates are estimated to be on the order of one mutation per 10^3–10^7 nucleotides resulting in approximately one error per replicated genome [1,41]. In contrast the mutation rate in *E. coli*, where cellular polymerases benefit from error-correcting mechanisms, is on the order of one mutation per 10^9–10^{10} nucleotides [1]. The large error rate results in the high genetic variability of the HCV viruses and provides a molecular basis for the rapid development of resistance to therapies.

6. NS5B Conformational Changes during the Replication Cycle

One characteristic unique to viral RdRps is their "closed-hand" shape. This terminology started to be used because their X-ray structures appear to be more closed than the previously characterized DdDps, DdRps and RTs (called "open-hand") [7,14,38,42]. This "closed-hand" shape is characterized by the fingertips region, a hallmark of RdRps, that connects the fingers and palm domains on the back of the enzyme as well as by the so-called β-flap on the front of the enzyme (see Figure 3). The latter is specific to the Flaviviridae RdRps while the linker, or a variation of it, is common to most of the *de novo* initiating RdRps [40] (note, however, that it is not found in the Flavivirus RdRps). The linker (residues 531 to 570) connects the NS5B catalytic core (residues 1 to 530) with the C-terminus transmembrane anchor (residues 571 to 591). These last twenty-one C-terminal residues seem not to influence RNA synthesis *in vitro* [40]. Given that these residues are very hydrophobic, their removal facilitates expression and purification of the enzyme. Thus, most biochemical and all structural studies have been carried out with the so-called NS5B Δ21 enzyme variant in which these residues have been removed.

Most of the NS5B structures that have been reported are thought to be in the closed conformation. However, it has been observed that *de novo* initiation by NS5B *in vitro* does not only occur at the 3′ end of the template but also can take place at internal template sites [43,44] and on circular templates [31]. These facts suggest that in solution there is an equilibrium between the closed and open conformations. The existence of the open conformation is supported by the structure of NS5B from genotype 2a NS5B, [45] as well as the structure recently published by Mosley *et al.* [46]. The latter contains a variant of NS5B that lacks the β-flap in complex with primer-template RNA. Molecular dynamics simulations of Davis *et al.* [47] also indicate the occurrence of open NS5B conformations.

The closed conformation is thought to represent the initiation state of the polymerase. In this conformation the catalytic core only provides sufficient space for a single-stranded RNA template and the nucleotides required for *de novo* initiation of RNA synthesis, but is not wide enough to accommodate double-stranded RNA [40]. To transition to elongation a major conformational change is needed so the nascent RNA can egress. Primer-dependent RdRps undergo less dramatic conformational changes than *de novo*-initiating RdRps [14] because the thumb domain of primer-dependent RdRps is smaller, leaving enough room for the dsRNA product to exit. Transitioning to elongation in *de novo*-initiating RdRps thus requires the adoption of an open conformation [34,40,46,48]. To arrive at the open conformation the β-flap would need to be moved out of the way, the stabilizing GTP should unbind and also a rotation of the thumb domain should take place. This would position it further from the center of the enzyme, increasing the size of the template and duplex channels so the dsRNA can exit the enzyme [34,48]. If the C-terminal linker does act as an initiation platform together with the β-flap, this element would also need to move away from the template channel in the transition to elongation as described by Appleby *et al.* [37] (see Figure 2). It is worth noting that conformational changes have been reported in several RdRp structures [45,46,49]. These findings suggest that these enzymes exhibit considerable conformational variability, which is similar to observations made for other polymerases [50,51].

7. NS5B Inhibitors and Mechanisms of Action

There are two main classes of NS5B inhibitors: nucleoside inhibitors (NIs) and non-nucleoside inhibitors (NNIs) (see Figure 5). NIs bind in the active site and generally act as non-obligate terminators of RNA synthesis after being incorporated into the newly produced RNA strand. The advantages of NIs are that they have shown stronger antiviral activity, are able to inhibit multiple HCV genotypes and have a higher barrier to the emergence of drug resistance [48]. However, they have the potential to also affect host polymerases since they interact with an active site that has similar features among diverse types of polymerases. Sofosbuvir, the drug most recently approved for HCV treatment is in this group. NNIs are allosteric inhibitors that bind to sites other than the active site. NNIs are also promising, though they have not yet been used in a clinical setting. NNIs are attractive for use in future anti-HCV therapies due to the decreased likelihood that they will exhibit nonspecific side effects compared to NIs. However, HCV is more likely to become resistant to these inhibitors because there is typically not strong evolutionary pressure to maintain the amino acid sequence of NNI binding sites. We focus on NNIs in this review because the role of NIs as terminators of RNA synthesis is well understood. In contrast, although many structures with NNIs bound have been solved, their mechanism of action still remains to be elucidated.

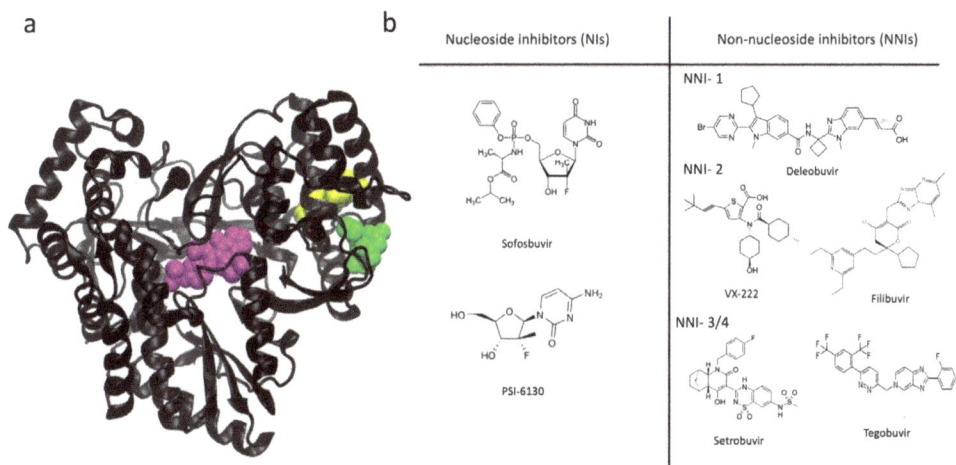

Figure 5. NS5B inhibitors. (**a**) The three allosteric sites of NS5B are highlighted with space filling representations of inhibitors that bind in these locations. Thumb site 1 (NNI-1) in yellow, thumb site 2 (NNI-2) in green, palm sites (NNI-3/4) in purple; (**b**) chemical structures of NIs and NNIs that are in clinical trials or have already been approved [52–59].

Four NNI sites have been identified: two in the thumb (NNI-1 and NNI-2) and two in the palm (NNI-3 and NNI-4) (see Figure 5). Brown and Thorpe [60] provide evidence that NNI-3 and NNI-4 are likely to be distinct regions within a single large pocket rather than two individual pockets. For this reason we use the nomenclature NNI-3/4 to denote both of these partially overlapping sites. Due to the fact that there are multiple distinct allosteric sites it may be possible to use multiple NNIs in combination with each other or with NIs in the effort to overcome resistance. NNIs are thought to inhibit NS5B by affecting the equilibrium distribution of conformational states required for normal catalytic activity of the enzyme [13,47]. Most of the NNIs that bind to the palm domain have been found to stabilize the β-flap via critical interactions with Tyr448 [61], fixing it in the closed, initiation-appropriate conformation and preventing these residues from moving out to allow the RNA double helix to egress [46]. NNI-2 ligands have also been suggested to prevent the occurrence of important conformational changes in NS5B [16,45,62]. Some studies have suggested that palm NNIs inhibit initiation while thumb NNIs inhibit an early phase of replication that occurs after initiation but before elongation starts [63–66]. Thus, the different allosteric sites may display distinct modes of action. Davis *et al.* [46] studied the mechanism of inhibition of allosteric inhibitors in the different allosteric sites. They found that inhibitors in the NNI-1 pocket seem to prevent enzyme function by reducing its overall stability and preventing it from stably adopting functional conformations. In contrast, NNI-2 inhibitors seem to reduce conformational sampling, preventing the transitions between conformational states that are required for NS5B to function. NNI-3 inhibitors were also observed to restrict conformational sampling, though the dominant mode of action of these molecules was predicted to result from blocking access of the RNA template (see Figure 6).

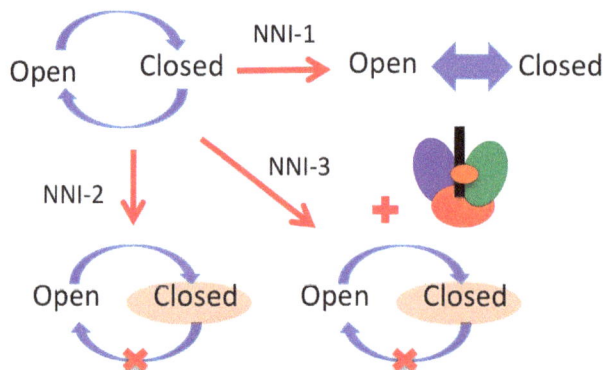

Figure 6. Mechanisms of inhibition for NNIs. NS5B must transition between open and closed states to perform replication (upper left). NNI-1 inhibitors have been observed to reduce enzyme stability. NNI-2 inhibitors have been shown to reduce conformational sampling, confining the enzyme in closed conformations. NNI-3 inhibitors mainly block access of the RNA template but also induce some restriction of conformational sampling. The RNA template is represented as a black rectangle and the inhibitor as an orange ellipse.

This may facilitate their use in combination therapies by degrading complementary functionalities in the enzyme. Understanding the molecular mechanisms by which small molecules in general and NNIs in particular inhibit the function of NS5B is essential for rationally design NS5B inhibitors. Such molecules may ultimately serve as a basis for more efficacious or cost-effective HCV therapies, either individually or in combination.

One informative example that illustrates the useful interplay between determining the roles of structural and functional elements of NS5B and understanding the efficacy of NNIs is provided by recent studies of Gilead pharmaceuticals. Boyce *et al.* [67] assessed the activities and biophysical properties of a number of NS5B variants using mutations and deletions in the enzyme C-terminus and β-flap, in concert with challenging the enzyme using diverse NNIs. Their observations suggest that ligands which bind to NNI-2 exhibit a unique inhibitory mechanism relative to other NNIs. Boyce *et al.* [67] discovered that NNI-2 ligands are most effective when both the C-terminus and β-flap of the enzyme are present. These inhibitors were found to stabilize NS5B in a closed conformation, consistent with simulation studies by Davis *et al.* [16,47]. Boyce *et al.* [67] found that interactions between the C-terminus and the β-flap were required for inhibition, but not for ligand binding. These authors determined that NNI-2 inhibitors exhibited decreased efficacy for truncated NS5B variants and suggested that while the C-terminus and β-flap do not alter the intrinsic interactions of NNI-2 ligands with the enzyme, they do play an important role in propagating the allosteric effects that result from inhibitors binding to distant enzyme locations. This finding is consistent with mutational data for NNI-2 inhibitors, which map viral resistance mutations to areas around the β-flap [68].

Simulation studies by Davis and Thorpe suggest that the enzyme C-terminus reduces conformational sampling in NS5B, likely eliminating transitions between the closed and open conformations necessary for the initiation and elongation phases of replication respectively [69]. These observations predict that enzymes without C-terminal residues should display increased activity, consistent with the findings of Boyce *et al.* [67]. Other studies from Davis *et al.* [16,47] indicate that an NNI-2 ligand can restrict conformational sampling even if the C-terminus is absent, stabilizing the enzyme in a very closed state. One might expect that this property could account for the inhibitory action of NNI-2 ligands without needing to invoke a role for the C-terminus as suggested by Boyce *et al.* [67]. However, there are several important considerations to be noted. First, the simulation studies examine the impact of binding a ligand to the enzyme and do not directly probe inhibition. The studies of Boyce *et al.* [67] indicate that binding affinities of NNI-2 ligands are not a good proxy for inhibition efficacy. Thus, observations in the simulation studies may not be explicitly linked to allosteric inhibition. Another consideration is that the simulation studies were not carried out with both the inhibitor and the C-terminus present.

It is possible that conformational restriction of the enzyme in the presence of both entities would be even more dramatic, consistent with the enhanced inhibition in Boyce *et al.* [67] measured in the presence of the C-terminus. Finally, different ligands were employed in each study and it could be that distinct inhibitors elicit different effects even though they bind to the same location. There is evidence that different NNI-2 ligands are able to alter the conformational distribution of NS5B to different extents [47]. Thus, it is possible that the enzyme C-terminus is only required for observing the inhibitory effects of certain NNI-2 ligands.

In contrast to NNI-2, Boyce *et al.* [67] observed that the potency of NNI-1 ligands was not affected by the presence of the C-terminus or β-flap. This observation suggests that these ligands possess a completely different mechanism of action compared to NNI-2 inhibitors. These authors noted that the presence of NNI-1 ligands lowered the melting temperature of NS5B, consistent with decreased stability of the enzyme. The decrease of NS5B stability in the presence of NNI-1 ligands was noted as well in other studies [70,71]. This finding is also consistent with results from simulations of Davis *et al.* [47] that suggest NNI-1 and NNI-2 ligands have distinct modes of action. In contrast to the stabilizing effect of NNI-2 ligands, it was observed that an NNI-1 ligand destabilized conformational sampling in NS5B, preventing the enzyme from stably occupying functional conformational states.

With regard to palm inhibitors, Boyce *et al.* [67] observed that such ligands display larger dissociation constants in NS5B constructs for which C-terminal residues were deleted, suggesting that the C-terminus facilitates binding to palm sites. Palm site inhibitors also demonstrate decreased potency in these deletion constructs, indicating that the C-terminus is needed for both binding and inhibition. In simulation studies Davis *et al.* [47] observed that NNI-3 ligands were able to bind to the enzyme without the C-terminus present and also restricted conformational sampling of NS5B in a similar manner to NNI-2 ligands. However, the conformations sampled when ligands were bound to NNI-3 tended to be more open in general than those induced by an NNI-2 ligand. These conformations may perturb the replication cycle to a reduced extent compared to NNI-2 or NNI-1 ligands. It is possible that in the presence of the C-terminus NNI-3 ligands elicit more dramatic changes in conformational sampling. Nonetheless, the authors concluded that the dominant inhibitory effect of palm ligands is likely due to direct obstruction of the RNA template channel (thus preventing the template from accessing the active site) rather than conformational restriction. This observation is consistent with previous predictions [72].

The findings of Boyce *et al.* [67] are important because they indicate the enzyme C-terminus plays a crucial role in modulating the efficacy of NNIs. The likely molecular basis of this observation can be readily understood by considering the schematic shown in Figure 2. In this figure it is apparent that the C-terminus acts as a "stopper" in the template channel, preventing elongation of the nascent RNA strand. Thus, both the C-terminus and β-flap need to be removed from the template channel before elongation can proceed. If the C-terminus is not present, the template channel cannot be effectively blocked and replication is less likely to be affected by presence of the inhibitor. This is a quite interesting result, as it points to the limitations of some inhibitor studies that may have been carried out *in vitro* using enzyme variants without the C-terminus. It is likely that any ligands employing the inhibitory mechanisms described by Boyce *et al.* [67] would not be identified in such studies. Thus, the role of NS5B regulatory elements in strongly modulating the efficacy of inhibitors must be taken into account when assessing ligand potency.

Studies such as those of Boyce *et al.* [67] or Davis and colleagues [16,47,69] may be useful to understand the differing susceptibility of different NS5B variants (and thus different HCV strains or genotypes) to the presence of diverse inhibitors. For example, in some viral genotypes the C-terminus might interact more strongly with the template channel than in others. One would anticipate that NNI-2 ligands would be more effective in inhibiting such enzyme variants. The studies reviewed in this article indicate that understanding the structure and function of NS5B provides powerful insight into the molecular mechanisms governing inhibition of this enzyme and the functional properties of other RdRps. For example, recent structural studies of the Influenza virus polymerase reveal a β-flap element similar to that which modulates the activity of NS5B and which may adopt a similarly important role in these enzymes [73,74]. We note that simulation studies are particularly helpful in this regard by allowing molecular mechanisms underlying the observed structure-function relationships to be elucidated [16,47,69].

Understanding the molecular mechanisms involved in inhibition by NNIs could facilitate the design and deployment of these molecules. The insights acquired may also be transferable to other polymerases to better understand the relationship between structure, function and dynamics in these enzymes. Due to the fact that individual NNIs can have distinct sites of binding, it should be possible to combine multiple NNIs such that their total inhibitory effect is enhanced relative to applying any given inhibitor on its own [60]. It may be beneficial to target complementary activities or distinct conformational states of the enzyme with an array of small molecules to degrade a wide spectrum of NS5B functionality in a therapeutic context. For example, it is possible that a large fraction of NS5B exists within the host cell in an auto-inhibited state with the C-terminus occupying the template channel. In this way, the virus can avoid negatively perturbing the host cell and facilitate evasion of the host immune response. One could envision targeting both actively replicating and auto-inhibited NS5B molecules with different inhibitors in order to more effectively degrade intracellular enzyme activity.

8. Summary

Flaviviridae viruses are (+) RNA viruses with RdRp polymerases that utilize the *de novo* mechanism for initiation. While *Flaviviridae* polymerases possess elements common to other RdRps such as the fingertips region, they are also unique in possessing the β-flap that may be used as an initiation platform during genome replication.

The important pathogen HCV is a member of the *Flaviviridae* family within the *Hepacivirus* genus and employs NS5B as the RdRp that replicates its genome. There are two key steps involved in the replication process: (1) the formation of the initial dinucleotide and (2) the transition from initiation to processive elongation. Structural elements of NS5B that likely have a crucial role in these steps are the C-terminal linker and the β-flap (see Figure 3). Initiation is also facilitated by the so-called "stabilizing GTP" in the active site (see Figure 2). Finally, a conformational change involving movement of the thumb and fingers domains to position them further apart has been observed to accompany the transition from initiation to elongation, resulting in an open-hand conformation. The linker and the β-flap may have dual roles: (1) acting as initiation platforms to stabilize formation of the first dinucleotide and (2) regulating the transition to elongation. These structural elements can prevent the enzyme from moving to the elongation stage and must be displaced to allow for processive elongation to take place.

Thus, the available evidence suggests that NS5B possesses an intrinsic capacity to be regulated via allosteric effectors including NNIs, the β-flap and the C-terminal linker. In addition, the role of these effectors seems to be strongly modulated by the specific context of the interaction. Understanding how these structural elements govern enzyme activity and how they interface with inhibitors is important for understanding the molecular mechanisms of allosteric inhibition in NS5B. Such knowledge paves the way for rational design of inhibitors and combination therapies both for NS5B and for the polymerases to which these insights can be generalized. This information may also be useful in designing enzymes with attenuated activity, as would be required if one sought to develop a strain of HCV that could serve as the basis for a vaccine. Attenuating HCV by degrading the activity of NS5B is one strategy that could prove useful in this regard. One potential drawback to such efforts is the high mutation rate of HCV that results from the error-prone nature of NS5B. However, it is possible that one could circumvent this issue by generating a polymerase that not only possesses reduced efficacy, but also displays increased fidelity and thus faithfully replicates the viral genome.

Viral polymerases and, specifically, RdRps share many common structural, functional and dynamic features. Thus, the knowledge obtained in understanding how NS5B functions may be transferable to polymerases from closely related viruses such as Dengue or West Nile virus, or even to other more distantly related polymerases such as reverse transcriptase from HIV and 3D-pol from poliovirus.

Conflicts of Interest: The authors declare no conflict of interest.

References

1. Choi, K.H. Viral polymerases. In *Viral Molecular Machines*; Springer Science: New York, NY, USA, 2012.
2. Ortin, J.; Parra, F. Structure and function of RNA replication. *Annu. Rev. Microbiol.* **2006**, *60*, 305–326.
3. Zhou, Y.; Ray, D.; Zhao, Y.; Dong, H.; Ren, S.; Li, Z.; Guo, Y.; Bernard, K.A.; Shi, P.-Y.; Li, H.; *et al.* Structure and function of flavivirus ns5 methyltransferase. *J. Virol.* **2007**, *81*, 3891–3903.
4. Zhou, D.; Chung, S.; Miller, M.; Grice, S.F.J.L.; Wlodawer, A. Crystal structures of the reverse transcriptase-associated ribonuclease h domain of xenotropic murine leukemia-virus related virus. *J. Struct. Biol.* **2012**, *177*, 638–645.
5. Knopf, C. Evolution of viral DNA-dependent DNA polymerases. *Virus Genes* **1998**, *16*, 47–58.
6. Baltimore, D. Expression of animal virus genomes. *Bacteriol. Rev.* **1971**, *35*, 235–241.
7. Shatskaya, G.S. Structural organization of viral RNA-dependent RNA polymerases. *Biochemistry* **2013**, *78*, 231–235.
8. Ollis, D.L.; Brick, P.; Hamlin, R.; Xuong, N.G.; Steitz, T.A. Structure of large fragment of *Escherichia coli* DNA polymerase i complexed with dtmp. *Nature* **1985**, *313*, 762–766.
9. McDonald, S.M. RNA synthetic mechanisms employed by diverse families of RNA viruses. *WIREs RNA* **2013**, *4*, 351–367.
10. Ferrer-Orta, C.; Verdaguer, N. RNA virus polymerases. In *Viral Genome Replication*; Cameron, C., Gotte, M., Raney, K.D., Eds.; Springer Science: New York, NY, USA, 2009.
11. Gao, G.; Orlova, M.; Georgiadis, M.M.; Hendrickson, W.A.; Goff, S.P. Conferring RNA polymerase activity to a DNA polymerase: A single residue in reverse transcriptase controls substrate selection. *Proc. Natl. Acad. Sci. USA* **1997**, *94*, 407–411.
12. Cameron, C.E.; Moustafa, I.M.; Arnold, J.J. Dynamics: The missing link between structure and function of the viral RNA-dependent RNA polymerase? *Curr. Opin. Struct. Biol.* **2009**, *19*, 768–774.
13. Ng, K.K.-S.; Arnold, J.J.; Cameron, C.E. *Structure and Function Relationships Ammong RNA-Dependent RNA Polymerases*; Springer-Verlag: Berlin, Germany; Heidelberg, Germany, 2008; Volume 320.
14. Subissi, L.; Decroly, E.; Selisko, B.; Canard, B.; Imbertl, I. A closed-handed affair: Positive-strand RNA virus polymerases. *Future Virol.* **2014**, *9*, 769–784.
15. Moustafa, I.M.; Shen, H.; Morton, B.; Colina, C.M.; Cameron, C.E. Molecular dynamics simulations of viral RNA polymerases link conserved and correlated motions of functional elements to fidelity. *J. Mol. Biol.* **2011**, *410*, 159–181.

16. Davis, B.; Thorpe, I.F. Thumb inhibitor binding eliminates functionally important dynamics in the hepatitis c virus RNA polymerase. *Proteins Struct. Funct. Bioinform.* **2013**, *81*, 40–52.

17. International Committee on Taxonomy of Viruses. Available online: http://www.Ictvonline.Org (accessed on 15 February 2015).

18. Gong, J.; Fang, H.; Li, M.; Liu, Y.; Yang, K.; Xu, W. Potential targets and their relevant inhibitors in anti-influenza fields. *Curr. Med. Chem.* **2009**, *16*, 3716–3739.

19. Malet, H.; Masse, N.; Selisko, B.; Romette, J.L.; Alvarez, K.; Guillemot, J.C.; Tolou, H.; Yap, T.L.; Vasudevan, S.; Lescar, J.; *et al.* The flavivirus polymerase as a target for drug discovery. *Antivir. Res.* **2008**, *2008*, 23–35.

20. Welsch, S.; Miller, S.; Romero-Brey, I. Composition and three-dimensional architecture of the dengue virus replication and assembly sites. *Cell Host Microbe* **2009**, *5*, 365–375.

21. Hsu, N.Y.; Ilnytska, O.; Belov, G. Viral reorganization of the secretory pathway generates distinct organelles for RNA replication. *Cell* **2010**, *141*, 799–811.

22. Zuckerman, A.J. Hepatitis viruses. In *Medical Microbiology*; Baron, S., Ed.; The University of Texas Medical Branch: Galveston, TX, USA, 1996.

23. Hansen, J.L.; Long, A.M.; Schultz, S.C. Structure of the RNA-dependent RNA polymerase of poliovirus. *Structure* **1997**, *5*, 1109–1122.

24. Lindenbach, B.D.; Tellinghuisen, T.L. Hepatitis C virus genome replication. In *Viral Genome Replication*; Cameron, C., Gotte, M., Raney, K.D., Eds.; Springer Science: New York, NY, USA, 2009.

25. Astier-Manifacier, S.; Cornuet, P. RNA-dependent RNA polymerase in chinese cabbage. *Biochim. Biophys. Acta* **1971**, *232*, 484–493.

26. Boege, F.; Heinz, L.S. RNA-dependent RNA polymerase from healthy tomato leaf tissue. *FEBS Lett.* **1980**, *121*, 91–96.

27. Cogoni., C.; Macino, G. Gene silencing in neurospora crassa requires a protein homologous to RNA-dependent RNA polymerase. *Nature* **1999**, *399*, 166–169.

28. Smardon, A.; Spoerke, J.M.; Stacey, S.C.; Klein, M.E.; Mackin, N.; Maine, E.M. Ego-1 is related to RNA-directed RNA polymerase and functions in germ-line development and RNA interference in c. Elegans. *Curr. Biol.* **2000**, *10*, 169–178.

29. Maida, Y.; Masutomi, K. RNA-dependent RNA polymerases in RNA silencing. *Biol. Chem.* **2011**, *392*, 299–304.

30. Rohayem, J.; Robel, I.; Jager, K.; Scheffler, U.; Rudolph, W. Protein-primed and *de novo* initiation of RNA synthesis by norovirus 3dpol. *J. Virol.* **2006**, *80*, 7060–7069.

31. Ranjith-Kumar, C.T.; Kao, C.C. Recombinant viral rdrps can initiate RNA synthesis from circular templates. *RNA* **2006**, *12*, 303–312.

32. Luo, G.; Hamatake, R.K.; Mathis, D.M.; Racela, J.; Rigat, K.L.; Lemm, J.; Colonno, R.J. *De novo* initiation of RNA synthesis by the RNA-dependent RNA polymerase (ns5b) of hepatitis C virus. *J. Virol.* **2000**, *74*, 851–863.

33. Kao, C.C.; Vecchio, A.M.D.; Zhong, W. *De novo* initiation of RNA synthesis by a recombinant flaviviridae RNA-dependent RNA polymerase. *Virology* **1999**, *253*, 1–7.

34. Harrus, D. Further insights into the roles of GTP and the C terminus of the hepatitis C virus polymerase in the initiation of RNA synthesis. *J. Biol. Chem.* **2010**, *285*, 32906–32918.

35. D'Abramo, C.M.; Deval, J.; Cameron, C.E.; Cellai, L.; Gotte, M. Control of template positioning during de novo initiation of RNA synthesis by the bovine viral diarrhea virus NS5B polymerase. *J. Biol. Chem.* **2006**, *281*, 24991–24998.

36. Bressanelli, S. Structural analysis of the hepatitis C virus RNA polymerase in complex with ribonucleotides. *J. Virol.* **2002**, *76*, 3482–3492.

37. Appleby, T.C.; Perry, J.K.; Murakami, E.; Barauskas, O.; Feng, J.; Cho, A.; Fox, D., III; Wetmore, D.R.; McGrath, M.E.; Ray, A.S.; *et al.* Structural basis for RNA replication by the hepatitis C virus polymerase. *Science* **2015**, *347*, 771–775.

38. Van Dijk, A.A.; Makeyev, E.V.; Bamford, D.H. Initation of viral RNA-dependent RNA polymerization. *J. Gen. Virol.* **2004**, *85*, 1077–1093.

39. Steitz, T. A mechanism for all polymerases. *Nature* **1998**, *391*, 231–232.

40. Lohmann, V. *Hepatitis C Virus: From Molecular Virology to Antiviral Therapy*; Springer-Verlag: Berlin, Germany; Heidelberg, Germany, 2013; Volume 369.

41. Drake, J.W. A constant rate of spontaneous mutation in DNA-based microbes. *Proc. Natl. Acad. Sci. USA* **1991**, *88*, 7160–7164.

42. Ferrer-Orta, C.; Arias, A.; Escarmi, C.; Verdaguer, N. A comparison of viral RNA-dependent RNA polymerases. *Curr. Opin. Struct. Biol.* **2006**, *16*, 27–34.

43. Binder, M.; Quinckert, D.; Bochkarova, O.; Klein, R.; Kezmic, N.; Bartenschalager, R.; Lohmann, V. Identification of determinants involved in initatiation of hepatitis c virus RNA synthesis by using intergenotipic chimeras. *J. Virol.* **2007**, *81*, 5270–5283.

44. Shim, J.H.; Larson, G.; Hong, J.Z. Selection of 3′ template bases and initatiting nucleotides by hepatitis c virus RNA by and ago2-miR-122 complex. *Proc. Natl. Acad. Sci. USA* **2002**, *109*, 941–946.

45. Biswal, B.K.; Cherney, M.M.; Wang, M.; Chan, L.; Yannopoulos, C.G.; Bilimoria, D.; Nicolas, O.; Bedard, J.; James, M.N. Crystal structures of the RNA-dependent RNA polymerase genotype 2A of hepatitis C virus reveal two conformations and suggest mechanisms of inhibition by non-nucleoside inhibitors. *J. Biol. Chem.* **2005**, *280*, 18202–18210.

46. Mosley, R.T. Structure of hepatitis C virus polymerase in complex with primer- template RNA. *J. Virol.* **2012**, *86*, 6503–6511.

47. Davis, B.C.; Brown, J.A.; Thorpe, I.F. Allosteric inhibitors have distinct effects, but also common modes of action, in the hcv polymerase. *Biophys. J.* **2015**, *108*, 1785–1795.

48. Caillet-Saguy, C.; Lim, S.P.; Shi, P.-Y.; Lescar, J.; Bressanelli, S. Polymerases of hepatitis C viruses and flaviviruses: Structural and mechanistic insights and drug development. *Antivir. Res.* **2014**, *105*, 8–16.

49. Choi, K.H.; Groarke, J.M.; Young, D.C.; Kuhn, R.J.; Smith, J.L.; Pevear, D.C.; Rossmann, M.G. The structure of the RNA-dependent RNA polymerase from bovine viral diarrhea virus establishes the role of GTP in de novo initiation. *Proc. Natl. Acad. Sci. USA* **2004**, *101*, 4425–4430.

50. Rothwell, P.J.; Waksman, G. Structure and mechanism of DNA polymerases. *Adv. Protein Chem.* **2005**, *71*, 401–440.

51. Doublie, S.; Sawaya, M.R.; Ellenberger, T. An open and closed case for all polymerases. *Structure* **1999**, *7*, R31–R35.

52. Wendt, A.; Adhoute, X.; Castellani, P.; Oules, V.; Ansaldi, C.; Benali, S.; Bourliere, M. Chronic hepatitis c: Future treatment. *Clin. Pharmacol.* **2014**, *6*, 1–17.

53. Larrey, D.; Lohse, A.W.; de Ledinghen, V.; Trepo, C.; Gerlach, T.; Zarski, J.P.; Tran, A.; Mathurin, P.; Thimme, R.; Arasteh, K.; *et al.* Rapid and strong antiviral activity of the non-nucleosidic NS5B polymerase inhibitor BI 207127 in combination with peginterferon α 2a and ribavirin. *J. Hepatol.* **2012**, *57*, 39–46.

54. Jacobson, I.; Pockros, P.J.; Lalezari, J.; Lawitz, E.; Rodriguez-Torres, M.; DeJesus, E.; Haas, F.; Martorell, C.; Pruitt, R.; Purohit, V.; *et al.* Virologic response rates following 4 weeks of filibuvir in combination with pegylated interferon α-2a and ribavirin in chronically-infected HCV genotype-1 patients. *J. Hepatol.* **2010**, *52*, S465–S465.

55. Rodriguez-Torres, M.; Lawitz, E.; Conway, B.; Kaita, K.; Sheikh, A.M.; Ghalib, R.; Adrover, R.; Cooper, C.; Silva, M.; Rosario, M.; *et al.* Safety antiviral activity of the HCV non-nucleoside polymerase inhibitor VX-222 in treatment-naive genotype 1 HCV-infected patients. *J. Hepatol.* **2010**, *52*, S14–S14.

56. Lawitz, E.; Rodriguez-Torres, M.; Rustgi, V.K. Safety and antiviral activity of ana 598 in combination with pegylated interferon α-2a plus ribavirin in treatment-naive genotype 1 chronic HCV patients. *J. Hepatol.* **2010**, *52*, 334A–335A.

57. Lawitz, E.; Jacobson, I.; Godofsky, E.; Foster, G.R.; Flisiak, R.; Bennett, M.; Ryan, M.; Hinkle, J.; Simpson, J.; McHutchison, J.; *et al.* A phase 2b trial comparing 24 to 48 weeks treatment with tegobuvir (GS-9190)/PEG/RBV to 48 weeks treatment with PEG/RBV for chronic genotype 1 HCV infection. *J. Hepatol.* **2011**, *54*, S181–S181.

58. Gane, E.J.; Stedman, C.A.; Hyland, R.H. Nucleotide polymerase inhibi- tor sofosbuvir plus ribavirin for hepatitis C. *N. Engl. J. Med.* **2013**, *368*, 34–44.

59. Wedemeyer, H.; Jensen, D.; Herring, R., Jr. Efficacy and safety of mericitabine in combination with PEG-IFN α-2a/RBV in G1/4 treatment naive HCV patients: Final analysis from the propel study. *J. Hepatol.* **2012**, *56*, S481–S482.

60. Brown, J.A.; Thorpe, I.F. Dual allosteric inhibitors jointly modulate protein structure and dynamics in the hepatitis c virus polymerase. *Biochemistry* **2015**, *54*, 4131–4141.

61. Pfefferkorn, J.A. Inhibitors of hcv ns5b polymerase. Part 1: Evaluation of the southern region of (2Z)-2-(benzoylamino)-3-(5-phenyl-2-furyl)acrylic acid. *Bioorg. Med. Chem. Lett.* **2005**, *15*, 2481–2486.

62. Wang, M. Non-nucleoside analogue inhibitors bind to an allosteric site on hcv ns5b polymerase. Crystal structures and mechanism of inhibition. *J. Biol. Chem.* **2003**, *278*, 9489–9495.

63. Ontoria, J.M.; Rydberg, E.H.; Carfi, A. Identification and biological evaluation of a series of 1*H*-benzo[*de*]isoquinoline-1,3(2*H*)-diones as hepatitis C virus NS5B polymerase inhibitors. *J. Med. Chem.* **2009**, *52*, 5217–5227.

64. Nyanguile, O.; Pauwels, F.; van den Broeck, W.; Boutton, C.W.; Quirynen, L.; Ivens, T.; van der Helm, L.; Vandercruyssen, G.; Mostmans, W.; Delouvroy, F.; *et al.* 1,5-Benzodiazepines, a novel class of hepatitis C virus polymerase nonnucleoside inhibitors. *Antimicrob. Agents Chemother.* **2008**, *52*, 4420–4431.

65. Nyanguile, O.; Devogelaere, B.; Fanning, G.C. 1a/1bsubtype profiling of nonnucleoside polymerase inhibitors of hepatitis C virus. *J. Virol.* **2010**, *84*, 2923–2934.

66. Tomei, L.; Altamura, S.; Migliaccio, G. Mechanism of action and antiviral activity of benzimidazole-based allosteric inhibitors of the hepatitis C virus RNA-dependent RNA polymerase. *J. Virol.* **2003**, *77*, 13225–13231.

67. Boyce, S.E.; Tirunagari, N.; Niedziela-Majka, A.; Perry, J.; Wong, M.; Kan, E.; Lagpacan, L.; Barauskas, O.; Hung, M.; Fenaux, M.; *et al.* Structural and regulatory elements of HCV NS5B polymerase—B-Loop and C-terminal tail—Are required for activity of allosteric thumb site II inhibitors. *PLoS ONE* **2014**, *9*, e84808.

68. Howe, A.Y.; Cheng, H.; Thompson, I.; Chunduru, S.K.; Herrmann, S. Molecular mechanism of a thumb domain hepatitis C virus nonnucleoside RNA-dependent RNA polymerase inhibitor. *Antimicrob. Agents Chemother.* **2006**, *50*, 4103–4113.

69. Davis, B.; Thorpe, I.F. Molecular simulations illuminate the role of regulatory components of the RNA polymerase from the hepatitis C virus in influencing protein structure and dynamics. *Biochemistry* **2013**, *52*, 4541–4552.

70. Ando, I.; Adachi, T.; Ogura, N.; Toyonaga, Y.; Sugimoto, K. Preclinical characterization of JTK-853, a novel nonnucleoside inhibitor of the hepatitis C virus RNA-dependent RNA polymerase. *Antimicrob. Agents Chemother.* **2012**, *56*, 4250–4256.

71. Caillet-Saguy, C.; Simister, P.C.; Bressanellli, S. An objective asessment of conformational variability in complexes of hepatitis C virus polymerase with non-nucleoside inhibitors. *J. Mol. Biol.* **2011**, *414*, 370–384.

72. Beaulieu, P. Recent advances in the development of NS5B polymerase inhibitors for the treatment of hepatitis C virus infection. *Expert Opin. Ther. Pat.* **2009**, *49*, 145–164.

73. Pflug, A.; Guilligay, D.; Reich, S.; Cusack, S. Structure of influenza a polymerase bound to the viral RNA promoter. *Nature* **2014**, *516*, 355–360.

74. Reich, S.; Guilligay, D.; Pflug, A.; Malet, H.; Berger, I.; Crepin, T.; Hart, D.; Lunardi, T.; Nanao, M.; Ruigrok, R.W.; *et al.* Structural insight into cap-snatching and RNA synthesis by influenza polymerase. *Nature* **2014**, *516*, 361–366.

Nucleobase but not Sugar Fidelity is Maintained in the Sabin I RNA-Dependent RNA Polymerase

Xinran Liu, Derek M. Musser, Cheri A. Lee, Xiaorong Yang, Jamie J. Arnold, Craig E. Cameron and David D. Boehr

Abstract: The Sabin I poliovirus live, attenuated vaccine strain encodes for four amino acid changes (*i.e.*, D53N, Y73H, K250E, and T362I) in the RNA-dependent RNA polymerase (RdRp). We have previously shown that the T362I substitution leads to a lower fidelity RdRp, and viruses encoding this variant are attenuated in a mouse model of poliovirus. Given these results, it was surprising that the nucleotide incorporation rate and nucleobase fidelity of the Sabin I RdRp is similar to that of wild-type enzyme, although the Sabin I RdRp is less selective against nucleotides with modified sugar groups. We suggest that the other Sabin amino acid changes (*i.e.*, D53N, Y73H, K250E) help to re-establish nucleotide incorporation rates and nucleotide discrimination near wild-type levels, which may be a requirement for the propagation of the virus and its efficacy as a vaccine strain. These results also suggest that the nucleobase fidelity of the Sabin I RdRp likely does not contribute to viral attenuation.

Reprinted from *Viruses*. Cite as: Liu, X.; Musser, D.M.; Lee, C.A.; Yang, X.; Arnold, J.J.; Cameron, C.E.; Boehr, D.D. Nucleobase but not Sugar Fidelity is Maintained in the Sabin I RNA-Dependent RNA Polymerase. *Viruses* **2015**, *7*, 5571–5586.

1. Introduction

Positive-strand RNA viruses are common causative agents of human disease, including the common cold, myocarditis, encephalitis, hepatitis, and paralytic poliomyelitis [1–7]. Poliovirus (PV) has been the subject of a largely-successful global eradication campaign [8,9]. These efforts have relied on the live attenuated oral poliovirus virus vaccine (OPV), and sustained in the developed world by inactivated poliovirus vaccine (IPV) [10]. Historically, OPV has been the favored treatment in the majority of countries, owing in part to its ease of use and lower cost [10]. However, most developed countries have transitioned to IPV because of OPV's risk of vaccine-associated paralytic poliomyelitis (VAPP) and vaccine-derived polioviruses (VDPV) [11]. OPV is generally comprised of three vaccine strains empirically developed by Albert Sabin and colleagues [12–14]. The three Sabin vaccine strains all have mutations located in the virus' internal ribosome entry site (IRES), which reduces the ability of PV to translate its RNA template within the neuronal cell [13,14]. Other mutations outside the IRES, including those within protein coding regions, may also contribute to viral attenuation [15–20]. The most abundant strain of the vaccine, Sabin I, has 57 mutations in its RNA sequence compared to the parental Mahoney strain, leading to 21 amino acid changes in viral proteins [21]. Four of these amino acid changes occur in the RNA-dependent RNA polymerase (RdRp) that is responsible for genome replication. We have previously shown that recombinant PV (*i.e.*, Mahoney background) encoding the Sabin-derived T362I mutation in the RdRp has a statistically significant reduction in viral virulence, likely because the T362I RdRp is a more error-prone polymerase than the "wild-type" (WT) enzyme [22]. These results are intriguing, and might suggest that the Sabin I RdRp also contributes to viral attenuation. Such findings would be intriguing in light of the suggestion that viruses encoding RdRp enzymes with altered fidelity might serve as live, attenuated vaccine candidates [23].

PV RdRp has a highly-conserved canonical cupped right-hand structure with palm, thumb, and finger subdomains (Figure 1) [24]. There are seven conserved structural motifs, five of which (A to E) are located in the palm subdomain [25,26]. The T362I amino acid change occurs on structural motif D, which we have proposed is important in phosphodiester bond formation and nucleotide discrimination [27,28]. More specifically, motif D contains a highly-conserved lysine residue (Lys359 in PV RdRp), which we have proposed acts as a general acid to protonate the β-phosphate of the incoming nucleotide to facilitate bond breakage between the α- and β-phosphates and create a better pyrophosphate leaving group [28,29]. We have suggested that the active-site loop containing Lys359 fluctuates between "closed" and "open" conformations in which Lys359 is positioned and out-of-position for catalysis, respectively. Our previous nuclear magnetic resonance (NMR) studies are consistent with this proposal [22,29,30]. We have shown that the Sabin-derived T362I amino acid substitution alters the motions of motif D, allowing the enzyme to fluctuate more readily into a closed conformation even in the presence of incorrect nucleotide, leading to a less faithful polymerase [22].

Other amino acid substitutions in the Sabin I RdRp may also change RdRp function. The Y73H substitution has been shown to interfere with the initiation of RNA synthesis [19], which might help to explain why PV encoding the Y73H substitution is attenuated [17,20]. The D53N, Y73H, and T362I substitutions also contribute to the temperature sensitivity of the Sabin I vaccine [15–17,31]. The Sabin substitutions may also affect each other. Our previous molecular dynamics (MD) simulations suggest that the T362I amino acid substitution induces different nanosecond timescale motions in distant parts of the enzyme, including around Asp53 [22]. We have also previously noted anti-correlated motions between motif D and the α-helix containing Tyr73 [32]. It is also noted that the helices containing Tyr73 and Lys250 are packed closely together (Figure 1). Altogether, these results suggest that there are structural and dynamic connections between the various Sabin sites, which may contribute to the function of the Sabin I RdRp.

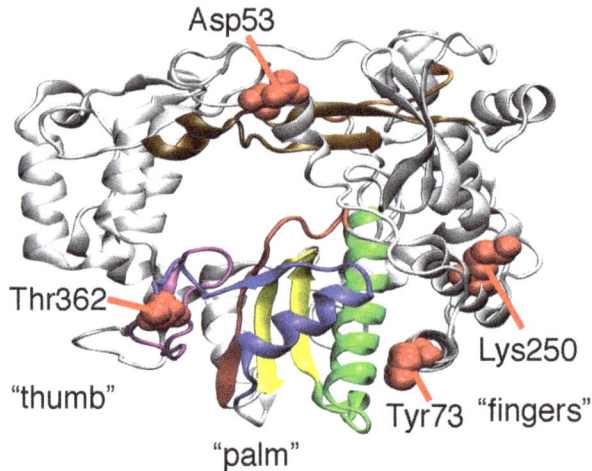

Figure 1. Locations of the amino acid substitutions in the Sabin I polymerase. The structure of the PV RdRp (PDB 1RA6 [33]) has fingers, palm, and thumb subdomains. Shown here is the "backside" of the protein, which allows easier visualization of the Sabin residues. The conserved structural motifs are colored (A, red; B, green; C, yellow; D, blue; E, purple; F, orange; G, brown). The locations of the amino acid residues changed in the Sabin I polymerase are indicated in red (*i.e.*, D53N, Y73H, K250E, and T362I).

In this article, we show that although the T362I substitution by itself lowers RdRp fidelity, the Sabin I RdRp, encoding all four substitutions, discriminates against nucleotides with incorrect nucleobases at the same level as wild-type (WT) enzyme. These results may suggest that there was evolutionary pressure during the selection of the Sabin I virus to maintain an optimal level of RdRp fidelity. In contrast, the Sabin I RdRp is less faithful when selecting against 2′-modified nucleotides, which would not be under the same selection pressure.

2. Results

2.1. Sabin PV RdRp Discriminates against Incorrect Nucleobases, but not against Incorrect Sugars, to the Same Extent as WT RdRp

Our previous studies indicated that the T362I substitution in the PV RdRp lowers nucleotide fidelity, likely because it alters the structural dynamics of the motif-D active-site loop [22], an important structural component in nucleotide discrimination [29]. Recombinant PV encoding the T362I substitution was also attenuated in a mouse model of PV [22]. These results suggested that the Sabin I RdRp may also have altered fidelity and that its function may contribute to the attenuation of the Sabin I vaccine strain. One way to probe polymerase fidelity is to perform single nucleotide incorporation assays using purified enzyme [34,35]. As such, we produced a modified RdRp enzyme encoding all four amino acid substitutions (*i.e.*, D53N/Y73H/K250E/T362I). To ensure that the Sabin I RdRp was amenable to the single nucleotide incorporation assays, we first examined its ability to interact with RNA. We have previously shown that the T362I substitution does not significantly weaken interactions between enzyme and RNA compared to WT RdRp [22].

The RNA template used in the kinetic assays was the symmetrical primer/template substrate (s/sU) that encodes for six complimentary base pairs and a four nucleotide overhang at the 5' end (*i.e.*, 5'-GCAUGGGCCC-3'), which has been used in previous kinetic and NMR analyses of the PV RdRp [22,29,34–36]. The s/sU RNA has a uracil as the first templating base in the RNA duplex. The rate and yield of competent RdRp s/sU complexes for the Sabin I RdRp was highly similar to that of WT PV RdRp (Figure 2). The dissociation rate constants for the RdRp-RNA complexes were also very similar to that of WT PV RdRp (Figure 2). Experiments with single-substituted variants (*i.e.*, D53N, Y73H, and K250E) also did not reveal substantial differences from the WT results. Any small differences we observed between WT and variant RdRp enzymes were deemed not sufficient to interfere with the single nucleotide incorporation assays.

We have previously investigated the nucleobase and sugar selectivity of the T362I variant using single nucleotide incorporation assays, which yield the maximal rate constant for nucleotide incorporation (k_{pol}) and the apparent dissociation constant for the incoming nucleotide ($K_{d,app}$) (Figure 3) [22,34,35]. These results indicated that the T362I variant had similar k_{pol} and $K_{d,app}$ values for correct nucleotide incorporation as WT RdRp, but had higher catalytic efficiency for incorrect nucleotide incorporation, including those nucleotides with an incorrect sugar (*i.e.*, 2'-dNTP) or incorrect nucleobase [22]. We performed similar experiments on the Sabin I variant (Table 1). It should be noted that the correct nucleotide was ATP in this case, since it is templated against U, and so nucleotides with incorrect sugar and nucleobase were 2'-dATP and GTP, respectively.

The Sabin I RdRp had higher rates of single nucleotide incorporation than WT enzyme using ATP, 2'-dATP, and GTP (Table 1; Figure 3). The fidelity of nucleotide incorporation can be expressed according to the kinetic experiments as $(k_{pol}/K_{d,app})_{correct}/(k_{pol}/K_{d,app})_{incorrect}$, where $(k_{pol}/K_{d,app})_{correct}$ and $(k_{pol}/K_{d,app})_{incorrect}$ are the second-order rate constants for correct (*i.e.*, ATP) and incorrect nucleotide (*i.e.*, using 2'-dATP or GTP) incorporation respectively (Table 1). Our results suggested that the Sabin I RdRp had a similar ability to discriminate against nucleotides with incorrect nucleobase (*i.e.*, $(k_{pol}/K_{d,app})_{ATP}/(k_{pol}/K_{d,app})_{GTP}$) as WT enzyme (Table 1). However, Sabin I RdRp had a reduced ability to discriminate against nucleotides with a $2'$-deoxyribose sugar (*i.e.*, $(k_{pol}/K_{d,app})_{ATP}/(k_{pol}/K_{d,app})_{2'-dATP}$) compared to WT enzyme. These results suggested that the other three amino acid substitutions in the Sabin I RdRp (*i.e.*, D53N, Y73H, K250E) may also impact RdRp fidelity.

2.2. The K250E Substitution is Unstable in Cell Culture

The nucleobase fidelity of the Sabin I RdRp was similar to that of WT enzyme (Table 1). Nonetheless, the Sabin amino acid substitutions may alter other functions of the RdRp to impact virus biology. To explore this idea, we attempted to encode recombinant PV (in Mahoney background) with the four substitutions occurring in the Sabin I polymerase. Unfortunately, this variant was not genetically stable and virus recovered from HeLa cells only retained the corresponding D53N/Y73H/T362I mutations. These results suggested that the K250E substitution was not stable in the Mahoney background, in the absence of other Sabin I mutations. Nonetheless, there was a possibility that the triple variant D53N/Y73H/T362I could be used in place of the Sabin variant for further biological characterization. Unfortunately for those studies, the D53N/Y73H/T362I variant had different sugar and nucleobase selectivities than the Sabin I variant (Table 1), and so we did not proceed with cell-based or mouse-based studies. There were also functional differences between the D53N/Y73H/T362I variant and T362I and WT RdRp enzymes (Table 1). Again, these results suggested that the other two substitutions, D53N and Y73H, had some effect on the rates and fidelity of nucleotide incorporation.

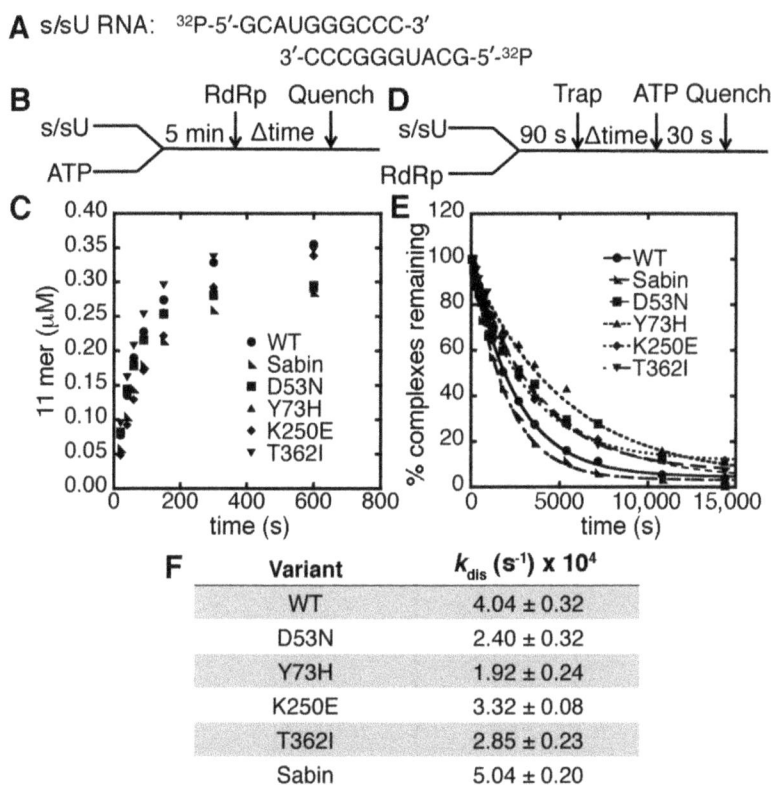

A s/sU RNA: ^{32}P-5'-GCAUGGGCCC-3'
                3'-CCCGGGUACG-5'-^{32}P

B

RdRp   Quench   **D**   Trap   ATP   Quench

s/sU ——⟍ 5 min ↓ Δtime ↓   s/sU ——⟍ 90 s ↓ Δtime ↓ 30 s ↓

ATP ——╱   RdRp ——╱

C

E

F

Variant	k_{dis} (s^{-1}) × 10^4
WT	4.04 ± 0.32
D53N	2.40 ± 0.32
Y73H	1.92 ± 0.24
K250E	3.32 ± 0.08
T362I	2.85 ± 0.23
Sabin	5.04 ± 0.20

Figure 2. The Sabin amino acid substitutions do not substantially change the association or dissociation of RdRp complexes. (**A**) The s/sU RNA is ^{32}P-labeled on the 5' end for the association and dissociation assays; (**B**) the experimental design for the RdRp-RNA-NTP assembly assay. RNA (0.5 μM duplex) and 500 μM ATP (*i.e.*, which will template against U) were pre-incubated for 5 min at 30 °C before the addition of 1 μM RdRp. Reactions were quenched at the indicated times by adding 25 mM EDTA; (**C**) comparisons of the RdRp-RNA-NTP assembly assay for WT (●), D53N (■), Y73H (▲), K250E (♦), T362I (▾) and Sabin (▲) RdRp. The Sabin I RdRp contains all four amino acid substitutions (*i.e.*, D53N, Y73H, K250E and T362I); (**D**) the experimental design for the RdRp-RNAA dissociation assay. RdRp (1 μM) and RNA (0.1 μM) were pre-incubated at 30 °C for 90 s before the addition of 100 μM unlabeled RNA (*i.e.*, "trap"). After the indicated times, the reaction buffer was mixed with 500 μM ATP and then quenched after 30 s by the addition of 25 mM EDTA; (**E**) the RdRp-RNA dissociation assays for WT (●), D53N (■), Y73H (▲), K250E (♦), T362I (▾) and Sabin (▲) RdRp. The lines represent the data fits to a single exponential function; (**F**) the RdRp-RNA dissociation rate constants derived from the data in panel (**E**). The dissociation rate constants for the RdRp variants are not substantially different from the WT enzyme.

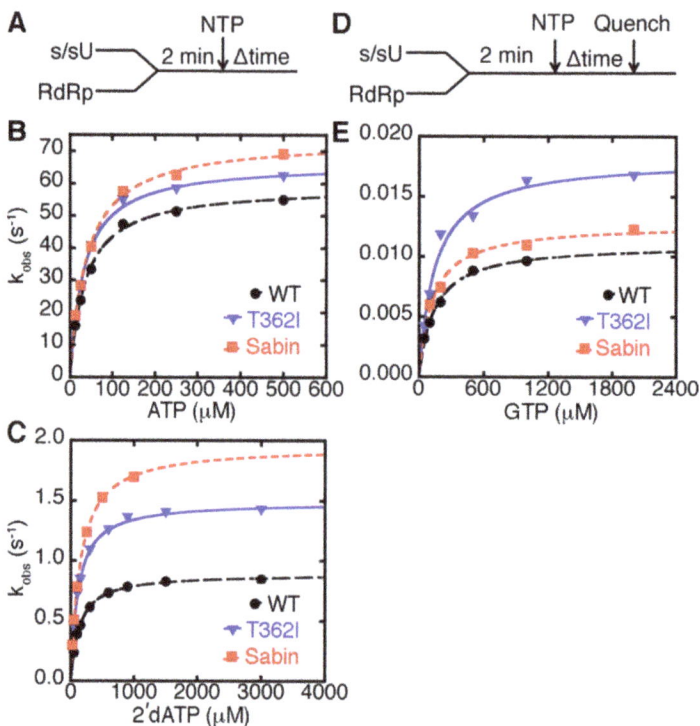

Figure 3. The Sabin I RdRp discriminates less against nucleotides with an incorrect 2′-deoxyribose sugar, but maintains nucleobase fidelity similar to WT enzyme (**A**) experimental design for the single nucleotide incorporation assay using ATP and 2′-dATP. RdRp (1 μM) was pre-incubated with s/sU RNA (1 μM), before being quickly mixed with equal volume of ATP or 2′-dATP with different concentrations. These reactions were monitored by fluorescence changes over time using a stopped-flow apparatus. Kinetic data for (**A**) AMP and (**B**) 2′-dAMP incorporation are plotted. The results for WT, T362I, and Sabin RdRp are shown in black, blue, and red, respectively. The lines represent data fit to a hyperbola function to give an apparent dissociation constant ($K_{d,app}$) and a maximal rate constant for nucleotide incorporation (k_{pol}); (**D**) the experimental design for the single nucleotide incorporation assay using GTP. RdRp (1 μM) was pre-incubated with s/sU RNA (1 μM) at room temperature for 3 min and then at the assay temperature of 30 °C for 2 min, before being quickly mixed with equal volume of GTP at different concentrations. In this case, RNA was ^{32}P labeled on 5′-end; and (**E**) kinetic data for GMP incorporation are plotted in black, blue, and red for WT, T362I and Sabin RdRp respectively. The lines represent data fit to a hyperbola function to yield $K_{d,app}$ and k_{pol} values.

2.3. D53N, Y73H and K250E PV RdRp Present Different Fidelities for Sugar and Nucleobase Selection

The functional differences between the Sabin I, D53N/Y73H/T362I, and T362I variants suggested that the D53N, Y73H, and K250E substitutions all affect RdRp function. As such, we determined kinetic values for the other single-substituted variants (Figure 1; Table 1). Although the changes induced by the single substitutions were relatively small, we note that small changes in RdRp fidelity can lead to biological effects (e.g., [37–40]). All three variants (*i.e.*, D53N, Y73H, and K250E) were a little less selective against nucleotides with incorrect sugars compared to WT RdRp. However, there were different effects for nucleobase selection. The nucleobase selectivities for the D53N, Y73H, and K250E variants were lower than, higher than, and similar to that of WT RdRp, respectively (Table 1).

Table 1. The Sabin amino acid substitutions induce small changes in RdRp catalytic rates and fidelity.

Variant	NTP	k_{pol} (s^{-1})	$K_{d,app}$ (μM)	$k_{pol}/K_{d,app}$ (μM$^{-1}\cdot$s^{-1})	$k_{pol,corr.}/$ $k_{pol,incorr.}$	$(k_{pol}/K_{d,app})_{corr.}/$ $(k_{pol}/K_{d,app})_{incorr.}$
WT		$5.9 \pm 0.1 \times 10^1$	36 ± 2	1.6	–	–
D53N		$6.2 \pm 0.1 \times 10^1$	39 ± 2	1.6	–	–
Y73H		$4.6 \pm 0.1 \times 10^1$	40 ± 2	1.2	–	–
K250E	ATP	$7.6 \pm 0.1 \times 10^1$	58 ± 3	1.3	–	–
T362I		$6.7 \pm 0.1 \times 10^1$	33 ± 2	2.0	–	–
D53N/T362I		$7.2 \pm 0.1 \times 10^1$	44 ± 2	1.6	–	–
D53N/Y73H/T362I		$8.4 \pm 0.2 \times 10^1$	47 ± 4	1.8	–	–
Sabin		$7.4 \pm 0.1 \times 10^1$	39 ± 2	1.9	–	–
WT		$8.9 \pm 0.1 \times 10^{-1}$	134 ± 4	6.7×10^{-3}	70	240
D53N		$9.3 \pm 0.2 \times 10^{-1}$	101 ± 9	9.3×10^{-3}	70	170
Y73H		$7.3 \pm 0.1 \times 10^{-1}$	117 ± 5	6.2×10^{-3}	60	190
K250E	2′-dATP	1.4 ± 0.0	174 ± 6	8.0×10^{-3}	50	160
T362I		1.5 ± 0.0	112 ± 4	1.3×10^{-2}	40	150
D53N/T362I		1.5 ± 0.0	132 ± 6	1.1×10^{-2}	50	150
D53N/Y73H/T362I		1.4 ± 0.0	101 ± 9	1.4×10^{-2}	60	130
Sabin		2.0 ± 0.0	145 ± 4	1.4×10^{-2}	40	140
WT		$1.1 \pm 0.1 \times 10^{-2}$	142 ± 15	7.7×10^{-5}	5400	21,000
D53N		$7.3 \pm 0.8 \times 10^{-3}$	91 ± 35	8.0×10^{-5}	8400	20,000
Y73H		$8.2 \pm 0.5 \times 10^{-3}$	154 ± 31	5.3×10^{-5}	5600	23,000
K250E	GTP	$9.9 \pm 0.8 \times 10^{-3}$	160 ± 41	6.1×10^{-5}	7700	21,000
T362I		$1.8 \pm 0.1 \times 10^{-2}$	149 ± 25	1.2×10^{-4}	3700	17,000
D53N/T362I		$1.2 \pm 0.1 \times 10^{-2}$	115 ± 21	1.0×10^{-4}	6000	16,000
D53N/Y73H/T362I		$1.1 \pm 0.1 \times 10^{-2}$	128 ± 12	8.6×10^{-5}	7600	21,000
Sabin		$1.3 \pm 0.1 \times 10^{-2}$	127 ± 16	1.0×10^{-4}	5700	19,000
WT	2′-C-methyl ATP	1.2 ± 0.0	160 ± 9	7.5×10^{-3}	50	210
Sabin		1.9 ± 0.0	129 ± 5	1.5×10^{-2}	40	130

2.4. Allosteric Effects among the Sabin Amino Acid Substitutions

The changes induced by the four single-substituted variants do not readily explain the kinetic results with the Sabin I variant, suggesting that there may be allosteric interactions between the amino acid substitutions. Another way of characterizing protein variants is to analyze the thermodynamic effects of the amino acid substitutions. Similar to the methods of Fersht and Mildvan [41–43], we determined $\Delta\Delta G$ values where $\Delta\Delta G_X = RT \ln((k_{pol}/K_{d,app})_X/(k_{pol}/K_{d,app})_{WT})$ where X is a particular enzyme variant (N.B. in our analysis, variants that have a lower $k_{pol}/K_{d,app}$ value than WT enzyme will yield a negative $\Delta\Delta G$ value). In the case of the Sabin I variant, there were small thermodynamic effects for AMP and GMP incorporation, but more substantial effects for 2′-dAMP incorporation (Figure 4a). This type of analysis also allows us to determine if the four amino acid substitutions were additive (*i.e.*, $\Delta\Delta G_{Sabin}$ = $\Delta\Delta G_{D53N} + \Delta\Delta G_{Y73H} + \Delta\Delta G_{K250E} + \Delta\Delta G_{T362I}$) or non-additive. Intriguingly, the amino acid substitutions were non-additive for AMP, 2′-dAMP, and GMP incorporation (Figure 4a). In fact, adding together the thermodynamic effects of the four single amino acid substitutions would predict a more catalytically-efficient Sabin enzyme (*i.e.*, $\Delta\Delta G$ is negative) for AMP and GMP incorporation, but less catalytically efficient for 2′-dAMP incorporation (*i.e.*, $\Delta\Delta G_{D53N} + \Delta\Delta G_{Y73H} + \Delta\Delta G_{K250E} + \Delta\Delta G_{T362I} >$ $\Delta\Delta G_{Sabin}$). However, it should be pointed out that these differences are smaller (*i.e.*, $|\Delta\Delta G_{Sabin} - (\Delta\Delta G_{D53N} + \Delta\Delta G_{Y73H} + \Delta\Delta G_{K250E} + \Delta\Delta G_{T362I})| \sim 0.2$–$0.3$ kcal/mol) than the kT value of 0.6 kcal/mol.

The results with the Sabin I RdRp suggested that there were small allosteric effects among the four amino acid substitutions. Our previous MD simulations with the T362I variant had indicated that the nanosecond timescale dynamics near the region encompassing Asp53 are different from what was observed in the WT enzyme [22]. To gain more insight into potential allosteric effects between these amino acid substitutions, we characterized the double variant D53N/T362I. The sugar selectivity of the double variant was similar to that of the T362I variant, and the nucleobase selectivity was reduced slightly compared to the two single variants D53N and T362I (Table 1). To more rigorously compare the effects of the single-substituted variants to the double-substituted variant, we also determined the thermodynamic effects, as we had for the Sabin variant. The effects of the single amino acid substitutions were non-additive for AMP, 2′-dAMP, and GMP incorporation (*i.e.*, $\Delta\Delta G_{D53N/T362I} < \Delta\Delta G_{D53N} + \Delta\Delta G_{T362I}$), which suggests negative cooperativity between the T362I and D53N substitutions (Figure 4a). It is also interesting to note that the double variant D53N/T362I had similar $\Delta\Delta G$ values for AMP and GMP incorporation as the Sabin variant.

A Sabin (D53N/Y73H/K250E/T362I)
D53N+Y73H+K250E+T362I
D53N/Y73H/T362I
D53N+Y73H+T362I
D53N/T362I
D53N+T362I

$\Delta\Delta G$ (kcal/mol)
0.8
0.6
0.4
0.2
0
-0.2
-0.4

ATP   2'dATP   GTP

B   ATP

WT $\xrightarrow{0.00}$ D53N/T362I   WT $\xrightarrow{0.07}$ D53N/Y73H/T362I

-0.17 $\searrow$ 0.07 $\downarrow$ 0.07   -0.13 $\searrow$ 0.10 $\downarrow$ 0.03

Y73H $\xrightarrow{0.24}$ D53N/Y73H/T362I   K250E $\xrightarrow{0.23}$ Sabin

C   2'dATP

WT $\xrightarrow{0.30}$ D53N/T362I   WT $\xrightarrow{0.44}$ D53N/Y73H/T362I

-0.05 $\searrow$ 0.44 $\downarrow$ 0.14   0.11 $\searrow$ 0.44 $\downarrow$ 0.00

Y73H $\xrightarrow{0.49}$ D53N/Y73H/T362I   K250E $\xrightarrow{0.33}$ Sabin

D   GTP

WT $\xrightarrow{0.16}$ D53N/T362I   WT $\xrightarrow{0.07}$ D53N/Y73H/T362I

-0.22 $\searrow$ 0.07 $\downarrow$ -0.09   -0.14 $\searrow$ 0.16 $\downarrow$ 0.09

Y73H $\xrightarrow{0.29}$ D53N/Y73H/T362I   K250E $\xrightarrow{0.30}$ Sabin

E 2'dATP   **F** GTP
80   10000
60   8000
40   6000
20   4000
                       2000

$k_{pol,correct}/k_{pol,incorrect}$

WT T362I D53N/T362I D53N/Y73H/T362I Sabin

WT T362I D53N/T362I D53N/Y73H/T362I Sabin

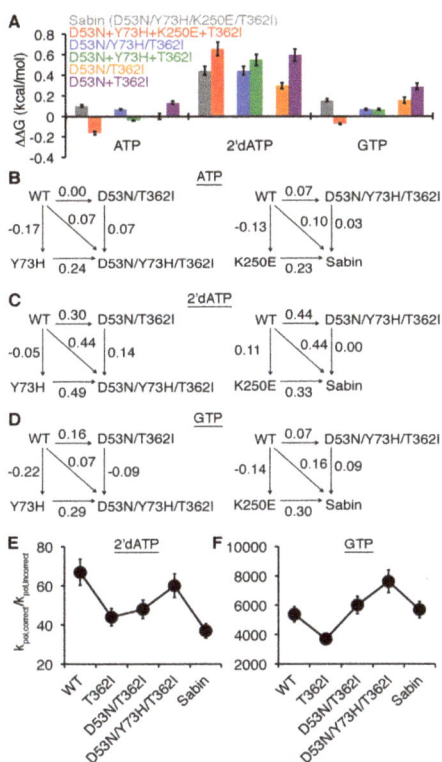

Figure 4. The Sabin I amino acid substitutions are cooperative. (**A**) Comparisons of single-substituted variants with multi-substituted variants. In this case, $\Delta\Delta G_X = RT \ln((k_{pol}/K_{d,app})_X/(k_{pol}/K_{d,app})_{WT})$ where X is a particular enzyme variant. The $\Delta\Delta G$ result for the Sabin variant with all four amino acid substitutions (D53N/Y73H/K250E/T362I) is in grey, and is compared to the sum of the effects for all four single variants in red (*i.e.*, $\Delta\Delta G$(D53N) + $\Delta\Delta G$(Y73H) + $\Delta\Delta G$(K250E) + $\Delta\Delta G$(T362I)). Likewise, the $\Delta\Delta G$ result for the triple variant D53N/Y73H/T362I is in blue and is compared to the sum of the effects for the three single variants in green. The $\Delta\Delta G$ result for the double variant D53N/T362I is in orange and is compared to the sum of the effects for the two single variants in purple. The $\Delta\Delta G$ values for the multi-variants are also compared using thermodynamics cycles when (**B**) ATP, (**C**) 2'-dATP, and (**D**) GTP are the incoming nucleotides (templated against U); all values are reported as kcal/mol. On the left, the $\Delta\Delta G$ values for the triple variant D53N/Y73H/T362I are compared to those of the double variant D53N/T362I and single variant Y73H. On the right, the $\Delta\Delta G$ values for Sabin variant with all four amino acid substitutions are compared to those of the triple variant D53N/Y73H/T362I and single variant K250E. (**E**) Sugar and (**F**) nucleobase selectivities are also compared between WT, T362I, D53N/T362I, D53N/Y73H/T362I, and the Sabin RdRp.

61

In the case of the triple variant D53N/Y73H/T362I, the effects of the D53N, Y73H, and T362I substitutions appeared to be additive (Figure 4a), which was surprising considering the non-additive results with the D53N/T362I and Sabin variants. To gain more insight, we constructed thermodynamic cycles to gauge the contributions of the Y73H and K250E substitutions to the triple and Sabin variants respectively (Figure 4b–d). Based on these thermodynamic cycles, there were small differences between $\Delta\Delta G_{D53N/Y73H/T362I}$ and the sum of $\Delta\Delta G_{Y73H}$ and $\Delta\Delta G_{D53N/T362I}$, with the largest difference being associated with 2'-dAMP incorporation. There were also small differences between $\Delta\Delta G_{Sabin}$ and the sum of $\Delta\Delta G_{K250E}$ and $\Delta\Delta G_{D53N/Y73H/T362I}$. Here, the largest difference was associated with GMP incorporation.

These results indicated that the Y73H and K250E substitutions provide small adjustments to the catalytic efficiency of the Sabin polymerase, which also has consequences for RdRp fidelity. To better visualize these changes, we compared the $k_{pol,correct}/k_{pol,incorrect}$ values between the T362I, D53N/T362I, D53N/Y73H/T362I, and Sabin variants (Figure 4e,f). It should be kept in mind that this series of variants does not necessarily recapitulate the order that these amino acid substitutions arose during the selection of the Sabin I virus. Nonetheless, this analysis provides some additional insight into a potential selection process in regards to the Sabin I polymerase, keeping in mind that other mutations in the Sabin I virus may have also played a role in regards to the polymerase changes (e.g., K250E is not stable in the Mahoney background). The results indicated that the Y73H substitution induced an increase in the relative rates of nucleotide misincorporation when comparing the D53N/T362I and D53N/Y73H/T362I variants. Interestingly, the Y73H effects appear to be opposed by the K250E substitution when comparing the D53N/Y73H/T362I and Sabin variants, such that the relative rate of the Sabin variant for misincorporation of nucleotide with incorrect nucleobase is very near that of WT RdRp, whereas the relative rate for misincorporation of nucleotide with incorrect sugar is near that of the T362I variant (Table 1, Figure 4).

2.5. Structural Dynamic Differences between the Sabin I RdRp and the Triple Variant D53/Y73H/T362I

We had previously shown that there is a correlation between RdRp fidelity and the conformational state of motif D, as reported on by the [*methyl*-^{13}C]Met ^{1}H-^{13}C HSQC NMR spectra [22,29,30]. In the NMR experiments, we first add 3'-dATP to s/sU RNA and enzyme, so that the nucleotide will become incorporated but lead to chain termination. Following passage through a de-salting column, another nucleotide is added to form a ternary complex between the enzyme, RNA, and the incoming nucleotide [29]. The correct nucleotide is considered to be UTP as it will basepair to A, which is the next templating base in the RNA. We have suggested that the peak position(s) of Met354 in motif D reports on whether motif D is in an open or closed conformation, which determines whether Lys359 is in a position to act as a general acid [29]. For WT enzyme, the ternary complexes bound with UTP and 2'-dUTP yield different [*methyl*-^{13}C] Met NMR spectra, providing fingerprints for the closed and open conformations, respectively (Figure 5a). We have also previously shown that Met354 in the T362I ternary complex bound with 2'-dUTP gives rise to two resonances, consistent with this region of the enzyme fluctuating between the open and closed conformations on the slow NMR timescale (Figure 5b); the Met6 and Met74 peak also provide evidence of conformational exchange. We have suggested that T362I RdRp has reduced nucleotide discrimination because the motif-D loop can more readily access a closed conformation, even in the presence of an incorrect nucleotide [22].

It was very interesting to note that there were differences in both sugar and nucleobase discrimination between the triple variant D53N/Y73H/T362I and the Sabin variant (Table 1; Figure 4). The Sabin I RdRp, in particular, had the lowest sugar discrimination among all variants tested (Table 1). To gain more insight into structural changes that might account for these functional behaviors, we collected [*methyl*-^{13}C] Met NMR spectra for the triple variant D53N/Y73H/T362I and Sabin variant. The ternary complexes bound with UTP yielded [*methyl*-^{13}C] Met spectra very similar to that of WT enzyme (Figure 5c,d). However, the triple variant D53N/Y73H/T362I and Sabin variant yielded small differences in the NMR spectrum for the ternary complex bound with 2'-dUTP (Figure 5c,d). In these cases, Met6 gave rise to two (or more) peaks, suggesting conformational exchange around this residue. We note that the ε-methyl group of Met6 packs against Phe59, which is located on the small α-helix that is N-capped by Asp53. Perhaps most informative is the difference between the Met354 peak(s) in the triple and Sabin variants compared to the T362I variant. There was no evidence for two Met354 resonances in the D53N/Y73H/T362I ternary complex bound with 2'-dUTP, suggesting that the enzyme was not substantially populating the closed conformation. For the Sabin variant, there was a very low intensity peak at the chemical shift position we expect for Met354 in the closed conformation. These results suggest that the Sabin 2'-dUTP ternary complex populates the closed conformation to a greater extent than the corresponding complexes for WT RdRp and the triple variant D53N/Y73H/T362I, but not nearly to the degree populated by the T362I variant.

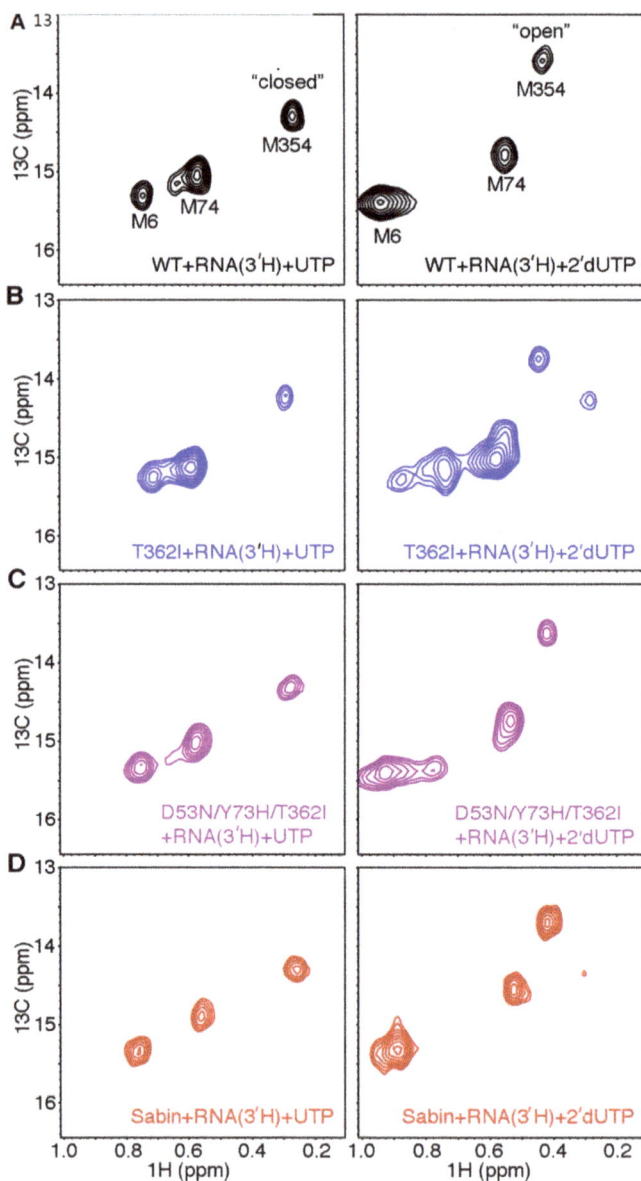

Figure 5. The Sabin I RdRp does not readily populate the closed conformation when incorrect nucleotide binds. [*Methyl*-^{13}C] Met ^{1}H-^{13}C HSQC spectra for (**A**) WT; (**B**) T362I; (**C**) D53N/Y73H/T362I; and (**D**) Sabin I RdRp when the enzyme is bound with s/s RNA lacking a 3′-OH group and (**left**) correct UTP nucleotide or (**right**) incorrect 2′-dUTP nucleotide. Spectra were collected at 293 K with 250 μM RdRp, 1000 μM s/s RNA, and 4 mM UTP or 8 mM 2′-dUTP.

2.6. The Sabin I RdRp is More Susceptible to 2′-Modified Nucleotides

It was interesting to note that the Sabin I RdRp more readily incorporates 2′-dAMP than WT RdRp, especially in light of the 2′ modified nucleotide derivatives that have been very successful in treating RNA virus infections, especially Hepatitis C virus [44,45]. As such, we characterized the ability of WT and Sabin RdRp to incorporate 2′-C-methylAMP (Table 1). Intriguingly, comparisons of the second-order rate constants $k_{pol}/K_{d,app}$ indicated that Sabin RdRp more readily incorporated 2′-C-methylAMP by a factor of two compared to WT RdRp. This result may suggest that Sabin I RdRp is more susceptible to incorporating 2′-modified nucleotides than WT enzyme.

3. Discussion

The Sabin live, attenuated vaccine has been a major component of the global efforts to eradicate poliovirus. It has been previously suggested that changes in the IRES are the major factor contributing to the attenuation of the virus [13,14]. Nonetheless, other mutations in the vaccine strains may also contribute to viral attenuation and efficacy of the vaccine [15–20,22]. We were especially interested in the RdRp amino acid changes encoded by the Sabin I strain. Our previous studies had shown that the T362I amino acid substitution induces a lower fidelity polymerase, and viruses encoding the T362I change showed a statistically significant decrease in viral pathogenesis [22]. We were thus interested in determining if there are similar functional changes in the Sabin I RdRp carrying all four amino acid changes (*i.e.*, D53N/Y73H/K250E/T362I). To our surprise, the Sabin I RdRp discriminates against nucleotides with incorrect nucleobases to the same extent as WT enzyme (Table 1; Figure 3), suggesting that the other three amino acid changes (D53N, Y73H and K250E) also modify RdRp function.

We also attempted to initiate cell-based assays with recombinant PV (*i.e.*, Mahoney background) encoding the Sabin I RdRp, but the K250E change was not genetically stable and virus recovered from HeLa cells only retained the D53N/Y73H/T362I changes. It should be kept in mind that the PV RdRp (also known as 3D) is found within other polyproteins, including 3CD, and changes in the 3D domain may have effects on the functions of these other proteins. In fact, 3CD is known to interact with the 5′ untranslated region (UTR) [46–48] and these interactions are important for regulating RNA synthesis and protein translation [48,49]. A change from a positively-charged residue (*i.e.*, Lys) to a negatively-charged residue (*i.e.*, Glu) at position 250 may disrupt interactions between 3CD and the 5′UTR. The loss of the K250E change in the Mahoney background may suggest that other elements in the Sabin I virus are important for retaining this mutation, or from a different perspective, the K250E change may have been necessary to compensate for other mutations in the Sabin I strain during the selection process, such as those in the 5′UTR.

The lack of selective pressure to maintain WT-levels of sugar discrimination may have allowed the Sabin I RdRp to drift towards a lower sugar selectivity, even lower than that of the T362I variant (Table 1; Figure 3). We had previously suggested that the lower fidelity of the T362I variant is likely because this substitution allows the motif-D active-site loop to fluctuate more readily into a "closed" conformation even in the presence of incorrect nucleotide [22]. However, this reasoning likely does not fully explain the lower sugar selectivity of the Sabin I RdRp. The equilibrium population of the "closed" state, according to the NMR experiments (Figure 5), is lower for the Sabin I RdRp compared to the T362I variant. It should be kept in mind, however, that these NMR experiments are under equilibrium conditions, and so the open-closed transition may still occur more rapidly in the Sabin I enzyme compared to WT RdRp, but not be reflected in the NMR spectra. Nonetheless, the NMR experiments suggest that the lower sugar fidelity of the Sabin I RdRp may owe to factors outside those of the motif-D active-site loop. The lower sugar selectivity of the Sabin I RdRp and RdRps associated with VAPP may serve as their Achilles heel. We have also shown that Sabin I RdRp more readily incorporated nucleotides with modified sugar groups (Table 1). These results suggest that Sabin I may be more sensitive to 2′-modified nucleotide analogs, which have been used to treat Hepatitis C and related viruses [44,45]. Considering that all variants we tested have lower sugar fidelity than WT RdRp, VAPP may also be more sensitive to this class of compounds, potentially providing an additional treatment option.

We have previously suggested that the small change in fidelity for the T362I variant likely contributes to viral attenuation [22]. In contrast, our results here suggest that fidelity of the Sabin I RdRp likely would not contribute to viral attenuation. These results would be consistent with previous studies that suggest that mutations in the IRES primarily contribute to viral attenuation in the Sabin strain [13,14]. Nonetheless, other RdRp functions might be impacted by the Sabin substitutions. It is already known that the Y73H change interferes with RdRp initiation [19]. The Sabin changes may also affect interactions with other viral or host proteins, including when the RdRp is found as a domain in other important viral polyproteins (e.g., 3CD). In these cases, the Sabin I substitutions may also be acting cooperatively to fine-tune these other functions.

4. Materials and Methods

4.1. Materials

[γ-^{32}P]ATP and [α-^{32}P]UTP (>7000Ci/mmol) was from VWR-MP Biomedical (Santa Ana, CA, USA); nucleoside 5′-triphosphates and 2′-deoxynucleoside 5′-triphosphates (all nucleotides were ultrapure solutions) were from GE Healthcare Bio-Sciences (Pittsburgh, PA, USA). 3′-Deoxyadenosine 5′-triphosphate (cordycepin) was from Trilink Biotechnologies (San Diego, CA, USA). All RNA oligonucleotides were from Dharmacon Research, Inc. (Boulder, CO, USA). T4 polynucleotide kinase was from New England Biolabs, Inc (Ipswich, MA, USA). [*Methyl*-^{13}C] methionine was from Cambridge Isotope Laboratories (Tewksbury, MA, USA). HisPur Ni-NTA resin was from ThermoFisher Scientific (Waltham, MA, USA). Q-Sepharose fast flow resin was from GE Healthcare Bio-Sciences. Polyethylenimine-cellulose thin layer chromatography (TLC) plates were from EM Science (Gibbstown, NJ, USA). QuickChange site-directed mutagenesis kit was from Stratagene (La Jolla, CA, USA). The plasmid DNA isolation miniprep kit was from Qiagen (Frederick, MD, USA). All other reagents were of the highest grade available from Sigma-Aldrich (St. Louis, MO, USA) or ThermoFisher.

4.2. Plasmid Construction

All RdRp variants including D53N, Y73H, K250E, D53N/T362I, D53N/Y73H/T362I, and D53N/Y73H/K250E/T362I (Sabin variant) were generated using the QuickChange site-directed mutagenesis kit using appropriate forward and reverse primers. Mutations were confirmed by DNA sequencing at the Nucleic Acid Facility at the Pennsylvania State University. It should be noted that all RdRp constructs contain two additional interface I amino acid substitutions (L446D and L455D) to reduce protein oligomerization.

4.3. Overexpression and Protein Purification

The overexpression and protein purification were conducted by following the procedure previously described [30,50–52].

4.4. Kinetic Assays

The kinetic assays including the active site titration assay, assembly assay and the dissociation assay were conducted as described previously [30,51]. Briefly, the reaction contained 50 mM HEPES pH 7.5, 10 mM 2-mercaptoethanol, 5 mM MgCl$_2$, and 60 μM ZnCl$_2$. The stopped flow experiments and the benchtop assays for nucleotide incorporation were conducted as described previously [22,29,34]. Reactions were incubated at 30 °C and quenched by the addition of an equal volume of ethylenediaminetetraacetic acid (EDTA) to a final concentration of 25 mM. Specific concentrations of RdRp, s/sU RNA, and nucleotide are indicated in the corresponding figure legend.

4.5. Determination of Kinetic Constants ($K_{d,app}$, and k_{pol}) for Nucleotide Incorporation Catalyzed by RdRp

Kinetic data analysis was conducted by following the procedure and equations previously described [22,29]. Briefly, the quenched product samples from kinetic assays were analyzed by 23% highly cross-linked denaturing polyacrylamide gel. The gel was visualized using a PhosphorImager and quantified using the ImageQuant software (GE Healthcare Bio-Sciences). The data were fit into different curves using Kaleidagraph (Synergy Software, Reading, PA, USA).

4.6. NMR Sample Preparation and Spectroscopy

NMR sample preparation followed procedures described previously using [*methyl*-^{13}C] Met-labeled PV RdRp [29,30]. ^{1}H, ^{13}C HSQC (heteronuclear single quantum coherence) NMR spectra were collected on a Bruker Avance III 600 MHz spectrometer equipped with a 5-mm "inverse detection" triple-resonance (^{1}H, ^{13}C, ^{15}N) single axis gradient TCI probe at 293 K. NMR samples generally contained 250 μM RdRp, 1000 μM s/s RNA, and 4 mM UTP, 8 mM 2'-dUTP or 16 mM ATP.

4.7. Construction of Mutated Viral cDNA Clones and Replicons

To introduce the four mutations, D53N, Y73H, K250E and T362I, into the RdRp/3D^{pol}-coding sequence of viral cDNA, pMovRA, overlap PCR was performed with oligonucleotides PV-3D-D53N-for (5'-GAT CCC AGG CTT AAG ACA AAT TTT GAG GAG GCA ATT TTC-3'), PV-3D-D53N-rev (5'-AGA AAA TTG CCT CCT CAA AAT TTG TCT TAA GCC TGG GAT C-3'), PV-3D-Y73H-for (5'-ATT ACT GAA GTG GAT GAG CAT ATG AAA GAG GCA GTA GAC-3'), PV-3D-Y73H-rev (5'-GTC TAC TGC CTC TTT CAT ATG CTC ATC CAC TTC AGT AAT-3'), PV-3D-K250E-for (5'-GCT TGG TTC GAG GCA CTA CAA ATG GTG CTT GAG AAA ATC GGA-3'), PV-3D-K250E-rev (5'-TCC GAT TTT CTC AAG CAC CAT TTC TAG TGC CTC GAA CCA AGC-3'), PV-3D-*Bgl*II-for (5'- GGC AAA GAA GTG GAG ATC TTG GAT GCC AAA GC-3'), PV-3D-*EcoRI*-*Apa*I-polyA-rev (5'-CGC TCA TCG ATG AAT TCG GGC CCT TTT TTT TTT TTT TTT TTT TCT CC-3') and the pMoV-3D-BPKN-I92T-T362I plasmid as template [22]. PCR products were purified and digested with *Bgl* II and *Eco* RI and the digested PCR product was ligated into pMovRA vector.

4.8. RNA Transcription

The pMo-3D-D53N-Y73H-K250E-T362I plasmid was linearized with *Apa*I and purified with Qiaex II suspension (Qiagen) by following the manufacturer's protocol. RNA was then transcribed from the linearized plasmid DNAs in a 20-μL reaction mixture containing 350 mM HEPES, pH 7.5, 32 mM magnesium acetate, 40 mM dithiothreitol (DTT), 2 mM spermidine, 28 mM nucleoside triphosphates, 0.025 μg/μL linearized DNA, and 0.025 μg/μL T7 RNA polymerase. The reaction mixture was incubated for 4 h at 37 °C, and magnesium pyrophosphate was removed by centrifugation for 2 min. The supernatant was transferred to a new tube, and RQ1 DNase (Promega; Madison, WI, USA) was used to remove the template. The RNA concentration was determined by measuring absorbance at 260 nm, assuming that an A_{260} of one was equivalent to 40 µg/mL, and the RNA quality was verified by 0.8% agarose gel electrophoresis.

4.9. Infectious Center Assays

HeLa cells were transfected by electroporation with 5 µg of viral RNA transcript and these cells were serially diluted and plated onto HeLa cell monolayers. Cells were allowed to adhere to the plate for 1 h at 37 °C and then the medium/PBS was aspirated. Cells were covered with 1X DMEM/F12 plus 10% fetal bovine serum and 1% agarose. After 2–4 days of incubation, the agarose overlay was removed and the cells were stained with crystal violet.

4.10. Virus Isolation, RNA Isolation, cDNA Synthesis, and Sequencing to Confirm the Presence of the Quadruple Mutation

HeLa cells were transfected by electroporation with 5 µg viral RNA transcript, added to HeLa cell monolayers and incubated at 37 °C. Upon cytopathic effect (CPE), viruses were harvested by three repeated freeze-thaw cycles, cell debris removed by centrifugation. Viral RNA was isolated with QiaAmp viral RNA purification kit (Qiagen), as recommended by the manufacturer. The 3Dpol cDNA was prepared from purified viral RNA by reverse transcription with MMuLV-RT (New England Biolabs) with Random Hexamer Primers (manufacturer). The resulting DNA product was then PCR amplified using SuperTaq DNA polymerase (Ambion; Naugtuck, CT, USA) and oligonucleotides PV-3D-*Bgl*II-for and PV-3D-*EcoRI*-*Apa*I-polyA-rev as primers. The presence of all four mutations was determined by sequencing of the nucleic acid obtained in second PCR step with oligonucleotides PV-3D-seq100-for (5'-GAA GGG GTG AAG GAA CCA G-3') and PV-3D-seq500-for (5'-AGG TTG AGC AGG GGA AA-3').

Acknowledgments: This work was supported by NIH grant R01 AI104878 to DDB and NIH grant R01 AI45818 to CEC.

Author Contributions: X.L. conducted the kinetic experiments, analyzed the data and helped write the manuscript; D.M.M. prepared protein samples for kinetic and NMR characterization; C.A.L. generated recombinant virus and performed cell-based assays; X.Y. performed NMR experiments; J.J.A. and C.E.C. helped design the kinetic experiments and gave scientific advice; D.D.B. designed experiments, conducted NMR experiments and wrote the manuscript. All authors contributed substantially to the present work, then read and approved the final manuscript.

Conflicts of Interest: The authors declare no conflict of interest.

References

1. Hayden, F.G. Antivirals for influenza: Historical perspectives and lessons learned. *Antiviral Res.* **2006**, *71*, 372–378.
2. Hayden, F.G. Respiratory viral threats. *Curr. Opin. Infect. Dis.* **2006**, *19*, 169–178.

3. Hayden, F.G. Antiviral resistance in influenza viruses—Implications for management and pandemic response. *N. Engl. J. Med.* **2006**, *354*, 785–788.

4. Howard, R.S. Poliomyelitis and the postpolio syndrome. *BMJ* **2005**, *330*, 1314–1318.

5. Kim, W.R. The burden of hepatitis C in the United States. *Hepatology* **2002**, *36*, S30–S34.

6. Racaniello, V.R. One hundred years of poliovirus pathogenesis. *Virology* **2006**, *344*, 9–16.

7. Weiss, S.R.; Navas-Martin, S. Coronavirus pathogenesis and the emerging pathogen severe acute respiratory syndrome coronavirus. *Microbiol. Mol. Biol. Rev.* **2005**, *69*, 635–664.

8. Modlin, J.; Wenger, J. Achieving and maintaining polio eradication—New strategies. *N. Engl. J. Med.* **2014**, *371*, 1476–1479.

9. Cochi, S.L.; Freeman, A.; Guirguis, S.; Jafari, H.; Aylward, B. Global polio eradication initiative: Lessons learned and legacy. *J. Infect. Dis.* **2014**, *210*, S540–S546.

10. Bandyopadhyay, A.S.; Garon, J.; Seib, K.; Orenstein, W.A. Polio vaccination: Past, present and future. *Future Microbiol.* **2015**, *10*, 791–808.

11. Dowdle, W.R.; de Gourville, E.; Kew, O.M.; Pallansch, M.A.; Wood, D.J. Polio eradication: The OPV paradox. *Rev. Med. Virol.* **2003**, *13*, 277–291.

12. Sabin, A.B.; Ramos-Alvarez, M.; Alvarez-Amezquita, J.; Pelon, W.; Michaels, R.H.; Spigland, I.; Koch, M.A.; Barnes, J.M.; Rhim, J.S. Live, orally given poliovirus vaccine. Effects of rapid mass immunization on population under conditions of massive enteric infection with other viruses. *JAMA* **1960**, *173*, 1521–1526.

13. Ochs, K.; Zeller, A.; Saleh, L.; Bassili, G.; Song, Y.; Sonntag, A.; Niepmann, M. Impaired binding of standard initiation factors mediates poliovirus translation attenuation. *J. Virol.* **2003**, *77*, 115–122.

14. Gromeier, M.; Bossert, B.; Arita, M.; Nomoto, A.; Wimmer, E. Dual stem loops within the poliovirus internal ribosomal entry site control neurovirulence. *J. Virol.* **1999**, *73*, 958–964.

15. Christodoulou, C.; Colbere-Garapin, F.; Macadam, A.; Taffs, L.F.; Marsden, S.; Minor, P.; Horaud, F. Mapping of mutations associated with neurovirulence in monkeys infected with Sabin 1 poliovirus revertants selected at high temperature. *J. Virol.* **1990**, *64*, 4922–4929.

16. Georgescu, M.M.; Tardy-Panit, M.; Guillot, S.; Crainic, R.; Delpeyroux, F. Mapping of mutations contributing to the temperature sensitivity of the Sabin 1 vaccine strain of poliovirus. *J. Virol.* **1995**, *69*, 5278–5286.

17. McGoldrick, A.; Macadam, A.J.; Dunn, G.; Rowe, A.; Burlison, J.; Minor, P.D.; Meredith, J.; Evans, D.J.; Almond, J.W. Role of mutations G-480 and C-6203 in the attenuation phenotype of Sabin type 1 poliovirus. *J. Virol.* **1995**, *69*, 7601–7605.

18. Omata, T.; Kohara, M.; Kuge, S.; Komatsu, T.; Abe, S.; Semler, B.L.; Kameda, A.; Itoh, H.; Arita, M.; Wimmer, E.; *et al.* Genetic analysis of the attenuation phenotype of poliovirus type 1. *J. Virol.* **1986**, *58*, 348–358.

19. Paul, A.V.; Mugavero, J.; Yin, J.; Hobson, S.; Schultz, S.; van Boom, J.H.; Wimmer, E. Studies on the attenuation phenotype of polio vaccines: Poliovirus RNA polymerase derived from Sabin type 1 sequence is temperature sensitive in the uridylylation of VPg. *Virology* **2000**, *272*, 72–84.

20. Tardy-Panit, M.; Blondel, B.; Martin, A.; Tekaia, F.; Horaud, F.; Delpeyroux, F. A mutation in the RNA polymerase of poliovirus type 1 contributes to attenuation in mice. *J. Virol.* **1993**, *67*, 4630–4638.

21. Nomoto, A.; Omata, T.; Toyoda, H.; Kuge, S.; Horie, H.; Kataoka, Y.; Genba, Y.; Nakano, Y.; Imura, N. Complete nucleotide sequence of the attenuated poliovirus Sabin 1 strain genome. *Proc. Natl. Acad. Sci. USA* **1982**, *79*, 5793–5797.

22. Liu, X.; Yang, X.; Lee, C.A.; Moustafa, I.M.; Smidansky, E.D.; Lum, D.; Arnold, J.J.; Cameron, C.E.; Boehr, D.D. Vaccine-derived mutation in motif D of poliovirus RNA-dependent RNA polymerase lowers nucleotide incorporation fidelity. *J. Biol. Chem.* **2013**, *288*, 32753–32765.

23. Vignuzzi, M.; Wendt, E.; Andino, R. Engineering attenuated virus vaccines by controlling replication fidelity. *Nat. Med.* **2008**, *14*, 154–161.

24. Hansen, J.L.; Long, A.M.; Schultz, S.C. Structure of the RNA-dependent RNA polymerase of poliovirus. *Structure* **1997**, *5*, 1109–1122.

25. O'Reilly, E.K.; Kao, C.C. Analysis of RNA-dependent RNA polymerase structure and function as guided by known polymerase structures and computer predictions of secondary structure. *Virology* **1998**, *252*, 287–303.

26. Poch, O.; Sauvaget, I.; Delarue, M.; Tordo, N. Identification of four conserved motifs among the RNA-dependent polymerase encoding elements. *EMBO J.* **1989**, *8*, 3867–3874.

27. Acosta-Hoyos, A.J.; Scott, W.A. The role of nucleotide excision by reverse transcriptase in HIV drug resistance. *Viruses* **2010**, *2*, 372–394.

28. Castro, C.; Smidansky, E.D.; Arnold, J.J.; Maksimchuk, K.R.; Moustafa, I.; Uchida, A.; Gotte, M.; Konigsberg, W.; Cameron, C.E. Nucleic acid polymerases use a general acid for nucleotidyl transfer. *Nat. Struct. Mol. Biol.* **2009**, *16*, 212–218.

29. Yang, X.; Smidansky, E.D.; Maksimchuk, K.R.; Lum, D.; Welch, J.L.; Arnold, J.J.; Cameron, C.E.; Boehr, D.D. Motif D of viral RNA-dependent RNA polymerases determines efficiency and fidelity of nucleotide addition. *Structure* **2012**, *20*, 1519–1527.

30. Yang, X.; Welch, J.L.; Arnold, J.J.; Boehr, D.D. Long-range interaction networks in the function and fidelity of poliovirus RNA-dependent RNA polymerase studied by nuclear magnetic resonance. *Biochemistry* **2010**, *49*, 9361–9371.

31. Toyoda, H.; Yang, C.F.; Takeda, N.; Nomoto, A.; Wimmer, E. Analysis of RNA synthesis of type 1 poliovirus by using an *in vitro* molecular genetic approach. *J. Virol.* **1987**, *61*, 2816–2822.

32. Moustafa, I.M.; Shen, H.; Morton, B.; Colina, C.M.; Cameron, C.E. Molecular dynamics simulations of viral RNA-dependent RNA polymerases link conserved and correlated motions of functional elements to fidelity. *J. Mol. Biol.* **2011**, *410*, 159–181.

33. Thompson, A.A.; Peersen, O.B. Structural basis for proteolysis-dependent activation of the poliovirus RNA-dependent RNA polymerase. *EMBO J.* **2004**, *23*, 3462–3471.

34. Arnold, J.J.; Cameron, C.E. Poliovirus RNA-dependent RNA polymerase (3Dpol): Pre-steady-state kinetic analysis of ribonucleotide incorporation in the presence of Mg^{2+}. *Biochemistry* **2004**, *43*, 5126–5137.

35. Arnold, J.J.; Gohara, D.W.; Cameron, C.E. Poliovirus RNA-dependent RNA polymerase (3Dpol): Pre-steady-state kinetic analysis of ribonucleotide incorporation in the presence of Mn^{2+}. *Biochemistry* **2004**, *43*, 5138–5148.

36. Arnold, J.J.; Cameron, C.E. Poliovirus RNA-dependent RNA polymerase (3Dpol) is sufficient for template switching *in vitro*. *J. Biol. Chem.* **1999**, *274*, 2706–2716.

37. Vignuzzi, M.; Stone, J.K.; Arnold, J.J.; Cameron, C.E.; Andino, R. Quasispecies diversity determines pathogenesis through cooperative interactions in a viral population. *Nature* **2006**, *439*, 344–348.

38. Korboukh, V.K.; Lee, C.A.; Acevedo, A.; Vignuzzi, M.; Xiao, Y.; Arnold, J.J.; Hemperly, S.; Graci, J.D.; August, A.; Andino, R.; *et al.* RNA virus population diversity, an optimum for maximal fitness and virulence. *J. Biol. Chem.* **2014**, *289*, 29531–29544.

39. Gnadig, N.F.; Beaucourt, S.; Campagnola, G.; Borderia, A.V.; Sanz-Ramos, M.; Gong, P.; Blanc, H.; Peersen, O.B.; Vignuzzi, M. Coxsackievirus B3 mutator strains are attenuated *in vivo*. *Proc. Natl. Acad. Sci. USA* **2012**, *109*, E2294–E2303.

40. Campagnola, G.; McDonald, S.; Beaucourt, S.; Vignuzzi, M.; Peersen, O.B. Structure-function relationships underlying the replication fidelity of viral RNA-dependent RNA polymerases. *J. Virol.* **2015**, *89*, 275–286.

41. Fersht, A.R. Dissection of the structure and activity of the tyrosyl-tRNA synthetase by site-directed mutagenesis. *Biochemistry* **1987**, *26*, 8031–8037.

42. Mildvan, A.S.; Weber, D.J.; Kuliopulos, A. Quantitative interpretations of double mutations of enzymes. *Arch. Biochem. Biophys.* **1992**, *294*, 327–340.

43. Serrano, L.; Horovitz, A.; Avron, B.; Bycroft, M.; Fersht, A.R. Estimating the contribution of engineered surface electrostatic interactions to protein stability by using double-mutant cycles. *Biochemistry* **1990**, *29*, 9343–9352.

44. Li, H.C.; Lo, S.Y. Hepatitis C virus: Virology, diagnosis and treatment. *World J. Hepatol.* **2015**, *7*, 1377–1389.

45. Noell, B.C.; Besur, S.V.; deLemos, A.S. Changing the face of hepatitis C management—The design and development of sofosbuvir. *Drug Des. Devel Ther.* **2015**, *9*, 2367–2374.

46. Gamarnik, A.V.; Andino, R. Interactions of viral protein 3CD and poly (rC) binding protein with the 5' untranslated region of the poliovirus genome. *J. Virol.* **2000**, *74*, 2219–2226.

47. Cornell, C.T.; Semler, B.L. Subdomain specific functions of the RNA polymerase region of poliovirus 3CD polypeptide. *Virology* **2002**, *298*, 200–213.

48. Toyoda, H.; Franco, D.; Fujita, K.; Paul, A.V.; Wimmer, E. Replication of poliovirus requires binding of the poly (rC) binding protein to the cloverleaf as well as to the adjacent C-rich spacer sequence between the cloverleaf and the internal ribosomal entry site. *J. Virol.* **2007**, *81*, 10017–10028.

49. Vogt, D.A.; Andino, R. An RNA element at the 5'-end of the poliovirus genome functions as a general promoter for RNA synthesis. *PLoS Pathog.* **2010**, *6*, e1000936.

50. Arnold, J.J.; Bernal, A.; Uche, U.; Sterner, D.E.; Butt, T.R.; Cameron, C.E.; Mattern, M.R. Small ubiquitin-like modifying protein isopeptidase assay based on poliovirus RNA polymerase activity. *Anal. Biochem.* **2006**, *350*, 214–221.
51. Arnold, J.J.; Cameron, C.E. Poliovirus RNA-dependent RNA polymerase (3Dpol). Assembly of stable, elongation-competent complexes by using a symmetrical primer-template substrate (sym/sub). *J. Biol. Chem.* **2000**, *275*, 5329–5336.
52. Gohara, D.W.; Ha, C.S.; Kumar, S.; Ghosh, B.; Arnold, J.J.; Wisniewski, T.J.; Cameron, C.E. Production of "Authentic" Poliovirus RNA-dependent RNA polymerase (3Dpol) by ubiquitin-protease-mediated cleavage in *Escherichia coli*. *Protein Expr. Purif.* **1999**, *17*, 128–138.

Conformational Ensemble of the Poliovirus 3CD Precursor Observed by MD Simulations and Confirmed by SAXS: A Strategy to Expand the Viral Proteome?

Ibrahim M. Moustafa, David W. Gohara, Akira Uchida, Neela Yennawar and Craig E. Cameron

Abstract: The genomes of RNA viruses are relatively small. To overcome the small-size limitation, RNA viruses assign distinct functions to the processed viral proteins and their precursors. This is exemplified by poliovirus 3CD protein. 3C protein is a protease and RNA-binding protein. 3D protein is an RNA-dependent RNA polymerase (RdRp). 3CD exhibits unique protease and RNA-binding activities relative to 3C and is devoid of RdRp activity. The origin of these differences is unclear, since crystal structure of 3CD revealed "beads-on-a-string" structure with no significant structural differences compared to the fully processed proteins. We performed molecular dynamics (MD) simulations on 3CD to investigate its conformational dynamics. A compact conformation of 3CD was observed that was substantially different from that shown crystallographically. This new conformation explained the unique properties of 3CD relative to the individual proteins. Interestingly, simulations of mutant 3CD showed altered interface. Additionally, accelerated MD simulations uncovered a conformational ensemble of 3CD. When we elucidated the 3CD conformations in solution using small-angle X-ray scattering (SAXS) experiments a range of conformations from extended to compact was revealed, validating the MD simulations. The existence of conformational ensemble of 3CD could be viewed as a way to expand the poliovirus proteome, an observation that may extend to other viruses.

Reprinted from *Viruses*. Cite as: Moustafa, I.M.; Gohara, D.W.; Uchida, A.; Yennawar, N.; Cameron, C.E. Conformational Ensemble of the Poliovirus 3CD Precursor Observed by MD Simulations and Confirmed by SAXS: A Strategy to Expand the Viral Proteome? *Viruses* **2015**, *7*, 5962–5986.

1. Introduction

Establishment of a bacterial or viral infection requires that the pathogen be able to evade intrinsic and innate immune defenses of the host cell [1]. Pathogenic bacteria that replicate within cells often encode as many as tens to hundreds "effector" proteins to evade intrinsic and innate immunity [2]. RNA viruses tend to have small genomes, typically encode no more than a dozen proteins in total. This smallness of genome size could be related to the fact that RNA viruses are highly susceptible to mutations [3,4]. RNA viruses developed a strategy to overcome the limited capacity of their encoding genomes and to maximize their functional usability [5,6]. Indeed, these viruses are as efficient as their bacterial counterparts at opposing cellular defense mechanisms.

One well known strategy of RNA viruses is the use of a virus-encoded protease(s) to inactivate host defense proteins, for example pathogen recognition receptors, their adaptors and/or signaling molecules that lead to expression of interferon [7]. These same proteases also enable RNA viruses to express the full complement of its proteome as a single polyprotein that is cleaved co- and/or post-translationally to yield a set of terminally processed proteins [8,9]. These terminally processed proteins often exhibit multiple functions, all of which are generally required for some aspect of virus multiplication [10]. Invariably, cleavage occurs in a manner that yields intermediates: dimeric, trimeric or longer fusion proteins (also referred to here as precursor proteins). Often these precursor proteins actually encode unique functions relative to the fully processed components, thus the use of a polyprotein can actually serve as a strategy and mechanism to expand the proteome [5,6].

Poliovirus (PV) and other picornaviruses, single-stranded RNA viruses with genome size 7–9 kb [11], use a polyprotein strategy to produce their proteins. The polyprotein is divided into three regions: P1, P2 and P3. The P3 region contains four proteins: 3A, 3B, 3C and 3D. In addition to the fully processed proteins, precursor proteins exist, and some of these accumulate in infected cells. One fusion protein that accumulates is 3CD, a multifunctional protein. PV 3CD consists of the amino acid sequences of both the viral protease 3C (aa 1–183) and the RNA-dependent RNA polymerase (RdRp) or 3D (aa 184–644) [12]. The 3CD protein exhibits protease and RNA-binding activities of 3C, although the specificity may differ [13,14], but the 3CD protein does not exhibit any of the RdRp activity of 3D [15]. In addition, 3CD exhibits numerous activities that are not present in 3C, 3D or the combination of the two individually (Figure 1) [10].

Both 3C and its 3CD precursor perform the majority of proteolytic cleavages of the 247 kDa polyprotein encoded by the poliovirus genome, playing a critical role in regulating the viral gene expression. Notably, 3CD was found to be more efficient (~100–1000 times) and specific than 3C in processing the P1-region of the polyprotein to yield the capsid proteins [13]. For RNA-binding activity, the 3C on its own and in the context of 3CD has been shown to bind the cis-acting replication elements (cre) of the viral RNA, forming important interactions required for genome replication; however, 3CD exhibits enhanced ability to form these interactions compared to 3C [16–22]. As shown in previous studies, 3CD is more efficient than 3C in recognizing the cloverleaf RNA at the 5′-end of the RNA genome, in association with host proteins, to form a ribonucleoprotein complex that is critical for virus replication [17,19,20]. Interactions between 3CD, but not 3C, and the *cre* at the 3'-end of the RNA genome have also been shown [20]. Additionally, 3CD binds to *oriI*, the internal *cre*, with an enhanced specificity compared to that of 3C in the VPg uridylylation reactions [21,22]— incorporation of UMP into VPg protein to form VPg-pUpU that serves as the protein primer for RNA synthesis.

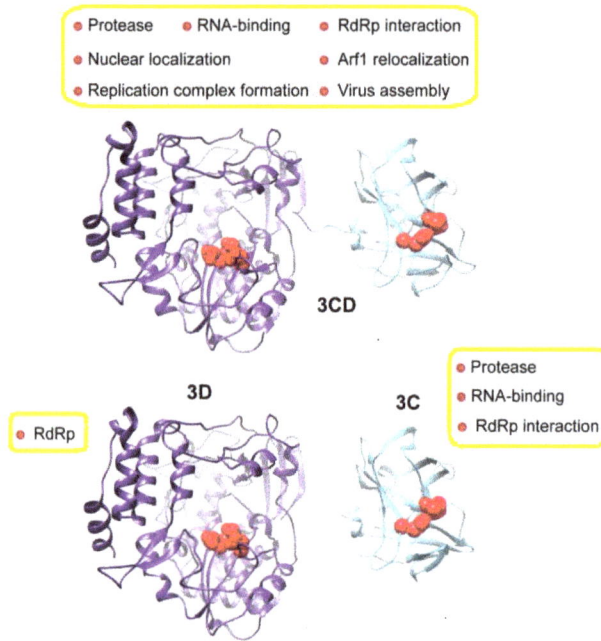

Figure 1. Crystal structure of 3CD is a composite of 3C and 3D proteins. Shown are the crystal structures of the precursor protein 3CD (PDB 2IJD) and the processed proteins 3D (1RA6, blue), and 3C (1L1N, cyan) proteins from poliovirus. The active-site residues of the protease (His-40, Glu-71, Cys-147) and polymerase (Asp-416, Asp-511, Asp-512, 3CD numbering) are shown as red spheres to indicate the relative orientation of the two domains. The different functions associated with the precursor and cleaved proteins are shown next to the structures.

Usually, polyproteins are considered "beads-on-a-string". For each protein (bead) to be cleaved, the viral protease would require an accessible cleavage site (string). However, with such arrangement, it is difficult to understand how unique function is created. PV 3CD represents a typical example. In the determined 3CD structure [23] (Figure 1), the 3C and 3D domains are arranged as beads-on-a-string, separated by a short linker (aa 180–186) to form an extended conformation with no interactions between the two domains. The precursor is a composite of the structures of 3C [24] and 3D proteins [25,26] (Figure 1). Of note, the linker region between the two domains in the crystal structure appeared weakly ordered, suggesting some flexibility in the 3CD structure [23]. Clearly, the 3CD crystal structure cannot explain how 3C-3D fusion in the precursor protein resulted in enhancing and modulating the protease activity of 3C and abrogating the polymerase function of 3D.

When the structure of PV 3CD was solved, showing a beads-on-a-string organization, the case has been made that dynamics is the missing link between structure and function of viral proteins [27]. Here we use molecular dynamics simulations of PV 3D to determine the extent to which precursor dynamics contributes to the unique activities of this protein relative to its processed components. These studies reveal a compact form of 3CD. The two domains collapse, forming an interface and masking the cleavage site. Interestingly, multiple orientations of the 3C–3D interface in 3CD were observed. The compact conformation of 3CD perturbed dynamics of the 3D domain, for example opening and closing of the nucleotide channel, thus explaining the absence of RdRp activity in the precursor. A single amino acid change in the interface produced a conformational ensemble distinct from the wild-type protein. Small angle X-ray scattering was used to demonstrate the existence of compact conformations in solution. We propose that the conformational dynamics of a viral precursor protein, coupled with the production of new ensembles by single amino acid substitutions may represent a strategy for viruses to expand their proteome/interactome. Studying interdomain interactions in these validated target precursor proteins in RNA viruses generates opportunities to develop new drugs against viral infections.

2. Materials and Methods

2.1. Molecular Dynamics Simulations

MD simulations were performed using Amber12 software suite [28] and parameters from amber99SB force field as described previously [29]. The starting coordinates for MD simulations were prepared from the crystal structure of 3CD (PDB 2IJD) [23]. The mutated residues in the crystal structure that were engineered into 3CD to facilitate crystallization were mutated back to wild-type residues. The crystal structure has two monomers 1 and 2 in the asymmetric unit that differ in the region encompassing residues 181–184 of the linker; monomer 1 is relatively more extended than monomer 2 by ~1 Å. Independent simulations for the two monomers were carried out. We also carried out a simulation on the 3CD crystal structure without modifying its sequence to wild-type protein. For all simulations, the prepared 3CD monomers were immersed in a truncated octahedral cell filled with TIP3P water molecules. The solvent molecules in the unit cell extended to at least 12 Å from any protein atoms; counterions were added to neutralize the system charge.

Simulations were conducted under NPT conditions of constant temperature (300 K) and pressure (1 atm); Berendsen thermostat was used to maintain the constant temperature. Simulations were conducted for a total of 100 ns (wild-type 3CD, monomer 1), 100 ns (WT 3CD, monomer 2), and 50 ns (mutant 3CD structure), using 1 fs integration time-step; SHAKE algorithm was employed to constrain bonds involving any hydrogen atoms. Periodic boundary conditions were employed to calculate non-bonded interactions using 9 Å cutoff distance. The Particle Mesh Ewald (PME) method was used to calculate electrostatic interactions. Snapshots of the simulated structure were taken at 1 ps interval. Analysis of the MD trajectories was done using PTRAJ and CPPTRAJ of the Amber suite [30].

To enhance the conformational sampling of 3CD, we carried out accelerated MD (aMD) simulations of WT 3CD (monomer 1) using the GPU version of pmemd in Amber 14 [31], modified to support the rotatable accelerated molecular dynamics-dual boost (RaMD-db) procedure described by Doshi and Hamelberg [32]. The structure was solvated in TIP3P water with 15 Å buffer, minimized and subjected to 10 ns of conventional MD using Amber ff14SB force field under NPT conditions with a 2 fs time-step. Boosting parameters for the RaMD-db run were calculated from the equilibration run according to the procedures described for traditional aMD in the Amber manual. The RaMD-db calculation was switched to constant temperature and volume (NVT) and run for an additional 170 ns of simulation time with snapshots saved every 10 ps. The resulting trajectory was analyzed using CPPTRAJ.

For free energy calculations of the interdomain interactions, we used SITRAJ program [33,34]. The program calculates ΔG for selected snapshots as the sum of (i) van der Waals interactions; (ii) Coulomb interactions; (iii) change in reaction field energy that is determined by solving Poisson-Boltzmann equation; and (iv) non-polar solvation energy that is proportional to the solvent accessible solvent area. The calculated ΔG is scaled by an empirical parameter as a crude treatment for entropy-enthalpy compensation [34,35]. The calculations were carried out on 5000 snapshots across the last 50 ns of the MD trajectory of the wild-type simulation (monomer 1). The selected snapshots were first clustered into six groups by PTRAJ using means algorithm [36] to make sure that ΔG is calculated over parts of the trajectory that have small fluctuations of the root-mean-square deviations (RMSD); which is recommended by SITRAJ developers. The estimated ΔGs for the different clusters were obtained by averaging of 530, 611, 1008, 645, 874, and 1332 calculations (corresponding to the number of snapshots in the clusters 1 to 6, in the same order). Knowing the limitations and caveats in this type of free energy calculations, we calculated ΔG for structures corresponding to the beads-on-a-string conformation of 3CD, where there is no interdomain interactions, and used it as a reference to calculate ΔΔG for the different clusters. The reported ΔΔG represents the favorable interdomain interaction in compact 3CD conformations sampled during simulations relative to the beads-on-a-string extended conformation shown by X-ray crystallography.

2.2. SAXS Experiments

2.2.1. Expression and Purification of 3CD

3CD protein was expressed using SUMO-fusion system. The poliovirus 3CD gene, containing the same series of mutations in the 3CD construct utilized for crystallization [23], was subcloned into the expression plasmid pET24-6His-SUMO and transformed into Rosetta(DE3) competent cells [37]. The cells containing the fusion plasmid were grown at 30 °C overnight in 100 mL of NZCYM medium supplemented with kanamycin at 25 μg/mL (K25), chloramphenicol at 20 μg/mL (C20), and dextrose at 0.4%. The overnight cell culture was used to inoculate 2 liters of NZCYM autoinducing medium supplemented with K25 and C20; cells were grown at 37 °C to an OD_{600} of ~1.0 and then growth continued at 25 °C for an additional 16 h before harvesting the cells. Cell pellets were washed once with buffer containing 10 mM Tris, pH 8.0, and 1 mM EDTA, suspended in lysis buffer [100 mM potassium phosphate, pH 8.0, 500 mM NaCl, 5 mM imidazole, 1.0 mM EDTA, 20% glycerol, 10 mM β-mercaptoethanol (β-ME), 2.8 μg/mL pepstatin A, and 2.0 μg/mL leupeptin], and disrupted by passage through a French pressure cell at 20,000 psi. Phenylmethanesulfonyl fluoride (PMSF) and Nonidet P-40 were added immediately to the cell lysate to final concentrations of 1.0 mM and 0.1%, respectively. Polyethyleneimine was added slowly to the cell lysate, to precipitate nucleic acid, at a concentration of 0.025%. The lysate was stirred at 4 °C for 30 min and then centrifuged at 25,000 rpm; the clear solution was retained, to which ammonium sulfate was added slowly to reach 60% saturation. The ammonium sulfate suspension was pelleted by centrifugation at 25,000 rpm for 30 min; the pellet, containing 3CD protein, was re-suspended in buffer A (50 mM Tris, pH 8.0, 500 mM NaCl, 10% glycerol, and 10 mM β-ME) containing 5 mM imidazole. The protein sample was loaded onto an Ni-nitrilotriacetic acid column (Ni-NTA, ~1 mL bed volume per 25 mg total protein) that was pre-equilibrated with buffer A containing 5 mM imidazole. The column was washed to baseline with the equilibrating buffer and eluted with a linear imidazole gradient (50 to 500 mM) in buffer A. Fractions containing 3CD protein, checked by SDS-PAGE, were pooled, Ulp1 protease was added to cleave the SUMO tag (~1 μg protease per 1.0 mg SUMO-fusion protein), and the sample was dialyzed overnight against buffer A containing 0.5 mM EDTA. The dialyzed sample was further fractionated on a HiLoad 26/60 Superdex 200 gel filteration column, equilibrated with buffer B [50 mM Tris, pH 7.5, 200 mM NaCl, 10% glycerol, 0.5 mM EDTA, and 2 mM dithiothreitol (DTT)]. Fractions containing 3CD protein were pooled together and concentrated using Vivaspin concentrator (MWCO 10,000) to a final concentration of 5 mg/mL.

2.2.2. SAXS Data Collection and Analysis

Samples of 3CD protein at concentrations of 0.54, 1.1, 2.2, and 4.3 mg/mL were prepared in a buffer containing (50 mM Tris, pH 7.5, 200 mM NaCl, 5% glycerol, 2 mM DTT, and 0.5 mM EDTA). Monodispersity of samples was checked by DLS. Synchrotron X-ray scattering data were collected at MacCHESS on the G1-line station. SAXS data were collected at 293 K using dual PILATUS 100K-S detector and a wavelength of 1.224 Å. The setup of the sample-to-detector distance facilitated simultaneous small- and wide-angle data recording, covering a momentum transfer range (q-range) of $0.01 < q < 0.8$ Å^{-1} [$q = \frac{4\pi\sin(\theta)}{\lambda}$, where 2θ is the scattering angle]. Exposure times of 1 min in fifteen 4-seconds frames were used for the measurements; this allowed monitoring for any radiation damage effect. No radiation damage was detected. The RAW software was used for initial data reduction and background subtraction [38]. The data at high concentrations showed some concentration dependence and a combined scattering curve was prepared from data recorded at low and high concentrations. The forward scattering $I(0)$ and the radius of gyration (R_g) were calculated using the Guinier approximation, which assumes that at very small angles ($q < 1.3/R_g$) the intensity is approximated as $I(q) = I(0)\exp\left[-\frac{(qR_g)^2}{3}\right]$. GNOM [39] was used to calculate the pair-distance distribution function P (r), from which the maximum particle dimension (D_{max}) and R_g were determined. The molecular mass was estimated using (i) comparison with lysozyme standard protein; (ii) Porod invariant [40]; and (iii) SAXS MoW web tool [41]. *Ab initio* low-resolution models were reconstructed using DAMMIN [42] for data in the range ($0.012 < q < 0.4$ Å^{-1}) and GASBOR [43] for data in the range ($0.012 < q < 0.5$ Å^{-1}). Ten models were generated from each program and averaged using DAMAVER [44]. The normalized spatial discrepancy parameter (NSD) obtained from DAMAVER indicated the similarity between models used for average calculations. NSD values ≤ 1.0 are expected for similar models. The theoretical scattering profiles of the constructed models were calculated and fitted to experimental scattering data using CRYSOL [45] and FoXS [46]. The 3CD structures from X-ray and MD simulations were fitted into the SAXS model using SUPCOMB [47].

To evaluate the conformational flexibility of the two domains in the protein, Ensemble Optimization Method (EOM 2.0) was used [48,49]. This method assumes the existence of a mixture of conformations in solution; the average scattering of the mixture fits the experimental data. In EOM, a random pool of 10,000 conformers of 3CD was generated; in these conformers the linker residues were allowed to have random-coil conformations and 3C-domain assumed different conformations relative to a fixed 3D-domain. The theoretical scattering was calculated for each generated model by CRYSOL. Using genetic algorithm in GAJOE, 100 sub-ensembles with varying numbers of conformers were selected. The distribution of R_g of the initial pool was compared with the corresponding distributions of the 100 sub-ensembles and the one with the best discrepancy (χ-value) was reported as the best solution. Information about conformers constituting the selected best sub-ensemble and their contributions to scattering was obtained. The EOM calculations were repeated three times; in each the same results were found.

2.2.3. Dynamic Light Scattering

Dynamic light scattering (DLS) experiments of the expressed and purified 3CD protein (2.0 mg/mL) were performed using Viscotec 802 instrument at 293 K. The DLS data were processed by OmniSIZE 3.0 software to get an estimate for the hydrodynamic radius R_h and polydespercity of the protein sample. The estimated R_h was 38 Å, and the estimated polydespersity was ~22.5%; see Figure S1.

3. Results

3.1. MD Simulations Revealed Domain Motions and Formation of Interdomain Interactions in 3CD

Conformational dynamics of 3CD was investigated using MD simulations as described in the Experimental section §2.1. MD simulations permit the study of time-dependent structural changes of proteins, providing detailed information on protein conformations that are relevant to function. The structure of the wild-type (WT, monomer 1) 3CD was immersed in a box containing water solvent and counterions were added to neutralize charges on the protein. The whole system was subjected to a 100 ns all-atom MD simulation with structural snapshots stored every 1 ps. Analysis of the trajectory of the MD simulation yielded detailed information on the dynamics of 3CD.

To obtain information on the extent of structural changes of 3CD during simulations, we calculated the root-mean-square deviations (RMSD) of all snapshots in the MD trajectory relative to the starting coordinates. The calculations were carried out utilizing backbone atoms of all residues in the protein and backbone atoms of the individual 3C and 3D domains (Figure 2). Clear differences were observed between the RMSD calculated using the 3CD precursor protein and that calculated for the individual domains. In the case of the precursor protein, the RMSD showed a dramatic increase during the first 10 ns of the simulation, reached an average of ~8 Å after 20 ns and continued fluctuating around this average for the rest of the MD trajectory. For 3C-domain, the calculated RMSD fluctuated around an average of ~2 Å throughout the entire trajectory. For 3D-domain, a gradual increase in RMSD values was observed during the first half of the MD trajectory, reaching a value of ~4 Å and fluctuating around this average value for the remainder of the simulation. This large discrepancy in RMSD values of the individual domains compared to that of the precursor protein suggested large domain movements during the simulation. Also, the larger RMSD values of the 3D-domain compared to that of 3C-domain (~4 Å $vs.$ ~2 Å) indicated that 3D-domain is more dynamic than 3C-domain.

The large domain movements in 3CD indicated by RMSD analysis (Figure 2) was readily observable by visual inspection of the MD trajectory using the molecular graphics software CHIMERA [50]. The 3C and 3D domains evidently approached each other early during simulations to form a compact conformation. To assess this observed conformational change in a more quantitative manner, the radius of gyration (R_g), which describes the root-mean-square distance of the protein atoms from their common center of mass, was calculated as a function of time. The R_g calculations were carried out for the full-length 3CD protein as well as for the individual 3C and 3D domains (Figure 3A); only backbone atoms were used in the calculations. For 3C-domain, the calculated R_g remained almost constant throughout the simulation with an average value of 14.7 ± 0.06 Å. The calculated R_g of the 3D-domain showed more variation than that of the 3C-domain, fluctuating around an average value of 23.4 ± 0.2 Å. For 3CD, the calculated R_g exhibited substantial variation; it decreased by approximately 8% from 30.1 Å at the starting time of simulation to an average of 27.8 ± 0.6 Å over the last 50 ns of the simulation. The decreasing R_g of the full-length 3CD without significant changes in the R_g values of the individual 3C and 3D domains is consistent with the large domain movements hinted by RMSD analysis and that the two domains are getting closer during the simulation. Calculating the average structure of 3CD using structural snapshots from the last 50 ns of the MD trajectory clearly revealed a compact conformation with interactions between 3C and 3D domains relative to their extended conformation in the crystal structure (Figure 3B). The 3C-domain moved approaching the back of the 3D-domain, towards the NTP channel of the polymerase.

To determine whether the arrangement of the 3C and 3D domains in the average MD structure corresponds to a single unique conformation or to multiple conformations sampled during simulations, we performed cluster analysis. The structural snapshots of the last 50 ns of the MD trajectory can be grouped into six clusters (1 to 6). Representative conformations of the clusters are shown in Figure 4A. Superpositioning of conformations of the different clusters using backbone atoms of the 3D-domain revealed relatively large conformational changes for residues of the 3C-domain. Also, small conformational changes for residues of the thumb (aa 563–644) in the polymerase domain were noticed. Inspection of these different conformations showed that they are similar to the compact conformation of the average structure (Figure 3B). Apparently, in all conformations, the same faces of the 3C and 3D domains are involved in interdomain interactions. Nevertheless, the details of these interactions vary from one conformation to another, suggesting a dynamic interface. Thus, a characteristic feature of the conformational changes observed in the simulations of 3CD is a large movement of the 3C-domain towards the 3D-domain with variations in the relative positions of the two domains. It is worthy to note that the collapsing of the two domains in 3CD structure to form a compact conformation occurred early during the first 10 ns of the simulation.

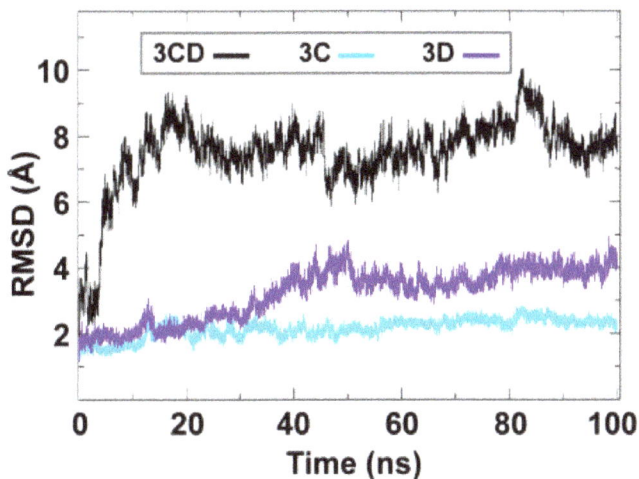

Figure 2. Root-mean-square deviations (RMSD) analysis of the molecular dynamics simulation (MD) trajectory reveals large domain movements relative to crystal structure. The RMSD for the backbone atoms of the full-length 3CD (black), of the domains 3C (cyan) and 3D (blue) are plotted as a function of time. The RMSD calculated for 3C and 3D domains showed much lower values than that calculated for 3CD. The dramatic increase in RMSD values for the full-length 3CD during the first 10 ns of the simulations suggested a large domain motion of the precursor protein. Moreover, the lower RMSD values of 3C-domain compared to that of 3D-domain indicated less dynamics of 3C-domain compared to 3D-domain.

To obtain information on the nature of motions of the two domains in 3CD and their correlations, we analyzed the correlations among positional fluctuations of Cα atoms in the simulated structure, or what is known as dynamic cross-correlation map (DCCM). In this analysis, residues that move in the same direction appear positively correlated whereas residues moving in opposite directions appear negatively correlated. Figure 4B shows the calculated DCCM for residues of 3CD using the last 50 ns of the MD trajectory. The residues of the 3C-domain exhibited strong intradomain positive correlations (red color in the calculated map), which can be interpreted as an indication of a collective movement of the whole domain. Also, the thumb residues (aa 563–644) of the polymerase domain showed strong positive correlations with each other, suggesting their collective movement as well. These collective movements of the thumb residues and 3C-domain are in agreement with the conformational changes observed by cluster analysis (Figure 4A).

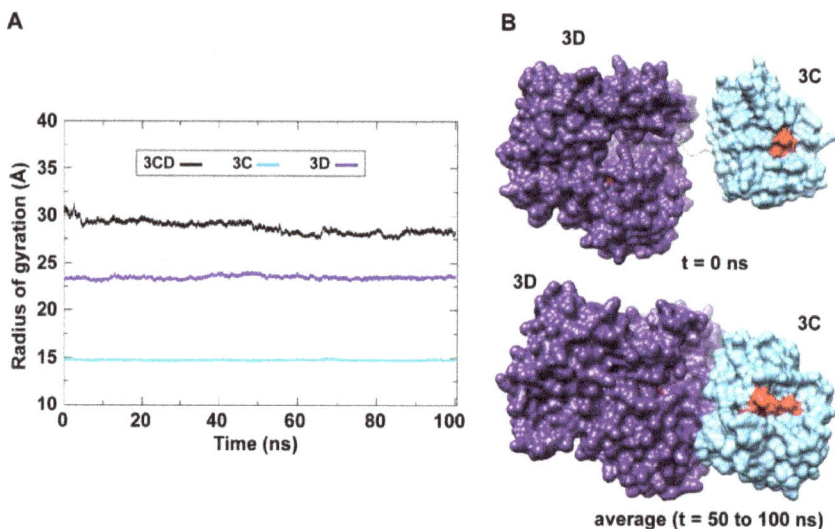

Figure 3. MD simulations reveal interdomain interactions in 3CD. (**A**) The radius of gyration (R_g) of the full-length 3CD protein (black) and the individual domains 3C (cyan) and 3D (blue) are plotted as a function of time. The R_g of 3CD substantially decreased during the simulation, whereas the R_g of 3C-domain remained almost constant and the R_g of 3D-domain showed very small variation. The decrease in R_g values of 3CD reflects the development of the interactions between 3C and 3D domains during simulations. (**B**) Shown are the 3CD structure at the starting time of the simulation ($t = 0$) and the average structure calculated over the last 50 ns of the trajectory of monomer 1. The structures are rendered as surface, the 3C and 3D domains are colored cyan and blue, respectively. The active-site residues of the protease (His-40, Glu-71, Cys-147) and polymerase (Asp-416, Asp-511, Asp-512) are indicated by red color to show the relative orientation of the two domains. The starting 3CD structure adopts an extended conformation with no interdomain interactions in contrast to that observed in the simulated average structure.

Of note, the intradomain correlations of 3D-domain differ significantly from the correlations reported previously in the MD simulation study of the cleaved polymerase [29] and protease [51]. The characteristic negative correlations of residues surrounding the template-nascent RNA duplex channel, the template channel and the NTP channel in the 3D^{pol}, which indicated the expansion and contraction of these channels of the polymerase, are lacking in the 3D-domain of the precursor protein;

additionally, the positive correlations among the functional motifs in the palm and fingers subdomains are less pronounced in the precursor protein compared to the mature polymerase. Inspecting the interdomain correlations revealed that residues of the 3C-domain were negatively correlated (blue color in the map) with regions of the fingers and palm of the 3D-domain encompassing residues 186–190, 238–246, 360–370, and 423–434. These regions of the 3D-domain appeared to be on the same side or near the interface with the 3C-domain. The negative correlations suggest that they move in a direction opposite to that of the 3C-domain movement. Thus, from DCCM analysis it can be concluded that the 3C-domain moves as a whole unit, or what is called "rigid-body" motion, towards a unique face of the 3D-domain. Further analysis of the MD trajectory using principal component analysis (PCA), which filters the noise from major motions in MD simulations as described in our previous study [29], revealed that the rigid-body movement of 3C-domain towards the polymerase domain is the major motion observed during simulations (Figure S2).

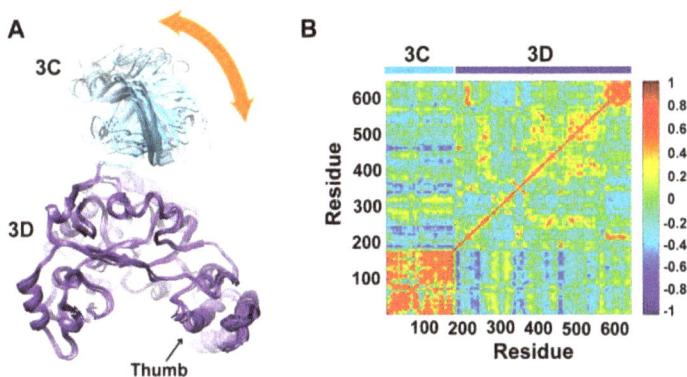

Figure 4. Conformational changes of 3CD during simulations. (**A**) Representative conformations of the clusters (1–6) sampled during the last 50 ns of wild-type (WT) 3CD simulation of monomer 1, superimposed using Cα atoms of 3D-domain. The positioning of 3C-domain (cyan) relative to 3D-domain (blue) varies among the different conformations, resulting in small variations at the interface between the two domains as indicated by the orange arrow. (**B**) Dynamic Cross-Correlation Map (DCCM) analysis of the last 50 ns, calculated for Cα atoms of all residues in 3CD. Residues of 3C-domain exhibited strong correlations, indicating rigid-body motion of the domain. Also, residues of the thumb (aa 563–644) in the 3D-domain showed strong correlations; which is consistent with the slightly larger deviations observed for the thumb residues shown in (**A**).

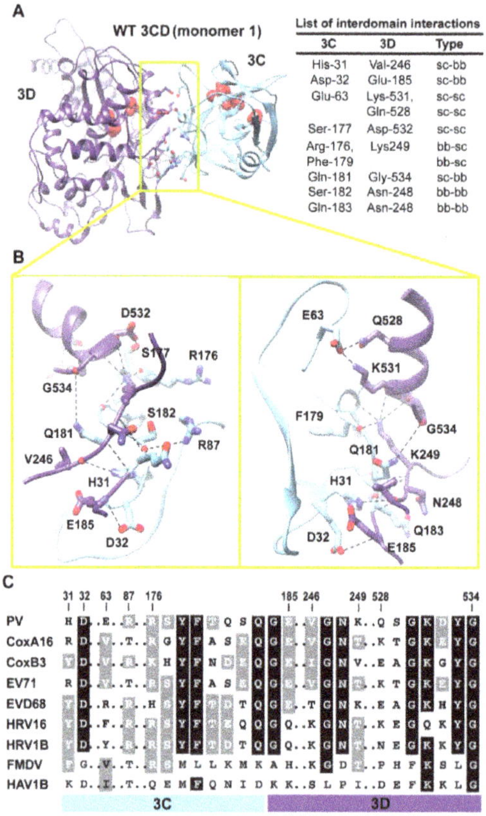

Figure 5. Interactions between 3C and 3D domains in the compact conformation of 3CD (monomer 1). (**A**) Shown is the simulated WT 3CD with interface residues displayed as sticks; the 3C and 3D domains are colored cyan and blue, respectively. The active-site residues of the protease (His-40, Glu-71, Cys-147) and polymerase (Asp-416, Asp-511, Asp-512) are shown as red spheres to help identifying the relative orientation of the two domains. The list of interdomain interactions is shown at the right of the panel; three types of interactions are indicated (sidechain-sidechain, sc-sc; sidechain-backbone, sc-bb/bb-sc; backbone-backbone, bb-bb). (**B**) Two close-up views of the interface marked by the yellow box in (**A**). Interacting residues are shown as sticks and their C-atoms are colored according to the corresponding domains. Interface residues are engaged in many H-bonding interactions and salt bridges, depicted as black dashed-lines. (**C**) Shown is part of the sequence alignment of 3CD proteins from picornaviruses, including poliovirus (PV), coxsackievirus A16 (CoxA16), coxsackievirus B3 (CoxB3), enterovirus 71 (EV71), enterovirus D68 (EVD68), human rhinovirus 16 (HRV16), human rhinovirus 1B (HRV1B), foot-and-mouth disease virus (FMDV), and hepatitis A virus 1B (HAV1B). The residues shown in the alignment are those involved in the interdomain interactions: 31–32, 63, 87, 176–185, 246–249, and 528–534. Many of the interface residues are highly conserved.

86

3.2. Interface Residues and Stability of the Interdomain Interactions

Next, we analyzed the conformations visited during simulations of the WT 3CD to obtain information on residues involved in the interdomain interactions. The 3C and 3D domains bury ~500 Å^2 of surface area between them. Residues mediating the interactions between the two domains are shown and listed in Figure 5A,B. The interface is formed by amino acids from the 3C-domain (residues 31, 32 and 63), the linker (residues 176–185), and from the 3D-domain (residues 246–248 of the fingers subdomain and residues 528–534 of motif D in the palm subdomain of the polymerase). The contact surface between the two domains involves electrostatic and weak hydrogen bonding interactions made by backbone and sidechain atoms. The linker region is sandwiched between the two domains to form an intricate network of interactions. The sidechain of Ser-177 forms a hydrogen bond with that of Asp-532. The carbonyl oxygen atoms of four residues, two from each domain, including Arg-176, Phe-179, Asp-532 and Tyr-533, form an oxyanion hole that accommodates the positively charged sidechain of Lys-249. The sidechain of Gln-181 is engaged in hydrogen bonding interaction with the backbone of Gly-534. The backbone atoms of Ser-182 mediate hydrogen bonding interactions with both backbone and sidechain atoms of Asn-248 and Arg-87, respectively. The backbone atoms of Asn-248 in turn are hydrogen bonded to the backbone atoms of Gln-183. The backbone nitrogen atom of Glu-185 forms a hydrogen bond with the sidechain of Asp-32. The residue preceding Asp-32, His-31, mediates a hydrogen bonding interaction between its sidechain and the carbonyl oxygen atom of Val-246. Finally, Glu-63 forms a salt bridge with Lys-531 and a hydrogen bond with the sidechain of Gln-528.

Interestingly, many of the residues at the interface are linked to functions. In the 3C-domain, His-31, Asp-32, Glu-63 and Arg-87 have been shown by NMR to be involved in RNA-binding by 3C [16,51]. Mutations of His-31 and Asp-32 in 3C impair the RNA recognition activity [17]. The residue Arg-176 is also known to be implicated in RNA recognition [24]. In addition, Arg-176 and Phe-179 are involved in peptide binding [51]. In the 3D-domain, Lys-249 (Lys-66 in PV RdRp numbering) is nearby Lys-61 of 3D that was previously shown to be critical for polymerase function [52]. Residues 528–534 correspond to motif D of the polymerase that is also known to be critical for polymerase function [53,54]. Therefore, the interactions between the two domains could be linked to the functional differences between the precursor protein 3CD and its cleaved products.

When we evaluated whether the interdomain interactions in PV 3CD exist in the related 3CD proteins from other picornaviruses (Figure 5C), we found that many of these interactions are conserved but some are not. Interactions in PV 3CD that involve backbone atoms and sidechains of residues at positions 31, 32, 87,177, 181, 249 and 532 are likely to exist in the related proteins; residues at these positions are either conserved or substituted by residues capable of making similar interactions. However, interactions mediated by Glu-63 are predicted to be unique for the poliovirus protein. Our interpretation of this observation is that picornaviral 3CD proteins likely adopt compact conformations in which the two domains interact with each other.

After examining the interface between 3C and 3D domains in the simulated WT 3CD and showing that the contact surface between the two domains involves many interactions, we evaluated the stability of the interdomain interactions in the compact conformation relative to the extended conformation seen in crystal structure. To do so, we carried out the free energy calculations using SIETRAJ as described in the Experimental section 2.1. The relative free energy ($\Delta\Delta G$) of 3C-3D interactions were calculated for the different conformations visited during simulations; see Table 1. The compact conformations with 3C-3D interdomain interactions were found to be more stable than the extended conformation with non-interacting domains by an average of -7.4 ± 0.89 kcal/mol. Considering the number of interactions between the two domains described above, the estimated stabilization energy seemed reasonable.

Table 1. Relative free energy of interdomain interactions in WT 3CD.

Conformation	$\Delta\Delta G$ [a] (kcal/mol)
Crystal structure [b]	0
Cluster 1	-7.73 ± 0.89
Cluster 2	-8.55 ± 0.98
Cluster 3	-6.81 ± 0.85
Cluster 4	-6.25 ± 0.74
Cluster 5	-8.26 ± 0.90
Cluster 6	-6.84 ± 0.98
average	-7.41 ± 0.89

[a] errors correspond to standard deviations; [b] used as a reference for the free energy calculations.

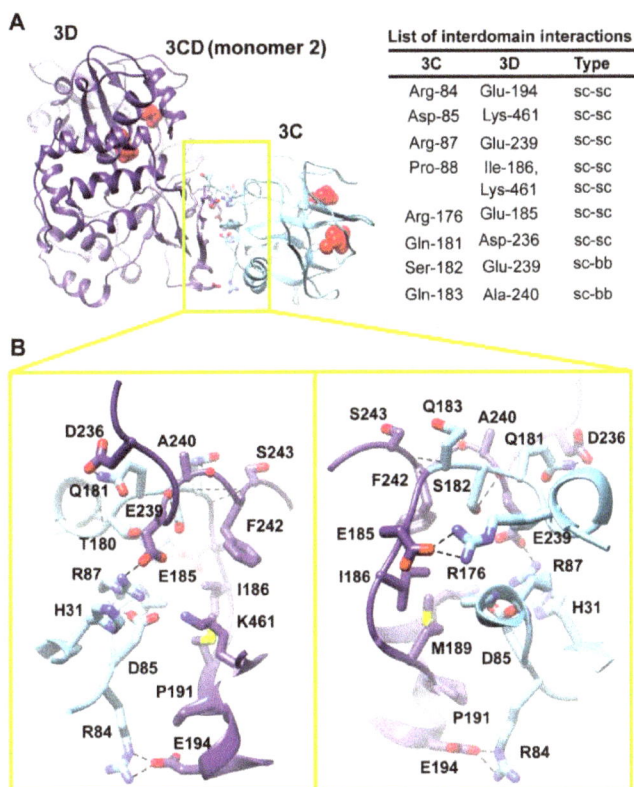

Figure 6. Interface in the compact conformation of 3CD (monomer 2). (**A**) The simulated 3CD is shown with interface residues displayed as sticks; the 3C and 3D domains are colored cyan and blue; respectively. The active-site residues of the protease (His-40, Glu-71, Cys-147) and polymerase (Asp-416, Asp-511, Asp-512) are shown as red spheres to help identifying the relative orientation of the two domains. Interdomain interactions are listed at the right of the panel; two types of interactions are indicated (sidechain-sidechain, sc-sc; sidechain-backbone, sc-bb). (**B**) Two close-up views of the interface marked by the yellow box in (**A**). Interacting residues are shown as sticks and their C-atoms are colored according to the corresponding domains. Interface residues are engaged in many H-bonding (black dashed-lines), hydrophobic and electrostatic interactions. The contact surface between the two domains in the mutant 3CD is distinct from that of monomer 1 shown in Figure 5.

89

Two monomers were present in the 3CD crystal structure with slightly different conformations of the linker. In order to assess the impact of starting conformation on the outcome of the simulation, we performed a second simulation using monomer 2. Similar to monomer 1 simulation, analysis of the MD trajectory revealed a 3CD molecule forming a compact conformation with a surface area of ~460 $Å^2$ buried between 3C and 3D domains. Interestingly, a new interface in monomer 2 was revealed and appeared to be different from that observed in monomer 1. The 3C-domain moved ~20 Å and rotated ~60° relative to the conformation observed for monomer 1, now reaching the upper part of the fingers in the 3D-domain (Figure S3A,B). The new interface is formed by the loop residues Arg-84, Asp-85, Arg-87 and Pro-88 projected from 3C-domain that interact with residues Ile-186, Glu-194, Glu-239 and Lys-461 from the fingers of 3D-domain (Figure 6). The interface residues from 3D-domain, in turn, interact with residues Arg-176, Gln-181, Ser-182 and Gln-183 at the C-terminus of the 3C-domain. Examination of the interface residues showed that residues at positions 84, 85, 87 and 186 are completely conserved in the related picornaviral proteins; the remaining residues showed a high level of conservation. The site of the 3D-domain contributing to the interface in monomer 2 is adjacent and non-overlapping to that presented in monomer 1. Also, the loop residues 84–88, which are known to play a role in RNA binding [16] in 3C-domain, are adjacent to the 31–32 site that is part of the interface in monomer 1 (Figure 5). The linker residues 176–183 contribute to the contact surfaces in the two monomers. The interfaces in the two monomers are predominately formed by electrostatic and hydrogen bonding interactions. Nevertheless, in monomer 2 there is a hydrophobic contribution to the interface from the interaction between Pro-88 from 3C-domain and Ile-186 from 3D-domain (Figure 6). Thus, it can be argued that 3CD is capable of assuming different compact conformations that could serve distinct functions at different stages of the viral lifecycle.

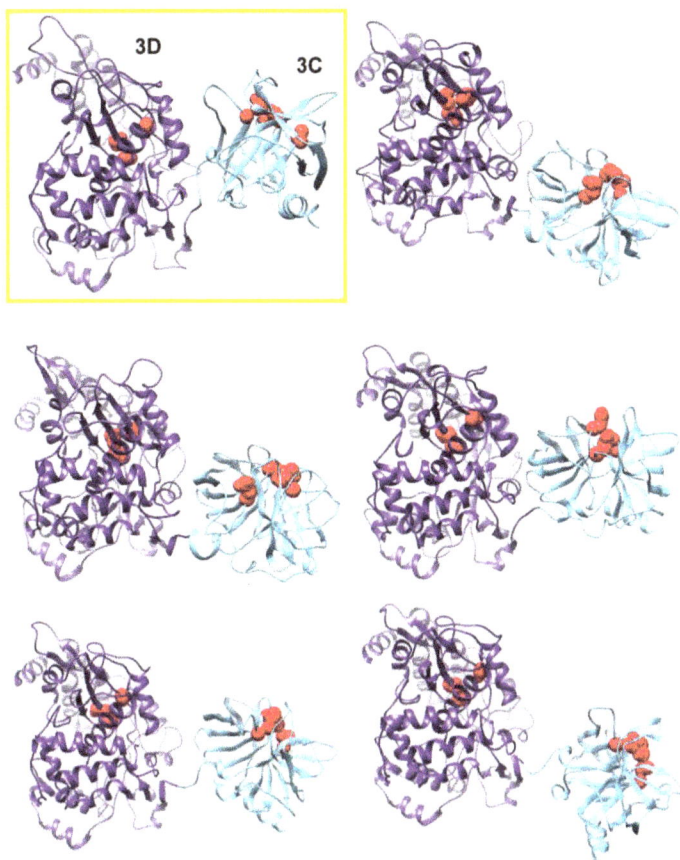

Figure 7. Accelerated MD simulations reveal dynamic interface between 3C and 3D domains. Shown are five representative conformations of 3CD sampled during the accelerated MD simulations (including both compact and extended conformations), revealing the dynamic nature of the interface between the two domains of the protein. The interface in the compact conformation from the conventional MD is shown in a yellow box for comparison. The 3C and 3D domains are colored cyan and blue, respectively; active-site residues of the protease (His-40, Glu-71, Cys-147) and polymerase (Asp-416, Asp-511, Asp-512) are shown as red spheres to help identifying the relative orientations of the two domains.

The observation of two unique orientations of the 3C-domain relative to the 3D-domain in the two simulated 3CD monomers begged the question: do additional conformations exist? To address this question we performed a 170 ns accelerated MD simulation (aMD) as described in the Experimental section §2.1 that enhanced sampling of the conformational space of 3CD. Similar to the above two simulations, the 3CD assumed compact conformations with interfaces that overlap with the interfaces observed in the conventional MD simulations (Figure 7). The switching among different conformations is shown in movie S1. The buried surface area between the two domains was in the range of ~500 to ~650 Å^2 in the different compact conformations sampled during the aMD simulation. Interestingly, in the accelerated simulation the 3CD has been seen transiently visiting relatively extended conformations similar to that observed in the crystal structure. The boosting energy in the aMD helped to overcome the interdomain interactions that exist in the compact conformations. From the results of MD simulations it can be concluded that 3CD may exist in solution as a mixture of many compact and extended conformations.

3.3. Mutated 3CD has Altered Interface Compared to the WT Protein

Crystal structures of 3C and 3D proteins revealed numerous packing interactions. Mutagenesis was used to prohibit these interactions during crystallization of 3CD. These mutations are E55A, D58A, E63A, C147A, L629D and R638D; see Figure S4. Because some of these substitutions were near the interface of the two domains, this structure provided an opportunity to evaluate the impact of these mutations on the outcome of the simulations. We performed a 50 ns MD simulation of the mutant 3CD and analyzed it in the same manner described for the WT protein; see Experimental section §2.1. Similar to WT simulations, RMSD and radius of gyration analyses indicated large domain movements (Figure S5). Analysis of the conformations sampled during simulations of the mutant protein suggested interdomain interactions between 3C and 3D domains. The interface between the domains in the mutant was found to be different from that observed in the WT protein (Figure S3C).

The interdomain interactions found in the mutant 3CD are displayed in Figure 8. The surface area buried between the two domains is ~700 Å^2, larger than that observed in the WT protein. Comparison of the interdomain interface of the mutant with that shown in Figures 5 and 6 for WT protein highlighted interesting similarities and differences. Whereas some residues of the 3D-domain at the interface in WT were also present in the mutant (Glu-185, Asp-236, Glu-239, Val-246, Lys-249 and Lys-531), interface residues from the 3C-domain in the mutant were totally different from that in the WT protein. Specifically, the substitution at position-63 in the mutant disrupted the interaction between Glu-63 and Lys-531 observed in the WT protein. The sidechain of Lys-531 in the mutant is engaged in hydrogen bonding interaction with backbone atoms of Gly-129; the aliphatic moiety participates in hydrophobic interactions with that of Asn-69. The sidechain of Asn-69 forms a bridge between the two domains via hydrogen bond formation with the carboxylate of Glu-71. In the vicinity of these interactions, the sidechain of Gln-131 forms a hydrogen bond with the carbonyl oxygen of Ser-529. Additionally, the interactions mediated by the linker region (residues 176–183) in WT are lost in the mutant. The only linker residue that is part of the contact surface in the mutant is Glu-185 of the 3D-domain. It forms a salt bridge with Lys-60, which is hydrogen bonded to the backbone oxygen of Val-246. The environment of Lys-249 in WT is lost in the mutant where it is hydrogen bonded to the backbone oxygen of Ala-61. Lastly, new interactions in the mutant mediated by Glu-45 and Lys-78 have no equivalent in WT. The residue Glu-45 forms two salt bridges with Lys-350 and Lys-355 and a hydrogen bond with the sidechain of Gln-353. Lys-78 forms a salt bridge with Asp-236 and is located within range to participate in electrostatic interactions with Glu-239. We conclude that a single substitution of Glu-63 with alanine at the interface was sufficient to remodel the interface to a completely new conformation.

3.4. Experimental SAXS Data Support more than one Conformation for 3CD

The MD simulations revealed a compact conformation with interdomain interactions for 3CD and transiently sampled the extended beads-on-a-string conformation reported in the crystal structure. To see if the two conformations exist and to gain more insight into the solution state, we performed small-angle X-ray scattering (SAXS) experiments; see the Experimental section §2.2. SAXS technique is ideal to explore the large domain movement suggested for 3CD by MD simulations [55].

Experimental data summary is given in Table 2. SAXS data collected at four different concentrations in the range of 0.54 to 4.3 mg/mL are shown in Figure 9A. Parameters that characterize the size of the protein, radius of gyration (R_g) and maximum particle dimension (D_{max}), were determined from data collected at low concentrations (0.54 and 1.1 mg/mL), for which no concentration dependence was observed. Data at very small angles showed linear correlations (Figure 9B) that satisfied Guinier approximation ($qR_g < 1.3$), from which an R_g value of 32.5 Å was obtained. The R_g parameter was also determined from the interatomic pair-distance distribution function P(r), computed by GNOM, which takes into account the entire scattering curve, not only the very small-angle portion (Figure 9C). The real-space R_g obtained from P(r) was determined to be 33.48 ± 1.53 Å, which is in good agreement with that obtained from Guinier approximation. It should be noted that the R_g from SAXS accounts for both protein atoms and solvent molecules in the hydration shell [56]. It is therefore larger than the R_g calculated from structure coordinates utilizing backbone atoms of the protein during MD simulations. The R_g from SAXS agrees well with the estimated hydrodynamic radius (R_h) from DLS with an R_g/R_h ratio of 0.86, which suggests a globular shape of 3CD. For comparison, an R_g/R_h value of ~0.8 is characteristic for globular proteins, the ratio increases as the molecules deviate from globular to elongated shapes, reaching ~1.4 for denatured proteins [57]. The D_{max} parameter was derived from P(r) with a value of 125 Å. The estimated size parameters are reasonable for a molecule with the size of 3CD, indicating a monomeric state of the protein in solution. In addition, the molecular masses derived from three different methods (76.9 ± 3.8, 71.5 ± 2.0 and 72.8 ± 2.9 kDa) were consistent with the calculated molecular mass of a monomer (71.92 kDa); see Table 2.

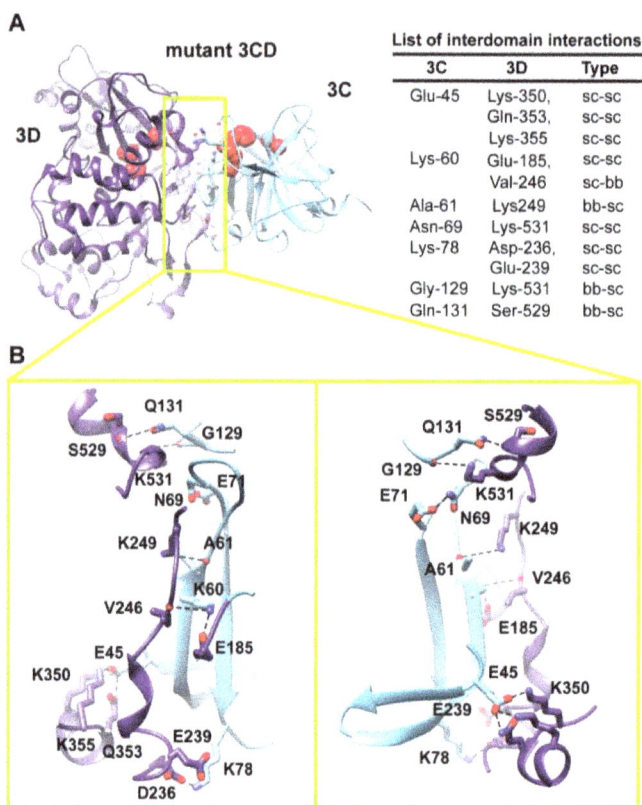

List of interdomain interactions

3C	3D	Type
Glu-45	Lys-350,	sc-sc
	Gln-353,	sc-sc
	Lys-355	sc-sc
Lys-60	Glu-185,	sc-sc
	Val-246	sc-bb
Ala-61	Lys249	bb-sc
Asn-69	Lys-531	sc-sc
Lys-78	Asp-236,	sc-sc
	Glu-239	sc-sc
Gly-129	Lys-531	bb-sc
Gln-131	Ser-529	bb-sc

Figure 8. MD simulation reveals altered interface in the mutant 3CD. (**A**) The simulated mutant 3CD is shown with interface residues displayed as sticks; the 3C and 3D domains are colored cyan and blue; respectively. The active-site residues of the protease (His-40, Glu-71, Cys-147) and polymerase (Asp-416, Asp-511, Asp-512) are shown as red spheres to help identifying the relative orientation of the two domains. Interdomain interactions are listed at the right of the panel; two types of interactions are indicated (sidechain-sidechain, sc-sc; sidechain-backbone, sc-bb/bb-sc). (**B**) Two close-up views of the interface marked by the yellow box in (**A**). Interacting residues are shown as sticks and their C-atoms are colored according to the corresponding domains. Interface residues are engaged in many H-bonding and salt bridges, depicted as black dashed-lines. The contact surface between the two domains in the mutant 3CD is different from that observed in the WT protein shown in Figures 5 and 6.

We used the combined scattering curve to construct low-resolution SAXS models using DAMMIN and GASBOR programs; see the Experimental section §2.2. Ten independent models were generated from each program and averaged. Some variations among the models were observed as indicated by NSD values that represent their similarity (1.211 ± 0.041 for DAMMIN models and 1.438 ± 0.049 for GASBOR models), which may reflect the flexibility of the protein. The average DAMMIN model is shown in Figure 10A. The model can reasonably fit the two conformations of 3CD: the one observed in the crystal structure and that revealed by MD simulations (Figure 10B). Of note, the structure from simulations showed slightly better fitting (NSD = 0.67) than the crystal structure (NSD = 0.96). Nevertheless, the precise orientation of the protein inside the SAXS envelope could not be determined without ambiguity. Moreover, fitting the calculated scattering profiles of the crystal structure and that of the average MD structure to the experimental curve appeared to be less than optimal with χ values of 1.9 and 2.9 for the crystal structure and the average MD structure, respectively (Figure 10C). This was interpreted to mean that no single conformation can satisfactorily fit the scattering data.

Table 2. Small-angle X-ray scattering (SAXS) data analysis for 3CD protein.

Data Collection	
Instrument	G1-line station at CHESS, dual Pilatus 100K-S detector
Beam diameter (μm)	250 × 250
Wavelength (Å)	1.244
q-range (Å^{-1})	0.006–0.800
Exposure time (s)	15 × 4
Concentration range (mg mL^{-1})	0.54–4.3
Temperature (K)	293
Structural parameters	
R_g [real-space R_g from $P(r)$] (Å)	33.48 ± 1.53
R_g (from Guinier) (Å)	32.5
D_{max} (Å)	125
Molecular-mass determination	
Molecular mass M_r (Da)	
From Porod volume ($V_P/1.6$)	76,875 ± 3750
From Lysozyme standard	71,492 ± 1953
From SAXS MoW	72,750 ± 2900
Calculated from sequence	71,920
Software employed	
Primary data reduction	RAW
Data processing	GNOM
Ab initio analysis	DAMMIN, GASBOR
Validation and averaging	DAMAVER
Conformational flexibility	EOM
Computation of model scattering	CRYSOL, FoXS
Fitting structure to SAXS model	SUPCOMB

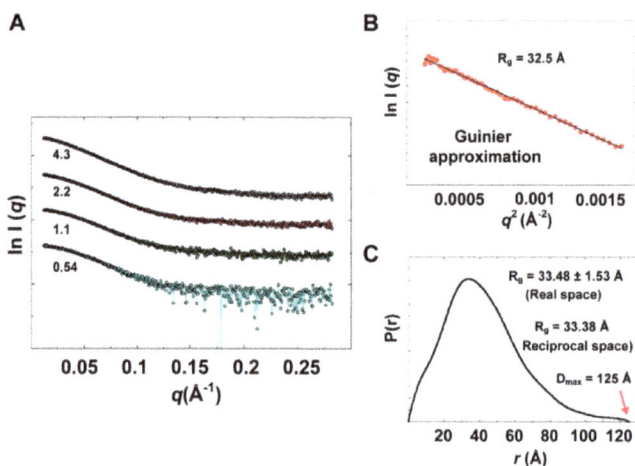

Figure 9. Experimental solution scattering data. (**A**) Shown are plots of the scattering intensity from SAXS data collected at different concentrations of 3CD protein: 0.54 (cyan), 1.1 (green), 2.2 (red) and 4.3 (grey) mg/mL. (**B**) Guinier plot of data at very small angles is shown with a linear regression satisfying the approximation $q < 1.3/R_g$. (**C**) The pair-distance distribution function P(r) calculated by GNOM is shown. The estimated maximum particle dimension (D_{max}) and R_g from P(r) are indicated on the plot. The R_g determined from Gunier approximation is in good agreement with that calculated by GNOM.

To assess the possibility of the presence of multiple conformations for 3CD in solution, we employed the Ensemble Optimization Method (EOM). This method assumes that a mixture of different conformers co-exist in solution and finds the best sub-ensemble out of a randomly generated ensemble consisting of a large number of conformers that best fits the experimental data. From an ensemble of 3CD conformers in which the 3C-domain and the linker residues were allowed to adopt different conformations, EOM selected a sub-ensemble that fits the scattering data better than any single conformation with a χ-value of 1.38 (Figure 11A). Comparison of the R_g histogram of the initial pool covering the range of 26–38 Å (corresponding to compact-to-extended conformations) with that of the selected sub-ensemble indicated the presence of both extended and compact conformations in solution (Figure 11B). The selected sub-ensemble has a bimodal distribution that could result from switching between the two conformations. The first peak centered at an R_g of ~31 Å corresponds to the compact conformation, and the second peak centered at an R_g of ~33 Å corresponds to the extended conformation. The conformer corresponding to the first peak accounts for 43% of the total scattering and that corresponding to the second peak accounts for 57%. Thus, both compact and extended conformations are well represented in solution. Of note, the entire initial pool is not represented in the selected sub-ensemble, suggesting a limited conformational space of the protein with the molecule fluctuating around its compact and extended conformations.

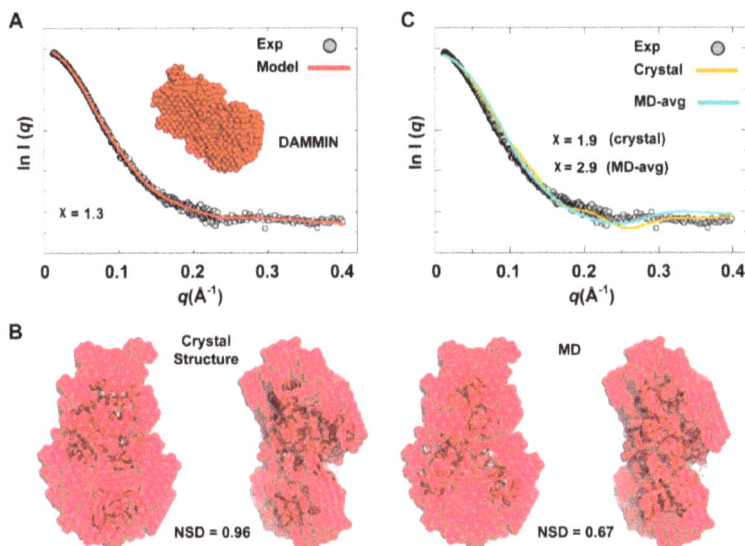

Figure 10. *Ab initio* SAXS model and fitting of single-conformation to experimental scattering data. (**A**) Shown are the average SAXS model constructed using DAMMIN (red spheres) and the fitting of the reference model (red line) to experimental SAXS data (grey). (**B**) The crystal and average MD structures of 3CD, represented as cartoons, are fitted to the DAMMIN model that is shown in A (red transparent surface); two different views are displayed for each fitting. The average MD structure fits the SAXS model slightly better (NSD = 0.67) than the crystal structure (NSD = 0.96). In both cases, however, the exact orientation of the structures cannot be determined without ambiguity. (**C**) The calculated scattering profiles of the crystal and average MD structures are compared to experimental data. The agreement between theoretical profiles and experimental data is relatively poor as indicated by the χ-values, suggesting that no single conformation can satisfactorily fit the experimental data.

In conclusion, the SAXS experiments are consistent with that the conformation of 3CD protein observed in the crystal structure and that revealed by MD simulations co-exist in solution in almost equal proportions. Furthermore, fluctuations around each conformation, similar to what was found in simulations (Figure 4A), could be inferred from the SAXS data.

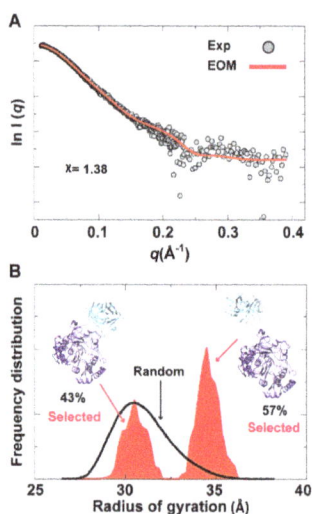

Figure 11. Ensemble Optimization Method (EOM) analysis suggested the presence of two conformations for 3CD in solution. (**A**) The scattering profile calculated from the selected EOM sub-ensemble (red) is in good agreement with the experimental data (grey) as indicated by the low χ-value. (**B**) Shown is the EOM radius of gyration distribution (R_g) of the initial random pool (black line) and the selected sub-ensemble (red shades) for 3CD; the 3C-domain adopts different conformations relative to the 3D-domain in the sub-ensemble. The selected sub-ensemble corresponds to two conformers: an extended conformer that has no interdomain interactions and a compact conformer with interdomain interactions; the extended one is slightly preferred, contributing 57% to the total scattering. The two conformers are displayed in the inset.

4. Discussion

Biochemical studies of the PV 3CD protein demonstrate quite convincingly that the protein is not merely the sum of its parts (Figure 1). Although 3CD exhibits the 3C-encoded protease activity, the catalytic efficiency and specificity are demonstrably different [13,14]. Even more striking, however, is the fact that 3CD fails to exhibit any of the $3D^{pol}$-encoded polymerase activity [15]. Observations such as these suggested that the structure of 3CD would be more than two "beads on a string." However, such an organization was observed when the crystal structure of 3CD was reported [23]. The current study was designed to determine if there was more to learn about 3CD than the crystal structure was telling us.

Starting from the crystal structure of 3CD, with modifications to reconstruct the wild-type sequence, molecular dynamics simulations revealed a collapse of the 3C and 3D domains to yield a conformation that could no longer be considered "beads on a string" (Figures 2 and 3). Interestingly, the simulation suggested that the 3C domain can move relative to the 3D domain to create multiple, unique interfaces (Figure 4), although one interface predominated the first trajectory (Figure 5). Performing the same experiment using 3CD coordinates from the second monomer in the unit cell produced the same results as above with respect to the collapse of the 3C and 3D domains (Figure 6). However, the predominating interface observed in this second trial was only partially overlapping with that observed in the first trial. Accelerated molecular dynamics simulations revealed an even wider range of possibilities for the conformations of 3C relative to 3D achievable for 3CD, including the extended conformation observed crystallographically (Figure 7). That the conformational ensemble of 3CD observed computationally exists in solution was shown by using small-angle X-ray scattering experiments (Figures 9–11). If each conformation of 3CD exhibits some unique function: catalytic activity or ability to interact specifically with a viral or host factor, then such a conformational ensemble may represent a strategy to expand the viral proteome.

In addition to revealing the conformational ensemble of the 3CD protein, the molecular dynamics simulations provided insight into the unique specificity/activity of 3CD relative to its domains in isolation. Regarding the proteolytic activity, the organization of the proteolytic active site in the context of 3CD when compared to 3C alone showed no significant difference (Figure 12A) [51]. However, residues of 3C that contribute to substrate binding are more dynamic in the precursor than in the processed protein (Figure 12A). The amino terminus was much more dynamic in the processed protein than in the precursor protein (Figure 12A). Indeed, these differences in dynamics appear as differences in structure when processed and precursor forms are evaluated by X-ray crystallography [23,24]. Of note, it has been shown before that binding of the peptide substrate is accompanied by conformational rearrangements within binding site [24]. Furthermore, portions of the 3D-domain should be able to interact directly with polyprotein substrate (Figure 12A). For the differences in RNA-binding between 3CD and 3C, MD simulations provided plausible insights. Residues of 3C that are implicated in stabilizing the conformation of the RNA recognition site [24] are engaged in interface interactions in the context of 3CD (Figures 5 and 6). These interactions may contribute to the differences in RNA binding activity of 3CD relative to 3C. Additionally, the drastic differences in dynamics of the amino terminus described above in the precursor and processed proteins may also contribute to the differences in RNA binding as these residues are known to be implicated in RNA binding activity [16].

In terms of the missing polymerase activity, it was clear going into this study that the amino terminus penetrates the fingers sub-domain of the polymerase and contributes to the organization of conserved structural motif A, which, in turn, contributes to the organization of conserved structural motif C [25]. Addition of a single amino acid to the amino terminus of the 3D protein reduces polymerase activity by 50–200 fold [37]. Interestingly, in the context of 3CD, the NTP channel is closed (Figure 12B). In the isolated 3D protein, this channel oscillates between open and closed but is most often observed in the open configuration (Figure 12B) [29,58]. This difference may also contribute to the complete absence of polymerase activity associated with 3CD.

We suggest that picornaviruses have evolved 3CD to be devoid of polymerase activity. First, the orthologous protein in the Calicivirus family, Pro-Pol, actually exhibits robust polymerase activity and is thought to be the active form of the enzyme in cells [59,60]. In this system, the role of the amino terminus has been replaced by the addition of an arginine residue to the fingers sub-domain [61]. Second, members of the Flavivirus family encode a methyltranferase domain followed by the polymerase domain [62,63]. In addition, the L (large) proteins of negative strand RNA viruses are all multi-functional, multi-domain proteins exhibiting polymerase activity [64,65].

We speculate that the observation made here for PV 3CD extends to all viral polyprotein precursor proteins. Structures of other precursor or multi-domain viral proteins have been solved (Figure 13) [62,63,66]. These are clearly compact structures with interdomain interactions. While solution information is not available for the alphavirus nsP23 precursor, SAXS experiments have been performed for the flavivirus NS5 protein [67]. These experiments suggest a conformational ensemble for NS5 as well. There is a clear need to place greater emphasis on the dynamics and conformational sampling of viral proteins, both fully processed and precursor forms, in the future.

Our laboratory has been quite intrigued by the concept that RNA viruses need to create mechanisms to expand their functional proteome [68]. This notion emerged when considering hepatitis C virus (HCV). HCV encodes fewer than one dozen proteins but is capable of precluding clearance by the human body as long as the liver can tolerate infection, usually many decades. We have provided very compelling evidence that in the case of HCV the use of an intrinsically disordered protein whose structure and dynamics can be regulated by phosphorylation is the strategy this virus uses to expand its functional proteome. Acute RNA viruses likely require more function than they encode using traditional approaches but not as many as persistent viruses. Use of the conformational ensemble is quite elegant. Even more elegant is the ability to make this ensemble sensitive to single amino acid substitutions. RNA viruses exist as a population of genetic variants, with the possibility for each amino acid to be sampled at each position of each protein. Our studies show that a single amino acid substitution in the interdomain interface of a precursor protein can completely change the architecture of the interface and the ensemble of conformations present and/or populated (Figure 8). Whether or not the conformational ensemble of multi-domain viral proteins as suggested here is a trait unique to this class of viral proteins, or cellular proteins as well, merits further investigation.

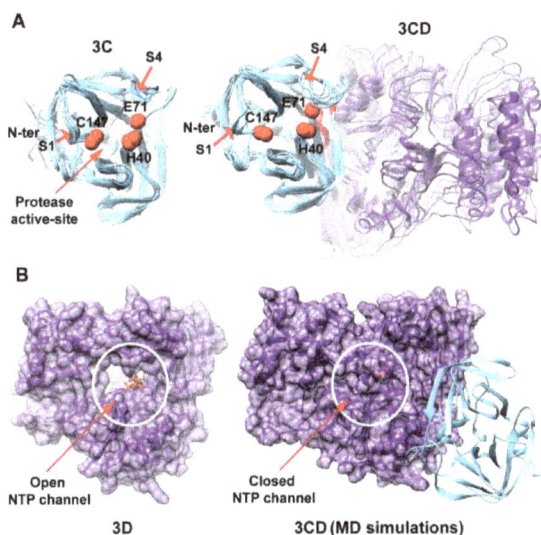

Figure 12. Influence of interdomain interactions on 3CD functions. (**A**) An ensemble of 3C conformations (left) obtained from MD simulations of 3C alone [51] and an ensemble of 3CD conformations (right) are shown. The catalytic triads of the protease active sites are indicated by red spheres; interface residues from 3C-domain in the precursor protein are colored red. MD simulations revealed that the N-terminal residues (aa 1–13), implicated in RNA binding, are much less dynamic in the context of 3CD than 3C alone. Whereas residues implicated in substrate binding or catalysis, including residues of pocket S1 (Thr-142, Gly-145 and Gln-146) and pocket S4 (Leu-125, Leu-127 and Phe-170), are more dynamic in the context of 3CD than 3C alone. These differences in dynamics may contribute to the unique specificity/activity of 3C relative to its precursor protein. Note that the 3D domain in 3CD can participate directly in interactions with polyprotein substrates, which may also contribute to the observed difference between the precursor and processed 3C protein. (**B**) Shown are the crystal structure of PV RdRp with the incoming NTP bound at the active site displayed as sticks (3D, PDB 3OL7 [26]), left, and the average 3CD structure from MD simulations, right. The two structures are oriented using the same view of the polymerase. The 3C and 3D domains are colored cyan and blue, respectively, and the polymerase domain is rendered as transparent surface. The NTP channel is circled and indicated by red arrow. In contrast to 3D structure in which the NTP channel is open, in 3CD the channel is closed and the nucleotide has no access to the active site of the polymerase. Also, interface residues in 3CD maintain interactions with the residues lining the NTP channel, interfering with its opening/closing motion. Additionally, the interdomain interactions in 3CD interfere with functionally important motions of the conserved structural motifs. All these differences in dynamics of 3D alone compared to that in the context of 3CD may contribute to the lack of polymerase activity in the precursor protein.

101

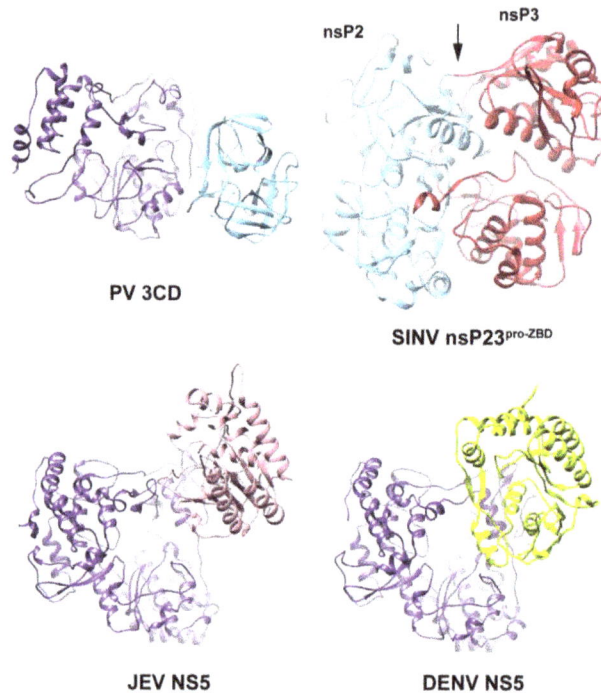

Figure 13. Interdomain interactions in PV 3CD and related proteins. Shown are the simulated structure of PV 3CD, structure of the polyprotein nsP23$^{pro-ZBD}$ from Sindbis virus (SINV, PDB 4GUA), structures of NS5 proteins from Japanese Encephalitis Virus (JEV NS5, PDB 4K6M) and Dengue Virus (DENV NS5, PDB 4V0R). The polymerase domains are colored blue and the N-terminal domains (protease in PV and MTase in NS5) are shown in cyan, pink and yellow. The nsP2 and nsP3 domains are colored cyan and red, respectively. In the four proteins, interfaces between the N- and C-terminal domains exist. The interface in 3CD is a reminiscent of the interfaces observed in the above related proteins. In the structure of nsP23$^{pro-ZBD}$, extensive interdomain interactions between the nsP2 protease and the nsP2/3 domain are maintained. To process this polyprotein, a conformational change is required to expose the cleavage site, suggesting that multiple conformations may exist. In the JEV NS5 structure, the interactions between the N-terminal MTase and the C-terminal polymerase domains are mostly hydrophobic; whereas in DENV NS5 the interdomain interactions are mostly polar. This difference in the nature of the interface suggested that JEV NS5 may possess multiple conformations similar to what has been shown by SAXS for DENV NS5. An interesting difference between 3CD and NS5 is that the linker residues in 3CD (7 amino acids long) are sandwiched between the two domains whereas the linkers in NS5 are not. Note that the linker of JEV NS5 is longer than 3CD (10 amino acids long) and that of DENV NS5 is shorter than 3CD (4 amino acids long). Also, the linker residues in NS5 proteins are less conserved than the linker residues in 3CD.

5. Conclusions

The use of a polyprotein for gene expression has always been considered a strategy to encode more functions into the limited coding capacity of a virus. Here we suggest that conformational sampling of interdomain interactions in polyprotein processing intermediates may further increase the viral proteome. The impact of mutation at the interdomain interfaces can be transformative to the conformational ensemble and hence the proteome, providing a possible explanation for the requirement of genetic diversity of an RNA virus for optimal fitness.

Supplementary Materials: Supplementary Materials: The following are available online at http://www.mdpi.com/1999-4915/7/11/2919, Figure S1: DLS analysis of PV 3CD protein, Figure S2: PCA analysis reveal major motions of 3CD, Figure S3: MD simulations reveal multiple interfaces in WT and mutant 3CD, Figure S4: Mutations of 3CD in the crystal structure, Figure S5: Analysis of the 50 ns MD simulations of mutant PV 3CD, and Video S1: Conformational Ensemble of the Poliovirus 3CD Precursor Observed by MD Simulations and Confirmed by SAXS: A Strategy to Expand the Viral Proteome?; DOI: 10.5281/zenodo.34538.

Acknowledgments: We thank Richard Gillian at MacCHESS for his assistance and support with SAXS data collection. This work was completed with financial support provided by grant AI053531 from NIAID, NIH and by the Eberly Family Endowment to CEC.

Author Contributions: C.E.C. and I.M. designed research; A.U. made the construct; N.Y. helped with SAXS data collection; I.M. did the SAXS experiments and the MD simulations, D.G. did the aMD; I.M. and C.E.C. analyzed the data and wrote the paper.

Conflicts of Interest: The authors declare no conflict of interest.

References

1.   Hornef, M.W.; Wick, M.J.; Rhen, M.; Normark, S. Bacterial strategies for overcoming host innate and adaptive immune responses. *Nat. Immunol.* **2002**, *3*, 1033–1040.
2.   Bhavsar, A.P.; Guttman, J.A.; Finlay, B.B. Manipulation of host-cell pathways by bacterial pathogens. *Nature* **2007**, *449*, 827–834.
3.   Belshaw, R.; Pybus, O.G.; Rambaut, A. The evolution of genome compression and genomic novelty in RNA viruses. *Genome Res.* **2007**, *17*, 1496–1504.
4.   Holmes, E.C. Error thresholds and the constraints to RNA virus evolution. *Trends Microbiol.* **2003**, *11*, 543–546.
5.   Richards, O.; Ehrenfeld, E. Poliovirus RNA replication. In *Picornaviruses*; Springer: Berlin Heidelberg, Germany, 1990; pp. 89–119.
6.   Wimmer, E.; Hellen, C.U.; Cao, X. Genetics of poliovirus. *Annu. Rev. Genet.* **1993**, *27*, 353–436.
7.   Nan, Y.; Nan, G.; Zhang, Y.-J. Interferon Induction by RNA Viruses and Antagonism by Viral Pathogens. *Viruses* **2014**, *6*, 4999–5027.
8.   Hellen, C.U.; Kraeusslich, H.G.; Wimmer, E. Proteolytic processing of polyproteins in the replication of RNA viruses. *Biochemistry* **1989**, *28*, 9881–9890.
9.   Palmenberg, A.C. Proteolytic processing of picornaviral polyprotein. *Annu. Rev. Microbiol.* **1990**, *44*, 603–623.

10. Cameron, C.E.; Suk Oh, H.; Moustafa, I.M. Expanding knowledge of P3 proteins in the poliovirus lifecycle. *Future Microbiol.* **2010**, *5*, 867–881.

11. Ehrenfeld, E.; Domingo, E.; Roos, R.P. *The Picornaviruses*; American Society for Microbiology Press: Washington, DC, USA, 2010.

12. Porter, D.C.; Ansardi, D.C.; Lentz, M.R.; Morrow, C.D. Expression of poliovirus P3 proteins using a recombinant vaccinia virus results in proteolytically active 3CD precursor protein without further processing to 3C pro and 3D pol. *Virus Res.* **1993**, *29*, 241–254.

13. Parsley, T.B.; Cornell, C.T.; Semler, B.L. Modulation of the RNA binding and protein processing activities of poliovirus polypeptide 3CD by the viral RNA polymerase domain. *J. Biol. Chem.* **1999**, *274*, 12867–12876.

14. Ypma-Wong, M.F.; Dewalt, P.G.; Johnson, V.H.; Lamb, J.G.; Semler, B.L. Protein 3CD is the major poliovirus proteinase responsible for cleavage of the P1 capsid precursor. *Virology* **1988**, *166*, 265–270.

15. Harris, K.; Reddigari, S.; Nicklin, M.; Hämmerle, T.; Wimmer, E. Purification and characterization of poliovirus polypeptide 3CD, a proteinase and a precursor for RNA polymerase. *J. Virol.* **1992**, *66*, 7481–7489.

16. Amero, C.; Arnold, J.; Moustafa, I.; Cameron, C.; Foster, M. Identification of the oriI-binding site of poliovirus 3C protein by nuclear magnetic resonance spectroscopy. *J. Virol.* **2008**, *82*, 4363–4370.

17. Andino, R.; Rieckhof, G.E.; Achacoso, P.L.; Baltimore, D. Poliovirus RNA synthesis utilizes an RNP complex formed around the 5′-end of viral RNA. *EMBO J.* **1993**, *12*, 3587–3598.

18. BLAIR, W.S.; PARSLEY, T.B.; BOGERD, H.P.; TOWNER, J.S.; SEMLER, B.L.; CULLEN, B.R. Utilization of a mammalian cell-based RNA binding assay to characterize the RNA binding properties of picornavirus 3C proteinases. *Rna* **1998**, *4*, 215–225.

19. Gamarnik, A.V.; Andino, R. Interactions of viral protein 3CD and poly (rC) binding protein with the 5′ untranslated region of the poliovirus genome. *J. Virol.* **2000**, *74*, 2219–2226.

20. Harris, K.S.; Xiang, W.; Alexander, L.; Lane, W.S.; Paul, A.V.; Wimmer, E. Interaction of poliovirus polypeptide 3CDpro with the 5′ and 3′ termini of the poliovirus genome. Identification of viral and cellular cofactors needed for efficient binding. *J. Biol. Chem.* **1994**, *269*, 27004–27014.

21. Pathak, H.B.; Arnold, J.J.; Wiegand, P.N.; Hargittai, M.R.; Cameron, C.E. Picornavirus Genome Replication Assembly and Organization of the Vpg Uridylylation Ribonucleoprotein (Initiation) Complex. *J. Biol. Chem.* **2007**, *282*, 16202–16213.

22. Paul, A.V.; Rieder, E.; Kim, D.W.; van Boom, J.H.; Wimmer, E. Identification of an RNA hairpin in poliovirus RNA that serves as the primary template in the *in vitro* uridylylation of VPg. *J. Virol.* **2000**, *74*, 10359–10370.

23. Marcotte, L.L.; Wass, A.B.; Gohara, D.W.; Pathak, H.B.; Arnold, J.J.; Filman, D.J.; Cameron, C.E.; Hogle, J.M. Crystal structure of poliovirus 3CD protein: Virally encoded protease and precursor to the RNA-dependent RNA polymerase. *J. Virol.* **2007**, *81*, 3583–3596.

24. Mosimann, S.C.; Cherney, M.M.; Sia, S.; Plotch, S.; James, M.N. Refined X-ray crystallographic structure of the poliovirus 3C gene product. *J. Mol. Biol.* **1997**, *273*, 1032–1047.

25. Thompson, A.A.; Peersen, O.B. Structural basis for proteolysis-dependent activation of the poliovirus RNA-dependent RNA polymerase. *EMBO J.* **2004**, *23*, 3462–3471.

26. Gong, P.; Peersen, O.B. Structural basis for active site closure by the poliovirus RNA-dependent RNA polymerase. *Proc. Natl. Acad. Sci. USA* **2010**, *107*, 22505–22510.

27. Cameron, C.E.; Moustafa, I.M.; Arnold, J.J. Dynamics: The missing link between structure and function of the viral RNA-dependent RNA polymerase? *Curr. Opin. Struct. Biol.* **2009**, *19*, 768–774.

28. Case, D.A.; Cheatham, T.E.; Darden, T.; Gohlke, H.; Luo, R.; Merz, K.M.; Onufriev, A.; Simmerling, C.; Wang, B.; Woods, R.J. The Amber biomolecular simulation programs. *J. Comput. Chem.* **2005**, *26*, 1668–1688.

29. Moustafa, I.M.; Shen, H.; Morton, B.; Colina, C.M.; Cameron, C.E. Molecular dynamics simulations of viral RNA polymerases link conserved and correlated motions of functional elements to fidelity. *J. Mol. Biol.* **2011**, *410*, 159–181.

30. Roe, D.R.; Cheatham, T.E., III. PTRAJ and CPPTRAJ: Software for processing and analysis of molecular dynamics trajectory data. *J. Chem. Theory Comput.* **2013**, *9*, 3084–3095.

31. Case, D.A.; Berryman, J.T.; Betz, R.M.; Cenrutti, T.E.; Cheatham, T.E., III; Darden, T.A.; Duke, R.E.; Giese, T.J.; Gohlke, H.; Goetz, A.W.; *et al.* *AMBER 2015*; University of California: San Francisco, CA, USA, 2015.

32. Doshi, U.; Hamelberg, D. Achieving rigorous accelerated conformational sampling in explicit solvent. *J. Phys. Chem. Lett.* **2014**, *5*, 1217–1224.

33. Cui, Q.; Sulea, T.; Schrag, J.D.; Munger, C.; Hung, M.-N.; Naïm, M.; Cygler, M.; Purisima, E.O. Molecular dynamics—Solvated interaction energy studies of protein–protein interactions: The MP1-p14 scaffolding complex. *J. Mol. Biol.* **2008**, *379*, 787–802.

34. Naïm, M.; Bhat, S.; Rankin, K.N.; Dennis, S.; Chowdhury, S.F.; Siddiqi, I.; Drabik, P.; Sulea, T.; Bayly, C.I.; Jakalian, A. Solvated interaction energy (SIE) for scoring protein-ligand binding affinities. 1. Exploring the parameter space. *J. Chem. Inf. Model.* **2007**, *47*, 122–133.

35. Chen, W.; Chang, C.-E.; Gilson, M.K. Calculation of cyclodextrin binding affinities: Energy, entropy, and implications for drug design. *Biophys. J.* **2004**, *87*, 3035–3049.

36. Shao, J.; Tanner, S.W.; Thompson, N.; Cheatham, T.E. Clustering molecular dynamics trajectories: 1. Characterizing the performance of different clustering algorithms. *J. Chem. Theory Comput.* **2007**, *3*, 2312–2334.

37. Gohara, D.W.; Ha, C.S.; Kumar, S.; Ghosh, B.; Arnold, J.J.; Wisniewski, T.J.; Cameron, C.E. Production of "authentic" poliovirus RNA-dependent RNA polymerase (3D pol) by ubiquitin–protease-mediated cleavage in Escherichia coli. *Protein Expr. Purif.* **1999**, *17*, 128–138.

38. Nielsen, S.; Toft, K.N.; Snakenborg, D.; Jeppesen, M.G.; Jacobsen, J.; Vestergaard, B.; Kutter, J.P.; Arleth, L. BioXTAS RAW, a software program for high-throughput automated small-angle X-ray scattering data reduction and preliminary analysis. *J. Appl. Crystallogr.* **2009**, *42*, 959–964.

39. Svergun, D. Determination of the regularization parameter in indirect-transform methods using perceptual criteria. *J. Appl. Crystallogr.* **1992**, *25*, 495–503.

40. Petoukhov, M.V.; Franke, D.; Shkumatov, A.V.; Tria, G.; Kikhney, A.G.; Gajda, M.; Gorba, C.; Mertens, H.D.; Konarev, P.V.; Svergun, D.I. New developments in the ATSAS program package for small-angle scattering data analysis. *J. Appl. Crystallogr.* **2012**, *45*, 342–350.

41. Fischer, H.; Oliveira Neto, M.D.; Napolitano, H.; Polikarpov, I.; Craievich, A. Determination of the molecular weight of proteins in solution from a single small-angle X-ray scattering measurement on a relative scale. *J. Appl. Crystallogr.* **2009**, *43*, 101–109.

42. Svergun, D. Restoring low resolution structure of biological macromolecules from solution scattering using simulated annealing. *Biophys. J.* **1999**, *76*, 2879–2886.

43. Svergun, D.I.; Petoukhov, M.V.; Koch, M.H. Determination of domain structure of proteins from X-ray solution scattering. *Biophys. J.* **2001**, *80*, 2946–2953.

44. Volkov, V.V.; Svergun, D.I. Uniqueness of ab initio shape determination in small-angle scattering. *J. Appl. Crystallogr.* **2003**, *36*, 860–864.

45. Svergun, D.; Barberato, C.; Koch, M. CRYSOL-a program to evaluate X-ray solution scattering of biological macromolecules from atomic coordinates. *J. Appl. Crystallogr.* **1995**, *28*, 768–773.

46. Schneidman-Duhovny, D.; Hammel, M.; Sali, A. FoXS: A web server for rapid computation and fitting of SAXS profiles. *Nucleic Acids Res.* **2010**, *38* (Suppl. 2), W540–W544.

47. Kozin, M.B.; Svergun, D.I. Automated matching of high-and low-resolution structural models. *J. Appl. Crystallogr.* **2001**, *34*, 33–41.

48. Bernadó, P.; Mylonas, E.; Petoukhov, M.V.; Blackledge, M.; Svergun, D.I. Structural characterization of flexible proteins using small-angle X-ray scattering. *J. Am. Chem. Soc.* **2007**, *129*, 5656–5664.

49. Tria, G.; Mertens, H.D.; Kachala, M.; Svergun, D.I. Advanced ensemble modelling of flexible macromolecules using X-ray solution scattering. *IUCrJ* **2015**, *2*, 207–217.

50. Pettersen, E.F.; Goddard, T.D.; Huang, C.C.; Couch, G.S.; Greenblatt, D.M.; Meng, E.C.; Ferrin, T.E. UCSF Chimera—A visualization system for exploratory research and analysis. *J. Comput. Chem.* **2004**, *25*, 1605–1612.

51. Chan, Y.M.; Moustafa, I.M.; Yennawar, N.; Arnold, J.J.; Cameron, C.E.; Boehr, D.D. Conformational adaptation in the multifunctional picornaviral 3C protease. **2015**, in preparation.

52. Richards, O.C.; Baker, S.; Ehrenfeld, E. Mutation of lysine residues in the nucleotide binding segments of the poliovirus RNA-dependent RNA polymerase. *J. Virol.* **1996**, *70*, 8564–8570.

53. Castro, C.; Smidansky, E.D.; Arnold, J.J.; Maksimchuk, K.R.; Moustafa, I.; Uchida, A.; Götte, M.; Konigsberg, W.; Cameron, C.E. Nucleic acid polymerases use a general acid for nucleotidyl transfer. *Nat. Struct. Mol. Biol.* **2009**, *16*, 212–218.

54. Yang, X.; Smidansky, E.D.; Maksimchuk, K.R.; Lum, D.; Welch, J.L.; Arnold, J.J.; Cameron, C.E.; Boehr, D.D. Motif D of viral RNA-dependent RNA polymerases determines efficiency and fidelity of nucleotide addition. *Structure* **2012**, *20*, 1519–1527.

55. Doniach, S. Changes in biomolecular conformation seen by small angle X-ray scattering. *Chem. Rev.* **2001**, *101*, 1763–1778.

56. Svergun, D.; Richard, S.; Koch, M.; Sayers, Z.; Kuprin, S.; Zaccai, G. Protein hydration in solution: Experimental observation by x-ray and neutron scattering. *Proc. Natl. Acad. Sci. USA* **1998**, *95*, 2267–2272.

57. Receveur-Bréchot, V.; Durand, D. How random are intrinsically disordered proteins? A small angle scattering perspective. *Curr. Protein Pept. Sci.* **2012**, *13*, 55–75.

58. Moustafa, I.M.; Korboukh, V.K.; Arnold, J.J.; Smidansky, E.D.; Marcotte, L.L.; Gohara, D.W.; Yang, X.; Sánchez-Farrán, M.A.; Filman, D.; Maranas, J.K. Structural Dynamics as a Contributor to Error-prone Replication by an RNA-dependent RNA Polymerase. *J. Biol. Chem.* **2014**, *289*, 36229–36248.

59. Belliot, G.; Sosnovtsev, S.V.; Chang, K.-O.; Babu, V.; Uche, U.; Arnold, J.J.; Cameron, C.E.; Green, K.Y. Norovirus proteinase-polymerase and polymerase are both active forms of RNA-dependent RNA polymerase. *J. Virol.* **2005**, *79*, 2393–2403.

60. Wei, L.; Huhn, J.S.; Mory, A.; Pathak, H.B.; Sosnovtsev, S.V.; Green, K.Y.; Cameron, C.E. Proteinase-polymerase precursor as the active form of feline calicivirus RNA-dependent RNA polymerase. *J. Virol.* **2001**, *75*, 1211–1219.

61. Arnold, J.J.; Vignuzzi, M.; Stone, J.K.; Andino, R.; Cameron, C.E. Remote site control of an active site fidelity checkpoint in a viral RNA-dependent RNA polymerase. *J. Biol. Chem.* **2005**, *280*, 25706–25716.

62. Lu, G.; Gong, P. Crystal structure of the full-length Japanese encephalitis virus NS5 reveals a conserved methyltransferase-polymerase interface. *PLoS Pathog.* **2013**, *9*, e1003549.

63. Zhao, Y.; Soh, T.S.; Zheng, J.; Chan, K.W.K.; Phoo, W.W.; Lee, C.C.; Tay, M.Y.; Swaminathan, K.; Cornvik, T.C.; Lim, S.P. A crystal structure of the dengue virus NS5 protein reveals a novel inter-domain interface essential for protein flexibility and virus replication. *PLoS Pathog.* **2015**, *11*, e1004682–e1004682.

64. Gerlach, P.; Malet, H.; Cusack, S.; Reguera, J. Structural Insights into Bunyavirus Replication and Its Regulation by the vRNA Promoter. *Cell* **2015**, *161*, 1267–1279.

65. Liang, B.; Li, Z.; Jenni, S.; Rahmeh, A.A.; Morin, B.M.; Grant, T.; Grigorieff, N.; Harrison, S.C.; Whelan, S.P. Structure of the L protein of vesicular stomatitis virus from electron cryomicroscopy. *Cell* **2015**, *162*, 314–327.

66. Shin, G.; Yost, S.A.; Miller, M.T.; Elrod, E.J.; Grakoui, A.; Marcotrigiano, J. Structural and functional insights into alphavirus polyprotein processing and pathogenesis. *Proc. Natl. Acad. Sci. USA* **2012**, *109*, 16534–16539.

67. Bussetta, C.C.; Choi, K.H. Dengue virus nonstructural protein 5 adopts multiple conformations in solution. *Biochemistry* **2012**, *51*, 5921–5931.

68. Cordek, D.G.; Croom-Perez, T.J.; Hwang, J.; Hargittai, M.R.; Subba-Reddy, C.V.; Han, Q.; Lodeiro, M.F.; Ning, G.; McCrory, T.S.; Arnold, J.J. Expanding the proteome of an RNA virus by phosphorylation of an intrinsically disordered viral protein. *J. Biol. Chem.* **2014**, *289*, 24397–24416.

Chapter 2:
Coordinated Interactions of Viral Proteins and Nucleic Acids

Coordinated DNA Replication by the Bacteriophage T4 Replisome

Erin Noble, Michelle M. Spiering and Stephen J. Benkovic

Abstract: The T4 bacteriophage encodes eight proteins, which are sufficient to carry out coordinated leading and lagging strand DNA synthesis. These purified proteins have been used to reconstitute DNA synthesis *in vitro* and are a well-characterized model system. Recent work on the T4 replisome has yielded more detailed insight into the dynamics and coordination of proteins at the replication fork. Since the leading and lagging strands are synthesized in opposite directions, coordination of DNA synthesis as well as priming and unwinding is accomplished by several protein complexes. These protein complexes serve to link catalytic activities and physically tether proteins to the replication fork. Essential to both leading and lagging strand synthesis is the formation of a holoenzyme complex composed of the polymerase and a processivity clamp. The two holoenzymes form a dimer allowing the lagging strand polymerase to be retained within the replisome after completion of each Okazaki fragment. The helicase and primase also form a complex known as the primosome, which unwinds the duplex DNA while also synthesizing primers on the lagging strand. Future studies will likely focus on defining the orientations and architecture of protein complexes at the replication fork.

Reprinted from *Viruses*. Cite as: Noble, E.; Spiering, M.M.; Benkovic, S.J. Coordinated DNA Replication by the Bacteriophage T4 Replisome. *Viruses* **2015**, *7*, 3186–3200.

1. Introduction

Bacteriophages were first discovered in the early 20th century due to their ability to kill bacteria [1]. Apart from their therapeutic uses, bacteriophages were found to encode proteins that carried out many of the same basic processes that are found in eukaryotic cells. The T4 bacteriophage, which infects *Escherichia coli*, is one of the best-studied viruses in this group. Its double-stranded DNA genome encodes all of the proteins necessary to carry out viral DNA replication in the infected cell. The components of the T4 replisome can be purified and used to reconstitute DNA replication *in vitro*. This system has been well characterized as a model for DNA replication at a fork [2–4]. The T4 replisome consists of eight proteins, which together catalyze coordinated leading and lagging strand synthesis (Figure 1). These proteins are similar in structure and function to their eukaryotic homologues [5]. Studies on the T4 system have contributed greatly to the understanding of DNA replication and paved the way for current studies on human and yeast DNA replication. This review will cover the current understanding of T4 DNA replication and highlight areas where recent research has yielded new mechanistic insight into functioning of the T4 replisome. For more detail on other prokaryotic model systems, see recent reviews highlighting studies of the T7 bacteriophage and *E. coli* replisomes [6–8].

2. T4 Replication Fork Components

T4 replication can be initiated via several different pathways [9]. Two specialized structures, R-loops and D-loops, have been shown to be important. R-loops form at T4 origin sites where an RNA primer is synthesized. D-loops are formed by the recombination machinery and are used to initiate origin-independent DNA synthesis. These two mechanisms of DNA replication initiation of have been reviewed elsewhere [10].

Synthesis of the T4 genomic DNA is accomplished by a holoenzyme complex composed of the gp43 polymerase and the gp45 sliding clamp [11–13]. On the leading strand, DNA synthesis is carried out continuously by one holoenzyme complex. On the lagging strand, DNA is synthesized in the opposite direction of the progression of the replication fork. Multiple priming events allow a second holoenzyme complex to carry out DNA synthesis discontinuously in 1 to 2 kb fragments known as Okazaki fragments. While there is no available crystal structure for the T4 gp43, the structure for the RB69 bacteriophage gp43 has been solved alone and as part of a binary and ternary complex [14–16]. The two proteins are 62% identical and 74% similar and thus, the proteins are likely very similar in topology. The RB69 structure reveals five conserved domains in a configuration similar to that of the eukaryotic B family polymerases. The *N*-terminus contains a 3' to 5' exonuclease active site. This truncated exonuclease domain from T4 gp43 has been isolated and the structure solved [12]. The catalytic activity of this domain is independent from the rest of the polymerase, as it retains full exonuclease activity *in vitro* [17]. The *C*-terminus of RB69 gp43 is organized into conserved finger, palm, and thumb domains, which catalyze DNA polymerization 5' to 3' [15].

The T4 sliding clamp, gp45, is a ring-shaped, trimeric protein that serves as a processivity factor for the polymerase [18,19]. The inner diameter of the ring is about 35 Å, which is large enough to accommodate duplex DNA. Unlike clamps in other systems, the T4 clamp exists in solution as a partially open ring with one of the three subunit interfaces disrupted [20–22]. Once loaded onto DNA, the interior of the clamp interacts with the DNA phosphate backbone through a number of basic residues and anchors the polymerase to the DNA [19]. gp43 has a C-terminal PIP box domain that mediates the interaction of the polymerase and the sliding clamp [23].

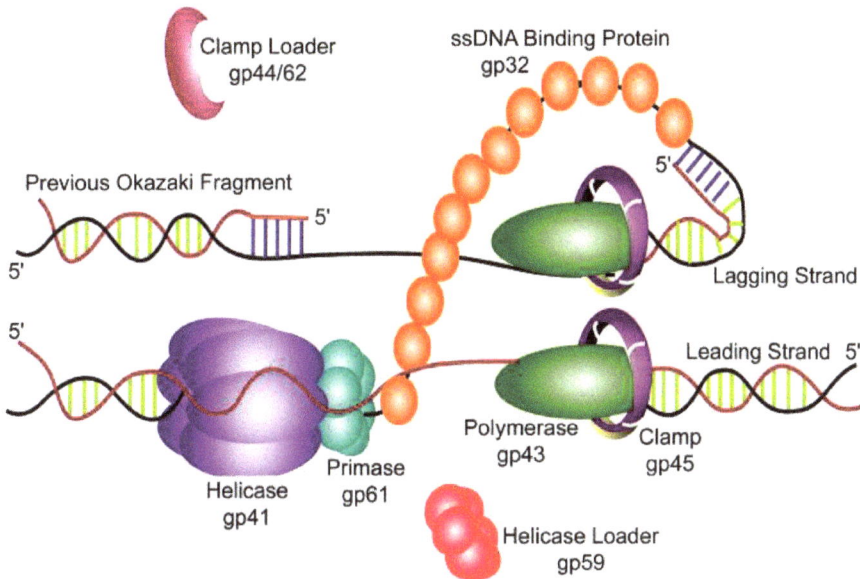

Figure 1. A model of the T4 bacteriophage DNA replisome. Replication of T4 genomic DNA is accomplished by a replication complex composed of eight proteins. The helicase (gp41) and primase (gp61) interact to form the primosome with the assistance of the helicase loader (gp59). The primosome complex encircles the lagging strand DNA, unwinding duplex DNA while synthesizing RNA primers for use by the lagging strand polymerase (gp43). DNA synthesis on both strands is catalyzed by a holoenzyme complex formed by a polymerase (gp43) and a trimeric processivity clamp (gp45). The clamp is loaded onto the DNA by the clamp loader complex (gp44/62). The leading and lagging strand holoenzymes interact to form a dimer. Single-stranded DNA formed by the helicase is coated with single-stranded DNA-binding protein (gp32).

The circular gp45 clamp is loaded onto the DNA by a clamp loader complex. In T4, four gp44 subunits associate with one gp62 subunit forming the gp44/62 clamp loader [24]. Each gp44 subunit binds ATP and the complex has a strong DNA-dependent ATPase activity [25,26]. The clamp loader is a member of the AAA+ family of ATPases, but unlike other enzymes of this type, clamp loaders are pentameric rather than hexameric. This asymmetry results in a gap that allows the clamp loader to specifically recognize the primer-template junction when loading a clamp [27,28].

The T4 helicase, gp41, forms a hexamer upon binding GTP or ATP [29]. This active form of the helicase hydrolyzes GTP/ATP to move along single-stranded DNA [30,31]. Electron microscopy has revealed that there are two forms of the hexameric gp41, a symmetric ring and a gapped asymmetric ring [32]. The "open" ring is thought to be important for the loading of the helicase onto DNA [29]. As part of the replication fork, gp41 unwinds the double stranded DNA by traveling 5′ to 3′, encircling the lagging strand while excluding the leading strand [33]. The preferred substrate for the helicase is a forked DNA with both 5′ and 3′ single-stranded DNA regions, suggesting the protein interacts with both the leading and lagging strands [33,34]. T4 also encodes two other helicases, UvsW and Dda. Both accessory helicases have been suggested to have roles in replication initiation, recombination, and repair (see review [35]).

Priming on the lagging strand is catalyzed by the gp61 primase, which interacts with gp41 to form the primosome [36]. This primosome synthesizes pentaribonucleotides from 5′-GTT-3′ priming sites. The 3′-T is necessary for priming but is not used to template the primer; the resulting primers have the sequence 5′-pppACNNN-3′ [37]. At high concentrations *in vitro*, gp61 alone can synthesize some RNA primers, but they are typically dimers primed from a 5′-GCT-3′ site [37,38]. In the presence of gp41, the rate of primer synthesis increases and shifts to pentaribonucleotide products primed from 5′-GTT-3′ sites, which is the priming site used *in vivo* [38,39]. gp61 alone is monomeric, but in the presence of gp41 and/or DNA, it oligomerizes into a hexameric ring [32,40].

Exposed single-stranded DNA is bound by gp32, which is necessary for DNA replication *in vivo*. It has many functions including preventing the formation of DNA secondary structure, protecting DNA from nuclease digestion, and stimulation of the gp43 synthesis rate and processivity [41–43]. A crystal structure of gp32 in complex with DNA reveals three domains. The N-terminus binds other gp32 monomers allowing for oligomerization, the C-terminus mediates interactions with other proteins such as the T4 polymerase, and the core domain binds single-stranded DNA [44].

In vivo a helicase loader, gp59, is required for origin-dependent initiation of replication [45]. In the presence of gp32, the helicase cannot efficiently load onto the DNA fork without the addition of gp59. gp59 interacts with gp41 stoichiometrically and helps to displace gp32, allowing the helicase to load [46]. gp59 is thought to mediate loading by inducing a conformational change in gp41 that promotes DNA binding [47]. It is unclear if gp59 dissociates or remains as part of the replication complex [48,49]. Binding events between gp43 and gp59 have been observed using single-molecule FRET [50].

3. Holoenzyme Formation

The gp43 polymerase alone can only copy short stretches of single-stranded DNA without dissociating [51]. The gp45 sliding clamp is a homotrimeric ring that allows gp43 to catalyze processive DNA synthesis. It is loaded onto DNA by gp44/62 with the clamp loader specifically recognizing the free 3′ end of the primer-template junction. As the clamp is partially open in solution, the function of the T4 clamp loader is to stabilize the open clamp and direct it onto DNA in the correct orientation. Crystal structures of the clamp/clamp loader complex, both with and without DNA, have provided detailed insight into how loading occurs [52]. The clamp loader has a low affinity for the clamp until the binding of ATP through an AAA+ module in each of the gp44 subunits. ATP binding causes the clamp loader subunits to adopt a spiral conformation that can bind to the clamp and open it further, allowing it to be loaded onto DNA. The opening of the clamp occurs in two planes. Movement of ~9 Å in the plane of the ring allows single-stranded DNA to pass through the gap, while an out-of-plane shift of ~23 Å results in a twisted conformation of the clamp, aligning it with the helical structure of the DNA. DNA binding stimulates the ATPase activity of the clamp loader and the hydrolysis of ATP in each of the four gp44 subunits [24,53]. This hydrolysis triggers a change in the conformation of the clamp loader, which closes the clamp around the DNA.

Once the clamp is closed around the DNA, it must be bound by the polymerase to form the holoenzyme. This process has been characterized using a FRET-based assay to monitor clamp loading and holoenzyme assembly. The clamp and clamp loader complex rapidly bind to the DNA after ATP binding. In the absence of the gp43 polymerase, the clamp and clamp loader remain as a complex and dissociate from the DNA together. In the presence of the polymerase, a functional holoenzyme forms in three kinetically distinct steps. The first corresponds to the hydrolysis of ATP and the dissociation of the clamp loader. The subsequent two steps involve slower conformational changes leading to the formation of a stable complex. The dissociation of the clamp in the presence of the polymerase is significantly slower than the clamp alone [54]. This stable holoenzyme complex is then able to efficiently carry out processive DNA synthesis on the leading strand and discontinuous DNA synthesis on the lagging strand.

4. Holoenzyme Processivity

The holoenzyme on the leading strand synthesizes DNA in the same direction as the movement of the replication fork. *In vivo*, the T4 genome can be synthesized within 15 min [55]. The half-life of the holoenzyme complex has been measured as 11 min as part of a moving fork and about 6 min on a small, defined DNA fork structure [56,57]. Given the half-life of the holoenzyme and the speed of synthesis, it is possible that the entire T4 genome could be synthesized by a single holoenzyme on the leading strand. While this highly processive holoenzyme would be advantageous on the leading strand, the lagging strand is synthesized discontinuously and the holoenzyme must repeatedly dissociate and rebind for synthesis of each Okazaki fragment.

A more recent study probing the processivity of the T4 holoenzyme confirmed the long half-life during replication using a standard dilution experiment [58]. However, it was found that an inactive mutant of the polymerase (D408N) was able to rapidly displace the wild-type polymerase and inhibit DNA synthesis. This inhibition occurred on both the leading and lagging strands. These results suggest that although the polymerase will not readily dissociate on its own, it can be actively displaced by a second polymerase without affecting DNA synthesis. The exchange process was termed dynamic processivity and is thought to be mediated through interactions with gp45 [58]. The C-terminus of gp43 is essential for polymerase binding to the clamp, but its deletion does not affect DNA polymerization [23]. When polymerase containing this deletion was used as a trap, it could no longer displace the replicating polymerase [58]. As the clamp is trimeric, it is hypothesized that multiple polymerases could bind and facilitate the exchange. This "toolbelt" model for the clamp has been suggested in other systems as well, with numerous proteins involved in DNA replication and repair also containing clamp binding domains [59,60]. In the T7 system, where there is no sliding clamp, the exchange process has been shown to be mediated by an interaction between the polymerase and the helicase [61]. It is thought that the helicase can bind multiple polymerases facilitating exchange on the leading strand and recycling on the lagging strand.

5. Coupling of Helicase and Polymerase for Leading Strand Synthesis

While both gp41 helicase and gp43/gp45 holoenzyme can function independently *in vitro* to unwind duplex DNA, the two enzymes work best when their activities are combined. The helicase alone is significantly slower and less processive than the replication fork, and the holoenzyme is very inefficient at strand displacement synthesis [33,62]. Together, the helicase and holoenzyme are able to efficiently carry out leading strand synthesis [63]. In the presence of a macromolecular crowding reagent, only gp43 and gp41 are needed, indicating the clamp does not play a role [64]. While the functional coupling between the two proteins has been clearly demonstrated, there is no evidence of a physical interaction between gp43 and gp41 [65,66]. One study also found that the T4 polymerase could be replaced with another processive polymerase and still carry out strand displacement synthesis, but could not be replaced with a low processivity polymerase [65]. This suggests that each enzyme is stabilized on the DNA replication fork by the activity of the other, with the helicase providing single-stranded DNA that the polymerase then traps.

In the T7 system, it was reported that nucleotide incorporation by the polymerase provided the driving force to stimulate helicase activity, but a detailed mechanism for helicase-polymerase coupling was not described [67]. A more recent single-molecule study of the coupling in the T4 system used magnetic tweezers to monitor both coupled and uncoupled activity [68]. A DNA hairpin was tethered to a glass slide with a magnetic bead on the other end. Force was applied to destabilize the duplex and assist enzymes in opening the hairpin. At low force, where the duplex of the hairpin is stable, the helicase moved at 6 times slower than its maximal translocation rate and showed sequence dependent pausing. As higher force was applied, the rate of helicase activity increased dramatically. Additionally, at low helicase concentrations, significant helicase slippage was observed involving the reannealing of tens to hundreds of base pairs. This fits with the passive model of helicase activity previously demonstrated, in which the helicase is not efficient in destabilizing duplex DNA and relies on transient fraying of base pairs to move forward [69].

The T4 holoenzyme was found to have very low strand displacement activity at low force and mainly exhibited exonuclease activity [68]. When higher forces destabilized the duplex, the holoenzyme was able to replicate the hairpin at maximal speeds. At moderate forces, the holoenzyme exhibited pausing and stalling. The proportion of holoenzymes observed synthesizing DNA, pausing, or degrading DNA was highly dependent on the force used. This indicates that at higher forces the holoenzyme is able to stay in the polymerization mode, while lower forces shift the holoenzyme to the exonuclease mode. When pausing and exonuclease events were excluded from analysis, the holoenzyme activity fits with a model of a strongly active motor. The basis for collaborative coupling then emerges in a model where the helicase provides the single-stranded DNA for the holoenzyme, but also prevents the fork regression pressure from switching the polymerase into the exonuclease mode. As the holoenzyme is kept in its highly processive polymerization mode, it stimulates the activity of the helicase and prevents slippage backwards [68].

6. Coordination of Helicase and Priming on the Lagging Strand

The leading and lagging strands are thought to be synthesized at the same net rate, despite the need for repeated priming and extension events on the lagging strand [4,70,71]. Priming is catalyzed by a gp61-gp41 complex known as the primosome. Both priming and DNA unwinding activity are stimulated when both proteins are present [34,38,39]. There is strong biochemical evidence for the interaction of the hexameric gp41 helicase and oligomeric gp61 primase [34,36,72,73]. Importantly, a gp61-gp41 fusion protein has been shown to have close to wild-type priming and helicase activity and can successfully catalyze coordinated leading and lagging strand synthesis [74].

This tight coordination of activity is clear, despite the fact that the helicase travels 5' to 3' unwinding duplex DNA while the primase synthesizes primers 3' to 5' on the same strand. There are three models for how this coupling can occur. The first model suggests that the helicase, and possibly the whole replisome, pauses while the primers are being synthesized. In the second model, primase subunits dissociate from the helicase and are left behind to synthesize primers. In the third model, coupling is accomplished by the formation of priming loops wherein the lagging strand folds back allowing for priming. The loop is then released after the primer is synthesized.

By observing helicase and priming activity on DNA hairpins using magnetic tweezers, the role of the three models in the T4 primosome could be directly observed [74]. In the T7 system, both pausing of the primosome [75] and priming loops have been reported [76]. The T4 study yielded no evidence of pausing of the T4 primosome. However, clear evidence of both primase disassembly and looping were seen in these experiments, indicating that there are two different mechanisms used by T4 to couple the helicase and primase (Figure 2). While primase disassembly was the predominant mode, in the case where the primase and helicase were fused only the looping mechanism was seen.

7. Recycling of the Lagging Strand Polymerase

The trombone model was proposed to explain the coordination of leading and lagging strand synthesis with the two polymerases synthesizing in opposite directions. In this model, the lagging strand DNA loops out during the formation of each Okazaki fragment [4]. These loops have been visualized in electron micrographs of T4 replication products [48]. The lagging strand polymerase is retained as part of the replisome after completing synthesis of each Okazaki fragment [4]. It dissociates from the DNA, but then rapidly binds the next primer to continue synthesis. This recycling of the lagging strand polymerase is supported by numerous studies. While the clamp, clamp loader, primase, and gp32 have all been shown to exchange with proteins in solution during replication, the polymerase is resistant to dilution [77–80]. The size of the Okazaki fragments is also independent of polymerase concentration [4,58]. Importantly, the leading and lagging strand polymerases interact in the presence of DNA, which provides a mechanism for tethering the lagging strand polymerase to the replisome [66].

While the holoenzyme on the leading strand is highly processive, on the lagging strand it must repeatedly dissociate. The trigger for the dissociation of the lagging strand polymerase has not clearly been defined despite a number of studies. Several models have been proposed with two gaining the most support and evidence suggests that both play a role during replication [81]. The collision model proposes that the lagging strand polymerase dissociates after colliding with the end of the previous Okazaki fragment, and this stimulates the primase to synthesize a new primer [62,82]. However, it has been also shown that dissociation of the lagging strand polymerase can occur before reaching the previous Okazaki fragment leaving single-stranded DNA gaps [81]. To account for this observation, the signaling model has been proposed where recycling is triggered by the synthesis of a new primer and the timing controlled by gp61 [80,81,83]. Recently, additional signals have been proposed to regulate this recycling in other replication systems such *E. coli* and T7. These new triggers include tension induced dissociation of the polymerase [84], primer availability [85], and a third polymerase [86]. While it has been shown that a third T4 polymerase does not seem to play a role in Okazaki fragment synthesis [87], the nature of the signal for recycling is still unknown.

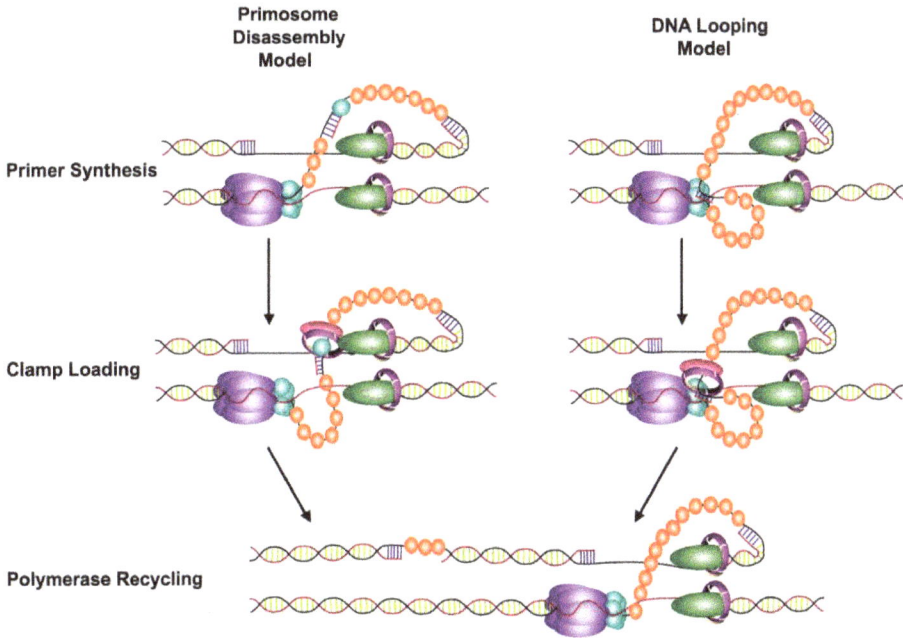

Figure 2. The two models of primosome activity used by T4 to initiate lagging strand synthesis. The helicase (gp41) and primase (gp61) interact as stacked rings encircling the lagging strand. This complex unwinds duplex DNA while synthesizing pentaribonucleotide RNA primers for use by the lagging strand polymerase (gp43). Primer synthesis occurs while the helicase continues to unwind DNA in the opposite direction. Two models have been proposed to accommodate these coupled activities. In the primosome disassembly model (shown left), one of the primase subunits dissociates from the primosome complex and remains with the newly synthesized primer. In the DNA looping model (shown right), the excess DNA unwound by the helicase during primer synthesis loops out allowing the primase to stay intact. In both models, the clamp loader (gp44/62) loads a clamp (gp45) onto the newly synthesized primer. The lagging strand polymerase is then signaled to release and recycle to the new primer.

8. Future Directions

The major unanswered questions concerning T4 DNA replication involve understanding the dynamics and organization of the proteins at the replication fork. While the protein complexes involved in replication, the primosome, holoenzyme, and single-stranded DNA binding protein, have been extensively studied, their orientations and spatial juxtapositions at the fork are unclear. According to the trombone model, the two polymerases are thought to interact in opposite orientations, but how these proteins assemble at the fork has not been demonstrated. It is also not known how the polymerases at the replication fork are able to readily exchange with polymerases in solution. It is possible that the polymerase and clamp transiently separate yielding the dynamic processivity that has been observed. Another area of uncertainty is the trigger for recycling of the lagging strand polymerase. A number of possible signals have been suggested but none of these models have been proven. Fluorescence resonance energy transfer (FRET) and single-molecule experiments will likely play an important role in resolving these uncertainties and more clearly defining the organization and coordination of the T4 replisome.

Acknowledgments: This work is supported by National Institutes of Health (NIH) grant GM013306 (Stephen J. Benkovic). Erin Noble is supported by the NIH under Award Number F32GM110857.

Author Contributions: All authors contributed to the conception and editing of this review. Erin Noble wrote the manuscript.

Conflicts of Interest: The authors declare no conflict of interest.

References

1. Sulakvelidze, A.; Alavidze, Z.; Morris, J.G. Bacteriophage therapy. *Antimicrob. Agents Chemother.* **2001**, *45*, 649–659.
2. Nossal, N.G. Protein-protein interactions at a DNA replication fork: Bacteriophage T4 as a model. *FASEB J.* **1992**, *6*, 871–878.
3. Liu, C.; Burke, R.; Hibner, U.; Barry, J.; Alberts, B. Probing DNA Replication Mechanisms with the T4 Bacteriophage *in Vitro* System. In *Cold Spring Harbor Symposia on Quantitative Biology*; Cold Spring Harbor Laboratory Press: Cold Spring Harbor, NY, USA, 1979; pp. 469–487.
4. Alberts, B.; Barry, J.; Bedinger, P.; Formosa, T.; Jongeneel, C.; Kreuzer, K. Studies on DNA Replication in the Bacteriophage T4 *in Vitro* System. In *Cold Spring Harbor Symposia on Quantitative Biology*; Cold Spring Harbor Laboratory Press: Cold Spring Harbor, NY, USA, 1983; pp. 655–668.
5. Mueser, T.C.; Hinerman, J.M.; Devos, J.M.; Boyer, R.A.; Williams, K.J. Structural analysis of bacteriophage T4 DNA replication: A review in the virology journal series on bacteriophage T4 and its relatives. *Virol. J.* **2010**, *7*.

6. Lee, S.-J.; Richardson, C.C. Choreography of bacteriophage T7 DNA replication. *Curr. Opin. Chem. Biol.* **2011**, *15*, 580–586.

7. Hamdan, S.M.; Richardson, C.C. Motors, switches, and contacts in the replisome. *Annu. Rev. Biochem.* **2009**, *78*, 205–243.

8. Van Oijen, A.M.; Loparo, J.J. Single-molecule studies of the replisome. *Annu. Rev. Biophys.* **2010**, *39*, 429–448.

9. Mosig, G.; Colowick, N.; Gruidl, M.E.; Chang, A.; Harvey, A.J. Multiple initiation mechanisms adapt phage T4 DNA replication to physiological changes during T4's development. *FEMS Microb. Rev.* **1995**, *17*, 83–98.

10. Kreuzer, K.N.; Brister, J.R. Initiation of bacteriophage T4 DNA replication and replication fork dynamics: A review in the virology journal series on bacteriophage T4 and its relatives. *Virol. J.* **2010**, *7*.

11. Benkovic, S.J.; Valentine, A.M.; Salinas, F. Replisome-mediated DNA replication. *Annu. Rev. Biochem.* **2001**, *70*, 181–208.

12. Reddy, M.K.; Weitzel, S.E.; von Hippel, P.H. Assembly of a functional replication complex without ATP hydrolysis: A direct interaction of bacteriophage T4 gp45 with T4 DNA polymerase. *Proc. Natl. Acad. Sci. USA* **1993**, *90*, 3211–3215.

13. Sexton, D.J.; Berdis, A.J.; Benkovic, S.J. Assembly and disassembly of DNA polymerase holoenzyme. *Curr. Opin. Chem. Biol.* **1997**, *1*, 316–322.

14. Franklin, M.C.; Wang, J.; Steitz, T.A. Structure of the replicating complex of a pol α family DNA polymerase. *Cell* **2001**, *105*, 657–667.

15. Wang, J.; Sattar, A.A.; Wang, C.; Karam, J.; Konigsberg, W.; Steitz, T. Crystal structure of a pol α family replication DNA polymerase from bacteriophage RB69. *Cell* **1997**, *89*, 1087–1099.

16. Shamoo, Y.; Steitz, T.A. Building a replisome from interacting pieces: Sliding clamp complexed to a peptide from DNA polymerase and a polymerase editing complex. *Cell* **1999**, *99*, 155–166.

17. Lin, T.-C.; Karam, G.; Konigsberg, W.H. Isolation, characterization, and kinetic properties of truncated forms of T4 DNA polymerase that exhibit 3′–5′exonuclease activity. *J. Biol. Chem.* **1994**, *269*, 19286–19294.

18. Bruck, I.; O'Donnell, M. The ring-type polymerase sliding clamp family. *Genome Biol.* **2001**, *2*.

19. Moarefi, I.; Jeruzalmi, D.; Turner, J.; O'Donnell, M.; Kuriyan, J. Crystal structure of the DNA polymerase processivity factor of T4 bacteriophage. *J. Mol. Biol.* **2000**, *296*, 1215–1223.

20. Soumillion, P.; Sexton, D.J.; Benkovic, S.J. Clamp subunit dissociation dictates bacteriophage T4 DNA polymerase holoenzyme disassembly. *Biochemistry* **1998**, *37*, 1819–1827.

21. Alley, S.C.; Shier, V.K.; Abel-Santos, E.; Sexton, D.J.; Soumillion, P.; Benkovic, S.J. Sliding clamp of the bacteriophage T4 polymerase has open and closed subunit interfaces in solution. *Biochemistry* **1999**, *38*, 7696–7709.

22. Millar, D.; Trakselis, M.A.; Benkovic, S.J. On the solution structure of the T4 sliding clamp (gp45). *Biochemistry* **2004**, *43*, 12723–12727.

23. Berdis, A.J.; Soumillion, P.; Benkovic, S.J. The carboxyl terminus of the bacteriophage T4 DNA polymerase is required for holoenzyme complex formation. *Proc. Natl. Acad. Sci. USA* **1996**, *93*, 12822–12827.

24. Jarvis, T.; Paul, L.; Hockensmith, J.; von Hippel, P. Structural and enzymatic studies of the T4 DNA replication system. II. Atpase properties of the polymerase accessory protein complex. *J. Biol. Chem.* **1989**, *264*, 12717–12729.

25. Jarvis, T.; Newport, J.; von Hippel, P. Stimulation of the processivity of the DNA polymerase of bacteriophage T4 by the polymerase accessory proteins. The role of atp hydrolysis. *J. Biol. Chem.* **1991**, *266*, 1830–1840.

26. Rush, J.; Lin, T.; Quinones, M.; Spicer, E.; Douglas, I.; Williams, K.; Konigsberg, W. The 44p subunit of the T4 DNA polymerase accessory protein complex catalyzes ATP hydrolysis. *J. Biol. Chem.* **1989**, *264*, 10943–10953.

27. Bowman, G.D.; O'Donnell, M.; Kuriyan, J. Structural analysis of a eukaryotic sliding DNA clamp-clamp loader complex. *Nature* **2004**, *429*, 724–730.

28. Simonetta, K.R.; Kazmirski, S.L.; Goedken, E.R.; Cantor, A.J.; Kelch, B.A.; McNally, R.; Seyedin, S.N.; Makino, D.L.; O'Donnell, M.; Kuriyan, J. The mechanism of ATP-dependent primer-template recognition by a clamp loader complex. *Cell* **2009**, *137*, 659–671.

29. Dong, F.; Gogol, E.P.; von Hippel, P.H. The phage T4-coded DNA replication helicase (gp41) forms a hexamer upon activation by nucleoside triphosphate. *J. Biol. Chem.* **1995**, *270*, 7462–7473.

30. Young, M.C.; Schultz, D.E.; Ring, D.; von Hippel, P.H. Kinetic parameters of the translocation of bacteriophage T4 gene 41 protein helicase on single-stranded DNA. *J. Mol. Biol.* **1994**, *235*, 1447–1458.

31. Liu, C.; Alberts, B. Characterization of the DNA-dependent gtpase activity of T4 gene 41 protein, an essential component of the t4 bacteriophage DNA replication apparatus. *J. Biol. Chem.* **1981**, *256*, 2813–2820.

32. Norcum, M.T.; Warrington, J.A.; Spiering, M.M.; Ishmael, F.T.; Trakselis, M.A.; Benkovic, S.J. Architecture of the bacteriophage T4 primosome: Electron microscopy studies of helicase (gp41) and primase (gp61). *Proc. Natl. Acad. Sci. USA* **2005**, *102*, 3623–3626.

33. Venkatesan, M.; Silver, L.; Nossal, N. Bacteriophage T4 gene 41 protein, required for the synthesis of RNA primers, is also a DNA helicase. *J. Biol. Chem.* **1982**, *257*, 12426–12434.

34. Richardson, R.W.; Nossal, N. Characterization of the bacteriophage T4 gene 41 DNA helicase. *J. Biol. Chem.* **1989**, *264*, 4725–4731.

35. Perumal, S.K.; Raney, K.D.; Benkovic, S.J. Analysis of the DNA translocation and unwinding activities of T4 phage helicases. *Methods* **2010**, *51*, 277–288.

36. Zhang, Z.; Spiering, M.M.; Trakselis, M.A.; Ishmael, F.T.; Xi, J.; Benkovic, S.J.; Hammes, G.G. Assembly of the bacteriophage T4 primosome: Single-molecule and ensemble studies. *Proc. Natl. Acad. Sci. USA* **2005**, *102*, 3254–3259.

37. Cha, T.; Alberts, B. Studies of the DNA helicase-RNA primase unit from bacteriophage T4. A trinucleotide sequence on the DNA template starts rna primer synthesis. *J. Biol. Chem.* **1986**, *261*, 7001–7010.

38. Hinton, D.; Nossal, N. Bacteriophage T4 DNA primase-helicase. Characterization of oligomer synthesis by T4 61 protein alone and in conjunction with T4 41 protein. *J. Biol. Chem.* **1987**, *262*, 10873–10878.

39. Cha, T.A.; Alberts, B.M. Effects of the bacteriophage T4 gene 41 and gene 32 proteins on rna primer synthesis: The coupling of leading-and lagging-strand DNA synthesis at a replication fork. *Biochemistry* **1990**, *29*, 1791–1798.

40. Yang, J.; Xi, J.; Zhuang, Z.; Benkovic, S.J. The oligomeric T4 primase is the functional form during replication. *J. Biol. Chem.* **2005**, *280*, 25416–25423.

41. Huberman, J.A.; Kornberg, A.; Alberts, B.M. Stimulation of t4 bacteriophage DNA polymerase by the protein product of T4 gene 32. *J. Mol. Biol.* **1971**, *62*, 39–52.

42. Huang, C.; Hearst, J.; Alberts, B. Two types of replication proteins increase the rate at which T4 DNA polymerase traverses the helical regions in a single-stranded DNA template. *J. Biol. Chem.* **1981**, *256*, 4087–4094.

43. Huang, C.-C.; Hearst, J.E. Pauses at positions of secondary structure during *in vitro* replication of single-stranded fd Bacteriophage DNA by T4 DNA polymerase. *Anal. Biochem.* **1980**, *103*, 127–139.

44. Shamoo, Y.; Friedman, A.M.; Parsons, M.R.; Konigsberg, W.H.; Steitz, T.A. Crystal structure of a replication fork single-stranded DNA binding protein (T4 gp32) complexed to DNA. *Nature* **1995**, *376*, 362–366.

45. Dudas, K.C.; Kreuzer, K.N. Bacteriophage T4 helicase loader protein gp59 functions as gatekeeper in origin-dependent replication *in vivo*. *J. Biol. Chem.* **2005**, *280*, 21561–21569.

46. Ishmael, F.T.; Alley, S.C.; Benkovic, S.J. Assembly of the Bacteriophage T4 helicase architecture and stoichiometry of the gp41-gp59 complex. *J. Biol. Chem.* **2002**, *277*, 20555–20562.

47. Delagoutte, E.; von Hippel, P.H. Mechanistic studies of the t4 DNA (gp41) replication helicase: Functional interactions of the c-terminal tails of the helicase subunits with the T4 (gp59) helicase loader protein. *J. Mol. Biol.* **2005**, *347*, 257–275.

48. Chastain, P.D.; Makhov, A.M.; Nossal, N.G.; Griffith, J. Architecture of the replication complex and DNA loops at the fork generated by the bacteriophage t4 proteins. *J. Biol. Chem.* **2003**, *278*, 21276–21285.

49. Nossal, N.G.; Makhov, A.M.; Chastain, P.D.; Jones, C.E.; Griffith, J.D. Architecture of the Bacteriophage T4 replication complex revealed with nanoscale biopointers. *J. Biol. Chem.* **2007**, *282*, 1098–1108.

50. Zhao, Y.; Chen, D.; Yue, H.; Spiering, M.M.; Zhao, C.; Benkovic, S.J.; Huang, T.J. Dark-field illumination on zero-mode waveguide/microfluidic hybrid chip reveals T4 replisomal protein interactions. *Nano Lett.* **2014**, *14*, 1952–1960.

51. Mace, D.C.; Alberts, B.M. T4 DNA polymerase: Rates and processivity on single-stranded DNA templates. *J. Mol. Biol.* **1984**, *177*, 295–311.

52. Kelch, B.A.; Makino, D.L.; O'Donnell, M.; Kuriyan, J. How a DNA polymerase clamp loader opens a sliding clamp. *Science* **2011**, *334*, 1675–1680.

53. Berdis, A.J.; Benkovic, S.J. Role of adenosine 5′-triphosphate hydrolysis in the assembly of the bacteriophage T4 DNA replication holoenzyme complex. *Biochemistry* **1996**, *35*, 9253–9265.

54. Perumal, S.K.; Ren, W.; Lee, T.-H.; Benkovic, S.J. How a holoenzyme for DNA replication is formed. *Proc. Natl. Acad. Sci. USA* **2013**, *110*, 99–104.

55. Mathews, C.K. *Bacteriophage T4*; Wiley Online Library: Washington, DC, USA, 1983.

56. Kaboord, B.F.; Benkovic, S.J. Accessory proteins function as matchmakers in the assembly of the T4 DNA polymerase holoenzyme. *Curr. Biol.* **1995**, *5*, 149–157.

57. Schrock, R.D.; Alberts, B. Processivity of the gene 41 DNA helicase at the bacteriophage T4 DNA replication fork. *J. Biol. Chem.* **1996**, *271*, 16678–16682.

58. Yang, J.; Zhuang, Z.; Roccasecca, R.M.; Trakselis, M.A.; Benkovic, S.J. The dynamic processivity of the T4 DNA polymerase during replication. *Proc. Natl. Acad. Sci. USA* **2004**, *101*, 8289–8294.

59. Maga, G.; Hübscher, U. Proliferating cell nuclear antigen (PCNA): A dancer with many partners. *J. Cell Sci.* **2003**, *116*, 3051–3060.

60. Maul, R.W.; Scouten Ponticelli, S.K.; Duzen, J.M.; Sutton, M.D. Differential binding of *Escherichia coli* DNA polymerases to the β-sliding clamp. *Mol. Microbial.* **2007**, *65*, 811–827.

61. Johnson, D.E.; Takahashi, M.; Hamdan, S.M.; Lee, S.-J.; Richardson, C.C. Exchange of DNA polymerases at the replication fork of bacteriophage T7. *Proc. Natl. Acad. Sci. USA* **2007**, *104*, 5312–5317.

62. Hacker, K.J.; Alberts, B.M. The rapid dissociation of the T4 DNA polymerase holoenzyme when stopped by a DNA hairpin helix. A model for polymerase release following the termination of each okazaki fragment. *J. Biol. Chem.* **1994**, *269*, 24221–24228.

63. Cha, T.-A.; Alberts, B.M. The bacteriophage t4 DNA replication fork. Only DNA helicase is required for leading strand DNA synthesis by the DNA polymerase holoenzyme. *J. Biol. Chem.* **1989**, *264*, 12220–12225.

64. Dong, F.; Weitzel, S.E.; Von Hippel, P.H. A coupled complex of T4 DNA replication helicase (gp41) and polymerase (gp43) can perform rapid and processive DNA strand-displacement synthesis. *Proc. Natl. Acad. Sci. USA* **1996**, *93*, 14456–14461.

65. Delagoutte, E.; von Hippel, P.H. Molecular mechanisms of the functional coupling of the helicase (gp41) and polymerase (gp43) of bacteriophage T4 within the DNA replication fork. *Biochemistry* **2001**, *40*, 4459–4477.

66. Ishmael, F.T.; Trakselis, M.A.; Benkovic, S.J. Protein-protein interactions in the bacteriophage T4 replisome the leading strand holoenzyme is physically linked to the lagging strand holoenzyme and the primosome. *J. Biol. Chem.* **2003**, *278*, 3145–3152.

67. Stano, N.M.; Jeong, Y.-J.; Donmez, I.; Tummalapalli, P.; Levin, M.K.; Patel, S.S. DNA synthesis provides the driving force to accelerate DNA unwinding by a helicase. *Nature* **2005**, *435*, 370–373.

68. Manosas, M.; Spiering, M.M.; Ding, F.; Croquette, V.; Benkovic, S.J. Collaborative coupling between polymerase and helicase for leading-strand synthesis. *Nucl. Acids Res.* **2012**, *40*, 6187–6198.

69. Lionnet, T.; Spiering, M.M.; Benkovic, S.J.; Bensimon, D.; Croquette, V. Real-time observation of bacteriophage T4 gp41 helicase reveals an unwinding mechanism. *Proc. Natl. Acad. Sci. USA* **2007**, *104*, 19790–19795.

70. Salinas, F.; Benkovic, S.J. Characterization of bacteriophage T4-coordinated leading-and lagging-strand synthesis on a minicircle substrate. *Proc. Natl. Acad. Sci. USA* **2000**, *97*, 7196–7201.

71. Yang, J.; Trakselis, M.A.; Roccasecca, R.M.; Benkovic, S.J. The application of a minicircle substrate in the study of the coordinated T4 DNA replication. *J. Biol. Chem.* **2003**, *278*, 49828–49838.

72. Jing, D.; Beechem, J.M.; Patton, W.F. The utility of a two-color fluorescence electrophoretic mobility shift assay procedure for the analysis of DNA replication complexes. *Electrophoresis* **2004**, *25*, 2439–2446.

73. Jing, D.H.; Dong, F.; Latham, G.J.; von Hippel, P.H. Interactions of bacteriophage t4-coded primase (gp61) with the t4 replication helicase (gp41) and DNA in primosome formation. *J. Biol. Chem.* **1999**, *274*, 27287–27298.

74. Manosas, M.; Spiering, M.M.; Zhuang, Z.; Benkovic, S.J.; Croquette, V. Coupling DNA unwinding activity with primer synthesis in the bacteriophage T4 primosome. *Nat. Chem. Biol.* **2009**, *5*, 904–912.

75. Lee, J.-B.; Hite, R.K.; Hamdan, S.M.; Xie, X.S.; Richardson, C.C.; Van Oijen, A.M. DNA T4 primase acts as a molecular brake in DNA replication. *Nature* **2006**, *439*, 621–624.

76. Pandey, M.; Syed, S.; Donmez, I.; Patel, G.; Ha, T.; Patel, S.S. Coordinating DNA replication by means of priming loop and differential synthesis rate. *Nature* **2009**, *462*, 940–943.

77. Kadyrov, F.A.; Drake, J.W. Conditional coupling of leading-strand and lagging-strand DNA synthesis at bacteriophage T4 replication forks. *J. Biol. Chem.* **2001**, *276*, 29559–29566.

78. Trakselis, M.A.; Roccasecca, R.M.; Yang, J.; Valentine, A.M.; Benkovic, S.J. Dissociative properties of the proteins within the bacteriophage T4 replisome. *J. Biol. Chem.* **2003**, *278*, 49839–49849.

79. Trakselis, M.A.; Alley, S.C.; Abel-Santos, E.; Benkovic, S.J. Creating a dynamic picture of the sliding clamp during T4 DNA polymerase holoenzyme assembly by using fluorescence resonance energy transfer. *Proc. Natl. Acad. Sci. USA* **2001**, *98*, 8368–8375.

80. Nelson, S.W.; Kumar, R.; Benkovic, S.J. Rna primer handoff in bacteriophage T4 DNA replication the role of single-stranded DNA-binding protein and polymerase accessory proteins. *J. Biol. Chem.* **2008**, *283*, 22838–22846.

81. Yang, J.; Nelson, S.W.; Benkovic, S.J. The control mechanism for lagging strand polymerase recycling during bacteriophage T4 DNA replication. *Mol. Cell* **2006**, *21*, 153–164.

82. Carver, T.E.; Sexton, D.J.; Benkovic, S.J. Dissociation of bacteriophage t4 DNA polymerase and its processivity clamp after completion of okazaki fragment synthesis. *Biochemistry* **1997**, *36*, 14409–14417.

83. Tougu, K.; Marians, K.J. The interaction between helicase and primase sets the replication fork clock. *J. Biol. Chem.* **1996**, *271*, 21398–21405.

84. Kurth, I.; Georgescu, R.E.; O'Donnell, M.E. A solution to release twisted DNA during chromosome replication by coupled DNA polymerases. *Nature* **2013**, *496*, 119–122.

85. Yuan, Q.; McHenry, C.S. Cycling of the *E. coli* lagging strand polymerase is triggered exclusively by the availability of a new primer at the replication fork. *Nucl. Acids Res.* **2014**, *42*, 1747–1756.

86. Geertsema, H.J.; van Oijen, A.M. A single-molecule view of DNA replication: The dynamic nature of multi-protein complexes revealed. *Curr. Opin. Struct. Biol.* **2013**, *23*, 788–793.

87. Chen, D.; Yue, H.; Spiering, M.M.; Benkovic, S.J. Insights into okazaki fragment synthesis by the T4 replisome the fate of lagging-strand holoenzyme components and their influence on Okazaki fragment size. *J. Biol. Chem.* **2013**, *288*, 20807–20816.

HIV Rev Assembly on the Rev Response Element (RRE): A Structural Perspective

Jason W. Rausch and Stuart F. J. Le Grice

Abstract: HIV-1 Rev is an ~13 kD accessory protein expressed during the early stage of virus replication. After translation, Rev enters the nucleus and binds the Rev response element (RRE), a ~350 nucleotide, highly structured element embedded in the *env* gene in unspliced and singly spliced viral RNA transcripts. Rev-RNA assemblies subsequently recruit Crm1 and other cellular proteins to form larger complexes that are exported from the nucleus. Once in the cytoplasm, the complexes dissociate and unspliced and singly-spliced viral RNAs are packaged into nascent virions or translated into viral structural proteins and enzymes, respectively. Rev binding to the RRE is a complex process, as multiple copies of the protein assemble on the RNA in a coordinated fashion *via* a series of Rev-Rev and Rev-RNA interactions. Our understanding of the nature of these interactions has been greatly advanced by recent studies using X-ray crystallography, small angle X-ray scattering (SAXS) and single particle electron microscopy as well as biochemical and genetic methodologies. These advances are discussed in detail in this review, along with perspectives on development of antiviral therapies targeting the HIV-1 RRE.

Reprinted from *Viruses*. Cite as: Rausch, J.W.; Le Grice, S.F.J. HIV Rev Assembly on the Rev Response Element (RRE): A Structural Perspective. *Viruses* **2015**, *7*, 3053–3075.

1. Introduction

Replication of retroviruses and transposition of endogenous retroelements exploits a unique mechanism of post-transcriptional regulation as a means of exporting their full-length and incompletely-spliced mRNAs (which serve as the genomic RNA and the template for protein synthesis, respectively) to the cytoplasm. This is achieved through the concerted interaction of highly structured *cis*-acting regulatory elements in the RNA genome with a variety of obligate viral and host proteins. Examples of such regulatory RNAs include the constitutive transport element (CTE) of the simian retroviruses (MPMV, SRV-1, SRV-2) [1], the musD transport element (MTE) of murine retroelements (intracisternal A particles and musD) [2], the ~1.7 kb posttranscriptional element (PTE) of gammaretroviruses such as murine leukemia virus [3] and the L1-NXF1 element of human LINE-1 transposons [4]. Studies such as these, which have made a significant contribution to our understanding of cellular processes regulating the fate of RNA, resulted from seminal work performed over 30 years ago to understand nucleocytoplasmic

RNA transport in human immunodeficiency virus (HIV). In particular, disrupting the HIV-1 genome in the immediate vicinity of the trans-activator of transcription (tat) open reading frame (ORF) had no effect on its expression levels, but inhibited expression of the gag, pol and env genes, resulting in a severe replication defect [5]. This defect could be corrected by inclusion of an overlapping ORF encoding the regulator of expression of viral proteins, or Rev, a small accessory protein containing both nuclear export and localization signals (NES and NLS, respectively).

The notion that Rev was involved in RNA transport rather than modifying splicing arose from observations of Malim *et al.* [5], who elegantly showed that cytoplasmic expression of the non-spliceable HIV-1 *env* gene was also subject to Rev control. At the same time, these authors identified the Rev response element (RRE), an ~350 nt highly-structured *cis*-acting RNA within the *env* gene (nucleotides 7709–8063), as the target of Rev. The principles of the Rev/RRE axis are outlined schematically in Figure 1. Early in the virus life cycle, Rev and additional HIV regulatory proteins are translated from completely spliced RNAs that are exported from the nucleus in a manner analogous to cellular mRNAs. Subsequently, the arginine-rich NLS facilitates entry of Rev into the nucleus, where it interacts with the RRE in unspliced and incompletely-spliced HIV RNAs. Rev initially binds to stem-loop IIB, a purine rich RNA secondary structure motif (*vide infra*), after which several additional Rev molecules assemble along the RRE to generate the Rev-RRE complex. This complex then recruits Crm1 and other host proteins into a larger complex that is exported from the nucleus into the cytoplasm. The goal of this review is to provide an updated account of our understanding of the HIV Rev-RRE complex with respect to the recently-elucidated structures of the protein and RNA components. For a broader perspective, the reader is referred to excellent reviews from the Cullen [6], Hope [7], Malim [8], Daelemans [9] and Frankel [10] groups.

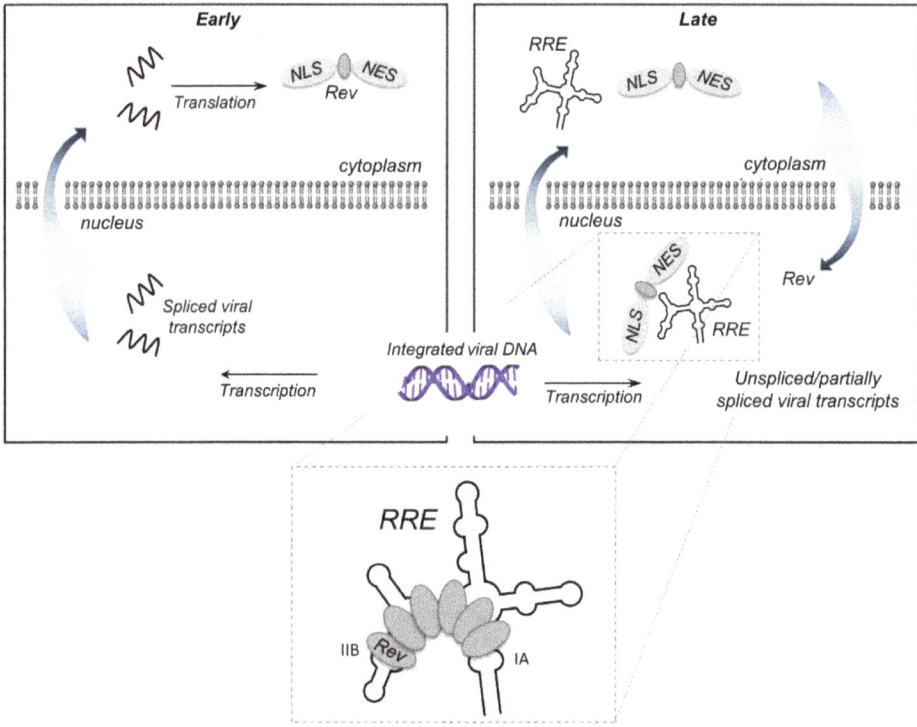

Figure 1. Rev-mediated nucleocytoplasmic transport of HIV-1 RNA containing the Rev RRE. RNA transport: In the early phase of the HIV life cycle, the genomic RNA transcript is completely spliced, generating RRE-free messages, which are transported to the cytoplasm *via* standard nuclear export pathways. One of these messages encodes Rev, which is imported into the nucleus *via* its nuclear localization sequence (NLS). The late phase of the viral life cycle is characterized by the expression of viral proteins encoded by the unspliced (9 kb) or partially spliced (4 kb) RRE-containing mRNAs. These large intron-containing RNAs are retained in the nucleus for splicing/degradation until a sufficient level of Rev accumulates, after which they are exported to the cytoplasm *via* a Rev-dependent export pathway. This involves assembly of the Rev-RRE complex and recruitment of host proteins CRM1 and Ran-GTP *via* the Rev nuclear export sequence (NES). Export of the Rev-RRE -CRM1/RanGTP complex to the cytoplasm provides mRNAs that are translated to produce the remaining viral proteins and full-length genomes that are packaged into the budding virion. Inset: Rev initially binds the IIB secondary structure motif, then cooperatively assembles along the RRE *via* a series of Rev-Rev and Rev-RNA interactions. Rev has also been reported to bind motif IA with high affinity.

129

2. Rev Structural Organization and RNA Binding

HIV-1 Rev is a 116 amino acid, ~13 kD protein generated by translation of a fully spliced 2 kb mRNA during the early phase of viral replication and organized as shown in Figure 2A [8]. The N-terminus of the protein assumes a helix-turn-helix configuration containing two functional domains: the nuclear localization signal and RNA binding domain (NLS/RBD) and the Rev multimerization domain. The NLS/RBD is housed within the distal portion of the helix-turn-helix motif. A stretch of amino acids within α-helix 2 in this domain contains several functionally important arginines and has thus been dubbed the arginine rich motif (ARM). The multimerization domain houses a number of hydrophobic amino acids located at opposite ends of the helix-turn-helix primary sequence, but proximal to each other in three dimensions (3D). Contacts among these hydrophobic residues stabilize the overall Rev structure, and their arrangement generates the two-sided interface for Rev multimerization. Outside of the helix-turn-helix motif, the C-terminus of Rev is intrinsically disordered; however, this segment of the protein contains a third, leucine-rich functional domain known as the nuclear export signal (NES). This is the effector domain of Rev, and is required for recruitment of Crm1 and other host proteins that facilitate nuclear export of the Rev-RRE complex.

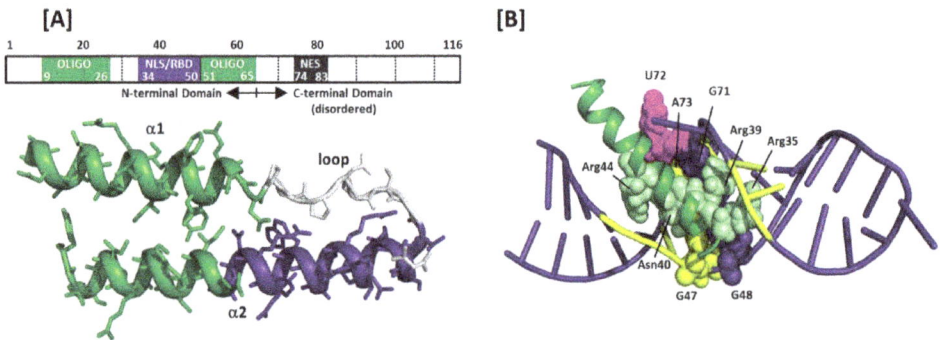

Figure 2. Rev structural organization and arginine-rich motif (ARM)/stem loop IIB complex: (**A**) Rev organization according to primary sequence and 3D structure. The bipartite oligomerization domain and the nuclear localization signal/RNA-binding domain (NLS/RBD) are depicted in green and blue, respectively. The C-terminal domain of Rev, which houses the nuclear export signal (NES), is intrinsically disordered; (**B**) ARM peptide in complex with stem-loop IIB model RNA. The ARM peptide binds in the RNA major groove widened by non-canonical base pairs (G47-A73, G48-G71) and an unpaired, unstacked uridine (U72). These nucleotides are space-filled in the model. Arginines that make specific contacts with nucleic acid bases, and the contacted nucleotides, are highlighted by space-filling and yellow coloration, respectively.

130

As its designation suggests, the NLS/RBD of HIV-1 Rev is important for both nuclear import of cytoplasmic Rev and binding to the RRE. Multiple reports indicate that initial Rev binding occurs at a purine rich segment in stem loop IIB, a substructure of the HIV-1 RRE for which Rev has been shown to bind with high affinity (*vide infra*) [11–15]. The first high-resolution picture of the Rev–stem loop IIB interaction was obtained from NMR structures of a Rev ARM peptide in complex with a short synthetic RNA engineered to stably recapitulate the IIB RNA substructure [16]. The averaged structure of the ARM-IIB complex is depicted in Figure 2B, wherein nucleotide numbering reflects that used for the artificial IIB construct in the file deposited in the Protein Data Bank (PDB). The IIB binding site is notable in that it contains two non-canonical base-pairs (G47-A73 and G48-G71) separated by an unstacked, bulged uridine (U72) [17,18]. This arrangement serves to widen the helical major groove at the site where the Rev peptide contacts the RNA [16,19,20]. Upon binding, the Rev ARM penetrates deeply into the stem loop IIB major groove [16,21–23], thereby inducing a conformational change that widens the groove even further [24–26]. Four ARM residues make base-specific contacts with nucleotides in stem loop IIB: Arg35 and Arg39 contact nucleotides U66, G67 and G70 on one side of the RNA major groove, while Asn40 and Arg44 contact U45, G46, G47 and A73 on the other. In addition, Thr34, located at the non-helical turn segment of the Rev helix-turn-helix motif, and six arginine residues (Arg38, Arg41–Arg43, Arg46 and Arg48) within the ARM, interact nonspecifically with the RNA sugar-phosphate backbone.

More recently, a crystallographic structure of a rev dimer in complex with an engineered RRE-like RNA was resolved that has greatly enhanced our understanding of both Rev interactions and how Rev binds RNA [27]. Unlike the NMR structure, the 47-nt RNA in the Rev dimer complex contains both the IIB binding site and an adjacent "junction site" engineered to resemble the three-way junction within stem loop II. This secondary site contains a non-canonical G-A base pair like that found in IIB, as well as an unpaired U. The unpaired U and adjacent A were shown to be essential for binding of the second Rev, as their removal results in a complex in which only a single Rev is bound. In the crystal structure, the ARMs of the dimerized Rev molecules bind along the RNA major grooves of the adjacent binding sites. Contacts at the IIB site closely resemble those observed in the NMR structure, with the notable exception of those involving Rev amino acid Asn40. At the junction site, all contacts except those involving Arg43 and Arg44 are with the RNA phosphate backbone, consistent with finding that binding at the second site is relatively sequence non-specific and requires only that the site contain an RNA bulge [28]. While the rotational positioning of the two Rev ARMs in the major grooves of the IIB and junctional binding sites appears to be similar, their linear positioning is displaced by approximately the length of an alpha helical turn. The structure also addresses how binding of the Rev dimer to adjacent sites on the RRE RNA affects the Rev dimer interface, which will be discussed below.

Another Rev binding site in RRE stem loop I has been identified using structural and biochemical techniques [29]. Like the IIB and junctional sites, the stem I site (designated site IA) is also comprised of a purine-rich bulge in which the major groove might be expected to be widened, flexible, and/or defined by non-canonical G-A base pairs. However, mutational analysis suggests that residues Arg38, Arg 41 and Arg 46 are crucial for binding at site IA, indicating a different rotational positioning of Rev in the RNA major groove and suggesting that the Rev-RRE interface is flexible and may be substantially different at distinct binding sites. Other segments of stem I have been implicated in Rev binding [30], but the Rev-RNA interactions at these sites have not been characterized structurally.

3. Structural Basis for Rev Oligomerization

Although oligomerization has been shown to be an important aspect of Rev function in virus replication [29,31,32], this property was first observed with recombinant protein. At low concentrations, Rev exists as monomer [11], dimer [33] or tetramer [11,34]. Above a critical concentration of ~80 ng/mL (~6 μM), however, Rev forms regular, unbranched filaments of indeterminate length [11,33,35,36]. The structural basis for filament formation resides in the two-sided multimerization interface in which a given Rev molecule can be flanked by additional Rev on either side.

The oligomerization domain of HIV-1 Rev is a flat, two-sided structure formed by juxtaposition of two α helical regions located at opposite ends of the Rev helix-turn-helix motif primary sequence. The domain has a polarity named for the sides of a coin, with the "heads" (H) and "tails" (T) faces interacting specifically with their equivalent counterparts in adjacent Rev molecules; i.e., "H/H" or "T/T" interfaces are preferentially formed in Rev oligomers [37]. Although the T/T Rev dimer is more stable than the H/H and is the form assumed by Rev dimers in solution and after initial binding to the RRE IIB substructure, a high resolution structure of a Rev dimer containing an H/H interface was the first to be resolved by X-ray crystallography [38]. Crystal formation was promoted and filament formation inhibited in this case by blocking the outer T faces of the Rev dimer with monoclonal F_{ab} fragments. Although present in the crystalized protein, the disordered C-terminal portion of Rev, including the NES, was unresolved.

In support of prior genetic and NMR structural data [16,37], a hydrophobic core comprising residues Leu12, Ile19, Leu22, Tyr23 of alpha-helix 1 and Trp45, Ile52, Ile59, Leu60 and Tyr63 of α-helix 2 were shown to be major contributors to Rev helical hairpin stability. Intermolecular interactions among an overlapping set of multimerization domain hydrophobic residues (Leu12, Leu13, Val16, Ile19,

Leu60 and Leu64) were shown to comprise much of the H/H interface, of which the roles of Leu12, Val16 and Leu60 were previously established genetically [37]. Because the H/H junction involves residues located at the extreme pronged end of the Rev helix-turn-helix, viewing the structured portion of the dimer from the intermolecular axis orthogonal to the plane of the interface demonstrates that the two molecules assume a "V-like" configuration relative to each other (*i.e.*, rather than one resembling an "X") (Figure 3A). Moreover, the helix-turn-helix motifs are arranged such that the two ARMs form an angle of approximately 140°. The spacing between NLS/RBDs and the opposing trajectories of the two motifs place their extreme termini approximately 8 nm apart, suggesting that the ARMs in such a dimer would be unlikely to bind the RRE in adjacent regions of the major groove on the same helix. Instead, a model based on superimposition of H/H Rev dimer crystal structure and the ARM/IIB NMR structure that suggests the H/H dimer may be best suited for linking either two separate RNAs or two RNA helices located in distant regions of the same RNA [38].

Details of the T/T Rev dimer interface were first resolved in a crystal structure in which H-surface residues Leu12 and Leu60 were mutated to suppress higher order multimerization [39] (Figure 3B). Despite these modifications, asymmetric units in these crystals contained four Rev molecules linked sequentially by T/T, H/H and T/T interfaces, where the H/H interaction matched that observed in the original dimer crystal structure almost exactly. Packing of hydrophobic residues on the T surfaces of adjacent Rev molecules formed an interface that buries over 1500 Å^2 of surface area. Leu18 and Ile55 form symmetric contacts between monomers at the T/T interface, are highly conserved, and essential for cooperative RNA binding and export [29,37,40]. Phe21, Leu22 and Ile59 are likewise present at the dimer interface, although these residues are also important for stabilizing monomeric Rev. From the vantage point of the oligomerization axis, the ordered portion of the T/T Rev dimer assumes a configuration resembling an asymmetric "X", with long and short NLS/RBD and multimerization domain protrusions flanking the Rev-Rev interface on either side, respectively. Moreover, at 120°, the ARM angle formed in the T/T dimer is narrower than observed in the H/H dimer structure. A molecular model of the T/T dimer in complex with a short IIB-like RNA suggests that both opposing ARMs could bind in the major groove on the same face of an RNA helix with contacts separated by approximately one A-form helical turn. It is further suggested that Rev-Rev and Rev-RNA interactions may propagate from the IIB initiation site along a contiguous region of the RRE, placing the Rev molecules—each with a disordered NES—projecting away from the RNA in a common direction like "tentacles of a jellyfish" [39]. The jellyfish model provides important insight into how multiple Revs may assemble on a comparatively large RRE, and will be discussed in more detail below. However, it is worth noting that in this initial manifestation of the model, the helical axis of the RNA was nearly perpendicular to the axis of Rev oligomerization—an arrangement that would likely be incompatible with the proposed coordination of multiple Revs binding at adjacent sites on the RRE.

How Rev-Rev and Rev-RNA interactions can occur concomitantly was largely explained by the aforementioned co-crystal structure in which a T/T Rev dimer binds a truncated RRE RNA at adjacent IIB and junctional sites [27]. ARM binding in the RNA major groove at the two sites appears to change the organization of the T/T interface relative to the naked Rev dimer, and a reciprocal effect can be observed in the distorted binding site major grooves. Among of the more important consequences of this reorganization are that the angle formed by the two Rev ARMs is reduced to ~50° and the Rev multimerization and RNA helical axes are nearly parallel (Figure 3C). Both of these observations are compatible with the jellyfish model of Rev assembly.

Relative to the naked dimer, the T/T interface of the Rev dimer-RNA complex is rotated around Ile55, substantially altering contacts among the hydrophobic residues of both T-surfaces. Although the interacting residues at the T/T interface are largely the same regardless of whether RNA is present, specific points of contact are altered considerably. Collectively, surface area buried at the T/T interface in the Rev dimer-RNA complex is reduced by ~33% (to ~1000 Å^2), suggesting that the dimer may be less stable in the presence of RNA. However, the energetic favorability of RNA binding likely compensates for reduced hydrophobic interactions, thereby rendering the Rev-dimer-RNA complex more stable than the naked dimer. The Phe21 residues that facilitated dimerization in the naked dimer are excluded from the T/T interface in the Rev-dimer RNA complex; moreover, reciprocal Gln51-Gln51 hydrogen bonding is observed only in the presence of RNA. Interestingly, introducing a Gln51Ala mutation into Rev appears to both impair Rev dimerization and reduce the affinity of Rev for the truncated RRE RNA approximately 30-fold. However, the effects of this mutation on the fully assembled Rev-RRE complex are relatively modest, suggesting that Gln51 hydrogen bonding may not be as important during later stages of Rev assembly.

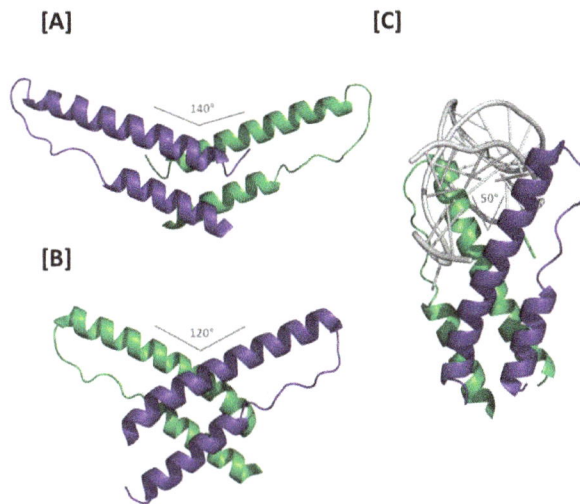

Figure 3. Crystal structures of Rev dimers: (**A**) Ribbon representation of the H/H Rev dimer. The angle formed by the two ARMs is 140°, which separates the apices of the two NLS/RBDs by approximately 8 nm and precludes binding at adjacent sites on an RNA helix; (**B**) T/T Rev dimer structure. In the absence of RNA, the ARM angle is 120°. Binding at adjacent RNA sites is conceivable, but higher order Rev multimerization would likely require a reduced ARM angle at H/H interfaces; (**C**) Structure of the T/T Rev dimer in complex with RNA. The two ARMs bind RNA at adjacent sites on the helix and are oriented at an angle of 50° relative to each other. The Rev oligomerization and RNA helical axes in this structure are roughly parallel, thereby facilitating consecutive binding and the higher order structures proposed in the jellyfish model of Rev assembly.

4. Secondary and 3D Structures of the RRE

Our structural understanding of Rev, Rev dimer and Rev–IIB interactions has been greatly facilitated by X-ray crystallography and NMR. Using the same approaches to study the structure of the HIV-1 RRE is problematic, however, given the size and flexibility of this highly structured RNA element. For the most part, experimental approaches have been restricted to using a number of enzymatic and chemical probing techniques to characterize the RRE secondary structure. Although these efforts sometimes produced differing secondary structural models for similar RRE variants, common structural features have been identified that help both in developing 3D models of the motif and for understanding how Rev assembles along this conserved segment of RNA.

Although the HIV-1 RRE is approximately 350 nt in length, the secondary structure of this RNA element was initially characterized using RNA folding prediction software together with enzymatic and chemical RNA probing experiments conducted using a truncated (∼235 nt), *in vitro* transcribed version of the RNA [41]. The sub-structure designations defined in this seminal work will also be used here. In the original model, the HIV-1 RRE RNA assumes a secondary structure comprised of five stems, stem loops or bifurcated stem-loops (I-V) arranged around a central 5-way junction (Figure 4A). Stem I is the longest of the stems/stem-loops, and is interrupted by a number of internal loops and bulges of varying size. Some of these internal loops are purine rich, and that most proximal to the central junction (stem IA) has been identified as a high-affinity Rev binding site [29]. Stem-loop II is bifurcated to form smaller substructures designated IIA, IIB and IIC. As previously noted, IIB contains a purine-rich internal segment characterized by non-canonical G-A and G-G base pairing and a widened major groove, and has been identified as the site at which Rev initially binds to the RRE to initiate assembly [31,32,42]. Moreover, according to the Rev dimer-RNA co-crystal structure [27], the adjacent stem-loop II junctional region serves as the binding site for the second Rev. Proceeding clockwise around the central junction in Figure 4A, stem loops III-V complete this 5-stem secondary structural model of the RRE.

A later 4-stem model obtained using enzymatic and chemical probing techniques suggests that stem loops III and IV, together with the intervening loop region, are organized to form a hybrid III/IV stem loop [32] (Figure 4B). While the central junction is compacted somewhat in this alternative folding, no additional differences between the 4- and 5-stem RRE secondary structures were proposed. Which of the two structures is assumed by the RRE in the context of HIV replication has been a subject of considerable debate, with multiple studies supporting both forms [2,43,44]. Recently, however, it has been demonstrated using native gel electrophoresis and in-gel SHAPE [45] that the HIV-1 RRE can assume either the 5- or the 4-stem conformation [46]. Moreover, supporting virological data suggest that the two forms may not be equivalent, *i.e.*, HIV-1 housing a mutant RRE that exclusively assumes the 5-stem conformation outgrows virus with a 4-stem RRE in growth competition experiments. This finding is in agreement with the observation that RRE61, an RRE mutant shown to assume a conformation resembling the 5-stem structure, confers resistance to the *trans*-dominant Rev mutant RevM10 [2].

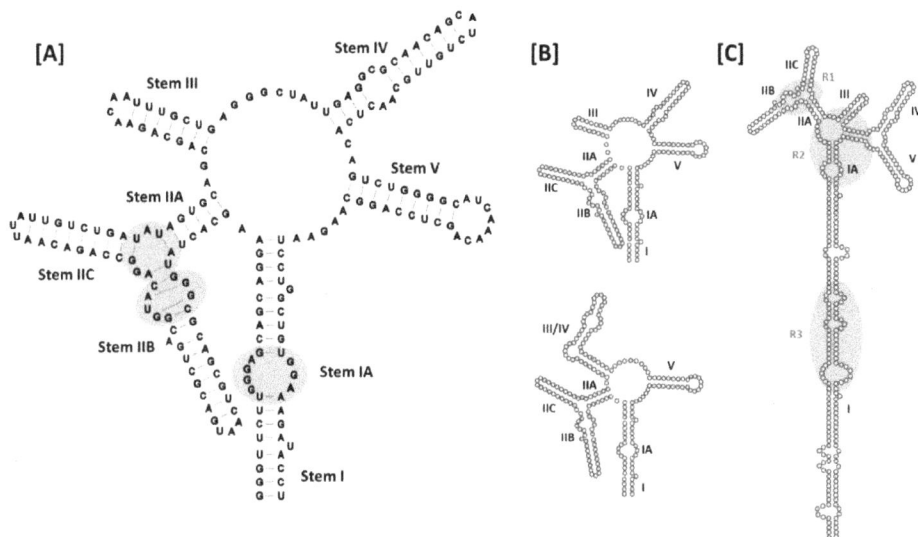

Figure 4. Secondary structures of HIV-1 RREs: (**A**) 5-stem structure. Sub-structure designations and base-pairing patterns, including non-canonical G-A and G-G base pairs in stem IIB, are indicated. Established Rev binding sites at IIB, the stem loop II junction and stem IA are also shown (gray ovals); (**B**) Comparison of the 5- and 4-stem structures. Differences between the two models are limited to the base pairing patterns of nucleotides comprising stem loops III and IV, and stem loop III/IV, in the respective structures. Due to space limitations, stem I has been truncated in the models presented in panels (**A–C**). Alternative 5-stem RRE structure proposed for an ARV-2/SF2 HIV-1 isolate. The entire RRE is shown. Stem I is truncated relative to the original 5-stem model, and there are more single-stranded nucleotides separating stem I and stem IIA. In addition, base pairing across the central junction separates stem loops IV and V from the rest of the alternative 5-stem structure. R1, R2 and R3 refer to Rev binding regions 1–3 identified for this RRE variant by time-resolved selective 2′-hydroxyl acylation analyzed by primer extension (SHAPE).

Recently, another secondary structural model was proposed for the RRE of an ARV-2/SF2 HIV-1 isolate (Figure 4C) that closely resembles the RRE configurations proposed for HIV-2 and SIVmac [30,43,47]. This model resembles the original 5-stem HIV-1 RRE structure in that stem-loops II, III, IV and V are identical in the two versions. However, in the more recent model, the loop region between I and IIA is larger than in original model, with a 6-basepair "bridge" across the central junction separating stem loops IV and V from the rest of the structure. Since nucleotides contributing to these alternative arrangements are base-paired at the

137

proximal terminus of stem I in the original structure, the stem I motif is four base pairs shorter in the newer model.

The modified 5-stem RRE structure is arguably the most compatible with the jellyfish model of Rev assembly, as single-stranded, purine rich, Rev binding sites at the IIB/stem II junction, central junction and proximal stem I internal loop are more evenly spaced than in other secondary structural models. However, the authors of this study also provide evidence for a heretofore unrecognized Rev binding region among the purine-rich internal loops located near the center of stem I, and propose a novel mechanism of Rev assembly that differs somewhat from the jellyfish model to explain binding at that site [30]. This model of assembly will be discussed in more detail in the following section.

While varying RRE sequence and secondary structure has been shown to affect RRE function, the nature and degree of these effects are not entirely predictable. For example, whereas large engineered changes such as deletion of stem loop II almost completely abolishes Rev binding in *in vitro* assays [14], disruption of stem loops III and IV (or III/IV) does not. The latter changes do, however, substantially impair Rev/RRE-mediated nuclear export function in cell culture [46]. Nuclear export of HIV-1 RNA is similarly reduced upon serial truncation of RRE stem I, although this process is gradual, and considerable function is retained in RRE variants as small as ~230 nt [32].

The effects of natural RRE sequence and structural variations in the context of viral replication can be more difficult to interpret. Since the RRE is embedded in the *env* gene, changes in RRE sequence may affect the functionality of Env as well as the nuclear export of HIV-1 RNA. Another consequence of this dual functionality may be that genetic flexibility is limited, making the RRE sequence a potentially promising target for antiviral therapies. Despite these seeming limitations, RRE sequences derived from clinical samples do exhibit a degree of genetic variation, and the effects of these sequence differences are not always easy to explain. For example, in one study, the function of patient-derived RRE variants appeared to be more affected by select single nucleotide polymorphisms than by more pronounced changes predicted to substantially alter RRE secondary structure [48]. In other work, select polymorphic RREs obtained from clinical isolates were shown to decrease Rev-dependent nuclear export 2–3-fold, although a corresponding effect on RRE structure was not established [49]. Although it is not clear how specific RRE mutations correlate with RRE structure and function, it has been suggested that mutational attenuation of RRE function could potentially serve as a natural means of down-regulating HIV-1 replication during the course of infection [46].

Thus far, molecular modeling and small angle X-ray scattering (SAXS) have been the only means of generating 3D models of the RRE. One option for the former approach is to use RNA Composer, a web-based RNA folding application that uses homology modeling and energy minimization to assemble a 3D RNA structure from primary sequence and an associated secondary structure map [50]. This software was used to generate the 3D model of the HIV-2 RRE depicted in Figure 5A [47]. While less is known about HIV-2 Rev assembly on its cognate RRE, alignment of the homologs of IIB, the central junction and stem loop I in this model suggest that the RRE 3D structures and mechanisms of Rev assembly may be similar between HIV-2 and HIV-1.

A combination of SAXS and molecular modeling was used to generate a 3D model of the HIV-1 RRE [51]. Using the 4-stem secondary structure map as a model, individual sub-structures were constructed (e.g., stem loop III/IV) and analyzed by SAXS. The molecular boundaries of the sub-structures were then mapped onto the global envelope of a truncated (233 nt) HIV-1 RRE, which was itself used to constrain and define the 3D organization of the RNA predicted by molecular modeling (Figure 5B). In the resulting structure, the HIV-1 RRE assumes an A-like configuration in which stem IIA and stem-loop III/IV are collinear, stem I is collinear with stem-loop V and the two extended motifs are centrally linked by a rigid, elongated 4-way junction. Moreover, whereas the loop regions of III/IV and V are proximally located, and perhaps contacting each other at the apex of the A-like configuration, IIB and the putative high affinity Rev binding site in stem I are opposite each other and separated by $\sim$55 Å. Because this distance approximates the separation between the ARMs of the Rev dimer in the absence of RNA [39], it was suggested that dimeric Rev spans from IIB to the high affinity site in I in the early stages of assembly [51]. In this model, multimerization then proceeds outward from the initiating dimer in both directions along the multimerization axis on one face of the A-like RRE RNA.

Subsequent SAXS analysis of the full length RRE extends the preceding model to include all of stem I [30]. These data suggest that an intact stem I folds back on itself so that the distal portion of the motif interacts with the proximal portion, and perhaps also with regions near the central junction. Interestingly, although A-like RRE SAXS envelopes are reported in both studies, probably in part because the former model was used to derive the latter, the 4-stem secondary structure used to model the former structure was not used in the more recent study. Instead, SHAPE analysis [52,53] determined that the RRE variant used in the latter work assumed the modified 5-stem structure with base-pairing across the central junction. While this secondary structural variation was used for subsequent SHAPE-based mapping of Rev binding sites on the RRE, the modified 5-stem RRE was not modeled into the A-like SAXS envelope in this study.

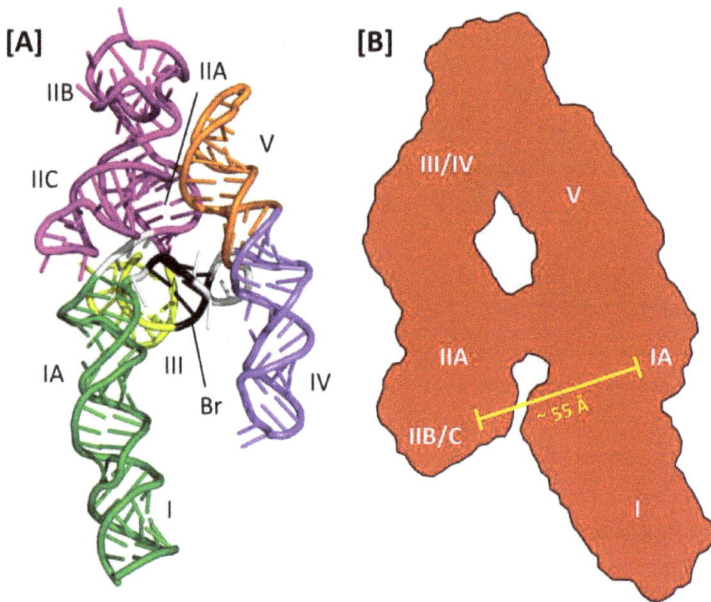

Figure 5. Three-dimensional models of the HIV-2 and HIV-1 RREs: (**A**) HIV-2 RRE model structure obtained using SHAPE and molecular modeling with RNA Composer. Substructures homologous to those reported for the HIV-1 RRE are indicated. The region of base pairing that bridges the gap between stem loops IV and V and the rest of the structure is also shown (Br, black ribbon); (**B**) A-like SAXS envelope obtained for a truncated HIV-1 RRE (233 nt). Sub-structure designations and positioning are determined by fitting an RRE molecular model into the SAXS envelope. The molecular model was generated using the RRE 4-stem secondary structure as a template. High-affinity Rev binding sites at IIB and IA are separated by approximately 55 Å.

5. Rev Assembly on the RRE

It is well established that multiple copies of HIV-1 Rev bind the RRE, and when this capacity is mutationally abolished, Rev-mediated nuclear export is adversely affected [32,42,54,55]. The stoichiometry of the saturated Rev-RRE complex, however, remains a subject of some debate. Reported binding ratios range from of 5:1 to 13:1 Rev/RRE, with an 8:1 ratio most often represented in early work in this area [11,32,33,39,56–63]. A study of particular note utilized surface plasmon resonance and a 244-nt RRE to measure Rev binding kinetics, with results indicating that up to 10 Rev molecules could bind the RRE before the complex became saturated [63]. Moreover, while the kinetics of the first four binding events suggest that Rev binding is sequential and specific, the subsequent Rev binding appeared to occur non-specifically. Most recently, a Rev/RRE ratio of 6:1 was determined for a

242-nt RRE using size exclusion chromatography [39]. The hexameric-Rev-RRE in this study was also shown to migrate as a discrete species by native polyacrylamide gel electrophoresis, suggesting structural uniformity, while complexes containing the Rev Leu18Gln/Leu60Arg multimerization mutant migrated as a broad, diffuse band to a position consistent with a Rev/RRE ratio of approximately 3–4:1.

Biochemical and biophysical experiments, together with single complex FRET measurements, demonstrate that Rev assembles on the RRE one molecule at time [2,64], binding initiates at stem-loop IIB [11,12,14,15,31] and the binding of successive Rev molecules is cooperative [29]. Nuclease protection analysis further indicates that Rev binding occurs principally on stem loop IIB and stem I of the RRE, as other structural motifs remained susceptible to nuclease cleavage in the presence of Rev [32]. Although a great deal is now known about how Rev interacts with itself and with RNA, the precise sequence of events required for Rev assembly, the positioning of individual Rev molecules on the RNA and the 3D structure of the saturated Rev-RRE complex remain unknown.

From the 3D structure of a 232-nt RRE obtained using molecular modeling in conjunction with SAXS, a model for Rev assembly was proposed in which 8 Rev molecules are coaxially arranged along one face of the A-like structure of the truncated RRE (Figure 6A) [51]. Support from this proposal comes primarily from observations that the high affinity Rev binding sites on IIB and stem-loop I are separated by ~55 Å, matching the approximate separation between distal portions of the ARMs in the T/T Rev dimer crystal structure [39]. However, the model also suggests direct binding of Rev to stem loops III/IV and V in higher order complexes and does not involve distal regions of stem I in Rev assembly. Both of these suppositions are inconsistent with prior nuclease protection analysis [32]. The high resolution Rev-dimer-RNA co-crystal structure [27], in which tandem NLS/RBSs bind at adjacent sites on the same helix likewise does not support the IIB-I bridging model of Rev assembly suggested by Wang and colleagues.

As noted previously, SAXS analysis of the full-length RRE suggests that the distal segment of stem I folds back on itself over one face of the A-like RRE structure [30]. Probing experiments further show highly variable sensitivity to chemical acylation at select regions of the RRE in the presence and absence of Rev. These area are designated Regions 1–3, and correspond, respectively, to (i) IIB and the stem loop II junction; (ii) the central junction and proximal purine rich internal loop of stem I; and (iii) a sequence of three adjacent, purine rich bulges located near the center of stem I. These regions of variable acylation sensitivity reportedly mark Rev binding sites, with Region 1 containing sites included in the Rev-dimer-RNA crystal structure [27].

Based on these data, a model for Rev assembly was presented in which Rev dimers bind sequentially at Regions 1–3 (Figure 6B). As in both previous models, assembly is proposed to initiate at the IIB nucleation site of Region 1. Moreover, as would be predicted by the jellyfish model, Rev binding at Regions 1 and 2 is temporally coupled, likely due to their spatial proximity. Unlike in the jellyfish model, however, this intriguing new model suggests that the 4-Rev complex subsequently undergoes an induced-fit conformational change to accommodate a third Rev dimer binding at Region 3, which is not adjacent to Regions 1 and 2 in the RNA secondary structure but is brought into proximity by tertiary RNA interactions. Specific Rev-Rev associations within the hexameric complex are not otherwise detailed in this model, and the relative positioning and orientation of individual Rev molecules in the fully assembled complex is likewise not predicted. Moreover, it is worth noting that in seeming contradiction to this model, a hexameric Rev-RRE complex has been reported to assemble on a 242-nt truncated RRE that lacks almost all of the Domain 3 Rev binding site(s) [29].

In the jellyfish model, Rev assembly extends from the IIB nucleation site through the RRE central junction and into the proximal segment of stem I, including stem IA (Figure 6C). Size exclusion chromatography suggests that coordinated assembly stops after six Rev molecules have been bound in the complex [29]. Rev-Rev and Rev-RNA interactions in such a complex would be consistent with the recent Rev-dimer RNA crystal structure [27], but a stable Rev hexamer would likely require similarly flexible H/H interface. More specifically, it would be necessary that the angles formed by the ARMs in adjacent Rev molecules are relatively narrow at both T/T and H/H interfaces, allowing for sequential binding at widened major grooves in adjacent segments of a contiguous RNA helix. Such an arrangement would be consistent with nuclease protection studies that suggest that Rev binds at IIB and along stem I [32], and also predict that the Rev effector domains align along the same face of the RRE.

Some RRE mutations have been shown to alter both RNA secondary structure and the capacity of the RRE to mediate nuclear export of HIV-1 RNA, yet do not appear to affect the binding kinetics or stoichiometry of Rev binding [2,46]. Such observations may be explained by aberrant Rev-RRE complexes that assemble with normal kinetics but where the relative orientations of Rev molecules are affected. Such subtle distinctions among complexes would be invisible in most biochemical assays, yet may substantially affect Rev-RRE function in a cellular context. In a recent study, the structure of a nuclear export complex containing Rev, the RRE and a Crm1 dimer was elucidated using single particle electron microscopy [65]. In this elegant work, the authors demonstrate that the Rev-RRE complex binds at the Crm1 dimer interface. It is suggested that each Crm1 subunit is bound by three NES domains of the hexameric Rev, although there is sufficient diversity among the complexes to allow for other binding stoichiometries. Crm1 binding appears to require a functional NES, as Rev bearing a dominant-negative M10 mutation [66] in the NES did not bind Crm1 under any condition. Although the resolution of these images is insufficient to definitively verify the jellyfish arrangement of Rev on the

RRE, the observed interactions are consistent with a hexameric Rev-RRE in which the NES domains are concentrated in a small area and oriented in a common direction.

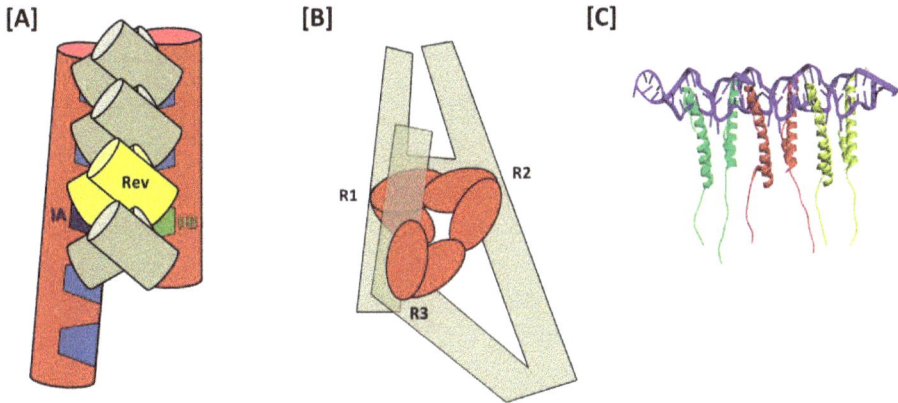

Figure 6. Models for assembly of HIV-1 Rev on the RRE: (**A**) "Bridging" model. The first two Rev in the complex bind to IIB and IA high affinity sites on the A-like RRE, as well as to each other, thereby forming a Rev dimer that "bridges" the distance separating two sub-structures. This proposal is based largely upon the observations that the distances separating the apices of the Rev arms in the T/T dimer crystal structure and the IIB and IA motifs modeled to fit the HIV-1 RRE SAXS envelope are both approximately 55 Å. After this initial Rev dimer-RRE complex is formed, assembly is proposed to propagate in both directions along the A-like RRE structure and involve stem loops III/IV, V and more distal portions of stem I; (**B**) Stem I "loop-back" model. Based on analyses using SAXS and time resolved SHAPE, it has been proposed that following relatively rapid assembly of Rev dimers at R1 and R2 (R1, Region 1—IIB and stem II junction; R2, Region 2—central junction and IA), the remaining dimer in the hexameric Rev complex assembles at R3 (Region 3—a series of purine-rich bulges in a more distant segment of stem I). RNA tertiary interactions bring the stem I terminus into proximity with the RRE central junction (even in the absence of Rev), and the last Rev dimer is accommodated in the fully assembled complex by an induced fit mechanism of conformational sampling; (**C**) Jellyfish model. Rev initially binds to the IIB high-affinity site, after which five additional Rev molecules assemble on the RRE *via* a series of consecutive T/T and H/H multimerization domain interactions. These Rev-Rev interactions are facilitated by concomitant Rev ARM binding at adjacent regions on the RRE RNA, such as is observed in the T/T Rev dimer-RNA crystal structure. The six C-terminal domains in this putative Rev-RRE complex would be expected to project in a common direction, which would in turn facilitate NES binding at the Crm1 dimer interface and promote nuclear export.

6. Therapeutic Targeting of the HIV-1 RRE

Based on its crucial role in HIV replication, interrupting RRE function clearly offers an attractive avenue for therapeutic intervention. The notion of using a transdominant negative version of the HIV-1 Rev protein harboring mutations in its nuclear export signal (RevM10), while disrupting export of Rev-dependent viral transcripts [67], was later shown to promote acquisition of resistance-conferring mutations [2]. Early attempts to target the HIV-1 RRE with small molecules have demonstrated the usefulness of neomycin B and related aminoglycosides as structural probes [68,69]. However, the therapeutic potential of this class of antagonist faces the challenges of specificity and poor cellular uptake [70]. Thus, alternative approaches to target the Rev/RRE interaction are warranted.

Both structural and mutagenic analyses have shown that the α-helical arginine-rich motif (ARM) of Rev mediates a critical interaction with the RRE by inserting into the major groove of an asymmetric bulge in IIB. Based on these observations, Mills *et al.* [71] synthesized a series of ARM-like conformationally-constrained (*i.e.*, α-helical) peptidomimetics to antagonize the Rev/RRE interaction. Of 15 candidate peptidomimetics examined, one bound the RRE with an affinity equivalent to that of the Rev ARM (Kd ~50 nM), while control experiments indicated that its unconstrained counterpart showed little specificity. An alternative approach has suggested targeting the RRE with an ARM peptide modified to incorporate reactive metal chelates [72,73]. This strategy envisions "catalytic" metallo-inhibitors, which, following destruction of their target biomolecule, would dissociate and circulate through the viral RNA population, and thus would not be required in saturating amounts to achieve maximum potency. Bifunctional metallo-inhibitors linking a high affinity targeting motif (the Rev ARM) with a metal chelate complex capable of damaging RNA in its vicinity *via* a variety of oxidative chemistries have been synthesized, and proof of concept was demonstrated by the ability of Cu^{2+}-Gly-Gly-His-ARM complexes to selectively cleave RRE RNA *in vitro* [74] and *in vivo* [72]. An extension of this strategy has investigated the oxidative properties of Rev ARM peptides linked to metal chelators such as tetraazacyclododecane-tetraacetic acid (DOTA), diethylenetriaminepentaacetic acid (DTPA), ethylenediaminetetraacetic acid (EDTA) and nitrilotriacetic acid (NTA) [75]. All complexes retained high affinity binding to RRE IIB (0.2–16 nM) and their Cu^{2+}-bound chelates efficiently induced RRE cleavage, with activity varying in the order Cu-NTA-Rev > Cu-DOTA-Rev > Cu-DTPA-Rev > Cu-EDTA-Rev. From a therapeutic perspective, oxidative damage of the RRE would result in dissociation of the metal chelate-ARM complex, which can be "reactivated" by reducing agents such as ascorbic acid or glutathione, whose concentrations are sufficiently high *in vivo*. Also, since the metal is extremely tightly chelated (Kd $\sim 10^{-15}$ M), the risk of toxicity due to metal ion leaching from the complex would be negligible.

Following initial reports that promoted multiple antigenic peptides for vaccine design [76], branched peptides comprising natural or unnatural amino acids, or combinations thereof, have been gaining attention as therapeutic agents, based on their potential for multivalent targeting of RNA and their resistance to proteolysis [77]. In screening a branched peptide boronic acid (BPBA) library whose boronic acid substituent was designed to mimic an acceptor for RNA 2′ OH groups and improve selectivity for RNA over DNA, Zhang *et al.* [78] identified BPBA1, a compound that bound RRE IIB with micromolar affinity and a 1:1 stoichiometry. Mutational studies supported the notion that the IIB tertiary structure was necessary for high affinity binding of BPBA1, while enzymatic footprinting highlighted several nuclease-insensitive regions in the presence of the branched peptide. Finally, studies with a fluorescent BPBA1 derivative suggest cellular uptake can be achieved, although suppression of HIV-1 replication remains to be established.

Of the currently available approaches, small molecules have superior delivery properties and can be "fine-tuned" through medicinal chemistry. With respect to the Rev/RRE interaction, several recent studies support their further development. By combining NMR spectroscopy and computational molecular dynamics, Steizer *et al.* have taken advantage of virtual screening to identify several compounds that antagonize the Tat/TAR interaction by interacting with nucleotides of both the apical loop and trinucleotide bulge [79]. Our laboratory has exploited small molecule microarrays (SMMs) with fluorescently-labeled, structured RNA motifs, where we identified a novel chemotype that binds the TAR hairpin with micromolar affinity, inhibits virus replication in culture and is not cytotoxic [80]. Finally, Informa, a computational approach to designing lead small molecules targeting RNA motifs based on sequence alone, has been used to identify a bioactive benzimidazole that induces apoptosis in cancer cells by sequestering the nuclease processing site of pre-micro RNA-96 [81]. Extending this approach to target the RRE would seem a logical next step.

Although this section has concentrated on small molecule inhibition of the Rev/RRE axis by targeting the viral *cis*-acting RNA, we should not rule out the possibility of targeting the HIV Rev protein to antagonize its interaction with host factors, thereby interrupting nucleocytoplasmic transport of unspliced and singly-spliced viral RNAs. In this respect, Campos *et al.* [82] have recently made the exciting discovery that the quinolin-2-amine ABX464 specifically targets Rev-dependent nuclear transport of HIV RNA without compromising cellular function. The lack of toxicity displayed by ABX464, combined with its ability to suppress viral load sustainably after treatment arrest, represent another powerful addition to the armament of HIV antivirals that will be needed until either a cross-clade vaccine or functional cure is achieved.

7. Summary and Perspective

The Rev-RRE interaction is vital to HIV replication and therefore constitutes an important axis for antiviral therapy. Although recent SAXS and crystallographic studies have greatly enhanced our understanding of these two viral components on a structural level, much remains to be learned—particularly regarding the details of Rev assembly and the overall structure of the nuclear export complex. Biochemical studies tell us that six Rev molecules cooperatively assemble on each RRE, and the resulting complex binds Crm1 at the dimer interface according to single particle electron microscopy. The latter study also suggests that while they share many common features, the structures of Rev_6-RRE-$Crm1_2$ complexes may not be completely uniform. Other factors, including the observed flexibility in the Rev T/T interface, potential flexibility in the H/H interface, and variability in RRE RNA sequence and secondary structure suggest this as well. It is conceivable, therefore, that select components of the "bridging", "loop-back" and jellyfish models of Rev assembly may all contribute to formation of the nuclear export complex in the context of the HIV-infected cell.

Acknowledgments: Stuart F. J. Le Grice and Jason W. Rausch were supported by the Intramural Research Program of the National Cancer Institute, National Institutes of Health, Department of Health and Human Services, USA.

Author Contributions: Jason W. Rausch and Stuart F. J. Le Grice co-wrote the manuscript.

Conflicts of Interest: The authors declare no conflict of interest.

References

1. Bray, M.; Prasad, S.; Dubay, J.W.; Hunter, E.; Jeang, K.T.; Rekosh, D.; Hammarskjold, M.L. A small element from the mason-pfizer monkey virus genome makes human immunodeficiency virus type 1 expression and replication Rev-independent. *Proc. Natl. Acad. Sci. USA* **1994**, *91*, 1256–1260.
2. Legiewicz, M.; Badorrek, C.S.; Turner, K.B.; Fabris, D.; Hamm, T.E.; Rekosh, D.; Hammarskjold, M.L.; le Grice, S.F. Resistance to RevM10 inhibition reflects a conformational switch in the HIV-1 Rev response element. *Proc. Natl. Acad. Sci. USA* **2008**, *105*, 14365–14370.
3. Pilkington, G.R.; Purzycka, K.J.; Bear, J.; Le Grice, S.F.; Felber, B.K. Gammaretrovirus mRNA expression is mediated by a novel, bipartite post-transcriptional regulatory element. *Nucl. Acids Res.* **2014**, *42*, 11092–11106.
4. Lindtner, S.; Felber, B.K.; Kjems, J. An element in the 3' untranslated region of human LINE-1 retrotransposon mRNA binds NXF1(TAP) and can function as a nuclear export element. *RNA* **2002**, *8*, 345–356.
5. Malim, M.H.; Hauber, J.; Le, S.Y.; Maizel, J.V.; Cullen, B.R. The HIV-1 Rev trans-activator acts through a structured target sequence to activate nuclear export of unspliced viral mRNA. *Nature* **1989**, *338*, 254–257.

6. Cullen, B.R. HIV-1 auxiliary proteins: Making connections in a dying cell. *Cell* **1998**, *93*, 685–692.

7. Hope, T.J. The ins and outs of HIV Rev. *Arch. Biochem. Biophys.* **1999**, *365*, 186–191.

8. Pollard, V.W.; Malim, M.H. The HIV-1 Rev protein. *Annu. Rev. Microbiol.* **1998**, *52*, 491–532.

9. Vercruysse, T.; Daelemans, D. HIV-1 Rev multimerization: Mechanism and insights. *Curr. HIV Res.* **2013**, *11*, 623–634.

10. Fernandes, J.; Jayaraman, B.; Frankel, A. The HIV-1 Rev response element: An RNA scaffold that directs the cooperative assembly of a homo-oligomeric ribonucleoprotein complex. *RNA Biol.* **2012**, *9*, 6–11.

11. Heaphy, S.; Finch, J.T.; Gait, M.J.; Karn, J.; Singh, M. Human immunodeficiency virus type 1 regulator of virion expression, Rev, forms nucleoprotein filaments after binding to a purine-rich "bubble" located within the Rev-responsive region of viral mrnas. *Proc. Natl. Acad. Sci. USA* **1991**, *88*, 7366–7370.

12. Iwai, S.; Pritchard, C.; Mann, D.A.; Karn, J.; Gait, M.J. Recognition of the high affinity binding site in Rev-response element RNA by the human immunodeficiency virus type-1 Rev protein. *Nucl. Acids Res.* **1992**, *20*, 6465–6472.

13. Malim, M.H.; Tiley, L.S.; McCarn, D.F.; Rusche, J.R.; Hauber, J.; Cullen, B.R. HIV-1 structural gene expression requires binding of the Rev trans-activator to its RNA target sequence. *Cell* **1990**, *60*, 675–683.

14. Olsen, H.S.; Nelbock, P.; Cochrane, A.W.; Rosen, C.A. Secondary structure is the major determinant for interaction of HIV Rev protein with RNA. *Science* **1990**, *247*, 845–848.

15. Tiley, L.S.; Malim, M.H.; Tewary, H.K.; Stockley, P.G.; Cullen, B.R. Identification of a high-affinity RNA-binding site for the human immunodeficiency virus type 1 Rev protein. *Proc. Natl. Acad. Sci. USA* **1992**, *89*, 758–762.

16. Battiste, J.L.; Mao, H.; Rao, N.S.; Tan, R.; Muhandiram, D.R.; Kay, L.E.; Frankel, A.D.; Williamson, J.R. A helix-RNA major groove recognition in an HIV-1 Rev peptide-RRE RNA complex. *Science* **1996**, *273*, 1547–1551.

17. Bartel, D.P.; Zapp, M.L.; Green, M.R.; Szostak, J.W. HIV-1 Rev regulation involves recognition of non-watson-crick base pairs in viral RNA. *Cell* **1991**, *67*, 529–536.

18. Battiste, J.L.; Tan, R.; Frankel, A.D.; Williamson, J.R. Binding of an HIV Rev peptide to Rev responsive element RNA induces formation of purine-purine base pairs. *Biochemistry* **1994**, *33*, 2741–2747.

19. Peterson, R.D.; Bartel, D.P.; Szostak, J.W.; Horvath, S.J.; Feigon, J. 1 h NMR studies of the high-affinity Rev binding site of the Rev responsive element of HIV-1 mRNA: Base pairing in the core binding element. *Biochemistry* **1994**, *33*, 5357–5366.

20. Williamson, J.R.; Battiste, J.L.; Mao, H.; Frankel, A.D. Interaction of HIV Rev peptides with the Rev response element RNA. *Nucl. Acids Symp. Ser.* **1995**, 46–48.

21. Tan, R.; Chen, L.; Buettner, J.A.; Hudson, D.; Frankel, A.D. RNA recognition by an isolated α helix. *Cell* **1993**, *73*, 1031–1040.

22. Tan, R.; Frankel, A.D. Costabilization of peptide and RNA structure in an HIV Rev peptide-Rre complex. *Biochemistry* **1994**, *33*, 14579–14585.

23. Ye, X.; Gorin, A.; Ellington, A.D.; Patel, D.J. Deep penetration of an α-helix into a widened RNA major groove in the HIV-1 Rev peptide-rna aptamer complex. *Nat. Struct. Biol.* **1996**, *3*, 1026–1033.

24. Hung, L.W.; Holbrook, E.L.; Holbrook, S.R. The crystal structure of the Rev binding element of HIV-1 reveals novel base pairing and conformational variability. *Proc. Natl. Acad. Sci. USA* **2000**, *97*, 5107–5112.

25. Ippolito, J.A.; Steitz, T.A. The structure of the HIV-1 rre high affinity rev binding site at 1.6 a resolution. *J. Mol. Biol.* **2000**, *295*, 711–717.

26. Peterson, R.D.; Feigon, J. Structural change in Rev responsive element RNA of HIV-1 on binding Rev peptide. *J. Mol. Biol.* **1996**, *264*, 863–877.

27. Jayaraman, B.; Crosby, D.C.; Homer, C.; Ribeiro, I.; Mavor, D.; Frankel, A.D. RNA-directed remodeling of the HIV-1 protein rev orchestrates assembly of the Rev-Rev response element complex. *eLife* **2014**, *3*, e04120.

28. Zemmel, R.W.; Kelley, A.C.; Karn, J.; Butler, P.J. Flexible regions of RNA structure facilitate co-operative Rev assembly on the Rev-response element. *J. Mol. Biol.* **1996**, *258*, 763–777.

29. Daugherty, M.D.; D'Orso, I.; Frankel, A.D. A solution to limited genomic capacity: Using adaptable binding surfaces to assemble the functional HIV Rev oligomer on RNA. *Mol. Cell* **2008**, *31*, 824–834.

30. Bai, Y.; Tambe, A.; Zhou, K.; Doudna, J.A. RNA-guided assembly of Rev-RRE nuclear export complexes. *eLife* **2014**, *3*, e03656.

31. Malim, M.H.; Cullen, B.R. HIV-1 structural gene expression requires the binding of multiple Rev monomers to the viral RRE: Implications for HIV-1 latency. *Cell* **1991**, *65*, 241–248.

32. Mann, D.A.; Mikaelian, I.; Zemmel, R.W.; Green, S.M.; Lowe, A.D.; Kimura, T.; Singh, M.; Butler, P.J.; Gait, M.J.; Karn, J. A molecular rheostat. Co-operative rev binding to stem I of the Rev-response element modulates human immunodeficiency virus type-1 late gene expression. *J. Mol. Biol.* **1994**, *241*, 193–207.

33. Wingfield, P.T.; Stahl, S.J.; Payton, M.A.; Venkatesan, S.; Misra, M.; Steven, A.C. HIV-1 REV expressed in recombinant escherichia coli: Purification, polymerization, and conformational properties. *Biochemistry* **1991**, *30*, 7527–7534.

34. Zapp, M.L.; Green, M.R. Sequence-specific rna binding by the HIV-1 Rev protein. *Nature* **1989**, *342*, 714–716.

35. Havlin, R.H.; Blanco, F.J.; Tycko, R. Constraints on protein structure in HIV-1 Rev and Rev-RNA supramolecular assemblies from two-dimensional solid state nuclear magnetic resonance. *Biochemistry* **2007**, *46*, 3586–3593.

36. Watts, N.R.; Misra, M.; Wingfield, P.T.; Stahl, S.J.; Cheng, N.; Trus, B.L.; Steven, A.C.; Williams, R.W. Three-dimensional structure of HIV-1 Rev protein filaments. *J. Struct. Biol.* **1998**, *121*, 41–52.

37. Jain, C.; Belasco, J.G. Structural model for the cooperative assembly of HIV-1 Rev multimers on the RRE as deduced from analysis of assembly-defective mutants. *Mol. Cell* **2001**, *7*, 603–614.

38. DiMattia, M.A.; Watts, N.R.; Stahl, S.J.; Rader, C.; Wingfield, P.T.; Stuart, D.I.; Steven, A.C.; Grimes, J.M. Implications of the HIV-1 Rev dimer structure at 3.2 a resolution for multimeric binding to the rev response element. *Proc. Natl. Acad. Sci. USA* **2010**, *107*, 5810–5814.

39. Daugherty, M.D.; Liu, B.; Frankel, A.D. Structural basis for cooperative rna binding and export complex assembly by HIV Rev. *Nat. Struct. Mol. Biol.* **2010**, *17*, 1337–1342.

40. Edgcomb, S.P.; Aschrafi, A.; Kompfner, E.; Williamson, J.R.; Gerace, L.; Hennig, M. Protein structure and oligomerization are important for the formation of export-competent HIV-1 Rev-RRE complexes. *Protein Sci.* **2008**, *17*, 420–430.

41. Kjems, J.; Brown, M.; Chang, D.D.; Sharp, P.A. Structural analysis of the interaction between the human immunodeficiency virus Rev protein and the Rev response element. *Proc. Natl. Acad. Sci. USA* **1991**, *88*, 683–687.

42. Huang, X.J.; Hope, T.J.; Bond, B.L.; McDonald, D.; Grahl, K.; Parslow, T.G. Minimal Rev-response element for type 1 human immunodeficiency virus. *J. Virol.* **1991**, *65*, 2131–2134.

43. Pollom, E.; Dang, K.K.; Potter, E.L.; Gorelick, R.J.; Burch, C.L.; Weeks, K.M.; Swanstrom, R. Comparison of SIV and HIV-1 genomic RNA structures reveals impact of sequence evolution on conserved and non-conserved structural motifs. *PLoS Pathogens* **2013**, *9*, e1003294.

44. Watts, J.M.; Dang, K.K.; Gorelick, R.J.; Leonard, C.W.; Bess, J.W., Jr.; Swanstrom, R.; Burch, C.L.; Weeks, K.M. Architecture and secondary structure of an entire HIV-1 RNA genome. *Nat.* **2009**, *460*, 711–716.

45. Kenyon, J.C.; Prestwood, L.J.; Le Grice, S.F.; Lever, A.M. In-gel probing of individual RNA conformers within a mixed population reveals a dimerization structural switch in the HIV-1 leader. *Nucl. Acids Res.* **2013**, *41*, e174.

46. Sherpa, C.; Rausch, J.W.; SF, J.L.G.; Hammarskjold, M.L.; Rekosh, D. The HIV-1 Rev response element (RRE) adopts alternative conformations that promote different rates of virus replication. *Nucl. Acids Res.* **2015**, *43*, 4676–4686.

47. Lusvarghi, S.; Sztuba-Solinska, J.; Purzycka, K.J.; Pauly, G.T.; Rausch, J.W.; Grice, S.F. The HIV-2 Rev-response element: Determining secondary structure and defining folding intermediates. *Nucl. Acids Res.* **2013**, *41*, 6637–6649.

48. Cunyat, F.; Beerens, N.; Garcia, E.; Clotet, B.; Kjems, J.; Cabrera, C. Functional analyses reveal extensive RRE plasticity in primary HIV-1 sequences selected under selective pressure. *PLoS ONE* **2014**, *9*, e106299.

49. Sloan, E.A.; Kearney, M.F.; Gray, L.R.; Anastos, K.; Daar, E.S.; Margolick, J.; Maldarelli, F.; Hammarskjold, M.L.; Rekosh, D. Limited nucleotide changes in the rev response element (RRE) during HIV-1 infection alter overall Rev-RRE activity and Rev multimerization. *J. Virol.* **2013**, *87*, 11173–11186.

50. Popenda, M.; Szachniuk, M.; Antczak, M.; Purzycka, K.J.; Lukasiak, P.; Bartol, N.; Blazewicz, J.; Adamiak, R.W. Automated 3D structure composition for large RNAs. *Nucl. Acids Res.* **2012**, *40*, e112.

51. Fang, X.; Wang, J.; O'Carroll, I.P.; Mitchell, M.; Zuo, X.; Wang, Y.; Yu, P.; Liu, Y.; Rausch, J.W.; Dyba, M.A.; *et al*. An unusual topological structure of the HIV-1 Rev response element. *Cell* **2013**, *155*, 594–605.

52. Lucks, J.B.; Mortimer, S.A.; Trapnell, C.; Luo, S.; Aviran, S.; Schroth, G.P.; Pachter, L.; Doudna, J.A.; Arkin, A.P. Multiplexed rna structure characterization with selective 2'-hydroxyl acylation analyzed by primer extension sequencing (shape-seq). *Proc. Natl. Acad. Sci. USA* **2011**, *108*, 11063–11068.

53. Merino, E.J.; Wilkinson, K.A.; Coughlan, J.L.; Weeks, K.M. Rna structure analysis at single nucleotide resolution by selective 2'-hydroxyl acylation and primer extension (shape). *J. Am. Chem. Soc.* **2005**, *127*, 4223–4231.

54. Holland, S.M.; Chavez, M.; Gerstberger, S.; Venkatesan, S. A specific sequence with a bulged guanosine residue(s) in a stem-bulge-stem structure of Rev-responsive element RNA is required for trans activation by human immunodeficiency virus type 1 Rev. *J. Virol.* **1992**, *66*, 3699–3706.

55. Kjems, J.; Sharp, P.A. The basic domain of rev from human immunodeficiency virus type 1 specifically blocks the entry of u4/u6.U5 small nuclear ribonucleoprotein in spliceosome assembly. *J. Virol.* **1993**, *67*, 4769–4776.

56. Brice, P.C.; Kelley, A.C.; Butler, P.J. Sensitive *in vitro* analysis of HIV-1 Rev multimerization. *Nucl. Acids Res.* **1999**, *27*, 2080–2085.

57. Cook, K.S.; Fisk, G.J.; Hauber, J.; Usman, N.; Daly, T.J.; Rusche, J.R. Characterization of HIV-1 Rev protein: Binding stoichiometry and minimal rna substrate. *Nucl. Acids Res.* **1991**, *19*, 1577–1583.

58. Daly, T.J.; Cook, K.S.; Gray, G.S.; Maione, T.E.; Rusche, J.R. Specific binding of HIV-1 recombinant Rev protein to the rev-responsive element *in vitro*. *Nature* **1989**, *342*, 816–819.

59. Daly, T.J.; Doten, R.C.; Rennert, P.; Auer, M.; Jaksche, H.; Donner, A.; Fisk, G.; Rusche, J.R. Biochemical characterization of binding of multiple HIV-1 Rev monomeric proteins to the Rev responsive element. *Biochemistry* **1993**, *32*, 10497–10505.

60. Holland, S.M.; Ahmad, N.; Maitra, R.K.; Wingfield, P.; Venkatesan, S. Human immunodeficiency virus Rev protein recognizes a target sequence in Rev-responsive element RNA within the context of rna secondary structure. *J. Virol.* **1990**, *64*, 5966–5975.

61. Jeong, K.S.; Nam, Y.S.; Venkatesan, S. Deletions near the *N*-terminus of HIV-1 Rev reduce RNA binding affinity and dominantly interfere with Rev function irrespective of the RNA target. *Arch. Virol.* **2000**, *145*, 2443–2467.

62. Pallesen, J.; Dong, M.; Besenbacher, F.; Kjems, J. Structure of the HIV-1 Rev response element alone and in complex with regulator of virion (Rev) studied by atomic force microscopy. *FEBS J.* **2009**, *276*, 4223–4232.

63. Van Ryk, D.I.; Venkatesan, S. Real-time kinetics of HIV-1 Rev-Rev response element interactions. Definition of minimal binding sites on RNA and protein and stoichiometric analysis. *J. Biol. Chem.* **1999**, *274*, 17452–17463.

64. Pond, S.J.; Ridgeway, W.K.; Robertson, R.; Wang, J.; Millar, D.P. HIV-1 Rev protein assembles on viral RNA one molecule at a time. *Proc. Natl. Acad. Sci. USA* **2009**, *106*, 1404–1408.

65. Booth, D.S.; Cheng, Y.; Frankel, A.D. The export receptor crm1 forms a dimer to promote nuclear export of HIV RNA. *eLife* **2014**, *3*, e04121.

66. Malim, M.H.; Bohnlein, S.; Hauber, J.; Cullen, B.R. Functional dissection of the HIV-1 Rev trans-activator—Derivation of a trans-dominant repressor of rev function. *Cell* **1989**, *58*, 205–214.

67. Bevec, D.; Dobrovnik, M.; Hauber, J.; Bohnlein, E. Inhibition of human immunodeficiency virus type 1 replication in human T cells by retroviral-mediated gene transfer of a dominant-negative Rev trans-activator. *Proc. Natl. Acad. Sci. USA* **1992**, *89*, 9870–9874.

68. Chittapragada, M.; Roberts, S.; Ham, Y.W. Aminoglycosides: Molecular insights on the recognition of RNA and aminoglycoside mimics. *Perspect. Med. Chem.* **2009**, *3*, 21–37.

69. Tok, J.B.; Dunn, L.J.; Des Jean, R.C. Binding of dimeric aminoglycosides to the HIV-1 Rev responsive element (RRE) rna construct. *Bioorganic Med. Chem. Lett.* **2001**, *11*, 1127–1131.

70. Ahn, D.G.; Shim, S.B.; Moon, J.E.; Kim, J.H.; Kim, S.J.; Oh, J.W. Interference of hepatitis C virus replication in cell culture by antisense peptide nucleic acids targeting the X-RNA. *J. Viral Hepat.* **2011**, *18*, e298–306.

71. Mills, N.L.; Daugherty, M.D.; Frankel, A.D.; Guy, R.K. An α-helical peptidomimetic inhibitor of the HIV-1 Rev-RRE interaction. *J. Am. Chem. Soc.* **2006**, *128*, 3496–3497.

72. Jin, Y.; Cowan, J.A. Cellular activity of Rev response element rna targeting metallopeptides. *J. Biol. Inorganic Chem.* **2007**, *12*, 637–644.

73. Jin, Y.; Lewis, M.A.; Gokhale, N.H.; Long, E.C.; Cowan, J.A. Influence of stereochemistry and redox potentials on the single- and double-strand DNA cleavage efficiency of Cu(ii) and Ni(ii) Lys-Gly-His-derived atcun metallopeptides. *J. Am. Chem. Soc.* **2007**, *129*, 8353–8361.

74. Jin, Y.; Cowan, J.A. Targeted cleavage of HIV Rev response element rna by metallopeptide complexes. *J. Am. Chem. Soc.* **2006**, *128*, 410–411.

75. Joyner, J.C.; Cowan, J.A. Target-directed catalytic metallodrugs. *Braz. J. Med. Biol. Res.* **2013**, *46*, 465–485.

76. Tam, J.P. Synthetic peptide vaccine design: Synthesis and properties of a high-density multiple antigenic peptide system. *Proc. Natl. Acad. Sci. USA* **1988**, *85*, 5409–5413.

77. Bryson, D.I.; Zhang, W.; McLendon, P.M.; Reineke, T.M.; Santos, W.L. Toward targeting RNA structure: Branched peptides as cell-permeable ligands to tar RNA. *ACS Chem. Biol.* **2012**, *7*, 210–217.

78. Zhang, W.; Bryson, D.I.; Crumpton, J.B.; Wynn, J.; Santos, W.L. Targeting folded RNA: A branched peptide boronic acid that binds to a large surface area of HIV-1 Rre RNA. *Organic Biomol. Chem.* **2013**, *11*, 6263–6271.

79. Stelzer, A.C.; Frank, A.T.; Kratz, J.D.; Swanson, M.D.; Gonzalez-Hernandez, M.J.; Lee, J.; Andricioaei, I.; Markovitz, D.M.; Al-Hashimi, H.M. Discovery of selective bioactive small molecules by targeting an rna dynamic ensemble. *Nat. Chem. Biol.* **2011**, *7*, 553–559.

80. Sztuba-Solinska, J.; Shenoy, S.R.; Gareiss, P.; Krumpe, L.R.; Le Grice, S.F.; O'Keefe, B.R.; Schneekloth, J.S., Jr. Identification of biologically active, HIV tar RNA-binding small molecules using small molecule microarrays. *J. Am. Chem. Soc.* **2014**, *136*, 8402–8410.

81. Velagapudi, S.P.; Gallo, S.M.; Disney, M.D. Sequence-based design of bioactive small molecules that target precursor micrornas. *Nat. Chem. Biol.* **2014**, *10*, 291–297.

82. Campos, N.; Myburgh, R.; Garcel, A.; Vautrin, A.; Lapasset, L.; Nadal, E.S.; Mahuteau-Betzer, F.; Najman, R.; Fornarelli, P.; Tantale, K.; *et al.* Long lasting control of viral rebound with a new drug ABX464 targeting Rev—Mediated viral RNA biogenesis. *Retrovirology* **2015**, *12*.

Translational Control of the HIV Unspliced Genomic RNA

Bárbara Rojas-Araya, Théophile Ohlmann and Ricardo Soto-Rifo

Abstract: Post-transcriptional control in both HIV-1 and HIV-2 is a highly regulated process that commences in the nucleus of the host infected cell and finishes by the expression of viral proteins in the cytoplasm. Expression of the unspliced genomic RNA is particularly controlled at the level of RNA splicing, export, and translation. It appears increasingly obvious that all these steps are interconnected and they result in the building of a viral ribonucleoprotein complex (RNP) that must be efficiently translated in the cytosolic compartment. This review summarizes our knowledge about the genesis, localization, and expression of this viral RNP.

Reprinted from *Viruses*. Cite as: Rojas-Araya, B.; Ohlmann, T.; Soto-Rifo, R. Translational Control of the HIV Unspliced Genomic RNA. *Viruses* **2015**, *7*, 4326–4351.

1. Introduction

Human Immunodeficiency virus type-1 (HIV-1) and type-2 (HIV-2) belong to the Lentivirus genus of the *Retroviridae* family and are the etiological agents of the Acquired Immunodeficiency Syndrome (AIDS) in humans [1]. Both viruses primarily infect cells of the immune system that express the CD4 receptor and one of the chemokine receptors CCR5 or CXCR4 that act as co-receptors for viral entry. The HIV replication cycle begins with the interactions between the surface glycoprotein gp120 with CD4 and one of the co-receptors in a process that induces conformational changes allowing insertion of the viral transmembrane protein gp41 in the host cell membrane to trigger fusion of both membranes and entry of the viral capsid into the host cell cytoplasm. Then, the positive single stranded RNA genome is converted into double stranded DNA by the virally encoded reverse transcriptase, which is located in the capsid. In association with viral and cellular proteins, viral DNA forms the so-called pre-integration complex (PIC), which is imported to the host cell nucleus in an active process orchestrated by the viral proteins capsid and integrase [2]. The latter then catalyzes integration of viral DNA into the host cell genome to establish what is known as the proviral state. Once integrated, the provirus can remain latent or undergo efficient gene expression in order to continue with late steps of the replication cycle. The full-length unspliced genomic RNA (hence referred as unspliced mRNA) has a dual function as it is both used as mRNA for the synthesis of Gag and Gag-Pol precursors and the genome that is incorporated into the viral particles. The structural protein Gag drives both packaging of the genomic RNA and assembly of newly synthesized viral particles, which will be maturated by the viral protease allowing initiation of a new replication cycle.

HIV gene expression relies on the host for transcription, RNA processing, nuclear export and translation, a series of complex processes that are assisted by at least, two major viral regulators namely Tat and Rev. HIV transcription relies both on the promoter sequences present in the viral 5′ long-terminal repeat (5′-LTR) region and the *trans*-activator viral protein Tat, which acts together with host cellular proteins including the RNA polymerase II and the pTEFb transcription factor [3–7]. Transcription from the provirus results in expression of the full-length unspliced mRNA, which is 9-kb long and encodes structural and enzymatic proteins (Gag and Gag-Pol). However, the presence of multiple splice donor and acceptor sites within the full-length mRNA supports alternative splicing which results in the generation of a complex pattern of viral mRNAs harboring the open reading frames of Vif, Vpr, Vpu/Env, Tat, Rev and Nef, which differ in their 5′ untranslated regions (5′-UTR) [8]. These transcripts are both incompletely (4-kb) and completely spliced (2-kb) and are used for expression of all remaining viral proteins. Several of these completely spliced transcripts coding for Tat, Rev, and Nef are produced during the early steps of infection [9–11]. Later on, the full-length unspliced mRNA together with further different 4-kb transcripts coding for Env/Vpu, Vif, Vpr, and Tat are then generated, exported and translated in the cytoplasm [9–12]. All these RNA processing events generate different viral mRNP complexes that will differ in the routes used to reach the host translational machinery. As such, while completely spliced transcripts are exported by the canonical nuclear export pathway, the unspliced and the 4-kb incompletely spliced transcripts require the binding of the virally encoded protein Rev to the *cis*-acting RNA element called the Rev responsive element (RRE) present in all of these intron-containing transcripts; this allows their export through the CRM1 pathway [13–20] (Figure 1).

As mentioned above, in addition to its nuclear function as a pre-mRNA template for the generation of the 2 and 4-kb transcripts, the 9-kb full-length unspliced mRNA plays two additional roles in the cytoplasm by serving both as a mRNA for viral protein production and as the packaged genome (Figure 1). In order to combine these different functions, the unspliced mRNA needs to overcome several structural and functional constraints that could affect cellular post-transcriptional events such as nuclear export and translation. In this review, we focus on how the virus has evolved to combine the building of a complex and specific mRNP on its mRNAs ensuring proper viral gene expression.

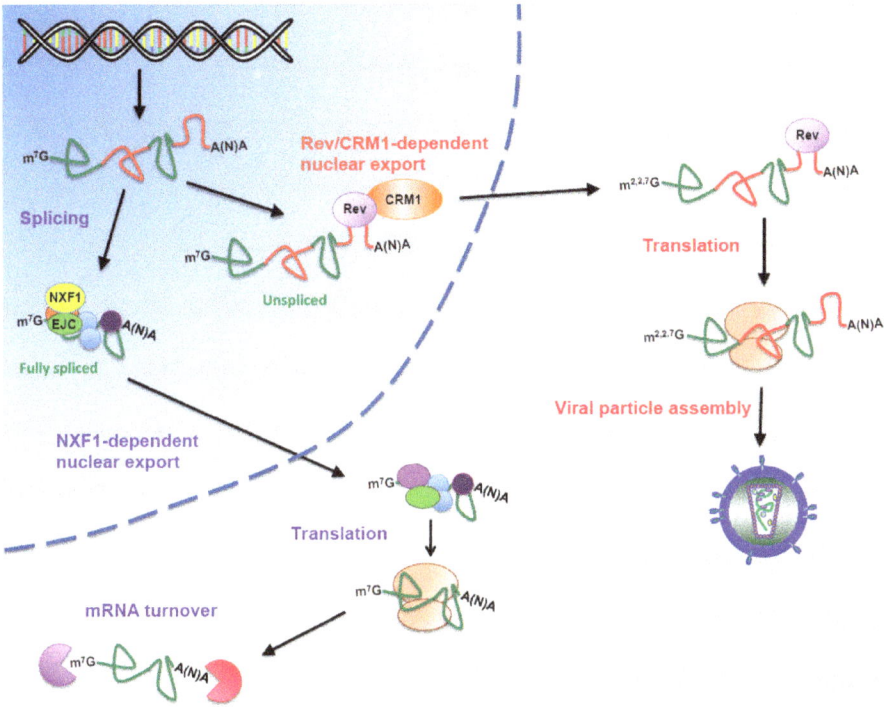

Figure 1. Post-transcriptional control on human immunodeficiency virus (HIV). Upon transcription, the capped and polyadenylated full-length genomic RNA is used as a template for the host mRNA processing machinery in order to generate fully spliced and partially spliced transcripts (partially spliced transcripts have been omitted for simplicity). In the nucleus, fully spliced transcripts form a classical messenger ribonucleoprotein complex (mRNP) together with host proteins such as the exon junction complex (EJC) and the mRNA export factor NXF1. In the cytoplasm, fully spliced mRNAs recruit the host translational apparatus for protein synthesis and later they are degraded by the mRNA turnover machinery. In the presence of the viral protein Rev, the unspliced genomic RNA (and partially spliced mRNAs) reaches the cytoplasm through the CRM1-dependent pathway avoiding the host cell surveillance mechanisms. During this journey to the cytoplasm, the unspliced genomic RNA forms a unique mRNP that favors its association with the host translational machinery. In contrast to the fully spliced transcripts, the unspliced genomic RNA does not undergo turnover as it is incorporated into viral particles.

2. Reaching the Cytoplasm Avoiding Surveillance Mechanisms

HIV-1 transcripts are synthesized by the RNA polymerase II and, consequently, are capped and polyadenylated by the host machinery [21–24]. As described above, many different viral mRNA species can be found in infected cells with at least four different 5′- and eight different 3′-splice sites being used during pre-mRNA processing [9–11]. However, the cellular splicing machinery must be inefficient in the usage of viral splice sites in order to ensure that appropriate pools of each subset of viral mRNAs can be produced in the nucleus [25–27]. The vast majority of cellular mRNAs are usually spliced to completion and thus all introns are removed during splicing [28]. This is the case for the viral 2-kb transcripts that are completely processed and can be exported to the cytoplasm through the nuclear export factor NXF1 [29–31]. However, nuclear export of mRNAs that harbor functional introns is quite unusual and they are often retained in the nucleus by the interaction with splicing factors until they are either spliced to completion or degraded [32–34]. In addition, viral intron-containing transcripts cannot be exported through the NXF1-dependent pathway due to surveillance mechanism ruled by, amongst others, the cellular protein Tpr [35–37]. As such, the 4-kb incompletely spliced and the 9-kb unspliced transcripts are retained and degraded in the host cell nucleus unless the viral protein Rev is present [25,38].

Rev is synthesized from a 2-kb completely spliced transcript and is essential for virus replication [39]. Although the step of the replication cycle in which Rev activity is the most important has been the subject of some controversy [39–41], there is no doubt that synthesis of the viral structural proteins Gag and Gag-Pol from the unspliced mRNA is dramatically reduced in the absence of Rev. Rev is a phosphoprotein of approximately 18-kDa that constantly shuttles between the nucleus and the cytoplasm but accumulates in the nucleus [39]. The N-terminal domain of the protein contains an arginine-rich motif that serves both as a nuclear localization signal (NLS) and as an RNA-binding domain (RBD) [39,42–50]. While the NLS allows recognition and nuclear import of Rev by Importin-β, the RBD allows the interaction with the Rev Responsive Element (RRE) which is present exclusively in the incompletely spliced and unspliced viral transcripts as it is located within the *env* gene [19,39,51]. The arginine-rich sequence is flanked from both sides by less defined sequences required for oligomerization [39,49,50]. The C-terminal domain contains the leucine-rich nuclear export signal (NES) that allows the interaction and nuclear export of the Rev-RRE complex with the karyopherin CRM1 (Chromosome maintenance-1) bound to Ran-GTP [16,52–55]. Recent structural studies have revealed that once bound to the RRE, the Rev protein oligomerizes in order to promote nuclear export [49,56] while CRM1 forms a dimer that favors nuclear export of the Rev-RRE complex [57]. Moreover, it was recently shown that Rev can interact with the nuclear cap-binding complex (CBC) component CBP80 and block NXF1 recruitment in order to specifically enter the nuclear export pathway through CRM1 [58]. In addition to CRM1-RanGTP and CBP80, Rev recruits several host proteins including eIF5A, hRIP, DDX3, DDX1, and Sam68 to promote nuclear export [59]. Thus, by using this alternative pathway, the viral protein Rev ensures the cytoplasmic accumulation of intron-containing transcripts and avoids NXF1-associated quality control mechanisms. This explains that despite the presence of introns, viral transcripts that do not undergo complete splicing are not substrates for non-sense mediated decay (NMD) [60,61].

After completion of their journey from the nucleus and through the nuclear pores, the viral transcripts must compete with cellular mRNAs in the cytoplasm to recruit the host translational machinery. In mammals, ribosome recruitment onto the mRNA occurs by two main mechanisms: the cap-dependent and the internal ribosome entry sites (IRES)-driven mechanisms [62,63] and HIV-1 has evolved strategies to use both [64].

3. An Overview on mRNA Translation Initiation in Eukaryotes

The vast majority of cellular mRNAs recruit ribosomes through a cap-dependent translation initiation mechanism. This process sequentially involves: (i) formation of a 43S pre-initiation complex; (ii) cap structure recognition and loading of the 43S pre-initiation complex onto the mRNA; (iii) ribosomal scanning of the 5'-UTR; (iv) initiation codon recognition and (v) joining of the 60S ribosomal subunit [62]. The 43S pre-initiation complex is composed of a recycled 40S small ribosomal subunit, an eIF2-GTP-tRNAi ternary complex (TC), eIF3, eIF1, eIF1A and probably eIF5 [62]. At the 5' end of the mRNA, the eIF4F holoenzyme binds to the cap-structure and unwinds local RNA structures assisted by eIF4B or eIF4H creating the landing pad for the 43S pre-initiation complex. The eIF4F multimeric complex is composed of the cap-binding protein eIF4E, the RNA helicase eIF4A, and the scaffold protein eIF4G [65,66]. eIF4E exhibits high affinity for the cap structure and interacts with eIF4G to mediate cap-dependent translation initiation by promoting assembly of eIF4F onto the capped mRNA. The DEAD-box protein eIF4A is an RNA helicase with ATP-dependent RNA unwinding activity [67,68]. Although the intrinsic helicase activity of eIF4A is weak, its inclusion into the eIF4F complex together with the binding of eIF4B and the related factor eIF4H strongly stimulates its enzymatic activity [69]. As mentioned above, the eIF4G scaffold protein associates with eIF4E and eIF4A to form the eIF4F holoenzyme that binds to the 5' end of capped mRNAs [70–73]. By further interacting with eIF3, eIF4G promotes attachment of the 43S pre-initiation complex onto the transcript to allow formation of a 48S pre-initiation complex [72,74–78]. Once attached, this complex immediately starts scanning in a 5' to 3' direction from the cap structure until it reaches an initiation codon, which often corresponds to the first AUG codon [79–81]. The ribosomal scanning model proposes that the translation initiation complex unwinds secondary structures present in the 5'-UTR and moves in the 5' to 3' direction in an ATP-dependent manner [82,83]. Thus, in addition to their role in

43S pre-initiation complex attachment, eIF4G, eIF4A, eIF4B (or eIF4H) also assist the scanning process [83,84]. Although the RNA helicase eIF4A and its associated factors eIF4B/eIF4H can support the unwinding process of the scanning pre-initiation complexes, it has been recently shown that additional RNA helicases can also be recruited [85]. As such, the related DExH box protein 29 (DHX29) binds the 40S small ribosomal subunits while RNA helicase A binds selected mRNAs and both are required for efficient scanning of mRNAs containing highly structured 5'-UTRs [86–88].

An alternative model of translation initiation has been described for mRNAs that harbor specific RNA sequences termed Internal Ribosome Entry Sites (IRES). These sequences are generally present in the 5'-UTR of the mRNA whose function is to recruit ribosomes for translation initiation in a cap-independent manner. IRES elements were first discovered in viral RNA genomes more than 25 years ago with the studies of picornavirus translation [89,90] and have now been characterized in many viral mRNAs including HCV, Pestiviruses, and Retroviruses [63]. Although IRES elements have also been described in near 100 cellular mRNAs their existence remains controversial mainly by the lack of essential controls discarding cryptic promoters and/or alternative splicing during the characterization process [91].

IRES elements promote the direct binding of the 43S pre-initiation complex and associated factors to the mRNA. However, the precise mechanism of IRES-mediated translation initiation is not completely understood. Although a classification of IRES elements by structural criteria is not possible due to the lack of any conserved sequence, viral IRES elements can be grouped based on a mechanistic and functional point of view involving: (i) the way by which the 43S pre-initiation complexes is recruited, e.g., whether it is assisted or not by eIFs; and (ii) the site where the 43S pre-initiation complex is positioned onto the mRNA, which can be close to the initiation codon or if it involves an additional step of scanning. Moreover, IRES elements can also be characterized by the requirement of diverse cellular accessory proteins denominated IRES *trans*-acting factors (ITAFs) for proper function.

IRES elements allow the selective translation of viral mRNAs under conditions in which global host translation is compromised. When faced by several stresses (such as viral infections, hypoxia, or heat shock) or particular cellular conditions (such as mitosis or apoptosis), Eukaryotic cells often respond by reducing the global rates of translation [92]. However, a significant fraction of cellular mRNAs was shown to remain associated to polysomes [93] and several of these are IRES-containing transcripts. This shows that the presence of the IRES element allows mRNAs to be translated under unfavorable conditions in which cap-dependent translation is slowed down or arrested [94–98].

4. Recruiting the Host Translational Machinery onto the Unspliced HIV Genomic RNA

The unspliced mRNA harbors a long (5'-UTR) organized in several RNA structures involved in many steps of the replication cycle [3,99–103]. Given the structure and complexity of the 5'-UTR, the mechanism by which translation initiation takes place on the HIV-1 genomic RNA has been the subject of debate for several years [104]. Indeed, it was initially shown that sequences derived from the 5'-UTR were inhibitory for translation [105–109]. Particularly, cell-free *in vitro* translation assays and *ex vivo* experiments using reporter genes suggested that the presence and folding of the TAR RNA motif, which is located at the very 5' end of the viral transcripts, exerted a negative effect on protein synthesis both by impeding ribosome recruitment and by activating the kinase PKR [105–108,110,111]. However, despite this incompatibility with ribosome recruitment by a cap-dependent ribosomal scanning mechanism, an IRES-driven mechanism on HIV-1 transcripts was rapidly discarded indicating that cap-dependent translation initiation was the major mechanism for ribosome recruitment [112].

5. Identification of a Cell Cycle-Dependent IRES

As mentioned above, initial attempts to identify sequences within viral 5'-UTR supporting IRES activity failed and the cap-dependent ribosomal scanning was proposed as the only mechanism to drive Gag synthesis [112]. However, a more detailed study revealed that an IRES element was indeed present within the 5'-UTR of the HIV-1 unspliced mRNA [113]. This IRES element was mapped to nucleotides 104 to 336 where it spans the primer-binding site (PBS), the dimerization site (DIS), the major splice donor (SD) and RNA motifs that are critical for encapsidation [113]. Interestingly, this IRES element was shown to be activated during the G2/M phase of the cell cycle [113]. This peculiarity not only explained why previous studies failed to detect IRES activity but also emphasized the physiological relevance that the use of an alternative mechanism of ribosome recruitment could have during viral replication. Indeed, during HIV-1 and other lentiviral infections, the viral protein Vpr induces a cell cycle arrest at the G2/M phase [114–117]. Although the G2/M phase is characterized by a strong inhibition of cap-dependent protein synthesis [95,118,119], HIV-1 viral gene expression was shown to continue during this phase of the cell cycle [120–122]. Although IRES elements were demonstrated to be able to drive efficiently protein synthesis in G2/M [96], other authors have proposed another alternative in which the translation of the HIV-1 unspliced mRNA was rather conducted by a eIF4E-independent, CBC-driven, cap-dependent mechanism during the G2/M arrest induced by Vpr [122]. Nevertheless, the ability of the 5'-UTR of HIV-1 transcripts to drive IRES-driven translation has now been evidenced by several groups in different experimental contexts and on different HIV-1 prototype strains [123–130]. Therefore, it is conceivable that the HIV-1 genomic RNA can use both strategies depending on some physiological conditions that remain to be found. Moreover, similar to poliovirus, the HIV-1 and HIV-2 proteases were shown to process translation initiation factors eIF4GI and PABP *in vitro* and *ex vivo* leading to the inhibition of cap-dependent ribosomal scanning with modest impact on viral unspliced mRNA translation [70,131–134]. However, processing of eIF4GI and PABP during viral infection was rather modest and occurred late during infection [131] and thus, the significance of these events in the course of viral replication remains to be demonstrated.

Protein synthesis from the HIV-1 and HIV-2 unspliced mRNAs presents an additional layer of complexity as IRES elements have also been characterized within the Gag coding region [64,135]. By using the HIV-1 Gag ORF lacking the viral 5′-UTR it was shown that this region was able to drive synthesis of full-length p55 Gag and a novel 40-kDa N-terminally truncated isoform of Gag (p40) initiated at an internal *in frame* AUG codon [136]. The presence of IRES elements downstream to the authentic initiation codon and the synthesis of N-terminally truncated isoforms of Gag were also characterized in other related lentiviruses such as HIV-2, SIV, and FIV indicating that the conservation of the mechanism is a common feature of the genus and could be important for replication [135,137–139]. In HIV-1 and HIV-2, these Gag isoforms are incorporated into viral particles despite the lack of a myristoylation site at their N-terminus, probably by protein-protein interactions with the full length Gag polyprotein; this suggests a role for these truncated isoforms in the replication cycle [136,138]. Although the molecular mechanisms controlling this process in the HIV-1 unspliced mRNA are not completely understood, an *in vitro* study revealed that the different modes of ribosome recruitment have different levels of requirements for eIF4F [140]. In the case of the HIV-2, it was shown that three IRES elements within the Gag coding region were able to directly recruit three independent 43S pre-initiation complexes [138,141–143].

6. Translation by a Cap-Dependent Mechanism

More recently, by using synthetic constructs Berkhout and co-workers demonstrated that cap-dependent ribosomal scanning occurs throughout the 5′-UTR of the HIV-1 unspliced mRNA [144]. Using similar approaches, other groups including ours demonstrated that the cap-dependent mechanism of translation initiation occurs both *in vitro* and *ex vivo* [109,121]. The ability of the 43S pre-initiation complex to scan through the highly structured 5′-UTR could be explained by the recruitment of the helicase RHA, which was shown to promote polysome association of the unspliced mRNA by interacting with a post-transcriptional control element (PCE) located at the 5′-UTR [88,145]. Although it is thought that RHA helicase activity contributes to the unwinding of secondary structures during ribosomal scanning, the involvement of other RNA helicases such as DHX29 has not yet been investigated and thus, cannot be discarded.

While the involvement of RHA shed light on how the 43S pre-initiation complex moves along the highly structured viral 5′-UTR, it was still unclear how the cap-structure could be recognized by the eIF4F complex in the presence of the TAR structure. Indeed, the 5′ end cap moiety of all HIV-1 transcripts is base-paired and embedded within the basis of the stem of the TAR RNA motif and thus, is likely to be inaccessible for the binding of the eIF4F complex and the ribosomal 43S subunit. Surprisingly, although the presence of TAR was shown to strongly interfere with translation initiation in the rabbit reticulocytes lysate (RRL), this was not the case in constructs expressed in living cells [109]. These data suggested that some specific host factor(s) that are absent or in limited concentration in the RRL can be used to overcome the structural constraint imposed by TAR. A likely candidate was found amongst one of the Rev cofactors namely the RNA helicase DDX3 [146]. DDX3 belongs to the DEAD-box family of proteins whose prototype member is the initiation factor eIF4A [147]. DEAD-box proteins are ATP-dependent RNA helicases that play pleiotropic functions within the cell by participating in all steps of RNA metabolism [148]. These proteins are thought to participate in RNA:RNA and RNA:protein remodeling or to act as RNA clamps for the assembly of large macromolecular complexes [148]. DDX3 was first proposed to be a host factor involved in Rev-dependent nuclear export [146]. By using a full-length reporter proviral DNA and viral infection, we were also able to show that DDX3 was required for translation of the unspliced genomic RNA both in HeLa and T-cells and this function required the ATP binding and ATPase activity of the enzyme [149,150]. We also reported that the molecular target for DDX3 was actually the TAR RNA motif, an observation recently validated by another group [151]. Interestingly, we observed that DDX3 was required to unwind TAR in cells and this functional interaction was necessary when the latter was at its original location (e.g., at the 5′ end of the HIV-1 transcript) but the dependence in DDX3 was abolished when the TAR motif was preceded by an unstructured spacer sequence [149]. These data suggested that DDX3 binds and unwinds TAR during a pre-translation initiation step that is necessary to remodel secondary structures in order to render the cap moiety accessible to the eIF4F holenzyme and the 43S complex [149] (Figure 2). In agreement with this, we could also show that DDX3 was bound to, at least, two additional and specific sites within the 5′-UTR of the HIV-1 genomic RNA [149]. These sites were located exclusively on RNA single stranded regions and could correspond to loading platforms for DDX3 as had been previously suggested [148]. In addition, an interaction between DDX3 and translation initiation factors eIF4GI and PABPC1 was also evidenced by biochemical assays as well as confocal microscopy [149]. Interestingly, we observed that the complex formed between the unspliced HIV-1 mRNA, DDX3, eIF4GI, and PABPC1 was assembled in localized cytoplasmic granules that resembled but were different from stress granules as they lacked the eIF4F components eIF4E, eIF4A, eIF4B, and the CBC component CBP80 [150] (Figure 2).

Another intriguing feature of the unspliced HIV-1 mRNA that could also influence its translation is the presence of a trimethylguanosine (TMG) cap structure [152]. A few years ago, the peroxisome proliferator-activated receptor-interacting protein with methyltransferase domain (PIMT) (the human homolog of the yeast cap hypermethylase TGS1) was shown to interact with Rev and this resulted in the hypermethylation of the 5′ cap structure of the HIV-1 unspliced mRNA [152]. It has been known for quite some time that TMG-capped mRNAs present reduced translational rates *in vitro* [153]. However, in the case of the HIV-1 genomic RNA, the latter is efficiently used for viral protein production and trimethylation of its cap was shown to be required in this process although the molecular mechanism underlying was not elucidated [152]. In light of our data showing the presence of a pre-initiation complex composed of DDX3/eIF4G/PABP, but lacking any of the major cap binding proteins CBC and eIF4E [150], a further investigation into the role of a TMG cap would be interesting. Indeed, as the affinity of the CBC and eIF4E for the TMG cap is largely reduced compared to the classical m^7G monomethylated cap [154], this could explain the exclusion of both eIF4E and CBC from this pre-initiation complex and could suggest that other TMG-bound cellular proteins may be recruited for initiation of HIV-1 unspliced mRNA translation.

7. Assembly of Unspliced mRNA-Containing Granules

Cellular mRNAs are in a dynamic equilibrium between polysomes and cytoplasmic granules such as stress granules and p-bodies [155,156]. While stress granules are sites of triage for mRNAs stalled in translation initiation as a response to cellular stress, p-bodies are sites intimately related to the mRNA decay machinery [155,156]. Both structures and/or some of their components have been shown to play pivotal roles during replication of several viruses and thus, it is not surprising that viruses have evolved different strategies to manipulate the assembly/disassembly of mRNA granules [157–159].

Although it was first proposed that HIV-1 translation could be negatively regulated by some components of p-bodies including APOBEC3G [160,161], there is new evidence showing that HIV-1 replication induces the disassembly of p-bodies [162]. Moreover, it was also recently shown that APOBEC3G activity on HIV-1 replication was independent of p-bodies [163] and that p-bodies components such as DDX6 and Argonaute 2 were rather involved in viral particle assembly independent of RNA packaging [164]. Therefore, further work is necessary to clarify the role of p-bodies in HIV-1 unspliced mRNA metabolism.

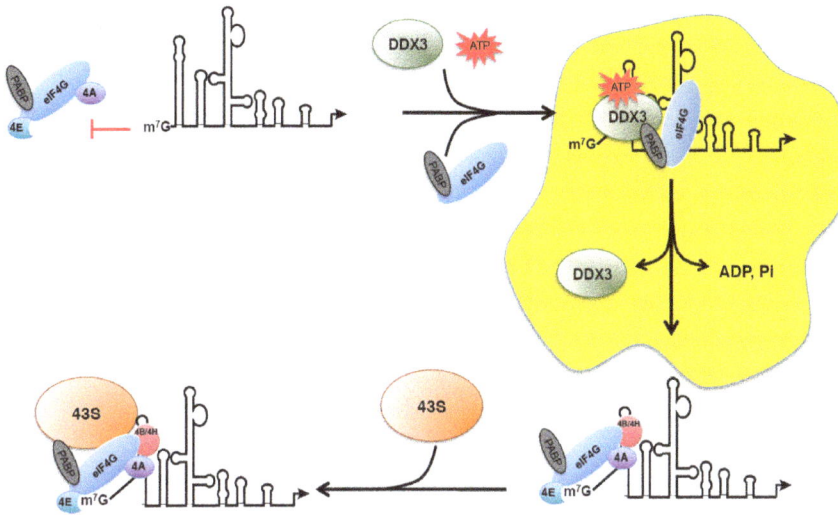

Figure 2. DDX3-mediated translation of the HIV-1 genomic RNA. In the absence of DDX3, the TAR RNA motif impedes binding of the eIF4F holoenzyme to the cap. Thus, DDX3 binds the viral 5′-UTR to nucleate formation of a pre-translation initiation complex that involves ATP-dependent unwinding of TAR and specific recruitment of translation initiation factors eIF4GI and PABP (and probably other unidentified cellular proteins). TAR unwinding renders the cap accessible for eIF4F binding and subsequent recruitment of the 43S pre-initiation complex. It is possible that such a pre-translation initiation step driven by DDX3 occurs compartmentalized in RNA granules (in yellow).

Interestingly, HIV-1 and HIV-2 have evolved completely opposite strategies to modulate and control the assembly of stress granules. As such, it was shown that HIV-1 has the ability to interfere with stress granule assembly induced by different types of stresses [162,165]. Indeed, the authors showed that the HIV-1 Gag protein has the ability to interfere with stress granules assembly through a direct interaction with eEF2 and G3BP1, two key factors required for assembly of these cytoplasmic structures [165]. Thus, it is possible that by doing so, the HIV-1 unspliced mRNA promotes the assembly of a pre-initiation complex with DDX3 and subset of eIFs in order to enter in translation initiation and associate with polysomes [150] (Figure 3A). Then, the HIV-1 unspliced mRNA is assembled into a Staufen1-dependent mRNP, which also contains the viral protein Gag and is required for RNA packaging [162] (Figure 3A).

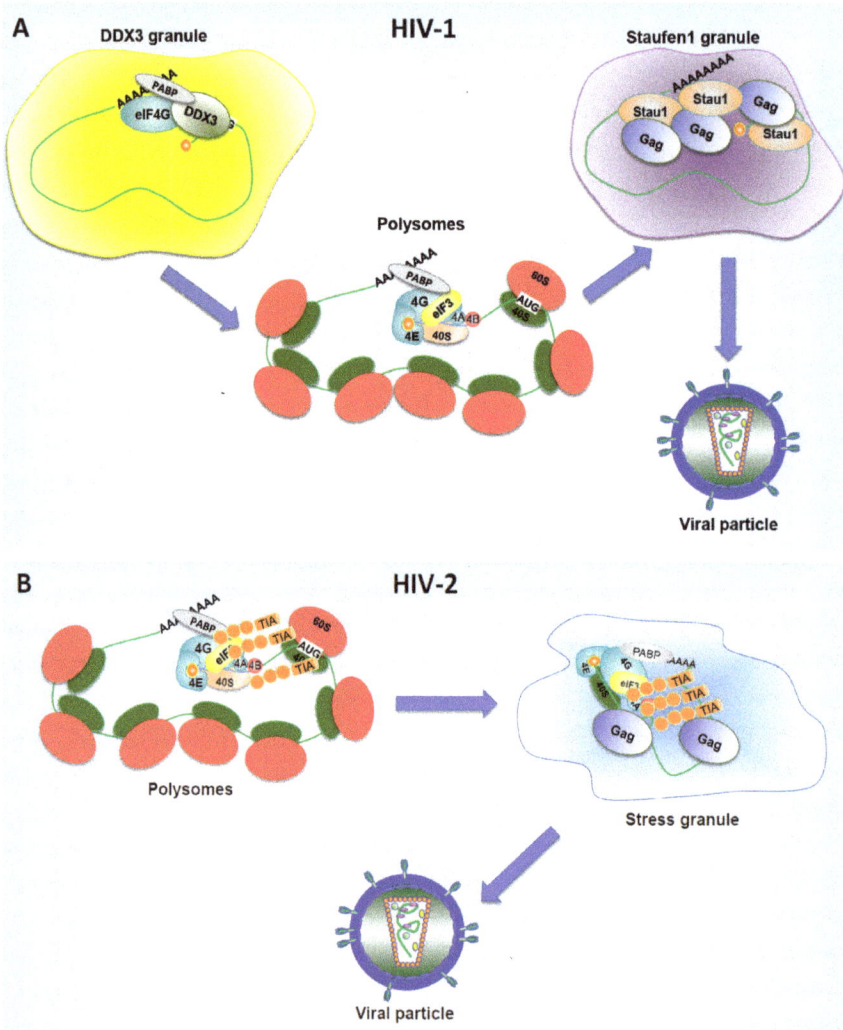

Figure 3. RNA granules assembled during HIV replication. (**A**) Polysome association of HIV-1 unspliced mRNA requires its previous assembly in DDX3-dependent granules together with eIF4GI, PABPC1 and probably other, yet, unidentified cellular proteins. Once translated, unspliced mRNA associates with the dsRNA-binding protein Staufen1 and the viral protein Gag in order to form another specific RNA granule (Staufen1 granule), which is required for viral particle assembly. This dynamic assembly of different RNA granules allows HIV-1 to coordinate genomic RNA translation and packaging; (**B**) The HIV-2 unspliced mRNA recruits the stress granule assembly factor TIAR to form a specific viral mRNP that accumulates in stress granules in the absence of active translation. The viral protein Gag also accumulates in stress granules suggesting that the transition from translation to RNA packaging could occur in these structures.

In sharp contrast with what was described for HIV-1, we showed that HIV-2 replication induces the spontaneous assembly of stress granules [166] (Figure 3B). Moreover, we observed that HIV-2 unspliced mRNA was directly associated with the stress granule assembly factor TIAR in order to form a specific viral mRNP [166] (Figure 3B). We have previously shown that ribosome recruitment onto the HIV-2 genomic RNA is very inefficient due to a strong interference imposed by the highly structured TAR RNA motif [109]. Thus, stress granules could serve as sites of storage for the viral genome while threshold levels of Gag required for RNA packaging are produced. Interestingly, the HIV-2 Gag polyprotein was also observed in stress granules indicating that the transition from translation to RNA packaging may occur in these structures [166] (Figure 3B).

8. Viral Proteins Promoting Translation

Some of the virally encoded proteins, namely Tat, Rev, and Gag have been involved in the control of viral mRNA translation (Figure 4). Initial studies carried out in the RRL and *Xenopus leavis* oocytes revealed that Tat was involved in the control of translation notably by counteracting the deleterious activation of PKR [107,167,168]. Indeed, secondary RNA structures constituting the TAR motif at the 5'-UTR were shown to activate the protein kinase R (PKR) leading to inhibition of translation [169,170]. Once activated, PKR phosphorylates the α subunit of eIF2 resulting in a global inhibition of translation initiation [171]. However, Tat binding to TAR and/or PKR could prevent activation of the kinase the phosphorylation of eIF2α [172]. In addition, it was shown that Tat is able to stimulate translation both *in vitro* and in living cells [129]. Moreover, binding of Tat to the 5'-UTR of the unspliced mRNA could stimulate the programmed-1 ribosomal frameshift [173]. More recently, Tat was shown to interact with DDX3 and remain associated to polysomes together with the unspliced mRNA further indicating its role in viral mRNA translation [151].

Another viral protein, Rev, was demonstrated to be required for association of the incompletely spliced mRNAs *vif, vpr, env,* and *vpu* into polysomes [174]. By using a Gag expression vector lacking the RRE, it was also shown that polysome association of the resulting gag mRNA was deficient either in the presence or absence of Rev, suggesting that the Rev-RRE interaction and not the presence of Rev *per se* is critical for ribosome recruitment [175]. Such a function of Rev in translation could be explained by an enhanced recruitment of PABPC1 to Rev-dependent mRNAs [176] or by direct binding to the loop-A of stem-loop 1 located within the packaging signal [177].

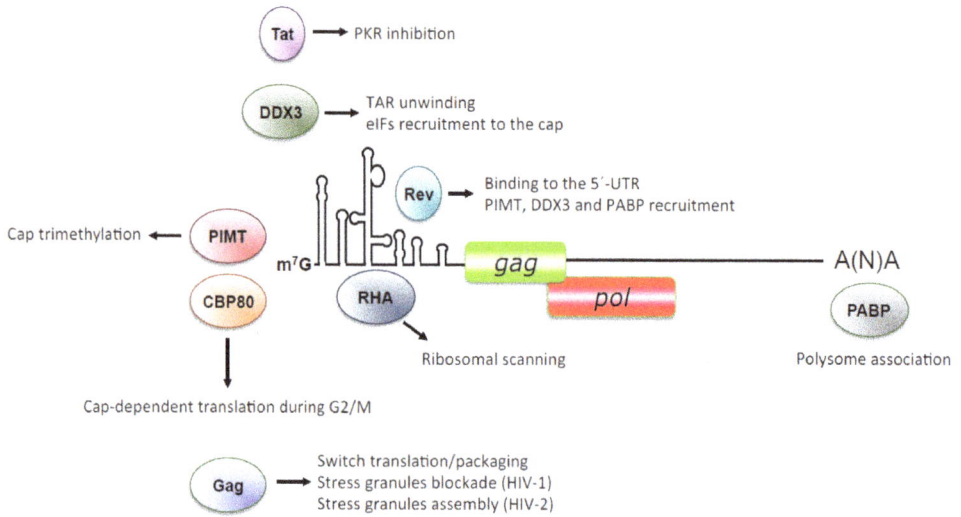

Figure 4. Translational control by host and viral proteins. Schematic representation of the panel of viral (Rev, Tat, Gag) and cellular (DDX3, PIMT, CBP80, RHA, and PABP) proteins required to assist translation initiation from the HIV-1 and HIV-2 unspliced mRNA.

Finally the Gag protein was shown to modulate its own translation by exerting a bimodal effect depending on its concentration [143,166,178]. As such, it was shown that HIV-1 and HIV-2 Gag stimulate translation at low concentrations to then inhibit protein synthesis. In the case of HIV-1, stimulation required the matrix domain while inhibition was dependent on the binding of nucleocapsid domain to the packaging signal [178]. In the case of HIV-2, we also observed a bimodal effect of Gag on translation with stimulation at low concentration and inhibition and higher concentrations [166]. Interestingly, we observed that such an effect of Gag on translation was concordant with the subcellular localization of the unspliced mRNA [166]. As such, we observed that both the unspliced mRNA and Gag localized diffusely in the cytoplasm at low concentrations of Gag while both components were assembled in stress granules at high concentrations of the viral protein [166].

9. Concluding Remarks

Post-transcriptional control of HIV-1 viral gene expression is regulated from the nucleus to the cytoplasm and involves many host and viral proteins throughout this process. It is amazing to realize that almost every step, from splicing, export, and translation, has its own regulatory pathway, which often differs from that used by cellular mRNAs. This results in the constitution of a unique viral RNP that reaches the cytoplasm to be translated by a spectrum of different mechanisms juggling with cap-dependent and cap-independent mechanisms of initiation. The great diversity of these means of expression confers to the virus several selective advantages such as preventing degradation of the unspliced mRNA by surveillance mechanisms in the nucleus and allowing selective translation under conditions that are not favorable for host gene expression. Evolution of diverse mechanisms for gene expression also allows the conciliation of the presence of multiple RNA structures in the 5'-UTR, that are required for genome replication, with the need of an efficient mechanism for viral protein synthesis. For instance, the TAR structure at the 5' end of the mRNA represents an essential element for transcription that would be severely inhibitory for ribosome binding and scanning unless it can be counteracted by the recruitment of the host RNA helicase DDX3 to assist pre-initiation complex formation. An interesting, promising new direction concerns the recent identification of compartmentalized cytoplasmic foci containing HIV-1 and HIV-2 viral RNPs. Although, the function of these foci are not fully characterized, they may serve as sites of storage to ensure an equilibrium between unspliced mRNA translation and its packaging into assembly virions. Interestingly, use of these cytoplasmic foci seems to be radically different between the two closely related human immunodeficiency viruses. A better understanding of this process may shed light on our understanding of the replication cycle of these two relatives. Such a specific and complex control of post-transcription gene expression in lentiviruses can point to new directions in the treatment of disease. As such, the targeting of essential host factors that are required for viral replication, such as DDX3 for instance, could bring new therapeutical approaches. Above all, due to great diversity of their strategies developed to express their genome, lentiviruses represent good paradigms for the studies on the control of post-transcriptional gene expression.

Acknowledgments: Research at the RSR laboratory is funded by the Comisión Nacional de Investigación en Ciencia y Tecnología (Conicyt) through the Fondecyt Initiation into Research Program (No 11121339) and the International Cooperation Program (DRI USA2013-0005). Research in the TO laboratory is financed by the ANRS.

Conflicts of Interest: The authors declare no conflict of interest.

References

1. Killian, M.S.; Levy, J.A. HIV/AIDS: 30 Years of progress and future challenges. *Eur. J. Immunol.* **2011**, *41*, 3401–3411.
2. Matreyek, K.A.; Engelman, A. Viral and cellular requirements for the nuclear entry of retroviral preintegration nucleoprotein complexes. *Viruses* **2013**, *5*, 2483–2511.
3. Arya, S.K. Human and simian immunodeficiency retroviruses: Activation and differential transactivation of gene expression. *AIDS Res. Hum. Retrovir.* **1988**, *4*, 175–186.

4. Berkhout, B.; Jeang, K.T. Trans activation of human immunodeficiency virus type 1 is sequence specific for both the single-stranded bulge and loop of the trans-acting-responsive hairpin: A quantitative analysis. *J. Virol.* **1989**, *63*, 5501–5504.

5. Cullen, B.R. Trans-activation of human immunodeficiency virus occurs via a bimodal mechanism. *Cell* **1986**, *46*, 973–982.

6. Gatignol, A. Transcription of HIV: Tat and cellular chromatin. *Adv. Pharmacol.* **2007**, *55*, 137–159.

7. Jones, K.A. HIV trans-activation and transcription control mechanisms. *New Biol.* **1989**, *1*, 127–135.

8. Tazi, J.; Bakkour, N.; Marchand, V.; Ayadi, L.; Aboufirassi, A.; Branlant, C. Alternative splicing: Regulation of HIV-1 multiplication as a target for therapeutic action. *FEBS J.* **2010**, *277*, 867–876.

9. Purcell, D.F.; Martin, M.A. Alternative splicing of human immunodeficiency virus type 1 mRNA modulates viral protein expression, replication, and infectivity. *J. Virol.* **1993**, *67*, 6365–6378.

10. Schwartz, S.; Felber, B.K.; Benko, D.M.; Fenyo, E.M.; Pavlakis, G.N. Cloning and functional analysis of multiply spliced mRNA species of human immunodeficiency virus type 1. *J. Virol.* **1990**, *64*, 2519–2529.

11. Schwartz, S.; Felber, B.K.; Pavlakis, G.N. Expression of human immunodeficiency virus type 1 vif and vpr mRNAs is Rev-dependent and regulated by splicing. *Virology* **1991**, *183*, 677–686.

12. Guerrero, S.; Batisse, J.; Libre, C.; Bernacchi, S.; Marquet, R.; Paillart, J.C. HIV-1 replication and the cellular eukaryotic translation apparatus. *Viruses* **2015**, *7*, 199–218.

13. Bogerd, H.P.; Echarri, A.; Ross, T.M.; Cullen, B.R. Inhibition of human immunodeficiency virus Rev and human T-cell leukemia virus Rex function, but not Mason-Pfizer monkey virus constitutive transport element activity, by a mutant human nucleoporin targeted to Crm1. *J. Virol.* **1998**, *72*, 8627–8635.

14. Dillon, P.J.; Nelbock, P.; Perkins, A.; Rosen, C.A. Function of the human immunodeficiency virus types 1 and 2 Rev proteins is dependent on their ability to interact with a structured region present in env gene mRNA. *J. Virol.* **1990**, *64*, 4428–4437.

15. Emerman, M.; Vazeux, R.; Peden, K. The rev gene product of the human immunodeficiency virus affects envelope-specific RNA localization. *Cell* **1989**, *57*, 1155–1165.

16. Fischer, U.; Huber, J.; Boelens, W.C.; Mattaj, I.W.; Luhrmann, R. The HIV-1 Rev activation domain is a nuclear export signal that accesses an export pathway used by specific cellular RNAs. *Cell* **1995**, *82*, 475–483.

17. Le, S.Y.; Malim, M.H.; Cullen, B.R.; Maizel, J.V. A highly conserved RNA folding region coincident with the Rev response element of primate immunodeficiency viruses. *Nucleic Acids Res.* **1990**, *18*, 1613–1623.

18. Lewis, N.; Williams, J.; Rekosh, D.; Hammarskjold, M.L. Identification of a cis-acting element in human immunodeficiency virus type 2 (HIV-2) that is responsive to the HIV-1 rev and human T-cell leukemia virus types I and II rex proteins. *J. Virol.* **1990**, *64*, 1690–1697.

19. Malim, M.H.; Hauber, J.; Le, S.Y.; Maizel, J.V.; Cullen, B.R. The HIV-1 Rev *trans*-activator acts through a structured target sequence to activate nuclear export of unspliced viral mRNA. *Nature* **1989**, *338*, 254–257.

20. Neville, M.; Stutz, F.; Lee, L.; Davis, L.I.; Rosbash, M. The importin-β family member Crm1p bridges the interaction between Rev and the nuclear pore complex during nuclear export. *Curr. Biol.* **1997**, *7*, 767–775.

21. Chiu, Y.L.; Coronel, E.; Ho, C.K.; Shuman, S.; Rana, T.M. HIV-1 Tat protein interacts with mammalian capping enzyme and stimulates capping of TAR RNA. *J. Biol. Chem.* **2001**, *276*, 12959–12966.

22. Chiu, Y.L.; Ho, C.K.; Saha, N.; Schwer, B.; Shuman, S.; Rana, T.M. Tat stimulates cotranscriptional capping of HIV mRNA. *Mol. Cell* **2002**, *10*, 585–597.

23. Shatkin, A.J. Capping of eucaryotic mRNAs. *Cell* **1976**, *9*, 645–653.

24. Karn, J.; Stoltzfus, C.M. Transcriptional and posttranscriptional regulation of HIV-1 gene expression. *Cold Spring Harb. Perspect. Med.* **2012**, *2*.

25. Chang, D.D.; Sharp, P.A. Regulation by HIV Rev depends upon recognition of splice sites. *Cell* **1989**, *59*, 789–795.

26. Lu, X.B.; Heimer, J.; Rekosh, D.; Hammarskjold, M.L. U1 small nuclear RNA plays a direct role in the formation of a rev-regulated human immunodeficiency virus *env* mRNA that remains unspliced. *Proc. Natl. Acad. Sci. USA* **1990**, *87*, 7598–7602.

27. Stutz, F.; Rosbash, M. A functional interaction between Rev and yeast pre-mRNA is related to splicing complex formation. *EMBO J.* **1994**, *13*, 4096–4104.

28. Han, J.; Xiong, J.; Wang, D.; Fu, X.D. Pre-mRNA splicing: Where and when in the nucleus. *Trends Cell Biol.* **2011**, *21*, 336–343.

29. Cullen, B.R. Nuclear mRNA export: Insights from virology. *Trends Biochem. Sci.* **2003**, *28*, 419–424.

30. Cullen, B.R. Nuclear RNA export. *J. Cell Sci.* **2003**, *116*, 587–597.

31. Gruter, P.; Tabernero, C.; von Kobbe, C.; Schmitt, C.; Saavedra, C.; Bachi, A.; Wilm, M.; Felber, B.K.; Izaurralde, E. TAP, the human homolog of Mex67p, mediates CTE-dependent RNA export from the nucleus. *Mol. Cell* **1998**, *1*, 649–659.

32. Legrain, P.; Rosbash, M. Some *cis*- and *trans*-acting mutants for splicing target pre-mRNA to the cytoplasm. *Cell* **1989**, *57*, 573–583.

33. Nakielny, S.; Dreyfuss, G. Nuclear export of proteins and RNAs. *Curr. Opin. Cell Biol.* **1997**, *9*, 420–429.

34. Stutz, F.; Izaurralde, E. The interplay of nuclear mRNP assembly, mRNA surveillance and export. *Trends Cell Biol.* **2003**, *13*, 319–327.

35. Coyle, J.H.; Bor, Y.C.; Rekosh, D.; Hammarskjold, M.L. The Tpr protein regulates export of mRNAs with retained introns that traffic through the Nxf1 pathway. *RNA* **2011**, *17*, 1344–1356.

36. Rajanala, K.; Nandicoori, V.K. Localization of nucleoporin Tpr to the nuclear pore complex is essential for Tpr mediated regulation of the export of unspliced RNA. *PLoS ONE* **2012**, *7*, e29921.

37. Porrua, O.; Libri, D. RNA quality control in the nucleus: The Angels' share of RNA. *Biochim. Biophys. Acta* **2013**, *1829*, 604–611.

38. Felber, B.K.; Hadzopoulou-Cladaras, M.; Cladaras, C.; Copeland, T.; Pavlakis, G.N. Rev protein of human immunodeficiency virus type 1 affects the stability and transport of the viral mRNA. *Proc. Natl. Acad. Sci. USA* **1989**, *86*, 1495–1499.

39. Pollard, V.W.; Malim, M.H. The HIV-1 Rev protein. *Annu. Rev. Microbiol.* **1998**, *52*, 491–532.

40. Groom, H.C.; Anderson, E.C.; Lever, A.M. Rev: Beyond nuclear export. *J. Gen. Virol.* **2009**, *90*, 1303–1318.

41. Blissenbach, M.; Grewe, B.; Hoffmann, B.; Brandt, S.; Uberla, K. Nuclear RNA export and packaging functions of HIV-1 Rev revisited. *J. Virol.* **2010**, *84*, 6598–6604.

42. Berger, J.; Aepinus, C.; Dobrovnik, M.; Fleckenstein, B.; Hauber, J.; Bohnlein, E. Mutational analysis of functional domains in the HIV-1 Rev trans-regulatory protein. *Virology* **1991**, *183*, 630–635.

43. Bohnlein, E.; Berger, J.; Hauber, J. Functional mapping of the human immunodeficiency virus type 1 Rev RNA binding domain: New insights into the domain structure of Rev and Rex. *J. Virol.* **1991**, *65*, 7051–7055.

44. Daly, T.J.; Cook, K.S.; Gray, G.S.; Maione, T.E.; Rusche, J.R. Specific binding of HIV-1 recombinant Rev protein to the Rev-responsive element *in vitro*. *Nature* **1989**, *342*, 816–819.

45. Hope, T.J.; McDonald, D.; Huang, X.J.; Low, J.; Parslow, T.G. Mutational analysis of the human immunodeficiency virus type 1 Rev transactivator: Essential residues near the amino terminus. *J. Virol.* **1990**, *64*, 5360–5366.

46. Kubota, S.; Siomi, H.; Satoh, T.; Endo, S.; Maki, M.; Hatanaka, M. Functional similarity of HIV-I [rev] and HTLV-I [rex] proteins: Identification of a new nucleolar-targeting signal in [rev] protein. *Biochem. Biophys. Res. Commun.* **1989**, *162*, 963–970.

47. Malim, M.H.; Bohnlein, S.; Hauber, J.; Cullen, B.R. Functional dissection of the HIV-1 Rev *trans*-activator—Derivation of a *trans*-dominant repressor of Rev function. *Cell* **1989**, *58*, 205–214.

48. Perkins, A.; Cochrane, A.W.; Ruben, S.M.; Rosen, C.A. Structural and functional characterization of the human immunodeficiency virus rev protein. *J. Acquir. Immune Defic. Syndr.* **1989**, *2*, 256–263.

49. Daugherty, M.D.; Booth, D.S.; Jayaraman, B.; Cheng, Y.; Frankel, A.D. HIV Rev response element (RRE) directs assembly of the Rev homooligomer into discrete asymmetric complexes. *Proc. Natl. Acad. Sci. USA* **2010**, *107*, 12481–12486.

50. DiMattia, M.A.; Watts, N.R.; Stahl, S.J.; Rader, C.; Wingfield, P.T.; Stuart, D.I.; Steven, A.C.; Grimes, J.M. Implications of the HIV-1 Rev dimer structure at 3.2 Å resolution for multimeric binding to the Rev response element. *Proc. Natl. Acad. Sci. USA* **2010**, *107*, 5810–5814.

51. Cochrane, A.W.; Chen, C.H.; Rosen, C.A. Specific interaction of the human immunodeficiency virus Rev protein with a structured region in the env mRNA. *Proc. Natl. Acad. Sci. USA* **1990**, *87*, 1198–1202.

52. Malim, M.H.; McCarn, D.F.; Tiley, L.S.; Cullen, B.R. Mutational definition of the human immunodeficiency virus type 1 Rev activation domain. *J. Virol.* **1991**, *65*, 4248–4254.

53. Venkatesh, L.K.; Chinnadurai, G. Mutants in a conserved region near the carboxy-terminus of HIV-1 Rev identify functionally important residues and exhibit a dominant negative phenotype. *Virology* **1990**, *178*, 327–330.

54. Weichselbraun, I.; Farrington, G.K.; Rusche, J.R.; Bohnlein, E.; Hauber, J. Definition of the human immunodeficiency virus type 1 Rev and human T-cell leukemia virus type I Rex protein activation domain by functional exchange. *J. Virol.* **1992**, *66*, 2583–2587.

55. Fornerod, M.; Ohno, M.; Yoshida, M.; Mattaj, I.W. CRM1 is an export receptor for leucine-rich nuclear export signals. *Cell* **1997**, *90*, 1051–1060.

56. Daugherty, M.D.; Liu, B.; Frankel, A.D. Structural basis for cooperative RNA binding and export complex assembly by HIV Rev. *Nat. Struct. Mol. Biol.* **2010**, *17*, 1337–1342.

57. Booth, D.S.; Cheng, Y.; Frankel, A.D. The export receptor Crm1 forms a dimer to promote nuclear export of HIV RNA. *eLife* **2014**, *3*.

58. Taniguchi, I.; Mabuchi, N.; Ohno, M. HIV-1 Rev protein specifies the viral RNA export pathway by suppressing TAP/NXF1 recruitment. *Nucleic Acids Res.* **2014**, *42*, 6645–6658.

59. Suhasini, M.; Reddy, T.R. Cellular proteins and HIV-1 Rev function. *Curr. HIV Res.* **2009**, *7*, 91–100.

60. Bohne, J.; Wodrich, H.; Krausslich, H.G. Splicing of human immunodeficiency virus RNA is position-dependent suggesting sequential removal of introns from the 5′ end. *Nucleic Acids Res.* **2005**, *33*, 825–837.

61. Ajamian, L.; Abrahamyan, L.; Milev, M.; Ivanov, P.V.; Kulozik, A.E.; Gehring, N.H.; Mouland, A.J. Unexpected roles for UPF1 in HIV-1 RNA metabolism and translation. *RNA* **2008**, *14*, 914–927.

62. Jackson, R.J.; Hellen, C.U.; Pestova, T.V. The mechanism of eukaryotic translation initiation and principles of its regulation. *Nat. Rev. Mol. Cell Biol.* **2010**, *11*, 113–127.

63. Balvay, L.; Soto Rifo, R.; Ricci, E.P.; Decimo, D.; Ohlmann, T. Structural and functional diversity of viral IRESes. *Biochim. Biophys. Acta* **2009**, *1789*, 542–557.

64. De Breyne, S.; Soto-Rifo, R.; Lopez-Lastra, M.; Ohlmann, T. Translation initiation is driven by different mechanisms on the HIV-1 and HIV-2 genomic RNAs. *Virus Res.* **2013**, *171*, 366–381.

65. Grifo, J.A.; Tahara, S.M.; Morgan, M.A.; Shatkin, A.J.; Merrick, W.C. New initiation factor activity required for globin mRNA translation. *J. Biol. Chem.* **1983**, *258*, 5804–5810.

66. Prevot, D.; Darlix, J.L.; Ohlmann, T. Conducting the initiation of protein synthesis: The role of eIF4G. *Biol. Cell* **2003**, *95*, 141–156.

67. Pause, A.; Methot, N.; Svitkin, Y.; Merrick, W.C.; Sonenberg, N. Dominant negative mutants of mammalian translation initiation factor eIF-4A define a critical role for eIF-4F in cap-dependent and cap-independent initiation of translation. *EMBO J.* **1994**, *13*, 1205–1215.

68. Pause, A.; Sonenberg, N. Mutational analysis of a DEAD box RNA helicase: The mammalian translation initiation factor eIF-4A. *EMBO J.* **1992**, *11*, 2643–2654.

69. Rogers, G.W., Jr.; Richter, N.J.; Lima, W.F.; Merrick, W.C. Modulation of the helicase activity of eIF4A by eIF4B, eIF4H, and eIF4F. *J. Biol. Chem.* **2001**, *276*, 30914–30922.

70. Prevot, D.; Decimo, D.; Herbreteau, C.H.; Roux, F.; Garin, J.; Darlix, J.L.; Ohlmann, T. Characterization of a novel RNA-binding region of eIF4GI critical for ribosomal scanning. *EMBO J.* **2003**, *22*, 1909–1921.

71. Korneeva, N.L.; Lamphear, B.J.; Hennigan, F.L.; Merrick, W.C.; Rhoads, R.E. Characterization of the two eIF4A-binding sites on human eIF4G-1. *J. Biol. Chem.* **2001**, *276*, 2872–2879.

72. Lamphear, B.J.; Kirchweger, R.; Skern, T.; Rhoads, R.E. Mapping of functional domains in eukaryotic protein synthesis initiation factor 4G (eIF4G) with picornaviral proteases. Implications for cap-dependent and cap-independent translational initiation. *J. Biol. Chem.* **1995**, *270*, 21975–21983.

73. Mader, S.; Lee, H.; Pause, A.; Sonenberg, N. The translation initiation factor eIF-4E binds to a common motif shared by the translation factor eIF-4 gamma and the translational repressors 4E-binding proteins. *Mol. Cell. Biol.* **1995**, *15*, 4990–4997.

74. Imataka, H.; Sonenberg, N. Human eukaryotic translation initiation factor 4G (eIF4G) possesses two separate and independent binding sites for eIF4A. *Mol. Cell. Biol.* **1997**, *17*, 6940–6947.

75. Korneeva, N.L.; Lamphear, B.J.; Hennigan, F.L.; Rhoads, R.E. Mutually cooperative binding of eukaryotic translation initiation factor (eIF) 3 and eIF4A to human eIF4G-1. *J. Biol. Chem.* **2000**, *275*, 41369–41376.

76. Ohlmann, T.; Rau, M.; Pain, V.M.; Morley, S.J. The C-terminal domain of eukaryotic protein synthesis initiation factor (eIF) 4G is sufficient to support cap-independent translation in the absence of eIF4E. *EMBO J.* **1996**, *15*, 1371–1382.

77. Rau, M.; Ohlmann, T.; Morley, S.J.; Pain, V.M. A reevaluation of the cap-binding protein, eIF4E, as a rate-limiting factor for initiation of translation in reticulocyte lysate. *J. Biol. Chem.* **1996**, *271*, 8983–8990.

78. Safer, B.; Kemper, W.; Jagus, R. Identification of a 48S preinitiation complex in reticulocyte lysate. *J. Biol. Chem.* **1978**, *253*, 3384–3386.

79. Kozak, M. Adherence to the first-AUG rule when a second AUG codon follows closely upon the first. *Proc. Natl. Acad. Sci. USA* **1995**, *92*, 2662–2666.

80. Kozak, M. Recognition of AUG and alternative initiator codons is augmented by G in position +4 but is not generally affected by the nucleotides in positions +5 and +6. *EMBO J.* **1997**, *16*, 2482–2492.

81. Wegrzyn, J.L.; Drudge, T.M.; Valafar, F.; Hook, V. Bioinformatic analyses of mammalian 5′-UTR sequence properties of mRNAs predicts alternative translation initiation sites. *BMC Bioinform.* **2008**, *9*.

82. Jackson, R.J. The ATP requirement for initiation of eukaryotic translation varies according to the mRNA species. *Eur. J. Biochem.* **1991**, *200*, 285–294.

83. Pestova, T.V.; Kolupaeva, V.G. The roles of individual eukaryotic translation initiation factors in ribosomal scanning and initiation codon selection. *Genes Dev.* **2002**, *16*, 2906–2922.

84. Poyry, T.A.; Kaminski, A.; Jackson, R.J. What determines whether mammalian ribosomes resume scanning after translation of a short upstream open reading frame? *Genes Dev.* **2004**, *18*, 62–75.

85. Parsyan, A.; Svitkin, Y.; Shahbazian, D.; Gkogkas, C.; Lasko, P.; Merrick, W.C.; Sonenberg, N. mRNA helicases: The tacticians of translational control. *Nat. Rev. Mol. Cell Biol.* **2011**, *12*, 235–245.

86. Parsyan, A.; Shahbazian, D.; Martineau, Y.; Petroulakis, E.; Alain, T.; Larsson, O.; Mathonnet, G.; Tettweiler, G.; Hellen, C.U.; Pestova, T.V.; *et al.* The helicase protein DHX29 promotes translation initiation, cell proliferation, and tumorigenesis. *Proc. Natl. Acad. Sci. USA* **2009**, *106*, 22217–22222.

87. Pisareva, V.P.; Pisarev, A.V.; Komar, A.A.; Hellen, C.U.; Pestova, T.V. Translation initiation on mammalian mRNAs with structured 5′ UTRs requires DExH-box protein DHX29. *Cell* **2008**, *135*, 1237–1250.

88. Hartman, T.R.; Qian, S.; Bolinger, C.; Fernandez, S.; Schoenberg, D.R.; Boris-Lawrie, K. RNA helicase A is necessary for translation of selected messenger RNAs. *Nat. Struct. Mol. Biol.* **2006**, *13*, 509–516.

89. Jang, S.K.; Krausslich, H.G.; Nicklin, M.J.; Duke, G.M.; Palmenberg, A.C.; Wimmer, E. A segment of the 5′ nontranslated region of encephalomyocarditis virus RNA directs internal entry of ribosomes during *in vitro* translation. *J. Virol.* **1988**, *62*, 2636–2643.

90. Pelletier, J.; Sonenberg, N. Internal initiation of translation of eukaryotic mRNA directed by a sequence derived from poliovirus RNA. *Nature* **1988**, *334*, 320–325.

91. Jackson, R.J. The current status of vertebrate cellular mRNA IRESs. *Cold Spring Harb. Perspect. Biol.* **2013**, *5*.

92. Yamasaki, S.; Anderson, P. Reprogramming mRNA translation during stress. *Curr. Opin. Cell Biol.* **2008**, *20*, 222–226.

93. Johannes, G.; Carter, M.S.; Eisen, M.B.; Brown, P.O.; Sarnow, P. Identification of eukaryotic mRNAs that are translated at reduced cap binding complex eIF4F concentrations using a cDNA microarray. *Proc. Natl. Acad. Sci. USA* **1999**, *96*, 13118–13123.

94. Holcik, M.; Sonenberg, N.; Korneluk, R.G. Internal ribosome initiation of translation and the control of cell death. *Trends Genet.* **2000**, *16*, 469–473.

95. Pyronnet, S.; Dostie, J.; Sonenberg, N. Suppression of cap-dependent translation in mitosis. *Genes Dev.* **2001**, *15*, 2083–2093.

96. Pyronnet, S.; Pradayrol, L.; Sonenberg, N. A cell cycle-dependent internal ribosome entry site. *Mol. Cell* **2000**, *5*, 607–616.

97. Spriggs, K.A.; Stoneley, M.; Bushell, M.; Willis, A.E. Re-programming of translation following cell stress allows IRES-mediated translation to predominate. *Biol. Cell* **2008**, *100*, 27–38.

98. Stoneley, M.; Willis, A.E. Cellular internal ribosome entry segments: Structures, trans-acting factors and regulation of gene expression. *Oncogene* **2004**, *23*, 3200–3207.

99. Baudin, F.; Marquet, R.; Isel, C.; Darlix, J.L.; Ehresmann, B.; Ehresmann, C. Functional sites in the 5′ region of human immunodeficiency virus type 1 RNA form defined structural domains. *J. Mol. Biol.* **1993**, *229*, 382–397.

100. Berkhout, B. Structure and function of the human immunodeficiency virus leader RNA. *Prog. Nucleic Acid Res. Mol. Biol.* **1996**, *54*, 1–34.

101. Paillart, J.C.; Dettenhofer, M.; Yu, X.F.; Ehresmann, C.; Ehresmann, B.; Marquet, R. First snapshots of the HIV-1 RNA structure in infected cells and in virions. *J. Biol. Chem.* **2004**, *279*, 48397–48403.

102. Rosen, C.A.; Sodroski, J.G.; Haseltine, W.A. The location of cis-acting regulatory sequences in the human T cell lymphotropic virus type III (HTLV-III/LAV) long terminal repeat. *Cell* **1985**, *41*, 813–823.

103. Watts, J.M.; Dang, K.K.; Gorelick, R.J.; Leonard, C.W.; Bess, J.W., Jr.; Swanstrom, R.; Burch, C.L.; Weeks, K.M. Architecture and secondary structure of an entire HIV-1 RNA genome. *Nature* **2009**, *460*, 711–716.

104. Yilmaz, A.; Bolinger, C.; Boris-Lawrie, K. Retrovirus translation initiation: Issues and hypotheses derived from study of HIV-1. *Curr. HIV Res.* **2006**, *4*, 131–139.

105. Geballe, A.P.; Gray, M.K. Variable inhibition of cell-free translation by HIV-1 transcript leader sequences. *Nucleic Acids Res.* **1992**, *20*, 4291–4297.

106. Parkin, N.T.; Cohen, E.A.; Darveau, A.; Rosen, C.; Haseltine, W.; Sonenberg, N. Mutational analysis of the 5′ non-coding region of human immunodeficiency virus type 1: Effects of secondary structure on translation. *EMBO J.* **1988**, *7*, 2831–2837.

107. SenGupta, D.N.; Berkhout, B.; Gatignol, A.; Zhou, A.M.; Silverman, R.H. Direct evidence for translational regulation by leader RNA and Tat protein of human immunodeficiency virus type 1. *Proc. Natl. Acad. Sci. USA* **1990**, *87*, 7492–7496.

108. Svitkin, Y.V.; Pause, A.; Sonenberg, N. La autoantigen alleviates translational repression by the 5′ leader sequence of the human immunodeficiency virus type 1 mRNA. *J. Virol.* **1994**, *68*, 7001–7007.

109. Soto-Rifo, R.; Limousin, T.; Rubilar, P.S.; Ricci, E.P.; Decimo, D.; Moncorge, O.; Trabaud, M.A.; André, P.; Cimarelli, A.; Ohlmann, T.; *et al.* Different effects of the TAR structure on HIV-1 and HIV-2 genomic RNA translation. *Nucleic Acids Res.* **2012**, *40*, 2653–2667.

110. Dorin, D.; Bonnet, M.C.; Bannwarth, S.; Gatignol, A.; Meurs, E.F.; Vaquero, C. The TAR RNA-binding protein, TRBP, stimulates the expression of TAR-containing RNAs *in vitro* and *in vivo* independently of its ability to inhibit the dsRNA-dependent kinase PKR. *J. Biol. Chem.* **2003**, *278*, 4440–4448.

111. Dugre-Brisson, S.; Elvira, G.; Boulay, K.; Chatel-Chaix, L.; Mouland, A.J.; DesGroseillers, L. Interaction of Staufen1 with the 5' end of mRNA facilitates translation of these RNAs. *Nucleic Acids Res.* **2005**, *33*, 4797–4812.

112. Miele, G.; Mouland, A.; Harrison, G.P.; Cohen, E.; Lever, A.M. The human immunodeficiency virus type 1 5' packaging signal structure affects translation but does not function as an internal ribosome entry site structure. *J. Virol.* **1996**, *70*, 944–951.

113. Brasey, A.; Lopez-Lastra, M.; Ohlmann, T.; Beerens, N.; Berkhout, B.; Darlix, J.L.; Sonenberg, N. The leader of human immunodeficiency virus type 1 genomic RNA harbors an internal ribosome entry segment that is active during the G2/M phase of the cell cycle. *J. Virol.* **2003**, *77*, 3939–3949.

114. Andersen, J.L.; Planelles, V. The role of Vpr in HIV-1 pathogenesis. *Curr. HIV Res.* **2005**, *3*, 43–51.

115. Elder, R.T.; Benko, Z.; Zhao, Y. HIV-1 VPR modulates cell cycle G2/M transition through an alternative cellular mechanism other than the classic mitotic checkpoints. *Front. Biosci.* **2002**, *7*, d349–d357.

116. Gemeniano, M.C.; Sawai, E.T.; Sparger, E.E. Feline immunodeficiency virus Orf-A localizes to the nucleus and induces cell cycle arrest. *Virology* **2004**, *325*, 167–174.

117. He, J.; Choe, S.; Walker, R.; di Marzio, P.; Morgan, D.O.; Landau, N.R. Human immunodeficiency virus type 1 viral protein R (Vpr) arrests cells in the G2 phase of the cell cycle by inhibiting p34cdc2 activity. *J. Virol.* **1995**, *69*, 6705–6711.

118. Fan, H.; Penman, S. Regulation of protein synthesis in mammalian cells. II. Inhibition of protein synthesis at the level of initiation during mitosis. *J. Mol. Biol.* **1970**, *50*, 655–670.

119. Tarnowka, M.A.; Baglioni, C. Regulation of protein synthesis in mitotic HeLa cells. *J. Cell. Physiol.* **1979**, *99*, 359–367.

120. Goh, W.C.; Rogel, M.E.; Kinsey, C.M.; Michael, S.F.; Fultz, P.N.; Nowak, M.A.; Hahn, B.H.; Emerman, M. HIV-1 Vpr increases viral expression by manipulation of the cell cycle: A mechanism for selection of Vpr *in vivo*. *Nat. Med.* **1998**, *4*, 65–71.

121. Monette, A.; Valiente-Echeverria, F.; Rivero, M.; Cohen, E.A.; Lopez-Lastra, M.; Mouland, A.J. Dual mechanisms of translation initiation of the full-length HIV-1 mRNA contribute to gag synthesis. *PLoS ONE* **2013**, *8*, e68108.

122. Sharma, A.; Yilmaz, A.; Marsh, K.; Cochrane, A.; Boris-Lawrie, K. Thriving under stress: Selective translation of HIV-1 structural protein mRNA during Vpr-mediated impairment of eif4e translation activity. *PLoS Pathog.* **2012**, *8*, e1002612.

123. Amorim, R.; Costa, S.M.; Cavaleiro, N.P.; da Silva, E.E.; da Costa, L.J. HIV-1 transcripts use IRES-initiation under conditions where Cap-dependent translation is restricted by poliovirus 2A protease. *PLoS ONE* **2014**, *9*, e88619.

124. Plank, T.D.; Whitehurst, J.T.; Kieft, J.S. Cell type specificity and structural determinants of IRES activity from the 5' leaders of different HIV-1 transcripts. *Nucleic Acids Res.* **2013**, *41*, 6698–6714.

125. Vallejos, M.; Deforges, J.; Plank, T.D.; Letelier, A.; Ramdohr, P.; Abraham, C.G.; Valiente-Echeverría, F.; Kieft, J.S.; Sargueil, B.; López-Lastra, M.; *et al.* Activity of the human immunodeficiency virus type 1 cell cycle-dependent internal ribosomal entry site is modulated by IRES *trans*-acting factors. *Nucleic Acids Res.* **2011**, *39*, 6186–6200.

126. Vallejos, M.; Carvajal, F.; Pino, K.; Navarrete, C.; Ferres, M.; Huidobro-Toro, J.P.; Sargueil, B.; López-Lastra, M. Functional and structural analysis of the internal ribosome entry site present in the mRNA of natural variants of the HIV-1. *PLoS ONE* **2012**, *7*, e35031.

127. Valiente-Echeverría, F.; Vallejos, M.; Monette, A.; Pino, K.; Letelier, A.; Huidobro-Toro, J.P.; Mouland, A.J.; López-Lastra, M. A cis-acting element present within the Gag open reading frame negatively impacts on the activity of the HIV-1 IRES. *PLoS ONE* **2013**, *8*, e56962.

128. Gendron, K.; Ferbeyre, G.; Heveker, N.; Brakier-Gingras, L. The activity of the HIV-1 IRES is stimulated by oxidative stress and controlled by a negative regulatory element. *Nucleic Acids Res.* **2011**, *39*, 902–912.

129. Charnay, N.; Ivanyi-Nagy, R.; Soto-Rifo, R.; Ohlmann, T.; Lopez-Lastra, M.; Darlix, J.L. Mechanism of HIV-1 Tat RNA translation and its activation by the Tat protein. *Retrovirology* **2009**, *6*.

130. Rivas-Aravena, A.; Ramdohr, P.; Vallejos, M.; Valiente-Echeverria, F.; Dormoy-Raclet, V.; Rodriguez, F.; Pino, K.; Holzmann, C.; Huidobro-Toro, J.P.; Gallouzi, I.E.; *et al.* The Elav-like protein HuR exerts translational control of viral internal ribosome entry sites. *Virology* **2009**, *392*, 178–185.

131. Alvarez, E.; Menendez-Arias, L.; Carrasco, L. The eukaryotic translation initiation factor 4GI is cleaved by different retroviral proteases. *J. Virol.* **2003**, *77*, 12392–12400.

132. Castello, A.; Franco, D.; Moral-Lopez, P.; Berlanga, J.J.; Alvarez, E.; Wimmer, E.; Carrasco, L. HIV-1 protease inhibits Cap- and poly(A)-dependent translation upon eIF4GI and PABP cleavage. *PLoS ONE* **2009**, *4*, e7997.

133. Ohlmann, T.; Prevot, D.; Decimo, D.; Roux, F.; Garin, J.; Morley, S.J.; Darlix, J.L. *In vitro* cleavage of eIF4GI but not eIF4GII by HIV-1 protease and its effects on translation in the rabbit reticulocyte lysate system. *J. Mol. Biol.* **2002**, *318*, 9–20.

134. Ventoso, I.; Blanco, R.; Perales, C.; Carrasco, L. HIV-1 protease cleaves eukaryotic initiation factor 4G and inhibits cap-dependent translation. *Proc. Natl. Acad. Sci. USA* **2001**, *98*, 12966–12971.

135. Chamond, N.; Locker, N.; Sargueil, B. The different pathways of HIV genomic RNA translation. *Biochem. Soc. Trans.* **2010**, *38*, 1548–1552.

136. Buck, C.B.; Shen, X.; Egan, M.A.; Pierson, T.C.; Walker, C.M.; Siliciano, R.F. The human immunodeficiency virus type 1 gag gene encodes an internal ribosome entry site. *J. Virol.* **2001**, *75*, 181–191.

137. Camerini, V.; Decimo, D.; Balvay, L.; Pistello, M.; Bendinelli, M.; Darlix, J.L.; Ohlmann, T. A dormant internal ribosome entry site controls translation of feline immunodeficiency virus. *J. Virol.* **2008**, *82*, 3574–3583.

138. Herbreteau, C.H.; Weill, L.; Decimo, D.; Prevot, D.; Darlix, J.L.; Sargueil, B.; Ohlmann, T. HIV-2 genomic RNA contains a novel type of IRES located downstream of its initiation codon. *Nat. Struct. Mol. Biol.* **2005**, *12*, 1001–1007.

139. Nicholson, M.G.; Rue, S.M.; Clements, J.E.; Barber, S.A. An internal ribosome entry site promotes translation of a novel SIV Pr55(Gag) isoform. *Virology* **2006**, *349*, 325–334.

140. De Breyne, S.; Chamond, N.; Decimo, D.; Trabaud, M.A.; Andre, P.; Sargueil, B.; Ohlmann, T. *In vitro* studies reveal that different modes of initiation on HIV-1 mRNA have different levels of requirement for eukaryotic initiation factor 4F. *FEBS J.* **2012**, *279*, 3098–3111.

141. Weill, L.; James, L.; Ulryck, N.; Chamond, N.; Herbreteau, C.H.; Ohlmann, T.; Sargueil, B. A new type of IRES within gag coding region recruits three initiation complexes on HIV-2 genomic RNA. *Nucleic Acids Res.* **2010**, *38*, 1367–1381.

142. Locker, N.; Chamond, N.; Sargueil, B. A conserved structure within the HIV gag open reading frame that controls translation initiation directly recruits the 40S subunit and eIF3. *Nucleic Acids Res.* **2011**, *39*, 2367–2377.

143. Ricci, E.P.; Herbreteau, C.H.; Decimo, D.; Schaupp, A.; Datta, S.A.; Rein, A.; Darlix, J.L.; Ohlmann, T. *In vitro* expression of the HIV-2 genomic RNA is controlled by three distinct internal ribosome entry segments that are regulated by the HIV protease and the Gag polyprotein. *RNA* **2008**, *14*, 1443–1455.

144. Berkhout, B.; Arts, K.; Abbink, T.E. Ribosomal scanning on the 5′-untranslated region of the human immunodeficiency virus RNA genome. *Nucleic Acids Res.* **2011**, *39*, 5232–5244.

145. Bolinger, C.; Sharma, A.; Singh, D.; Yu, L.; Boris-Lawrie, K. RNA helicase A modulates translation of HIV-1 and infectivity of progeny virions. *Nucleic Acids Res.* **2010**, *38*, 1686–1696.

146. Yedavalli, V.S.; Neuveut, C.; Chi, Y.H.; Kleiman, L.; Jeang, K.T. Requirement of DDX3 DEAD box RNA helicase for HIV-1 Rev-RRE export function. *Cell* **2004**, *119*, 381–392.

147. Soto-Rifo, R.; Ohlmann, T. The role of the DEAD-box RNA helicase DDX3 in mRNA metabolism. *Wiley Interdiscip. Rev. RNA* **2013**, *4*, 369–385.

148. Linder, P.; Jankowsky, E. From unwinding to clamping—The DEAD box RNA helicase family. *Nat. Rev. Mol. Cell Biol.* **2011**, *12*, 505–516.

149. Soto-Rifo, R.; Rubilar, P.S.; Limousin, T.; de Breyne, S.; Decimo, D.; Ohlmann, T. DEAD-box protein DDX3 associates with eIF4F to promote translation of selected mRNAs. *EMBO J.* **2012**, *31*, 3745–3756.

150. Soto-Rifo, R.; Rubilar, P.S.; Ohlmann, T. The DEAD-box helicase DDX3 substitutes for the cap-binding protein eIF4E to promote compartmentalized translation initiation of the HIV-1 genomic RNA. *Nucleic Acids Res.* **2013**, *41*, 6286–6299.

151. Lai, M.C.; Wang, S.W.; Cheng, L.; Tarn, W.Y.; Tsai, S.J.; Sun, H.S. Human DDX3 interacts with the HIV-1 Tat protein to facilitate viral mRNA translation. *PLoS ONE* **2013**, *8*, e68665.

152. Yedavalli, V.S.; Jeang, K.T. Trimethylguanosine capping selectively promotes expression of Rev-dependent HIV-1 RNAs. *Proc. Natl. Acad. Sci. USA* **2010**, *107*, 14787–14792.

177

153. Darzynkiewicz, E.; Stepinski, J.; Ekiel, I.; Jin, Y.; Haber, D.; Sijuwade, T.; Tahara, S.M. β-globin mRNAs capped with m^7G, $m_2^{2.7}(2)G$ or $m_3^{2.2.7}G$ differ in intrinsic translation efficiency. *Nucleic Acids Res.* **1988**, *16*, 8953–8962.

154. Worch, R.; Niedzwiecka, A.; Stepinski, J.; Mazza, C.; Jankowska-Anyszka, M.; Darzynkiewicz, E.; Cusack, S.; Stolarski, R. Specificity of recognition of mRNA 5′ cap by human nuclear cap-binding complex. *RNA* **2005**, *11*, 1355–1363.

155. Decker, C.J.; Parker, R. P-bodies and stress granules: Possible roles in the control of translation and mRNA degradation. *Cold Spring Harb. Perspect. Biol.* **2012**, *4*.

156. Stoecklin, G.; Kedersha, N. Relationship of GW/P-bodies with stress granules. *Adv. Exp. Med. Biol.* **2013**, *768*, 197–211.

157. Lloyd, R.E. Regulation of stress granules and P-bodies during RNA virus infection. *Wiley Interdiscip. Rev. RNA* **2013**, *4*, 317–331.

158. Reineke, L.C.; Lloyd, R.E. Diversion of stress granules and P-bodies during viral infection. *Virology* **2013**, *436*, 255–267.

159. Valiente-Echeverria, F.; Melnychuk, L.; Mouland, A.J. Viral modulation of stress granules. *Virus Res.* **2012**, *169*, 430–437.

160. Nathans, R.; Chu, C.Y.; Serquina, A.K.; Lu, C.C.; Cao, H.; Rana, T.M. Cellular microRNA and P bodies modulate host-HIV-1 interactions. *Mol. Cell* **2009**, *34*, 696–709.

161. Chable-Bessia, C.; Meziane, O.; Latreille, D.; Triboulet, R.; Zamborlini, A.; Wagschal, A.; Jacquet, J.M.; Reynes, J.; Levy, Y.; Saib, A.; *et al.* Suppression of HIV-1 replication by microRNA effectors. *Retrovirology* **2009**, *6*.

162. Abrahamyan, L.G.; Chatel-Chaix, L.; Ajamian, L.; Milev, M.P.; Monette, A.; Clement, J.F.; Song, R.; Lehmann, M.; DesGroseillers, L.; Laughrea, M.; *et al.* Novel Staufen1 ribonucleoproteins prevent formation of stress granules but favour encapsidation of HIV-1 genomic RNA. *J. Cell Sci.* **2010**, *123*, 369–383.

163. Phalora, P.K.; Sherer, N.M.; Wolinsky, S.M.; Swanson, C.M.; Malim, M.H. HIV-1 replication and APOBEC3 antiviral activity are not regulated by P bodies. *J. Virol.* **2012**, *86*, 11712–11724.

164. Reed, J.C.; Molter, B.; Geary, C.D.; McNevin, J.; McElrath, J.; Giri, S.; Klein, K.C.; Lingappa, J.R. HIV-1 Gag co-opts a cellular complex containing DDX6, a helicase that facilitates capsid assembly. *J. Cell Biol.* **2012**, *198*, 439–456.

165. Valiente-Echeverria, F.; Melnychuk, L.; Vyboh, K.; Ajamian, L.; Gallouzi, I.E.; Bernard, N.; Mouland, A.J. eEF2 and Ras-GAP SH3 domain-binding protein (G3BP1) modulate stress granule assembly during HIV-1 infection. *Nat. Commun.* **2014**, *5*.

166. Soto-Rifo, R.; Valiente-Echeverria, F.; Rubilar, P.S.; Garcia-de-Gracia, F.; Ricci, E.P.; Limousin, T.; Décimo, D.; Mouland, A.J.; Ohlmann, T. HIV-2 genomic RNA accumulates in stress granules in the absence of active translation. *Nucleic Acids Res.* **2014**, *42*, 12861–12875.

167. Braddock, M.; Thorburn, A.M.; Chambers, A.; Elliott, G.D.; Anderson, G.J.; Kingsman, A.J.; Kingsman, S.M. A nuclear translational block imposed by the HIV-1 U3 region is relieved by the Tat-TAR interaction. *Cell* **1990**, *62*, 1123–1133.

168. Braddock, M.; Powell, R.; Blanchard, A.D.; Kingsman, A.J.; Kingsman, S.M. HIV-1 TAR RNA-binding proteins control TAT activation of translation in Xenopus oocytes. *FASEB J.* **1993**, *7*, 214–222.

169. SenGupta, D.N.; Silverman, R.H. Activation of interferon-regulated, dsRNA-dependent enzymes by human immunodeficiency virus-1 leader RNA. *Nucleic Acids Res.* **1989**, *17*, 969–978.

170. Edery, I.; Petryshyn, R.; Sonenberg, N. Activation of double-stranded RNA-dependent kinase (DSL) by the TAR region of HIV-1 mRNA: A novel translational control mechanism. *Cell* **1989**, *56*, 303–312.

171. Williams, B.R. Signal integration via PKR. *Sci. STKE* **2001**, *2001*.

172. Clerzius, G.; Gelinas, J.F.; Gatignol, A. Multiple levels of PKR inhibition during HIV-1 replication. *Rev. Med. Virol.* **2011**, *21*, 42–53.

173. Charbonneau, J.; Gendron, K.; Ferbeyre, G.; Brakier-Gingras, L. The 5′ UTR of HIV-1 full-length mRNA and the Tat viral protein modulate the programmed-1 ribosomal frameshift that generates HIV-1 enzymes. *RNA* **2012**, *18*, 519–529.

174. Arrigo, S.J.; Chen, I.S. Rev is necessary for translation but not cytoplasmic accumulation of HIV-1 vif, vpr, and env/vpu 2 RNAs. *Genes Dev.* **1991**, *5*, 808–819.

175. Kimura, T.; Hashimoto, I.; Nishikawa, M.; Fujisawa, J.I. A role for Rev in the association of HIV-1 gag mRNA with cytoskeletal beta-actin and viral protein expression. *Biochimie* **1996**, *78*, 1075–1080.

176. Campbell, L.H.; Borg, K.T.; Haines, J.K.; Moon, R.T.; Schoenberg, D.R.; Arrigo, S.J. Human immunodeficiency virus type 1 Rev is required *in vivo* for binding of poly(A)-binding protein to Rev-dependent RNAs. *J. Virol.* **1994**, *68*, 5433–5438.

177. Groom, H.C.; Anderson, E.C.; Dangerfield, J.A.; Lever, A.M. Rev regulates translation of human immunodeficiency virus type 1 RNAs. *J. Gen. Virol.* **2009**, *90*, 1141–1147.

178. Anderson, E.C.; Lever, A.M. Human immunodeficiency virus type 1 Gag polyprotein modulates its own translation. *J. Virol.* **2006**, *80*, 10478–10486.

Flaviviral Replication Complex: Coordination between RNA Synthesis and 5′-RNA Capping

Valerie J. Klema, Radhakrishnan Padmanabhan and Kyung H. Choi

Abstract: Genome replication in flavivirus requires (−) strand RNA synthesis, (+) strand RNA synthesis, and 5′-RNA capping and methylation. To carry out viral genome replication, flavivirus assembles a replication complex, consisting of both viral and host proteins, on the cytoplasmic side of the endoplasmic reticulum (ER) membrane. Two major components of the replication complex are the viral non-structural (NS) proteins NS3 and NS5. Together they possess all the enzymatic activities required for genome replication, yet how these activities are coordinated during genome replication is not clear. We provide an overview of the flaviviral genome replication process, the membrane-bound replication complex, and recent crystal structures of full-length NS5. We propose a model of how NS3 and NS5 coordinate their activities in the individual steps of (−) RNA synthesis, (+) RNA synthesis, and 5′-RNA capping and methylation.

Reprinted from *Viruses*. Cite as: Klema, V.J.; Padmanabhan, R.; Choi, K.H. Flaviviral Replication Complex: Coordination between RNA Synthesis and 5′-RNA Capping. *Viruses* **2015**, *7*, 4640–4656.

1. Flavivirus Genome and Viral Non-Structural (NS) Proteins

Flaviviruses are positive (+) sense RNA viruses belonging to the family *Flaviviridae*, which also includes *Hepacivirus* and *Pestivirus*. The *Flavivirus* genus includes over 70 viruses, many of which cause arboviral diseases in humans, such as dengue (DENV), West Nile (WNV), tick-borne encephalitis (TBEV), and yellow fever virus. The 11 kb flaviviral RNA genome consists of a 5′-cap, a 5′-untranslated region (5′-UTR), a single open reading frame (ORF), and a 3′-UTR. The 5′- and 3′-UTRs contain conserved RNA secondary structures that are important for viral replication, including sequences that mediate long range 5′- and 3′-RNA interactions [1–8]. The viral ORF is translated into a polyprotein, C-prM-E-NS1-NS2A-NS2B-NS3-NS4A-NS4B-NS5, that is subsequently cleaved into individual proteins by viral and host proteases. Three structural proteins (C, prM, and E) form capsids and seven non-structural (NS) proteins (NS1, NS2A, NS2B, NS3, NS4A, NS4B, NS5) are involved in the assembly of the viral replication complex [9]. Among NS proteins, the functions of NS3 and NS5 in viral replication are well established. NS3 consists of an N-terminal serine protease and a C-terminal helicase. NS3 protease activity requires NS2B as a cofactor, and cleaves the viral polyprotein at several positions between NS proteins [10–14]. The NS3 helicase domain has helicase, RNA-stimulated nucleoside triphosphate hydrolase and 5'-RNA triphosphatase activities [15–19]. Helicase activity would be required for unwinding the double-stranded (ds) RNA intermediate formed during genome synthesis, and 5'-RNA triphosphatase activity is required for 5'-RNA cap formation [20,21]. NS5, the largest viral protein (103 kDa), consists of an N-terminal methyltransferase (MTase) and a C-terminal RNA-dependent RNA polymerase (RdRp). The RdRp is involved in viral genome replication and carries out both (−) and (+) strand RNA synthesis [22]. The NS5 MTase has RNA guanylyltransferase and methyltransferase activities necessary for 5'-RNA capping and cap methylations [23,24]. Little is known about functions of the membrane proteins NS2A, NS4A, and NS4B, but they are likely involved in membrane alterations and assembly of the viral replication complex on the cellular membrane [25–29]. NS1 may be involved in multiple steps in the viral life cycle, including viral replication [30–33]. NS1 exists as two forms, either a membrane-bound dimer in the viral replication complex or a secreted hexameric form with as-yet-unknown function [34,35]. More extensive reviews of NS protein functions can be found elsewhere [36,37].

2. Flavivirus Genome Replication

The flavivirus replication complex carries out RNA synthesis, RNA capping and RNA methylation steps to produce the genome with a type 1 cap structure (^{m7}GpppN^m-RNA) at its 5' end. RNA synthesis in flavivirus is semi-conservative and asymmetric. The (+) sense RNA, which is the same polarity as the viral genome, is predominantly formed over the (−) sense RNA; single-stranded (ss) RNA found in flavivirus-infected cells is (+) sense RNA, and (−) strand RNA is only detected in the dsRNA form [38]. When DENV or Kunjin virus-infected cells were incubated with a radiolabeled NTP (^{3}H-UTP or ^{32}P-labeled guanosine-5'-triphosphate (GTP)), three radiolabeled RNA products were identified: a double-stranded replicative form (RF) RNA, single-stranded genome length RNA, and the slowest migrating, replicative intermediate (RI) [22,39,40]. RNase treatment of the RI suggested that the RI likely contains growing nascent RNAs on the RF template, which displace a pre-existing strand of the same polarity (Figure 1A). One to ten nascent RNA strands are synthesized on one RF template at a time [29,39]. Based on these observations, a general scheme of RNA synthesis and the associated required enzymatic activities are outlined in Figure 1A. The genomic (+) sense RNA is first used as a template by NS5 RdRp to synthesize a complementary (−) sense RNA. The (−) strand RNA remains base-paired with the (+) strand RNA, resulting in a dsRNA intermediate. The (−) strand within the dsRNA intermediate then serves as the template to generate (+) sense RNA. NS3 helicase activity may be required to unwind the dsRNA. The nascent

(+) strand synthesized on the (−) RNA template displaces a pre-existing (+) strand and is released as a dsRNA product. The newly generated dsRNA is then recycled as a template to generate additional copies of (+) sense RNA.

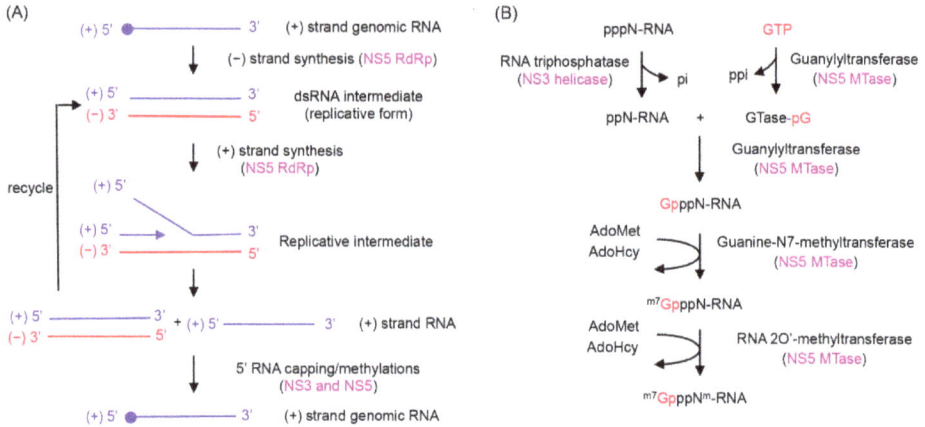

Figure 1. RNA synthesis by the flavivirus replication complex. (**A**) RNA replication by flaviviral NS3 and NS5 proteins. A (+) strand genomic RNA serves as a template to produce (−) strand RNA. The (−) strand RNA exists as a dsRNA intermediate (replicative form). The (−) strand within the dsRNA intermediate is then used as a template for (+) strand RNA synthesis. The dsRNA product is released and recycled for additional (+) strand synthesis. Flavivirus replication is asymmetric, and multiple copies of (+) strand RNA are synthesized from a (−) strand template. The (+) strand RNA is then capped and methylated by NS3 and NS5 to form (+) strand genomic RNA. The identities of the enzymes involved in each step are shown in purple. (**B**) 5′-RNA cap synthesis by flaviviral NS3 and NS5 proteins. The type 1 cap is formed on (+) strand RNA via the four sequential enzyme activities of RNA triphosphatase, guanylyltransferase, guanine-N7-MTase, and RNA 2′O-MTase. First, triphosphatase activity of the NS3 helicase releases the terminal phosphate from the 5′-triphosphate end of (+) strand RNA. A guanosine monophosphate (GMP) moiety from GTP is transferred to the 5′ end of the now-diphosphorylated RNA through guanylyltransferase activity of the NS5 MTase. The capped RNA is then methylated first at the N7 position of the guanine cap and subsequently at the ribose 2′-O position of the first RNA nucleotide. The MTase domain of NS5 carries out both methylations using S-adenosyl-L-methionine (AdoMet) as a methyl donor. AdoMet is converted to S-adenosyl-L-homosysteine (AdoHcy) during this process.

The (+) strand progeny RNA is subsequently capped at its 5′ end and methylated to form a type 1 cap (Figure 1B). The cap is shown to be present only on genomic RNA, and not on the dsRNA intermediate (RF form) in WNV-infected cells [38].

The RNA-capping process is likely to occur as the (+) strand RNA is synthesized during the initial stages of RNA synthesis (Figure 1), but little is known about how flavivirus coordinates RNA synthesis and 5′ end RNA capping. The RNA capping and methylation processes require three enzymatic activities. First, the 5′-triphosphate end of (+) RNA is converted into a 5′-diphosphate by the RNA triphosphatase activity of the NS3 helicase. Second, a GMP moiety from GTP is transferred to the 5′-diphosphate RNA by a guanylyltransferase (GTase). The NS5 MTase domain has weak guanylyltransferase activity that transfers a GMP cap from GTP to the 5′ end of (+) sense RNA [23]. Finally, the capped RNA is first methylated at the N7 position of guanine and then at the ribose 2′-OH position of the first nucleotide of the RNA. The MTase domain of NS5 functions as both the N7-MTase and the 2′O-MTase, and transfers a methyl group from the cofactor AdoMet to the substrate capped RNA in both reactions [24,41].

3. Flaviviral Replication Complex

Genome replication in flavivirus is carried out by a membrane-bound viral replication complex consisting of viral NS proteins, viral RNA and unidentified host proteins [9]. Membrane fractions prepared from the lysates of flavivirus-infected cells contained virus-specific RNA and proteins, and retained all of the RdRp activity [29]. Flaviviruses alter host cellular membrane structures, presumably to protect the viral RNA and replication proteins from triggering the host immune response and being degraded by cytoplasmic enzymes. Using electron microscopy and tomography, flavivirus (DENV, WNV, and TBEV)-infected cells have been shown to form spherical single-membrane vesicles of 80–100 nm in diameter via invagination of the ER membrane into the ER lumen [25,42,43] (Figure 2). This contrasts to the related hepatitis C virus, which shows the double membrane vesicle structures with varied sizes between 150–1000 nm [44,45]. Viral NS proteins NS1, NS2B, NS3, NS4A, NS4B and NS5 along with dsRNA are located in the membrane vesicles, suggesting that these vesicles are the sites of RNA replication [28,45]. The vesicles have a pore connecting the interior of the vesicle to the cytoplasm, which is thought to allow exchange of nucleotides and RNA product with the cytoplasm [25,42]. In addition, virus particles have been shown to bud out on ER membranes next to the replication vesicles, suggesting that the viral replication and encapsidation of the viral genome/virion assembly are likely coordinated.

Although the exact composition of the replication complex is not known, all flaviviral NS proteins have been shown to be a component of the replication complex. Using immunoprecipitation, yeast two-hybrid, and fluorescence resonance energy transfer (FRET) assays, interactions among NS proteins have been identified [27,46–48] (Figure 2). Membrane proteins NS2B, NS4A, and NS4B are likely involved in membrane alterations and/or anchoring the viral replication complex to the membrane [26]. NS2B interacts with the three other membrane proteins, NS2A, NS4A, and NS4B [47]. NS4A and NS4B have been proposed to interact with NS1 by genetic studies [49,50]. Oligomerization of NS4A, and dimerizations of NS4B and NS1 have also been reported [51,52]. NS3 itself does not have any membrane-association or transmembrane region. However, the active protease function of NS3 (in the N-terminal domain) requires the cofactor NS2B, thus NS3 localizes to the membrane as an NS3-NS2B complex [15,47]. NS3 has also been shown to interact with NS4B through its C-terminal helicase domain [53]. NS5 also does not have any membrane-associated region and interacts only with NS3. Thus, NS5 likely accumulates to the membrane via NS3-NS5 interactions. Unidentified host proteins are also involved in viral replication. It has been shown that DENV NS5 protein alone was not able to use a viral dsRNA intermediate (RF form) as a template for RNA synthesis. However, the addition of uninfected cell lysate to the NS5 reaction could restore polymerase activity, suggesting that host proteins are also involved in viral replication [40]. Consequently, an *in vitro* flaviviral replication system, which can use the viral genome to synthesize the methylated, capped RNA product, is currently unavailable.

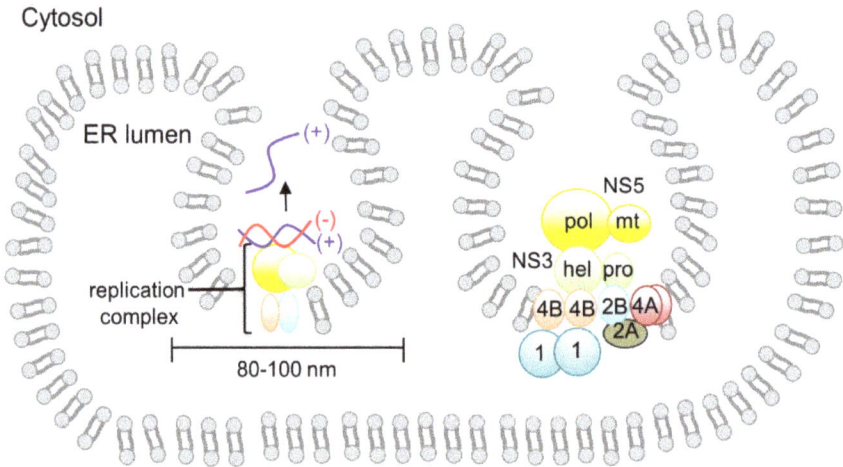

Figure 2. Flaviviral replication complex assembled on the cytoplasmic side of the ER membrane. The viral replication complex is associated with the virus vesicle (single membrane vesicle) formed by invagination of the ER membrane. Interactions between flaviviral NS proteins are indicated schematically. Viral membrane proteins NS2A, NS2B, NS4A, and NS4B form a scaffold for the assembly of NS3 and NS5 proteins. Oligomerization of NS4A and dimerization of NS4B are depicted [51,52]. NS3 interacts with NS2B through its protease domain and with NS4B through its helicase domain. NS5 interacts only with NS3. The NS1 dimer is located in the ER lumen, and associates with NS4A and NS4B.

4. Inter- and Intramolecular Coordination between NS5 and NS3 during Genome Replication

At the heart of the flaviviral replication complex are NS3 and NS5, which together account for all enzymatic activities required to amplify the RNA genome and to attach a type 1 cap to its $5'$ end (Figure 1). Other viral and host proteins in the viral replication complex would provide additional efficiency and specificity for viral replication. NS5 is involved in RNA synthesis, $5'$-RNA cap transfer, and cap methylations. The NS3 helicase domain is involved in dsRNA unwinding and removal of the γ-phosphate at the $5'$-RNA prior to RNA capping. However, how NS3 and NS5 activities are coordinated during viral replication is not known. It is possible that individual activities of the two proteins are modulated by mutual interaction with each other as well as with other proteins and viral RNA [23,54–56]. Interactions between NS3 and NS5 proteins from several flaviviral species have been shown using pull-down assays from infected cell extracts, *in vivo* fluorescent measurements, and yeast two-hybrid studies [46,47,57]. Their respective binding sites map to the C-terminal domain of NS3 (residues 303–618) and the RdRp domain of NS5 (residues 320–368, which also includes a nuclear localization sequence) [58–60].

The physical linkage between the MTase and RdRp domains within NS5 suggests that viral genome replication and $5'$-RNA capping may be coordinated between the two domains. Several genetic studies show that mutations in the MTase domain impact RNA replication by the RdRp domain, suggesting that the RdRp and MTase interact during viral genome replication [61–64]. In addition, a full-length DENV type 2 RNA containing an NS5 chimera, in which the MTase from DENV type 4 is fused to the RdRp from DENV type 2, cannot carry out replication although the MTase from DENV type 2 and 4 NS5 share high sequence identity ($\sim$70%) [65]. This suggests that sequence-specific interactions between the MTase and RdRp domains are necessary for viral replication. Consequently, there has been great interest in carrying out structural and biochemical studies to define whether and/or how coordination between these activities may occur, and how coordination may be exploited to design novel antiviral agents effective against flavivirus infection [66,67]. The individual MTase and RdRp domains from several flaviviral sources have been structurally characterized, but until recently there were no available structures of full-length flavivirus NS5. The crystal structures of full-length NS5 from Japanese encephalitis virus (JEV) and DENV were recently reported, and provided a first glimpse into how the MTase and RdRp domains interact.

5. Full-Length NS5 Structure and Function

The recently reported structures of full-length NS5 from JEV and DENV offer details of specific interactions between the MTase and RdRp domains and suggest

185

a structural mechanism by which their activities may be coordinated within the flavivirus replication complex [68,69]. We have recently determined the crystal structure of DENV NS5 as a dimer (in preparation), and will provide an overview of the three structures (Figure 3). Briefly, the MTase domain contains a central core structure characteristic of AdoMet-dependent MTases consisting of a 7-stranded β-sheet surrounded by 4 α-helices. The RdRp domain adopts the shape of a closed right hand and consists of palm, fingers, and thumb subdomains. The inner surfaces of the fingers and thumb subdomains form the entrance to a template-binding channel that leads to the active site in the palm subdomain. The locations of respective active sites in the MTase and the RdRp are indicated in Figure 3. Surprisingly, in both JEV and DENV full-length NS5 the active sites of the MTase and RdRp domains are located on opposite faces and do not interact with each other. The MTase domain sits "behind" the RdRp domain when viewed in its right hand orientation, opposite the dsRNA exit channel and close to the template-binding channel (Figure 3). In all three structures, the individual MTase and RdRp domains are nearly superimposable with one another. The major difference between the JEV and DENV NS5 structures is the relative orientation of the MTase and RdRp domains, primarily due to different conformations of the inter-domain linker (residues 263-272, DENV type 3 numbering). The MTase domains in JEV and DENV NS5 are rotated relative to one another by $\sim 100°$. A strictly conserved ^{260}Gly-Thr-Arg^{262} (GTR) motif, N-terminal to the inter-domain linker, is proposed to act as a hinge that allows movement of the MTase and RdRp domains relative to one another [68–70]. Accordingly, the relative orientation of the MTase and RdRp active sites are quite different in JEV and DENV NS5, and the domain interfaces involve different sets of MTase residues to interact with the same area of the RdRp. In the JEV NS5, the domain interface centers around a hydrophobic core consisting of residues P113, L115, and W121 from the MTase domain, and F467, F351, and P585 from the RdRp. In contrast, the domain interface in DENV NS5 is stabilized mainly by polar and electrostatic interactions involving MTase residues Q63, E252, and D256 to interact with relatively the same area of RdRp (Figure 3). Interestingly, the MTase residues involved in the inter-domain interactions in the JEV NS5 monomer (P113, P115, and W121 in DENV numbering) are present at the dimer interface in our DENV NS5 dimer structure. Thus, the same MTase region mediates intra- and inter-molecular interactions with the RdRp domain in JEV and DENV NS5, respectively, suggesting that NS5 may also function as a dimer (Figure 3). Taken together, comparison of the full-length NS5 structures from JEV and DENV shows that flavivirus NS5 can alter relative MTase and RdRp domain orientations and its monomer-dimer state by modulating its linker region. This may endow NS5 with the flexibility it needs to carry out multiple functions in the replication complex and interact with other proteins at different steps of viral replication (see below).

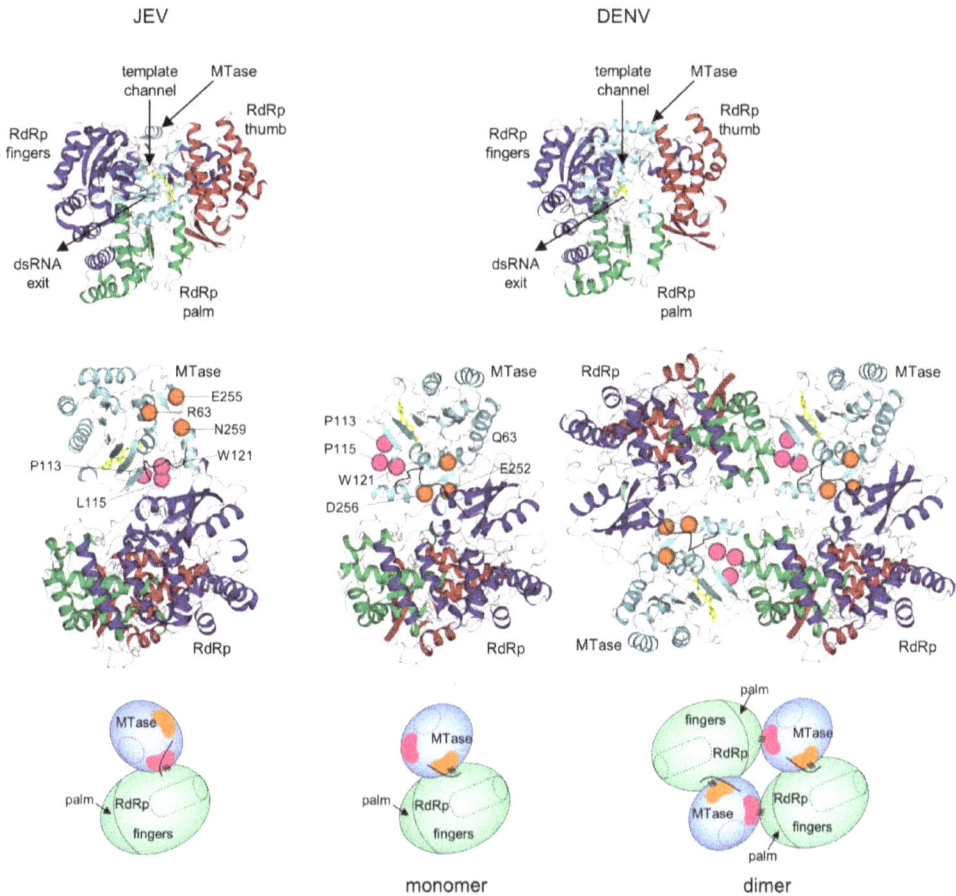

Figure 3. Comparison of full-length NS5 structures from JEV and DENV. Overall fold of JEV and DENV full-length NS5 (Protein Data Bank entries 4K6M and 4VOQ) are shown from the canonical right hand view of RdRp (top) and side view (bottom). Side views of both monomer and dimer forms of DENV NS5 are shown. The RdRp domain consists of fingers (green), palm (blue), and thumb (red) subdomains. The MTase domain (cyan) sits opposite the dsRNA exit channel of the RdRp and near the template-binding channel. The methyl donor used during MTase reaction, AdoMet, is shown in yellow. In each structure, the same region of the RdRp contacts the MTase. The MTase residues at the domain interface from JEV and DENV are indicated by pink and orange spheres, respectively. For comparison, corresponding residues are also mapped on the structures. Domain interactions are also indicated in schematics by a series of three lines below the side view. The MTase active site is indicated by dashed ovals, and the RdRp template channel is shown as a dashed tube.

6. Model of Flaviviral Replication

The multifunctional flavivirus NS5 carries out several reactions. These include synthesis of (−) strand RNA, synthesis of (+) strand RNA, addition of a guanine cap to the 5′ end of (+) strand RNA, and two methylation reactions to form a type 1 cap structure at the 5′ end of the nascent RNA. Consequently, NS5 interacts with other viral proteins and several different forms of viral RNA, including both single- and double-stranded, (+) and (−) strand, capped and uncapped, and methylated and unmethylated forms of capped RNA. Thus, NS5 likely needs to adopt multiple conformations during different stages of genome replication and processing, as suggested by the conformational variation observed for the structures of full-length NS5 from JEV and DENV. Below, we summarize viral replication steps in terms of coordination of NS3 and NS5 functions.

1. (−) strand RNA synthesis: (−) sense RNA synthesis is carried out by NS5 polymerase using the viral (+) RNA genome as a template. The function of NS3 does not seem to be required for (−) strand RNA synthesis, since NS5 (either the full-length protein or the polymerase domain) alone is capable of synthesizing RNA using viral subgenomic RNA. The (+) sense RNA genome contains cyclization sequences at the 5′- and 3′-UTR, and the cyclized genome serves as the template for (−) strand RNA synthesis [1–8,71]. NS5 polymerase recognizes a conserved RNA structure called stem loop A (SLA) within the 5′-UTR as a promoter, and initiates (−) strand RNA synthesis at the 3′ end of the genome. The nascent (−) RNA product exists as dsRNA, base-paired with the (+) strand RNA template [29].

2. (+) strand RNA synthesis: The viral replication complex uses the (−) strand RNA in the dsRNA intermediate as the template to synthesize multiple copies of (+) sense RNA. Both NS5 and NS3 activities are required for (+) RNA synthesis (Figure 4). NS3 helicase first unwinds the dsRNA intermediate into (+) and (−) sense RNAs. Following strand separation, the 3′ end of (−) sense RNA will bind the template-binding channel of NS5 RdRp and serve as a template for (+) strand RNA synthesis. Because (−) strand RNA only exists in the dsRNA form, specific RNA interactions, such as those between the cyclization sequences or stem loop structures necessary for (−) strand synthesis, are not required for (+) sense RNA synthesis [29,72]. Upon completion of nascent (+) sense RNA synthesis, a dsRNA product (consisting of the nascent (+) strand and the (−) strand template) will be released from the RdRp and recycled for another round of (+) strand RNA synthesis [29,36].

3. 5′-RNA capping and methylation of (+) sense RNA: The (+) sense RNA will be capped and methylated to form a type 1 cap at the 5′ end (Figure 1B). NS3 helicase hydrolyzes the 5′-terminal phosphate of (+) sense RNA and

188

converts it to a diphosphate using its RNA triphosphatase activity. The 5′-diphosphorylated (+) sense RNA then binds the NS5 MTase for capping and methylations. For 5′-RNA capping, NS5 MTase first reacts with GTP to form a covalently linked GMP-enzyme intermediate, and then transfers the GMP moiety to the 5′-diphosphate of the (+) strand RNA [23]. Next, the MTase methylates the N7 position of the guanine cap and the ribose 2′-OH position of the first nucleotide. Each methylation reaction requires distinct RNA sequences and lengths, and N7 cap methylation requires the presence of the stem loop A in the 5′-UTR of the viral genome [73]. Since the NS5 MTase carries out three separate reactions using one active site, the 5′ end of (+) RNA needs to dissociate and then re-associate with the MTase at each step [74].

4. Coordination of (+) RNA synthesis and 5′-RNA capping: The 5′-RNA capping and methylation steps are likely coupled with (+) sense RNA synthesis, but during which step of RNA synthesis 5′-RNA capping and methylation occur is not clear. Capping of 5′-RNA could occur on the nascent (+) strand RNA while it is being synthesized by the NS5 RdRp domain (co-transcriptional model in Figure 4A). After a short stretch of (+) sense RNA is synthesized, the 5′ end of the nascent (+) RNA could be dephosphorylated by NS3 helicase, and capped and methylated by NS5 MTase. The 5′-capped (+) sense RNA will then be continuously synthesized until the entire (−) strand is copied. Alternatively, the 5′-RNA capping could occur on the preexisting full-length (+) RNA that is separated from the (−) strand template RNA by NS3 helicase (post-transcriptional model in Figure 4A). Upon unwinding of dsRNA, NS3 helicase hydrolyzes the 5′-terminal phosphate of (+) sense RNA and NS5 MTase subsequently attaches the type 1 cap to the 5′ end. The consequential difference between the two mechanisms would be whether the (+) strand RNA in the dsRNA form is capped. In the co-transcriptional model, all (+) strand RNA, either ssRNA or dsRNA, would be capped and methylated because the nascent RNA is co-transcriptionally capped and methylated. In the post-transcriptional model, only displaced (+) strand RNA, and not the (+) sense RNA in the dsRNA form, would be capped and methylated. Since the cap is shown to be present only on genomic (+) sense RNA, and not on the dsRNA form in WNV-infected cell [29,38], it seems likely that 5′-capping occurs on the fully synthesized (+) sense RNA. In this case, upon completion of a cycle of nascent (+) sense RNA synthesis, a dsRNA product and a capped (+) strand RNA (identical to the viral genome) will be released from the RdRp and MTase domains of NS5 (Figure 4B).

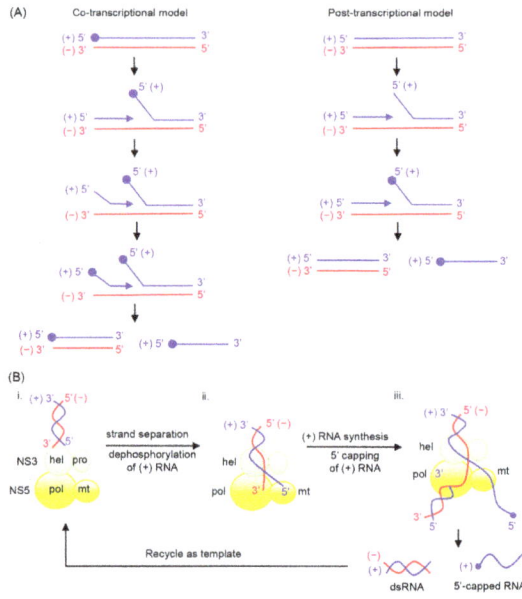

Figure 4. Possible mechanisms of coordination between RNA synthesis and 5'-RNA capping. (**A**) Two possible mechanisms are proposed for 5'-RNA capping for (+) sense RNA. In a co-transcriptional model, 5'-RNA capping occurs co-transcriptionally while nascent (+) sense RNA is being synthesized from dsRNA replicative form. After a short stretch of (+) sense RNA being synthesized, the 5' end of the nascent (+) RNA would be dephosphorylated by NS3 helicase, and capped and methylated by NS5 MTase. The cap structure is depicted with a closed circle. The 5'-capped (+) sense RNA would then be continuously synthesized until the entire (−) strand is copied. Upon completion of a cycle, a dsRNA containing a capped (+) strand and a capped (+) RNA are synthesized. In a post-transcriptional model, 5'-RNA capping occurs on the fully synthesized (+) strand RNA. Following strand separation of dsRNA, the (+) sense RNA could be dephoshphorylated by NS3 helicase, and capped and methylated by NS5 MTase. Upon completion of a cycle, a dsRNA and a capped (+) RNA are synthesized. (**B**) Coordination of RNA synthesis and 5'-RNA capping by flavivirus NS3 and NS5 in the post-transcriptional model. NS3 helicase unwinds the dsRNA intermediate into (+) and (−) sense RNA, and hydrolyzes the 5' end of (+) sense RNA by its 5'-RNA triphosphatase activity. (**ii**) The 3' end of the (−) strand RNA will enter the NS5 RdRp template-binding channel, and serve as a template for (+) strand RNA synthesis. The 5'-dephosphorylated (+) RNA enters the NS5 MTase active site for 5'-RNA capping and methylations. (**iii**) Upon completion of nascent (+) sense RNA synthesis, a dsRNA and a 5'-capped (+) strand RNA are released from the NS5 RdRp and MTase domains. The dsRNA product, consisting of the (−) strand template and nascent (+) strand, is then recycled for another round of (+) strand RNA synthesis.

190

7. Conclusions and Future Perspectives

The identification and visualization of subcellular membrane structures used by flavivirus as replication factories have provided insight into how flavivirus may orchestrate viral replication and virion assembly in infected cells. Additionally, crystal structures of the full-length NS5 have provided details of intra- and inter-molecular interactions that stabilize different forms of NS5. These forms may interact with RNA and protein components of the replication complex differently and thus serve as a platform to promote different steps during genome replication. Future studies will be geared toward understanding how these membrane vesicles are related to the assembly and function of the viral replication complex at the molecular level. In particular, how flaviviral and host proteins establish such elaborate membrane vesicles for replication, how flaviviral NS proteins are assembled in the membrane-bound viral replication complex, and how the viral replication complex carries out individual steps of RNA replication are of great interest.

Acknowledgments: This work was supported by NIH Research Grants AI087856 (to K.H.C. and R.P) and AI105985 (to K.H.C.).

Author Contributions: V.J.K., R.P. and K.H.C. wrote the paper.

Conflicts of Interest: The authors declare no conflict of interest.

References

1. Lodeiro, M.F.; Filomatori, C.V.; Gamarnik, A.V. Structural and functional studies of the promoter element for dengue virus RNA replication. *J. Virol.* **2009**, *83*, 993–1008.
2. Filomatori, C.V.; Lodeiro, M.F.; Alvarez, D.E.; Samsa, M.M.; Pietrasanta, L.; Gamarnik, A.V. A 5′ RNA element promotes dengue virus RNA synthesis on a circular genome. *Genes Dev.* **2006**, *20*, 2238–2249.
3. Ackermann, M.; Padmanabhan, R. De novo synthesis of RNA by the dengue virus RNA-dependent RNA polymerase exhibits temperature dependence at the initiation but not elongation phase. *J. Biol. Chem.* **2001**, *276*, 39926–39937.
4. You, S.; Falgout, B.; Markoff, L.; Padmanabhan, R. *In vitro* RNA synthesis from exogenous dengue viral rna templates requires long range interactions between 5′- and 3′-terminal regions that influence RNA structure. *J. Biol. Chem.* **2001**, *276*, 15581–15591.
5. Hahn, C.S.; Hahn, Y.S.; Rice, C.M.; Lee, E.; Dalgarno, L.; Strauss, E.G.; Strauss, J.H. Conserved elements in the 3′ untranslated region of flavivirus RNAs and potential cyclization sequences. *J. Mol. Biol.* **1987**, *198*, 33–41.
6. Khromykh, A.A.; Meka, H.; Guyatt, K.J.; Westaway, E.G. Essential role of cyclization sequences in flavivirus RNA replication. *J. Virol.* **2001**, *75*, 6719–6728.
7. Corver, J.; Lenches, E.; Smith, K.; Robison, R.A.; Sando, T.; Strauss, E.G.; Strauss, J.H. Fine mapping of a cis-acting sequence element in yellow fever virus RNA that is required for RNA replication and cyclization. *J. Virol.* **2003**, *77*, 2265–2270.

8. Lo, M.K.; Tilgner, M.; Bernard, K.A.; Shi, P.Y. Functional analysis of mosquito-borne flavivirus conserved sequence elements within 3′ untranslated region of West Nile virus by use of a reporting replicon that differentiates between viral translation and RNA replication. *J. Virol.* **2003**, *77*, 10004–10014.

9. Lindenbach, B.D.; Thiel, H.J.; Rice, C.M. *Flaviviridae: The Viruses and Their Replication*; Lippincott-Raven Publishers: Philadelphia, PA, USA, 2007; Volume 1, pp. 1101–1152.

10. Falgout, B.; Pethel, M.; Zhang, Y.M.; Lai, C.J. Both nonstructural proteins NS2B and NS3 are required for the proteolytic processing of dengue virus nonstructural proteins. *J. Virol.* **1991**, *65*, 2467–2475.

11. Yusof, R.; Clum, S.; Wetzel, M.; Murthy, H.M.; Padmanabhan, R. Purified NS2B/NS3 serine protease of dengue virus type 2 exhibits cofactor NS2B dependence for cleavage of substrates with dibasic amino acids *in vitro*. *J. Biol. Chem.* **2000**, *275*, 9963–9969.

12. Chambers, T.J.; Weir, R.C.; Grakoui, A.; McCourt, D.W.; Bazan, J.F.; Fletterick, R.J.; Rice, C.M. Evidence that the N-terminal domain of nonstructural protein NS3 from yellow fever virus is a serine protease responsible for site-specific cleavages in the viral polyprotein. *Proc. Natl. Acad. Sci. USA* **1990**, *87*, 8898–8902.

13. Jan, L.R.; Yang, C.S.; Trent, D.W.; Falgout, B.; Lai, C.J. Processing of japanese encephalitis virus non-structural proteins: NS2B-NS3 complex and heterologous proteases. *J. Gen. Virol.* **1995**, *76* (Pt 3), 573–580.

14. Clum, S.; Ebner, K.E.; Padmanabhan, R. Cotranslational membrane insertion of the serine proteinase precursor NS2B-NS3(pro) of dengue virus type 2 is required for efficient *in vitro* processing and is mediated through the hydrophobic regions of NS2B. *J. Biol. Chem.* **1997**, *272*, 30715–30723.

15. Li, H.; Clum, S.; You, S.; Ebner, K.E.; Padmanabhan, R. The serine protease and RNA-stimulated nucleoside triphosphatase and RNA helicase functional domains of dengue virus type 2 NS3 converge within a region of 20 amino acids. *J. Virol.* **1999**, *73*, 3108–3116.

16. Matusan, A.E.; Pryor, M.J.; Davidson, A.D.; Wright, P.J. Mutagenesis of the dengue virus type 2 NS3 protein within and outside helicase motifs: Effects on enzyme activity and virus replication. *J. Virol.* **2001**, *75*, 9633–9643.

17. Bartelma, G.; Padmanabhan, R. Expression, purification, and characterization of the RNA 5′-triphosphatase activity of dengue virus type 2 nonstructural protein 3. *Virology* **2002**, *299*, 122–132.

18. Benarroch, D.; Selisko, B.; Locatelli, G.A.; Maga, G.; Romette, J.L.; Canard, B. The RNA helicase, nucleotide 5′-triphosphatase, and RNA 5′-triphosphatase activities of dengue virus protein NS3 are Mg^{2+}-dependent and require a functional walker B motif in the helicase catalytic core. *Virology* **2004**, *328*, 208–218.

19. Wengler, G.; Wengler, G. The NS 3 nonstructural protein of flaviviruses contains an RNA triphosphatase activity. *Virology* **1993**, *197*, 265–273.

20. Xu, T.; Sampath, A.; Chao, A.; Wen, D.; Nanao, M.; Chene, P.; Vasudevan, S.G.; Lescar, J. Structure of the dengue virus helicase/nucleoside triphosphatase catalytic domain at a resolution of 2.4 a. *J. Virol.* **2005**, *79*, 10278–10288.

21. Wang, C.C.; Huang, Z.S.; Chiang, P.L.; Chen, C.T.; Wu, H.N. Analysis of the nucleoside triphosphatase, RNA triphosphatase, and unwinding activities of the helicase domain of dengue virus NS3 protein. *FEBS Lett.* **2009**, *583*, 691–696.

22. Chu, P.W.; Westaway, E.G. Replication strategy of kunjin virus: Evidence for recycling role of replicative form RNA as template in semiconservative and asymmetric replication. *Virology* **1985**, *140*, 68–79.

23. Issur, M.; Geiss, B.J.; Bougie, I.; Picard-Jean, F.; Despins, S.; Mayette, J.; Hobdey, S.E.; Bisaillon, M. The flavivirus NS5 protein is a true RNA guanylyltransferase that catalyzes a two-step reaction to form the RNA cap structure. *RNA* **2009**, *15*, 2340–2350.

24. Egloff, M.P.; Benarroch, D.; Selisko, B.; Romette, J.L.; Canard, B. An RNA cap (nucleoside-2′-o-)-methyltransferase in the flavivirus RNA polymerase NS5: Crystal structure and functional characterization. *EMBO J.* **2002**, *21*, 2757–2768.

25. Welsch, S.; Miller, S.; Romero-Brey, I.; Merz, A.; Bleck, C.K.; Walther, P.; Fuller, S.D.; Antony, C.; Krijnse-Locker, J.; Bartenschlager, R. Composition and three-dimensional architecture of the dengue virus replication and assembly sites. *Cell. Host Microbe* **2009**, *5*, 365–375.

26. Miller, S.; Kastner, S.; Krijnse-Locker, J.; Buhler, S.; Bartenschlager, R. The non-structural protein 4A of dengue virus is an integral membrane protein inducing membrane alterations in a 2k-regulated manner. *J. Biol. Chem.* **2007**, *282*, 8873–8882.

27. Westaway, E.G.; Mackenzie, J.M.; Kenney, M.T.; Jones, M.K.; Khromykh, A.A. Ultrastructure of kunjin virus-infected cells: Colocalization of NS1 and NS3 with double-stranded RNA, and of NS2b with NS3, in virus-induced membrane structures. *J. Virol.* **1997**, *71*, 6650–6661.

28. Mackenzie, J.M.; Khromykh, A.A.; Jones, M.K.; Westaway, E.G. Subcellular localization and some biochemical properties of the flavivirus kunjin nonstructural proteins NS2A and NS4A. *Virology* **1998**, *245*, 203–215.

29. Westaway, E.G.; Mackenzie, J.M.; Khromykh, A.A. Kunjin RNA replication and applications of kunjin replicons. *Adv. Virus Res.* **2003**, *59*, 99–140.

30. Youn, S.; Ambrose, R.L.; Mackenzie, J.M.; Diamond, M.S. Non-structural protein-1 is required for West Nile virus replication complex formation and viral RNA synthesis. *Virol. J.* **2013**, *10*, e339.

31. Mackenzie, J.M.; Jones, M.K.; Young, P.R. Immunolocalization of the dengue virus nonstructural glycoprotein NS1 suggests a role in viral RNA replication. *Virology* **1996**, *220*, 232–240.

32. Khromykh, A.A.; Sedlak, P.L.; Guyatt, K.J.; Hall, R.A.; Westaway, E.G. Efficient trans-complementation of the flavivirus kunjin NS5 protein but not of the NS1 protein requires its coexpression with other components of the viral replicase. *J. Virol.* **1999**, *73*, 10272–10280.

33. Lindenbach, B.D.; Rice, C.M. Trans-complementation of yellow fever virus NS1 reveals a role in early RNA replication. *J. Virol.* **1997**, *71*, 9608–9617.

34. Flamand, M.; Megret, F.; Mathieu, M.; Lepault, J.; Rey, F.A.; Deubel, V. Dengue virus type 1 nonstructural glycoprotein NS1 is secreted from mammalian cells as a soluble hexamer in a glycosylation-dependent fashion. *J. Virol.* **1999**, *73*, 6104–6110.

35. Winkler, G.; Maxwell, S.E.; Ruemmler, C.; Stollar, V. Newly synthesized dengue-2 virus nonstructural protein NS1 is a soluble protein but becomes partially hydrophobic and membrane-associated after dimerization. *Virology* **1989**, *171*, 302–305.

36. Roby, J.A.; Funk, A.; Khromykh, A.A. Flavivirus replication and assembly. In *Molecular Virology and Control of Flaviviruses*; Shi, P.Y., Ed.; Caister Academic Press: Norfolk, UK, 2012.

37. Brinton, M.A. Replication cycle and molecular biology of the West Nile virus. *Viruses* **2014**, *6*, 13–53.

38. Wengler, G.; Wengler, G.; Gross, H.J. Studies on virus-specific nucleic acids synthesized in vertebrate and mosquito cells infected with flaviviruses. *Virology* **1978**, *89*, 423–437.

39. Cleaves, G.R.; Ryan, T.E.; Schlesinger, R.W. Identification and characterization of type 2 dengue virus replicative intermediate and replicative form RNAs. *Virology* **1981**, *111*, 73–83.

40. Raviprakash, K.; Sinha, M.; Hayes, C.G.; Porter, K.R. Conversion of dengue virus replicative form RNA (RF) to replicative intermediate (RI) by nonstructural proteins NS-5 and NS-3. *Am. J. Trop. Med. Hyg.* **1998**, *58*, 90–95.

41. Ray, D.; Shah, A.; Tilgner, M.; Guo, Y.; Zhao, Y.; Dong, H.; Deas, T.S.; Zhou, Y.; Li, H.; Shi, P.Y. West Nile virus 5′-cap structure is formed by sequential guanine N-7 and ribose 2′-O methylations by nonstructural protein 5. *J. Virol.* **2006**, *80*, 8362–8370.

42. Miorin, L.; Romero-Brey, I.; Maiuri, P.; Hoppe, S.; Krijnse-Locker, J.; Bartenschlager, R.; Marcello, A. Three-dimensional architecture of tick-borne encephalitis virus replication sites and trafficking of the replicated RNA. *J. Virol.* **2013**, *87*, 6469–6481.

43. Gillespie, L.K.; Hoenen, A.; Morgan, G.; Mackenzie, J.M. The endoplasmic reticulum provides the membrane platform for biogenesis of the flavivirus replication complex. *J. Virol.* **2010**, *84*, 10438–10447.

44. Romero-Brey, I.; Merz, A.; Chiramel, A.; Lee, J.Y.; Chlanda, P.; Haselman, U.; Santarella-Mellwig, R.; Habermann, A.; Hoppe, S.; Kallis, S.; *et al.* Three-dimensional architecture and biogenesis of membrane structures associated with hepatitis C virus replication. *PLoS Pathog.* **2012**, *8*, e1003056.

45. Harak, C.; Lohmann, V. Ultrastructure of the replication sites of positive-strand RNA viruses. *Virology* **2015**, *479–480*.

46. Kapoor, M.; Zhang, L.; Ramachandra, M.; Kusukawa, J.; Ebner, K.E.; Padmanabhan, R. Association between NS3 and NS5 proteins of dengue virus type 2 in the putative RNA replicase is linked to differential phosphorylation of NS5. *J. Biol. Chem.* **1995**, *270*, 19100–19106.

47. Yu, L.; Takeda, K.; Markoff, L. Protein-protein interactions among West Nile non-structural proteins and transmembrane complex formation in mammalian cells. *Virology* **2013**, *446*, 365–377.

48. Chen, C.J.; Kuo, M.D.; Chien, L.J.; Hsu, S.L.; Wang, Y.M.; Lin, J.H. RNA-protein interactions: Involvement of NS3, NS5, and 3' noncoding regions of japanese encephalitis virus genomic RNA. *J. Virol.* **1997**, *71*, 3466–3473.

49. Youn, S.; Li, T.; McCune, B.T.; Edeling, M.A.; Fremont, D.H.; Cristea, I.M.; Diamond, M.S. Evidence for a genetic and physical interaction between nonstructural proteins NS1 and NS4B that modulates replication of West Nile virus. *J. Virol.* **2012**, *86*, 7360–7371.

50. Lindenbach, B.D.; Rice, C.M. Genetic interaction of flavivirus nonstructural proteins NS1 and NS4A as a determinant of replicase function. *J. Virol.* **1999**, *73*, 4611–4621.

51. Stern, O.; Hung, Y.F.; Valdau, O.; Yaffe, Y.; Harris, E.; Hoffmann, S.; Willbold, D.; Sklan, E.H. An N-terminal amphipathic helix in dengue virus nonstructural protein 4A mediates oligomerization and is essential for replication. *J. Virol.* **2013**, *87*, 4080–4085.

52. Zou, J.; Xie, X.; Lee le, T.; Chandrasekaran, R.; Reynaud, A.; Yap, L.; Wang, Q.Y.; Dong, H.; Kang, C.; Yuan, Z.; *et al.* Dimerization of flavivirus NS4B protein. *J. Virol.* **2014**, *88*, 3379–3391.

53. Umareddy, I.; Chao, A.; Sampath, A.; Gu, F.; Vasudevan, S.G. Dengue virus NS4B interacts with NS3 and dissociates it from single-stranded RNA. *J. Gen. Virol.* **2006**, *87*, 2605–2614.

54. Cui, T.; Sugrue, R.J.; Xu, Q.; Lee, A.K.; Chan, Y.C.; Fu, J. Recombinant dengue virus type 1 NS3 protein exhibits specific viral RNA binding and NTPase activity regulated by the NS5 protein. *Virology* **1998**, *246*, 409–417.

55. Yon, C.; Teramoto, T.; Mueller, N.; Phelan, J.; Ganesh, V.K.; Murthy, K.H.; Padmanabhan, R. Modulation of the nucleoside triphosphatase/RNA helicase and 5'-RNA triphosphatase activities of dengue virus type 2 nonstructural protein 3 (NS3) by interaction with NS5, the RNA-dependent RNA polymerase. *J. Biol. Chem.* **2005**, *280*, 27412–27419.

56. Liu, W.J.; Sedlak, P.L.; Kondratieva, N.; Khromykh, A.A. Complementation analysis of the flavivirus kunjin NS3 and NS5 proteins defines the minimal regions essential for formation of a replication complex and shows a requirement of NS3 in cis for virus assembly. *J. Virol.* **2002**, *76*, 10766–10775.

57. Johansson, M.; Brooks, A.J.; Jans, D.A.; Vasudevan, S.G. A small region of the dengue virus-encoded RNA-dependent RNA polymerase, NS5, confers interaction with both the nuclear transport receptor importin-beta and the viral helicase, NS3. *J. Gen. Virol.* **2001**, *82*, 735–745.

58. Tay, M.Y.; Saw, W.G.; Zhao, Y.; Chan, K.W.; Singh, D.; Chong, Y.; Forwood, J.K.; Ooi, E.E.; Gruber, G.; Lescar, J.; *et al.* The C-terminal 50 amino acid residues of dengue NS3 protein are important for NS3-NS5 interaction and viral replication. *J. Biol. Chem.* **2015**, *290*, 2379–2394.

59. Moreland, N.J.; Tay, M.Y.; Lim, E.; Rathore, A.P.; Lim, A.P.; Hanson, B.J.; Vasudevan, S.G. Monoclonal antibodies against dengue NS2B and NS3 proteins for the study of protein interactions in the flaviviral replication complex. *J. Virol. Methods* **2012**, *179*, 97–103.

60. Zou, G.; Chen, Y.L.; Dong, H.; Lim, C.C.; Yap, L.J.; Yau, Y.H.; Shochat, S.G.; Lescar, J.; Shi, P.Y. Functional analysis of two cavities in flavivirus NS5 polymerase. *J. Biol. Chem.* **2011**, *286*, 14362–14372.

61. Tan, C.S.; Hobson-Peters, J.M.; Stoermer, M.J.; Fairlie, D.P.; Khromykh, A.A.; Hall, R.A. An interaction between the methyltransferase and RNA dependent RNA polymerase domains of the West Nile virus NS5 protein. *J. Gen. Virol.* **2013**, *94*, 1961–1971.

62. Zhang, B.; Dong, H.; Zhou, Y.; Shi, P.Y. Genetic interactions among the West Nile virus methyltransferase, the RNA-dependent RNA polymerase, and the 5′ stem-loop of genomic RNA. *J. Virol.* **2008**, *82*, 7047–7058.

63. Malet, H.; Egloff, M.P.; Selisko, B.; Butcher, R.E.; Wright, P.J.; Roberts, M.; Gruez, A.; Sulzenbacher, G.; Vonrhein, C.; Bricogne, G.; *et al.* Crystal structure of the RNA polymerase domain of the West Nile virus non-structural protein 5. *J. Biol. Chem.* **2007**, *282*, 10678–10689.

64. Potisopon, S.; Priet, S.; Collet, A.; Decroly, E.; Canard, B.; Selisko, B. The methyltransferase domain of dengue virus protein NS5 ensures efficient RNA synthesis initiation and elongation by the polymerase domain. *Nucleic Acids Res.* **2014**, *42*, 11642–11656.

65. Teramoto, T.; Boonyasuppayakorn, S.; Handley, M.; Choi, K.H.; Padmanabhan, R. Substitution of NS5 N-terminal domain of dengue virus type 2 RNA with type 4 domain caused impaired replication and emergence of adaptive mutants with enhanced fitness. *J. Biol. Chem.* **2014**, *289*, 22385–22400.

66. Bussetta, C.; Choi, K.H. Dengue virus nonstructural protein 5 adopts multiple conformations in solution. *Biochemistry* **2012**, *51*, 5921–5931.

67. Takahashi, H.; Takahashi, C.; Moreland, N.J.; Chang, Y.T.; Sawasaki, T.; Ryo, A.; Vasudevan, S.G.; Suzuki, Y.; Yamamoto, N. Establishment of a robust dengue virus NS3-NS5 binding assay for identification of protein-protein interaction inhibitors. *Antivir. Res.* **2012**, *96*, 305–314.

68. Lu, G.; Gong, P. Crystal structure of the full-length japanese encephalitis virus NS5 reveals a conserved methyltransferase-polymerase interface. *PLoS Pathog.* **2013**, *9*, e1003549.

69. Zhao, Y.; Soh, T.S.; Zheng, J.; Chan, K.W.; Phoo, W.W.; Lee, C.C.; Tay, M.Y.; Swaminathan, K.; Cornvik, T.C.; Lim, S.P.; *et al.* A crystal structure of the dengue virus NS5 protein reveals a novel inter-domain interface essential for protein flexibility and virus replication. *PLoS Pathog.* **2015**, *11*, e1004682.

70. Li, X.D.; Shan, C.; Deng, C.L.; Ye, H.Q.; Shi, P.Y.; Yuan, Z.M.; Gong, P.; Zhang, B. The interface between methyltransferase and polymerase of NS5 is essential for flavivirus replication. *PLoS Negl. Trop. Dis.* **2014**, *8*, e2891.

71. You, S.; Padmanabhan, R. A novel *in vitro* replication system for dengue virus. Initiation of RNA synthesis at the 3′-end of exogenous viral RNA templates requires 5′- and 3′-terminal complementary sequence motifs of the viral RNA. *J. Biol. Chem.* **1999**, *274*, 33714–33722.

72. Nomaguchi, M.; Teramoto, T.; Yu, L.; Markoff, L.; Padmanabhan, R. Requirements for West Nile virus (−)- and (+)-strand subgenomic RNA synthesis *in vitro* by the viral RNA-dependent RNA polymerase expressed in escherichia coli. *J. Biol. Chem.* **2004**, *279*, 12141–12151.

73. Dong, H.; Ray, D.; Ren, S.; Zhang, B.; Puig-Basagoiti, F.; Takagi, Y.; Ho, C.K.; Li, H.; Shi, P.Y. Distinct RNA elements confer specificity to flavivirus RNA cap methylation events. *J. Virol.* **2007**, *81*, 4412–4421.

74. Dong, H.; Ren, S.; Zhang, B.; Zhou, Y.; Puig-Basagoiti, F.; Li, H.; Shi, P.Y. West Nile virus methyltransferase catalyzes two methylations of the viral RNA cap through a substrate-repositioning mechanism. *J. Virol.* **2008**, *82*, 4295–4307.

Modes of Human T Cell Leukemia Virus Type 1 Transmission, Replication and Persistence

Alexandre Carpentier, Pierre-Yves Barez, Malik Hamaidia, Hélène Gazon, Alix de Brogniez, Srikanth Perike, Nicolas Gillet and Luc Willems

Abstract: Human T-cell leukemia virus type 1 (HTLV-1) is a retrovirus that causes cancer (Adult T cell Leukemia, ATL) and a spectrum of inflammatory diseases (mainly HTLV-associated myelopathy—tropical spastic paraparesis, HAM/TSP). Since virions are particularly unstable, HTLV-1 transmission primarily occurs by transfer of a cell carrying an integrated provirus. After transcription, the viral genomic RNA undergoes reverse transcription and integration into the chromosomal DNA of a cell from the newly infected host. The virus then replicates by either one of two modes: (i) an infectious cycle by virus budding and infection of new targets and (ii) mitotic division of cells harboring an integrated provirus. HTLV-1 replication initiates a series of mechanisms in the host including antiviral immunity and checkpoint control of cell proliferation. HTLV-1 has elaborated strategies to counteract these defense mechanisms allowing continuous persistence in humans.

Reprinted from *Viruses*. Cite as: Carpentier, A.; Barez, P.-Y.; Hamaidia, M.; Gazon, H.; de Brogniez, A.; Perike, S.; Gillet, N.; Willems, L. Modes of Human T Cell Leukemia Virus Type 1 Transmission, Replication and Persistence. *Viruses* **2015**, *7*, 3603–3624.

1. Introduction

HTLV-1 infects approximately 5–10 million people worldwide mainly in subtropical areas [1]. In the vast majority of cases, HTLV-1 infection remains clinically silent. Among asymptomatic carriers, 3% to 5% will develop a leukemia/lymphoma (ATL) or a neurodegenerative disease (HAM/TSP) after long latent periods (40–60 years) [2]. ATL results from proliferation and accumulation of infected cells carrying an integrated proviral genome (here referred to as clones). HAM/TSP is associated with invasion of the central nervous system by infected cells, antiviral immunity, cytokine burst, and inflammation. Main clinical symptoms of HAM/TSP are urinary failures and paralysis of lower legs. Why infected subjects develop either ATL or HAM/TSP is currently unknown. There is no efficient treatment for HAM/TSP, except palliative attenuation of inflammation with corticosteroids. The leukemic form of ATL is initially responsive to general chemotherapy (CHOP) but almost invariably relapses after a few months. An antiviral therapy based on AZT combined with interferon yields 50% survival at five years [3]. Another type of treatment includes hematopoietic stem cell transplantation that yields, if successful, the best long-term survival rates [4–6]. An anti-CCR4 antibody is now in clinical use in Japan [7,8] and other promising approaches such as valproic acid are currently being investigated [8,9].

The HTLV-1 genome contains essential structural and enzymatic genes (Gag, Pro, Pol and Env) shared by all retroviral family members (reviewed by [10]). As a deltaretrovirus, HTLV-1 also encodes a series of accessory and regulatory proteins. Among these, the Tax oncoprotein and HTLV-1 basic leucine zipper factor (HBZ) play pivotal roles in the viral life cycle [11]. Here, we describe how these factors subvert cellular pathways to allow viral transmission, persistence, and replication.

2. Current Model of HTLV-1 Replication

HTLV-1 predominantly infects CD4+ T cells but also targets other cell types such as CD8+ T and B lymphocytes, dendritic cells (DCs), monocytes, and macrophages [12–14]. This pleiotropic pattern is permitted by the presence of membrane-associated receptors that interact with the viral envelope allowing efficient binding and entry. These include heparan sulfate proteoglycans (HSPGs), the glucose transporter 1 (GLUT-1) and neuropilin-1 (NRP-1) [15–18]. The mechanisms of receptor binding and virus entry have been reviewed elsewhere [19–21]. A number of studies have shown that cell-free infection is poorly efficient compared to cell-to-cell virus transfer (about 10,000 fold) [22,23], suggesting that HTLV-1 spread *in vivo* relies more on a cellular intermediate than on the virion itself. Whatever the route of infection used, the initial contact with HTLV-1 mainly occurs via breast feeding, sexual intercourse, and blood transfusion [24]. Except when contamination occurs by blood transfer, initial infection first requires interaction with oral, gastrointestinal, or cervical mucosa. Crossing of the mucosal barrier occurs by different mechanisms as schematized on Figure 1a. Although not formally demonstrated yet, HTLV-1 infected macrophages could transmigrate through an intact epithelium as observed for human immunodeficiency virus (HIV) [25,26]. Viral particles produced by HTLV-1 infected T-cells have been shown to cross the epithelium by transcytosis, *i.e.*, the transit of a virion incorporated into a vesicle from the apical to the basal surface of an epithelial cell [26,27]. Alternatively, HTLV-1 can also infect an epithelial cell and produce new virions that are then released from the basal surface [28]. Finally, HTLV-1 infected cells can directly bypass a disrupted mucosa [28].

Having crossed the epithelial barrier, HTLV-1 infects mucosal immune cells directly or via APCs such as DCs or macrophages. APCs can either undergo infection or transfer membrane bound extracellular virions to uninfected T-cells (trans-infection) [14]. Cell-to-cell transfer of HTLV-1 virions then potentially involves several non-exclusive mechanisms (reviewed in [28]): a virological synapse [29–31], cellular conduits [32], or extracellular viral assemblies [33,34]. Infection of resident cells occurs either in the mucosa or in secondary lymphoid organs.

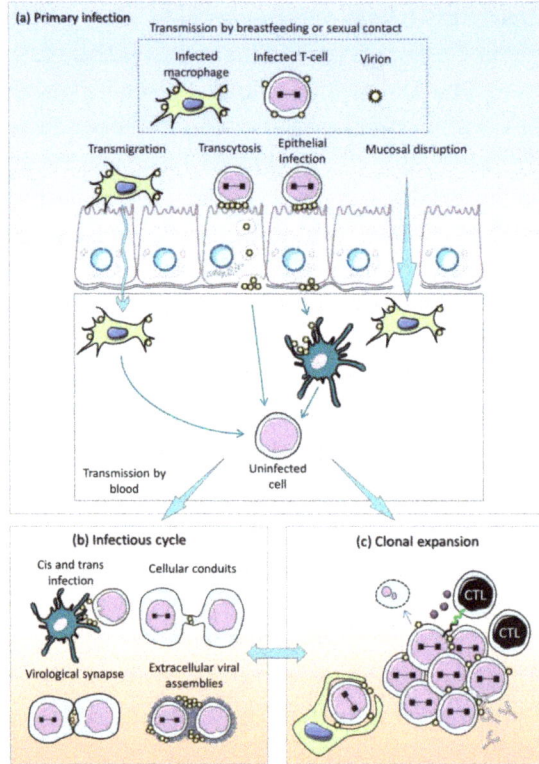

Figure 1. Model of HTLV-1 replication (**a**) HTLV-1 transmission occurs by breastfeeding, sexual intercourse, or blood transfusion. Except for blood transfer, initial infection requires crossing of the mucosal barrier by several mechanisms: (i) transmigration of HTLV-1 infected macrophages, (ii) transcytosis of viral particles, (iii) release of newly produced virions from the basal surface of infected epithelial cell, (iv) bypass of HTLV-1 infected cells through a damaged mucosa. HTLV-1 can then infect mucosal immune cells directly (cis-infection) or via antigen-presenting cells (APCs); (**b**) APCs can either become infected or transfer membrane-bound extracellular virions to T-cells (trans-infection). Cell-to-cell transfer of virions involves different non-exclusive mechanisms: a virological synapse, cellular conduits, or extracellular viral assemblies. Infection of resident cells occurs either in the mucosa or in secondary lymphoid organs. Soon after primary infection, HTLV-1 replicates by cell-to-cell infection (*i.e.*, the infectious cycle) or (**c**) by mitotic division of a cell containing an integrated provirus (clonal expansion). Since an antiviral immune response is quickly initiated, the efficacy of the infectious cycle is severely dampened down soon after infection.

Soon after primary infection, HTLV-1 attempts to expand by colonizing new targets by cell-to-cell transfer, reverse transcription of the viral RNA, integration of the provirus into the chromosome, expression of viral proteins and budding of new virions (the infectious cycle; Figure 1b). Another mode of replication involves mitotic division of a cell containing an integrated provirus (clonal expansion; Figure 1c). Recently, host restriction factors such as SAMHD1, APOBEC3 and miR-28-3p have been shown to limit HTLV-1 infection [35–37]. Since an antiviral immune response is also quickly initiated, the efficacy of the infectious cycle is severely attenuated soon after infection, although likely not completely abrogated later on. On the other side, clonal expansion and cell proliferation also require expression of viral factors such as Tax [38,39]. Survival of infected progeny cells therefore requires silencing of viral expression before immune-mediated destruction. This model is consistent with the following observations: (i) to block HTLV-1 infection, reverse transcriptase inhibitors (RTIs) must be administrated simultaneously with viral inoculation [40]; (ii) when used alone, RTIs do not reduce the proviral load in HTLV-1 infected subjects [41,42]; (iii) sustained T-cell proliferation in patients correlates with Tax expression [43], extending previous studies in BLV-infected animal models [44]; (iv) compared to HIV, the HTLV-1 genome undergoes limited variability [45], suggesting a replication mode by cellular DNA polymerase rather than by viral reverse transcriptase; (v) sequential high-throughput sequencing of proviral integration sites reveal a high clonal stability over years [46]. In this context, our recent study in BLV-infected cows also showed that most clones generated during primary infection are destroyed and replaced by others undergoing expansion [47].

Taken together, these data support a model of viral replication by cell-to-cell contact at the early stages of infection, followed by a sustained clonal proliferation counterbalancing the host immune response. Repetitive cycles of viral expression followed by transcriptional silencing continuously challenges the immune response thereby initiating inflammation and ultimately leading to HAM/TSP. By favoring emergence of sporadic mutations in the cell genome, unrestrained proliferation also paves the way to malignant transformation and development of ATL [43].

3. Tax and HBZ Are Two Main Drivers of Viral Replication

According to currently most accepted model, Tax and HBZ are believed to have the highest impact on viral replication and cell transformation, besides other components required to synthesize the viral particle. The modes of action of Tax and HBZ are remarkably pleiotropic and involve a variety of cell signaling pathways (CREB, NFkB and AKT; Figure 2). Tax inhibits tumor suppressors (p53, Bcl11B and TP53INP1 [48–50]) and activates cyclin-dependent kinases (CDKs) [51], both of these mechanisms leading to accelerated cell proliferation. In parallel, Tax attenuates the Mad1 spindle assembly checkpoint protein, induces genomic lesions and interferes with DNA repair thereby promoting aneuploidy [39,52,53]. Experimental evidence also shows that Tax drives tumor formation in transgenic mouse models, supporting its oncogenic potential [54–56]. Tax also induces genomic instability [39,57], generating somatic alterations [58] and promoting cell growth. However, expression of Tax alone fails to systematically immortalize human primary T cells [59], suggesting the involvement of other viral or cellular components. In particular, driver mutations affecting the CCR4 chemokine receptor have been identified in ~25% of ATL cases [60,61]. In about 50% of ATL cases, Tax is either inactivated by genetic mutation or transcriptionally silenced by hyper-methylation or deletion of the 5′-LTR [62–65]. Because of the strong immunogenicity of the Tax protein, it is possible that these mechanisms confer a selective advantage to HTLV-1-transformed T cells [66–69]. In comparison, HBZ triggers a less efficient immunity that is compatible with permanent expression throughout HTLV-1 infection [70,71]. Later in leukemogenesis, cell growth can thereby become independent of Tax and be promoted by HBZ. Indeed, HBZ is constitutively expressed throughout HTLV-1 infection [72,73], counteracts Tax-mediated viral and cellular pathways modulation (such as NF-κB, Akt and CREB) and stimulates cell proliferation [69,74] via apoptosis/senescence inhibition and cell cycle modulation [69,74]. This simplified model thus hypothesizes that Tax initiates transformation while HBZ is required to maintain the transformed phenotype if Tax expression is silenced [75]. Clinical data indicating that Tax mRNA expression allows estimating the risk of HAM/TSP development and that HBZ positively correlates with the severity of symptoms further supports a role of Tax and HBZ in pathogenesis [76,77].

3.1. Tax and HBZ Exert Opposite Functions in Signaling Pathways

Almost systematically, the activities of Tax on a series of cellular pathways are balanced by HBZ.

3.1.1. NF-κB

By controlling T lymphocyte activation and proliferation in response to diverse immune stimuli (such as antigens, cytokines or microbial components), the NF-κB pathway is a key player in regulation of immunity and inflammation [78]. HTLV-1 Tax activates the IKK complex through IKKγ/NEMO binding. Tax requires CADM1/TSLC1 for inactivation of the NF-kappaB inhibitor A20 and constitutive NF-κB signaling [79]. The subsequent translocation of p50/p65 complex into the nucleus activates transcription of NF-κB responsive genes [78,80]. Activation of the canonical NF-κB pathway by Tax requires IL17RB signaling [81]. On the other hand, Tax stimulates IKKα-dependent processing of p100 into p52 [78,80]. Tax also hijacks the cellular ubiquitin machinery to activate ubiquitin-dependent kinases and NF-κB signaling (reviewed in [82]). Tax thereby induces expression of a variety of growth promoting cytokines (such as IL-1, IL-6, TNF, and EGF [83,84]). Tax also upregulates antiapoptotic proteins: caspase-8 inhibitory protein c-FLIP [85,86] and members of the Bcl-2 family (Bcl-2, Bcl-xL, Mcl-1 and Blf-1) [87–90]. By activating the NF-κB pathway, Tax thus favors proliferation and survival of HTLV-1-infected T cells. On the contrary, HBZ suppresses the canonical NF-κB signaling pathway by inhibiting the activity of the RelA/p65 complex and thus mitigates excessive activation of NF-κB by Tax [91]. NF-κB activation by Tax is associated with an upregulation of $p21^{WAF1/CIP1}$ and $p27^{KIP1}$, leading to cellular senescence [92,93]. In HeLa cells, HBZ prevents Tax-induced senescence through down-regulation of NF-κB [92,94].

3.1.2. Akt

Tax promotes cell proliferation and survival through activator protein-1 (AP-1) and the phosphatidylinositol 3-kinase (PI3K)/Akt pathway [95]. Inhibition of Akt in HTLV-1-transformed cells decreases phosphorylated Bad and induces caspase-dependent apoptosis [96]. Stimulation of PI3K/Akt by Tax activates HiF-1 (hypoxia-inducible factor 1) [97], reduces expression of proapoptotic Bim and Bid and promotes IL-2 independent growth [98] and finally increases Bcl3 whose expression is associated with the growth of infected cells [99]. HBZ inhibits Tax-dependent activation of the PI3K/Akt pathway and downstream anti-apoptotic properties [100]. HBZ suppresses apoptosis by attenuating the function of FOXO3a and altering its localization [101]. Besides, the interaction of HBZ with AP-1 factors (cJun, JunB or MafB) results in the inhibition of their transcriptional activities via several mechanisms, such as sequestration into nuclear bodies or proteasomal degradation, and prevents the subsequent activation of AP-1 regulated genes [102–106].

3.1.3. CREB

Tax activates 5′-LTR-directed transcription by interacting with CREB, modulating its phosphorylation at Ser133 and connecting the histone acetyltransferase CBP (CREB-binding protein/p300) [107]. The ability of Tax to activate transcription via CREB is required to protect murine fibroblasts from serum-depletion-induced apoptosis. [108–110]. Tax modifies the phosphorylation state of CREB (i) by activating the upstream Akt kinase [111,112] or (ii) by decreasing the expression of PTEN phosphatase which is required for CREB dephosphorylation at Ser133 in the nucleus [111,113].

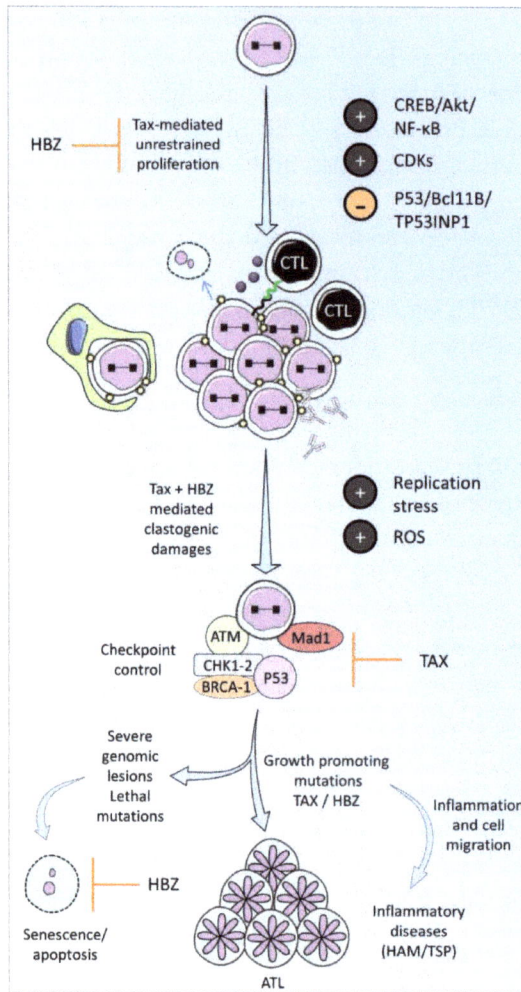

Figure 2. Tax and HBZ promote proliferation and persistence of the infected cell. Tax activates survival pathways (CREB/Akt/NFkB), promotes mitosis (CDKs), and inhibits tumor suppressors (p53, TP53INP1, Bcl11B). Tax-mediated growth-promoting activities are counteracted by HBZ, mitigating unrestrained proliferation. The host immune response further controls infected cell proliferation. Tax-induced proliferation creates replicative stress and generates reactive oxygen species (ROS). Tax interacts with the mitotic checkpoint control protein Mad1 thereby inducing clastogenic damage. Tax attenuates the DNA damage response (DDR) induced by unscheduled cell proliferation. Inhibition of the DDR allows cells to accumulate DNA lesions and stabilize mutations. If uncontrolled by senescence or cell death mechanisms, growth-promoting mutations pave the way to disease development.

HBZ represses viral transcription through interaction with the bZIP domain of CREB proteins and prevents their binding to the viral CRE elements [114,115]. HBZ interacts with the KIX domain of p300/CBP, competing for Tax binding and inhibiting the association of co-activators with the viral promoter [116]. HBZ modulates the occupancy of KIX domains of p300/CBP by modulating the activity of transcription factors, thereby influencing subsequent gene expression [117].

3.1.4. Wnt

Tax interacts with DAPLE (dishevelled-associating protein with a high frequency of leucine residues) to activate the canonical Wnt pathway. HBZ can suppress this activation by inhibiting DNA binding of TCF-1/LEF-1 transcription factors. On the other side, HBZ promotes transcription of WnT 5a, a key protein of the non-canonical Wnt pathway, by enhancing its promoter activity through transforming growth factor-beta (TGF-beta). Knockdown of Wnt5a represses proliferation and migration of ATL cells, pointing out the role of this pathway in HTLV-1 infected cell growth [118].

3.1.5. TGF-β/Smad

Tax represses TGF-β 1 signaling (i) by blocking the association of Smad proteins with Smad-binding elements, (ii) through its interaction with CREB-binding protein/p300 and (iii) via c-jun activation [119–121]. HBZ counteracts this effect and also interacts with Smad2/3 to enhance TGF-beta/Smad transcriptional responses in a p300-dependent manner, improving transcription of different genes, such as the FOXP3 mediator of regulatory T cells [122].

3.1.6. S Phase Entry and Cell Cycle Progression

Through interaction with cyclins and CDKs, Tax interferes with cell cycle progression by several mechanisms: (i) Tax stabilizes the cyclin D2/CDK4 complex and favors hyperphosphorylation of the retinoblastoma protein (Rb). Phosphorylated Rb frees E2F1 that activates transcription of genes required for G1/S transition; (ii) Tax represses cyclin-dependent kinases inhibitors (CKIs) such as members of INK4 family and KIP1; (iii) Tax interacts with and directs Rb to the proteasome for subsequent degradation; (iv) Tax activates the cyclin D1 transcription by enhancing p300 recruitment to the CRE site of cyclin D1 promoter through interaction with pCREB and TORC2 [123]. As a result, Tax favors S phase entry of HTLV-1 infected cells (reviewed in [51,75]). Tax also accelerates S phase progression by interaction with the replicative helicase (minichromosome maintenance complex, MCM2-7). Tax modulates the spatiotemporal program of replication origins through p300-dependent histone hyperacetylation, resulting in early firing of late replication origins. Tax also fires supplementary origins of replication accelerating S phase progression. This mechanism triggers replicative stress and genomic lesions, such as double strand breaks (DSBs) [39,57]. By modulating replication timing, Tax could also modulate the entire transcriptional landscape of infected cells. Indeed, the level of transcription at replication origins (ORC1 binding sites) correlates with replication timing [124].

In contrast to Tax, HBZ exerts a dual regulatory role in cell cycle progression. Indeed, HBZ interacts with CREB and inhibits transcription of cyclin D1 [125]. HBZ also binds activating transcription factor 3 (ATF3) that modulates expression of cell division cycle 2 (CDC2) and cyclin E2, thereby promoting proliferation of ATL cells [126]. Concomitantly, HBZ suppresses ATF3-induced p53 transcriptional activity. Moreover, the HBZ mRNA increases E2F1 gene transcription and promotes cell proliferation [69].

Together, this series of data on the signaling pathways illustrates the opposite functions of Tax and HBZ in finely tuned regulatory mechanisms of cell proliferation.

3.2. Cellular Checkpoints Control Unscheduled Proliferation

Cellular checkpoints act as a failsafe barrier against unrestrained cellular proliferation. Tax subverts the G1 restriction and the spindle mitotic checkpoints. In G1, the tumor-suppressor protein p53 is the main factor that controls the checkpoint. Although approximately 50% of cancers harbor a mutation in p53, this mechanism only appears in a small percentage of ATL patients. Instead, p53 is functionally inactivated in leukemic and HTLV-1 transformed cells [127]. It remains incompletely understood how Tax inactivates p53: (i) Tax competes with p53 in binding with CBP, thereby repressing p53 trans-activating function [128]; (ii) NF-kappaB p65 subunit is critical for Tax-induced p53 inactivation [129]; (iii) repression of p53 transcriptional activity by Tax is independent of NF-kB and CBP [130]; p53 is invalidated by wild-type p53-induced phosphatase 1 (Wip1) [131,132]. ATL cells are characterized by loss of spindle assembly checkpoint function [133] and aneuploidy [134]. Tax binding to Mad1 perturbs the organization of the spindle assembly and results in multinucleated cells [52]. Moreover, the direct interaction between Tax and the anaphase-promoting complex APC Cdc20 also explains the mitotic abnormalities in HTLV-1 infected cells [135]. Tax promotion of supernumerary centrosomes through recruitment of Ran and Ran-binding protein-1 is another mechanism contributing to leukemia [136].

3.3. Response to DNA Damage

By accelerating the replication-timing program, the Tax protein induces replicative stress and DSBs [39]. Tax expression generates reactive oxygen species (ROS) leading to oxidative and replication-dependent DSBs [53]. Tax-associated

DNA damages activate several phosphoproteins of the DDR pathway (H2AX, ATM, CHK1-2, P53, BRCA1), which in turn arrest the cell cycle transiently or lead to apoptosis and senescence. In presence of DNA damaging agents (e.g., UV irradiation), Tax inhibits the DDR machinery by sequestrating key signaling pathway components [137–144]. Induction of genomic lesions and inhibition of the DDR leads to proliferation in presence of DNA mutations, potentially to leukemogenesis.

HBZ induces DNA lesions through activation of miR-17 and miR-21 and downregulation of the DNA damage factor OBFC2A [145]. HBZ association with growth arrest and DNA damage gene 34 (GADD34) also deregulates the cellular responses to DNA damage [146].

3.4. DNA Repair Pathways

Besides modulating the DDR signaling pathway, Tax also directly interferes with the mechanisms of DNA repair. For example, Tax downregulates the expression of β-polymerase [147] and inhibits base excision repair (BER) [148]. Furthermore, Tax activates PCNA and interferes with nucleotide excision repair (NER) [149,150]. Tax decreases Ku80 gene transcription and interacts with Ku80 protein, interfering with non-homologous end joining (NHEJ) [151,152]. In Tax-expressing cells, DSBs are nevertheless preferentially repaired by error-prone NHEJ [153]. Another viral protein, p30, inhibits homologous recombination, shifting repair towards unfaithful pathways [154]. Whether HBZ also interferes with DNA damage repair mechanisms remains to be further clarified.

4. Conclusions

HTLV-1 persists and replicates by means of viral proteins, such as Tax and HBZ that finely tune cellular signaling pathways. Viral replication through the infectious and the mitotic routes requires viral expression and faces destruction by the host immune response. Expression of viral proteins creates genomic stress responsible for DNA lesions that initiate the DDR response. Imperfect repair of these errors stabilizes mutations that potentially drive oncogenesis.

Acknowledgments: This work was supported by the "Fonds National de la Recherche Scientifique" (FNRS), the Télévie, the Interuniversity Attraction Poles (IAP) Program "Virus-host interplay at the early phases of infection" BELVIR initiated by the Belgian Science Policy Office, the Belgian Foundation against Cancer (FBC), the Sixth Research Framework Programme of the European Union (project "The role of infections in cancer" INCA LSHC-CT-2005-018704), the "Neoangio" excellence program and the "Partenariat Public Privé", PPP INCA, of the "Direction générale des Technologies, de la Recherche et de l'Energie/DG06" of the Walloon government, the "Action de Recherche Concertée Glyvir" (ARC) of the "Communauté française de Belgique", the "Centre anticancéreux près ULg" (CAC), the "Subside Fédéral de Soutien à la Recherche Synbiofor and Agricultureislife" projects of Gembloux Agrobiotech (GxABT), the "ULg Fonds Spéciaux pour la Recherche", the "Plan Cancer" of the "Service Public Fédéral". A.C.; S.P. and A.B. are supported by grants of the Télévie. M.H. is a research fellow of the "Agriculture is life" project of GxABT. N.G. is supported by the IAP program. P.-Y. B. (FNRS research fellow), H.G. (post-doctoral researcher) and L.W. (Research Director) are members of the FNRS.

Author Contributions: A.C.; P.-Y.B. and M.H. drafted the manuscript. L.W. edited the manuscript. All authors corrected, edited and approved the text.

Conflicts of Interest: The authors declare no conflict of interest.

References

1. Gessain, A.; Cassar, O. Epidemiological Aspects and World Distribution of HTLV-1 Infection. *Front. Microbiol.* **2012**, *3*, e388.
2. Verdonck, K.; Gonzalez, E.; Van Dooren, S.; Vandamme, A.M.; Vanham, G.; Gotuzzo, E. Human T-lymphotropic virus 1: Recent knowledge about an ancient infection. *Lancet. Infect. Dis.* **2007**, *7*, 266–281.
3. Bazarbachi, A.; Plumelle, Y.; Carlos Ramos, J.; Tortevoye, P.; Otrock, Z.; Taylor, G.; Gessain, A.; Harrington, W.; Panelatti, G.; Hermine, O. Meta-analysis on the use of zidovudine and interferon-alfa in adult T-cell leukemia/lymphoma showing improved survival in the leukemic subtypes. *J. Clin. Oncol.* **2010**, *28*, 4177–4183.
4. Shiratori, S.; Yasumoto, A.; Tanaka, J.; Shigematsu, A.; Yamamoto, S.; Nishio, M.; Hashino, S.; Morita, R.; Takahata, M.; Onozawa, M.; *et al.* A retrospective analysis of allogeneic hematopoietic stem cell transplantation for adult T cell leukemia/lymphoma (ATL): Clinical impact of graft-versus-leukemia/lymphoma effect. *Biol. Blood Marrow Transpl.* **2008**, *14*, 817–823.
5. Ishida, T.; Hishizawa, M.; Kato, K.; Tanosaki, R.; Fukuda, T.; Takatsuka, Y.; Eto, T.; Miyazaki, Y.; Hidaka, M.; Uike, N.; *et al.* Impact of graft-versus-host disease on allogeneic hematopoietic cell transplantation for adult T cell leukemia-lymphoma focusing on preconditioning regimens: Nationwide retrospective study. *Biol. Blood Marrow Transpl.* **2013**, *19*, 1731–1739.
6. Ishida, T.; Hishizawa, M.; Kato, K.; Tanosaki, R.; Fukuda, T.; Taniguchi, S.; Eto, T.; Takatsuka, Y.; Miyazaki, Y.; Moriuchi, Y.; *et al.* Allogeneic hematopoietic stem cell transplantation for adult T-cell leukemia-lymphoma with special emphasis on preconditioning regimen: A nationwide retrospective study. *Blood* **2012**, *120*, 1734–1741.
7. Utsunomiya, A.; Choi, I.; Chihara, D.; Seto, M. Recent advances in the treatment of adult T-cell leukemia-lymphomas. *Cancer Sci.* **2015**, *106*, 344–351.
8. Yamauchi, J.; Coler-Reilly, A.; Sato, T.; Araya, N.; Yagishita, N.; Ando, H.; Kunitomo, Y.; Takahashi, K.; Tanaka, Y.; Shibagaki, Y.; *et al.* Mogamulizumab, an anti-CCR4 antibody, targets human T-lymphotropic virus type 1-infected CD8+ and CD4+ T cells to treat associated myelopathy. *J. Infect. Dis.* **2015**, *211*, 238–248.
9. Lezin, A.; Gillet, N.; Olindo, S.; Signate, A.; Grandvaux, N.; Verlaeten, O.; Belrose, G.; de Carvalho Bittencourt, M.; Hiscott, J.; Asquith, B.; *et al.* Histone deacetylase mediated transcriptional activation reduces proviral loads in HTLV-1 associated myelopathy/tropical spastic paraparesis patients. *Blood* **2007**, *110*, 3722–3728.

10. Katz, R.A.; Skalka, A.M. Generation of Diversity in Retroviruses. *Annu. Rev. Genet.* **1990**, *24*, 409–443.

11. Matsuoka, M.; Jeang, K.T. Human T-cell leukaemia virus type 1 (HTLV-1) infectivity and cellular transformation. *Nat. Rev. Cancer* **2007**, *7*, 270–280.

12. Macatonia, S.E.; Cruickshank, J.K.; Rudge, P.; Knight, S.C. Dendritic cells from patients with tropical spastic paraparesis are infected with HTLV-1 and stimulate autologous lymphocyte proliferation. *AIDS Res. Hum. Retrovir.* **1992**, *8*, 1699–1706.

13. Koyanagi, Y.; Itoyama, Y.; Nakamura, N.; Takamatsu, K.; Kira, J.; Iwamasa, T.; Goto, I.; Yamamoto, N. *In vivo* infection of human T-cell leukemia virus type I in non-T cells. *Virology* **1993**, *196*, 25–33.

14. Jones, K.S.; Petrow-Sadowski, C.; Huang, Y.K.; Bertolette, D.C.; Ruscetti, F.W. Cell-free HTLV-1 infects dendritic cells leading to transmission and transformation of CD4+ T cells. *Nat. Med.* **2008**, *14*, 429–436.

15. Manel, N.; Kim, F.J.; Kinet, S.; Taylor, N.; Sitbon, M.; Battini, J.L. The ubiquitous glucose transporter GLUT-1 is a receptor for HTLV. *Cell* **2003**, *115*, 449–459.

16. Jones, K.S.; Petrow-Sadowski, C.; Bertolette, D.C.; Huang, Y.; Ruscetti, F.W. Heparan sulfate proteoglycans mediate attachment and entry of human T-cell leukemia virus type 1 virions into CD4+ T cells. *J. Virol.* **2005**, *79*, 12692–12702.

17. Ghez, D.; Lepelletier, Y.; Lambert, S.; Fourneau, J.M.; Blot, V.; Janvier, S.; Arnulf, B.; van Endert, P.M.; Heveker, N.; Pique, C.; *et al.* Neuropilin-1 is involved in human T-cell lymphotropic virus type 1 entry. *J. Virol.* **2006**, *80*, 6844–6854.

18. Lambert, S.; Bouttier, M.; Vassy, R.; Seigneuret, M.; Petrow-Sadowski, C.; Janvier, S.; Heveker, N.; Ruscetti, F.W.; Perret, G.; Jones, K.S.; *et al.* HTLV-1 uses HSPG and neuropilin-1 for entry by molecular mimicry of VEGF165. *Blood* **2009**, *113*, 5176–5185.

19. Ghez, D.; Lepelletier, Y.; Jones, K.S.; Pique, C.; Hermine, O. Current concepts regarding the HTLV-1 receptor complex. *Retrovirology* **2010**, *7*, e99.

20. Jones, K.S.; Lambert, S.; Bouttier, M.; Benit, L.; Ruscetti, F.W.; Hermine, O.; Pique, C. Molecular aspects of HTLV-1 entry: Functional domains of the HTLV-1 surface subunit (SU) and their relationships to the entry receptors. *Viruses* **2011**, *3*, 794–810.

21. Hoshino, H. Cellular Factors Involved in HTLV-1 Entry and Pathogenicit. *Front. Microbiol.* **2012**, *3*, e222.

22. Derse, D.; Hill, S.A.; Lloyd, P.A.; Chung, H.; Morse, B.A. Examining human T-lymphotropic virus type 1 infection and replication by cell-free infection with recombinant virus vectors. *J. Virol.* **2001**, *75*, 8461–8468.

23. Mazurov, D.; Ilinskaya, A.; Heidecker, G.; Lloyd, P.; Derse, D. Quantitative comparison of HTLV-1 and HIV-1 cell-to-cell infection with new replication dependent vectors. *PLoS Pathog.* **2010**, *6*, e1000788.

24. Goncalves, D.U.; Proietti, F.A.; Ribas, J.G.; Araujo, M.G.; Pinheiro, S.R.; Guedes, A.C.; Carneiro-Proietti, A.B. Epidemiology, treatment, and prevention of human T-cell leukemia virus type 1-associated diseases. *Clin. Microbiol. Rev.* **2010**, *23*, 577–589.

25. Takeuchi, H.; Takahashi, M.; Norose, Y.; Takeshita, T.; Fukunaga, Y.; Takahashi, H. Transformation of breast milk macrophages by HTLV-I: Implications for HTLV-I transmission via breastfeeding. *Biomed. Res.* **2010**, *31*, 53–61.

26. Tugizov, S.M.; Herrera, R.; Veluppillai, P.; Greenspan, D.; Soros, V.; Greene, W.C.; Levy, J.A.; Palefsky, J.M. Differential transmission of HIV traversing fetal oral/intestinal epithelia and adult oral epithelia. *J. Virol.* **2012**, *86*, 2556–2570.

27. Martin-Latil, S.; Gnadig, N.F.; Mallet, A.; Desdouits, M.; Guivel-Benhassine, F.; Jeannin, P.; Prevost, M.C.; Schwartz, O.; Gessain, A.; Ozden, S.; *et al.* Transcytosis of HTLV-1 across a tight human epithelial barrier and infection of subepithelial dendritic cells. *Blood* **2012**, *120*, 572–580.

28. Pique, C.; Jones, K.S. Pathways of cell-cell transmission of HTLV-1. *Front. Microbiol.* **2012**, *3*, e378.

29. Igakura, T.; Stinchcombe, J.C.; Goon, P.K.; Taylor, G.P.; Weber, J.N.; Griffiths, G.M.; Tanaka, Y.; Osame, M.; Bangham, C.R. Spread of HTLV-I between lymphocytes by virus-induced polarization of the cytoskeleton. *Science* **2003**, *299*, 1713–1716.

30. Majorovits, E.; Nejmeddine, M.; Tanaka, Y.; Taylor, G.P.; Fuller, S.D.; Bangham, C.R. Human T-lymphotropic virus-1 visualized at the virological synapse by electron tomography. *PLoS ONE* **2008**, *3*, e2251.

31. Nejmeddine, M.; Negi, V.S.; Mukherjee, S.; Tanaka, Y.; Orth, K.; Taylor, G.P.; Bangham, C.R. HTLV-1-Tax and ICAM-1 act on T-cell signal pathways to polarize the microtubule-organizing center at the virological synapse. *Blood* **2009**, *114*, 1016–1025.

32. Van Prooyen, N.; Gold, H.; Andresen, V.; Schwartz, O.; Jones, K.; Ruscetti, F.; Lockett, S.; Gudla, P.; Venzon, D.; Franchini, G. Human T-cell leukemia virus type 1 p8 protein increases cellular conduits and virus transmission. *Proc. Natl. Acad. Sci. USA* **2010**, *107*, 20738–20743.

33. Jones, K.S.; Green, P.L. Cloaked virus slips between cells. *Nat. Med.* **2010**, *16*, 25–27.

34. Pais-Correia, A.M.; Sachse, M.; Guadagnini, S.; Robbiati, V.; Lasserre, R.; Gessain, A.; Gout, O.; Alcover, A.; Thoulouze, M.I. Biofilm-like extracellular viral assemblies mediate HTLV-1 cell-to-cell transmission at virological synapses. *Nat. Med.* **2010**, *16*, 83–89.

35. Sze, A.; Belgnaoui, S.M.; Olagnier, D.; Lin, R.; Hiscott, J.; van Grevenynghe, J. Host restriction factor SAMHD1 limits human T cell leukemia virus type 1 infection of monocytes via STING-mediated apoptosis. *Cell Host Microbe* **2013**, *14*, 422–434.

36. Ooms, M.; Krikoni, A.; Kress, A.K.; Simon, V.; Munk, C. APOBEC3A, APOBEC3B, and APOBEC3H haplotype 2 restrict human T-lymphotropic virus type 1. *J. Virol.* **2012**, *86*, 6097–6108.

37. Bai, X.T.; Nicot, C. miR-28-3p is a cellular restriction factor that inhibits human T cell leukemia virus, type 1 (HTLV-1) replication and virus infection. *J. Biol. Chem.* **2015**, *290*, 5381–5390.

38. Twizere, J.C.; Kruys, V.; Lefebvre, L.; Vanderplasschen, A.; Collete, D.; Debacq, C.; Lai, W.S.; Jauniaux, J.C.; Bernstein, L.R.; Semmes, O.J.; *et al.* Interaction of retroviral Tax oncoproteins with tristetraprolin and regulation of tumor necrosis factor-alpha expression. *J. Natl. Cancer Inst.* **2003**, *95*, 1846–1859.

39. Boxus, M.; Twizere, J.C.; Legros, S.; Kettmann, R.; Willems, L. Interaction of HTLV-1 Tax with minichromosome maintenance proteins accelerates the replication timing program. *Blood* **2012**, *119*, 151–160.

40. Miyazato, P.; Yasunaga, J.; Taniguchi, Y.; Koyanagi, Y.; Mitsuya, H.; Matsuoka, M. De novo human T-cell leukemia virus type 1 infection of human lymphocytes in NOD-SCID, common gamma-chain knockout mice. *J. Virol.* **2006**, *80*, 10683–10691.

41. Taylor, G.P.; Goon, P.; Furukawa, Y.; Green, H.; Barfield, A.; Mosley, A.; Nose, H.; Babiker, A.; Rudge, P.; Usuku, K.; *et al.* Zidovudine plus lamivudine in Human T-Lymphotropic Virus type-I-associated myelopathy: A randomised trial. *Retrovirology* **2006**, *3*, e63.

42. Trevino, A.; Parra, P.; Bar-Magen, T.; Garrido, C.; de Mendoza, C.; Soriano, V. Antiviral effect of raltegravir on HTLV-1 carriers. *J. Antimicrobial Chemother.* **2012**, *67*, 218–221.

43. Asquith, B.; Zhang, Y.; Mosley, A.J.; de Lara, C.M.; Wallace, D.L.; Worth, A.; Kaftantzi, L.; Meekings, K.; Griffin, G.E.; Tanaka, Y.; *et al.* In vivo T lymphocyte dynamics in humans and the impact of human T-lymphotropic virus 1 infection. *Proc. Natl. Acad. Sci. USA* **2007**, *104*, 8035–8040.

44. Debacq, C.; Asquith, B.; Kerkhofs, P.; Portetelle, D.; Burny, A.; Kettmann, R.; Willems, L. Increased cell proliferation, but not reduced cell death, induces lymphocytosis in bovine leukemia virus-infected sheep. *Proc. Natl. Acad. Sci. USA* **2002**, *99*, 10048–10053.

45. Ratner, L.; Philpott, T.; Trowbridge, D.B. Nucleotide sequence analysis of isolates of human T-lymphotropic virus type 1 of diverse geographical origins. *AIDS Res. Hum. Retrovir.* **1991**, *7*, 923–941.

46. Gillet, N.A.; Malani, N.; Melamed, A.; Gormley, N.; Carter, R.; Bentley, D.; Berry, C.; Bushman, F.D.; Taylor, G.P.; Bangham, C.R. The host genomic environment of the provirus determines the abundance of HTLV-1-infected T-cell clones. *Blood* **2011**, *117*, 3113–3122.

47. Gillet, N.A.; Gutierrez, G.; Rodriguez, S.M.; de Brogniez, A.; Renotte, N.; Alvarez, I.; Trono, K.; Willems, L. Massive depletion of bovine leukemia virus proviral clones located in genomic transcriptionally active sites during primary infection. *PLoS Pathog.* **2013**, *9*, e1003687.

48. Reid, R.L.; Lindholm, P.F.; Mireskandari, A.; Dittmer, J.; Brady, J.N. Stabilization of wild-type p53 in human T-lymphocytes transformed by HTLV-I. *Oncogene* **1993**, *8*, 3029–3036.

49. Takachi, T.; Takahashi, M.; Takahashi-Yoshita, M.; Higuchi, M.; Obata, M.; Mishima, Y.; Okuda, S.; Tanaka, Y.; Matsuoka, M.; Saitoh, A.; *et al.* Human T-cell leukemia virus type 1 Tax oncoprotein represses the expression of the BCL11B tumor suppressor in T-cells. *Cancer Sci.* **2015**, *106*, 461–465.

50. Yeung, M.L.; Yasunaga, J.; Bennasser, Y.; Dusetti, N.; Harris, D.; Ahmad, N.; Matsuoka, M.; Jeang, K.T. Roles for microRNAs, miR-93 and miR-130b, and tumor protein 53-induced nuclear protein 1 tumor suppressor in cell growth dysregulation by human T-cell lymphotrophic virus 1. *Cancer Res.* **2008**, *68*, 8976–8985.

51. Boxus, M.; Twizere, J.C.; Legros, S.; Dewulf, J.F.; Kettmann, R.; Willems, L. The HTLV-1 Tax interactome. *Retrovirology* **2008**, *5*, e76.

52. Jin, D.Y.; Spencer, F.; Jeang, K.T. Human T cell leukemia virus type 1 oncoprotein Tax targets the human mitotic checkpoint protein MAD1. *Cell* **1998**, *93*, 81–91.

53. Kinjo, T.; Ham-Terhune, J.; Peloponese, J.M., Jr.; Jeang, K.T. Induction of reactive oxygen species by human T-cell leukemia virus type 1 tax correlates with DNA damage and expression of cellular senescence marker. *J. Virol.* **2010**, *84*, 5431–5437.

54. Grossman, W.J.; Kimata, J.T.; Wong, F.H.; Zutter, M.; Ley, T.J.; Ratner, L. Development of leukemia in mice transgenic for the tax gene of human T-cell leukemia virus type I. *Proc. Natl. Acad. Sci. USA* **1995**, *92*, 1057–1061.

55. Hasegawa, H.; Sawa, H.; Lewis, M.J.; Orba, Y.; Sheehy, N.; Yamamoto, Y.; Ichinohe, T.; Tsunetsugu-Yokota, Y.; Katano, H.; Takahashi, H.; *et al.* Thymus-derived leukemia-lymphoma in mice transgenic for the Tax gene of human T-lymphotropic virus type I. *Nat. Med.* **2006**, *12*, 466–472.

56. Ohsugi, T.; Kumasaka, T.; Okada, S.; Urano, T. The Tax protein of HTLV-1 promotes oncogenesis in not only immature T cells but also mature T cells. *Nat. Med.* **2007**, *13*, 527–528.

57. Chaib-Mezrag, H.; Lemacon, D.; Fontaine, H.; Bellon, M.; Bai, X.T.; Drac, M.; Coquelle, A.; Nicot, C. Tax impairs DNA replication forks and increases DNA breaks in specific oncogenic genome regions. *Mol. Cancer* **2014**, *13*, 205.

58. Marriott, S.J.; Semmes, O.J. Impact of HTLV-I Tax on cell cycle progression and the cellular DNA damage repair response. *Oncogene* **2005**, *24*, 5986–5995.

59. Bellon, M.; Baydoun, H.H.; Yao, Y.; Nicot, C. HTLV-I Tax-dependent and -independent events associated with immortalization of human primary T lymphocytes. *Blood* **2010**, *115*, 2441–2448.

60. Nakagawa, M.; Schmitz, R.; Xiao, W.; Goldman, C.K.; Xu, W.; Yang, Y.; Yu, X.; Waldmann, T.A.; Staudt, L.M. Gain-of-function CCR4 mutations in adult T cell leukemia/lymphoma. *J. Exp. Med.* **2014**, *211*, 2497–2505.

61. Shannon, K.M. CCR4 drives ATLL jail break. *J. Exp. Med.* **2014**, *211*, 2485.

62. Furukawa, Y.; Kubota, R.; Tara, M.; Izumo, S.; Osame, M. Existence of escape mutant in HTLV-I tax during the development of adult T-cell leukemia. *Blood* **2001**, *97*, 987–993.

63. Koiwa, T.; Hamano-Usami, A.; Ishida, T.; Okayama, A.; Yamaguchi, K.; Kamihira, S.; Watanabe, T. 5′-long terminal repeat-selective CpG methylation of latent human T-cell leukemia virus type 1 provirus *in vitro* and *in vivo*. *J. Virol.* **2002**, *76*, 9389–9397.

64. Takeda, S.; Maeda, M.; Morikawa, S.; Taniguchi, Y.; Yasunaga, J.; Nosaka, K.; Tanaka, Y.; Matsuoka, M. Genetic and epigenetic inactivation of tax gene in adult T-cell leukemia cells. *Int. J. Cancer* **2004**, *109*, 559–567.

65. Taniguchi, Y.; Nosaka, K.; Yasunaga, J.; Maeda, M.; Mueller, N.; Okayama, A.; Matsuoka, M. Silencing of human T-cell leukemia virus type I gene transcription by epigenetic mechanisms. *Retrovirology* **2005**, *2*, e64.

66. Jacobson, S.; Shida, H.; McFarlin, D.E.; Fauci, A.S.; Koenig, S. Circulating CD8+ cytotoxic T lymphocytes specific for HTLV-I pX in patients with HTLV-I associated neurological disease. *Nature* **1990**, *348*, 245–248.

67. Kannagi, M.; Harada, S.; Maruyama, I.; Inoko, H.; Igarashi, H.; Kuwashima, G.; Sato, S.; Morita, M.; Kidokoro, M.; Sugimoto, M.; *et al.* Predominant recognition of human T cell leukemia virus type I (HTLV-I) pX gene products by human CD8+ cytotoxic T cells directed against HTLV-I-infected cells. *Int. Immunol.* **1991**, *3*, 761–767.

68. Kannagi, M.; Matsushita, S.; Harada, S. Expression of the target antigen for cytotoxic T lymphocytes on adult T-cell-leukemia cells. *Int. J. Cancer* **1993**, *54*, 582–588.

69. Satou, Y.; Yasunaga, J.; Yoshida, M.; Matsuoka, M. HTLV-I basic leucine zipper factor gene mRNA supports proliferation of adult T cell leukemia cells. *Proc. Natl. Acad. Sci. USA* **2006**, *103*, 720–725.

70. Macnamara, A.; Rowan, A.; Hilburn, S.; Kadolsky, U.; Fujiwara, H.; Suemori, K.; Yasukawa, M.; Taylor, G.; Bangham, C.R.; Asquith, B. HLA class I binding of HBZ determines outcome in HTLV-1 infection. *PLoS Pathog.* **2010**, *6*, e1001117.

71. Hilburn, S.; Rowan, A.; Demontis, M.A.; MacNamara, A.; Asquith, B.; Bangham, C.R.; Taylor, G.P. *In vivo* expression of human T-lymphotropic virus type 1 basic leucine-zipper protein generates specific CD8+ and CD4+ T-lymphocyte responses that correlate with clinical outcome. *J. Infect. Dis.* **2011**, *203*, 529–536.

72. Usui, T.; Yanagihara, K.; Tsukasaki, K.; Murata, K.; Hasegawa, H.; Yamada, Y.; Kamihira, S. Characteristic expression of HTLV-1 basic zipper factor (HBZ) transcripts in HTLV-1 provirus-positive cells. *Retrovirology* **2008**, *5*, e34.

73. Matsuoka, M.; Green, P.L. The HBZ gene, a key player in HTLV-1 pathogenesis. *Retrovirology* **2009**, *6*, e71.

74. Arnold, J.; Zimmerman, B.; Li, M.; Lairmore, M.D.; Green, P.L. Human T-cell leukemia virus type-1 antisense-encoded gene, Hbz, promotes T-lymphocyte proliferation. *Blood* **2008**, *112*, 3788–3797.

75. Matsuoka, M.; Jeang, K.T. Human T-cell leukemia virus type 1 (HTLV-1) and leukemic transformation: Viral infectivity, Tax, HBZ and therapy. *Oncogene* **2011**, *30*, 1379–1389.

76. Andrade, R.G.; Goncalves Pde, C.; Ribeiro, M.A.; Romanelli, L.C.; Ribas, J.G.; Torres, E.B.; Carneiro-Proietti, A.B.; Barbosa-Stancioli, E.F.; Martins, M.L. Strong correlation between tax and HBZ mRNA expression in HAM/TSP patients: Distinct markers for the neurologic disease. *J. Clin. Virol.* **2013**, *56*, 135–140.

77. Saito, M.; Matsuzaki, T.; Satou, Y.; Yasunaga, J.; Saito, K.; Arimura, K.; Matsuoka, M.; Ohara, Y. *In vivo* expression of the HBZ gene of HTLV-1 correlates with proviral load, inflammatory markers and disease severity in HTLV-1 associated myelopathy/tropical spastic paraparesis (HAM/TSP). *Retrovirology* **2009**, *6*, e19.

78. Sun, S.C.; Yamaoka, S. Activation of NF-kappaB by HTLV-I and implications for cell transformation. *Oncogene* **2005**, *24*, 5952–5964.

79. Pujari, R.; Hunte, R.; Thomas, R.; van der Weyden, L.; Rauch, D.; Ratner, L.; Nyborg, J.K.; Ramos, J.C.; Takai, Y.; Shembade, N. Human T-cell leukemia virus type 1 (HTLV-1) tax requires CADM1/TSLC1 for inactivation of the NF-kappaB inhibitor A20 and constitutive NF-kappaB signaling. *PLoS Pathog.* **2015**, *11*, e1004721.

80. Harhaj, E.W.; Harhaj, N.S. Mechanisms of persistent NF-kappaB activation by HTLV-I tax. *IUBMB Life* **2005**, *57*, 83–91.

81. Lavorgna, A.; Matsuoka, M.; Harhaj, E.W. A critical role for IL-17RB signaling in HTLV-1 tax-induced NF-kappaB activation and T-cell transformation. *PLoS Pathog.* **2014**, *10*, e1004418.

82. Lavorgna, A.; Harhaj, E.W. Regulation of HTLV-1 tax stability, cellular trafficking and NF-kappaB activation by the ubiquitin-proteasome pathway. *Viruses* **2014**, *6*, 3925–3943.

83. Karin, M. NF-kappaB as a critical link between inflammation and cancer. *Cold Spring Harb. Perspect. Biol.* **2009**, *1*.

84. Xiao, G.; Fu, J. NF-kappaB and cancer: A paradigm of Yin-Yang. *Am. J. Cancer Res.* **2011**, *1*, 192–221.

85. Krueger, A.; Fas, S.C.; Giaisi, M.; Bleumink, M.; Merling, A.; Stumpf, C.; Baumann, S.; Holtkotte, D.; Bosch, V.; Krammer, P.H.; *et al.* HTLV-1 Tax protects against CD95-mediated apoptosis by induction of the cellular FLICE-inhibitory protein (c-FLIP). *Blood* **2006**, *107*, 3933–3939.

86. Okamoto, K.; Fujisawa, J.; Reth, M.; Yonehara, S. Human T-cell leukemia virus type-I oncoprotein Tax inhibits Fas-mediated apoptosis by inducing cellular FLIP through activation of NF-kappaB. *Genes Cells: Devoted Mol. Cell. Mech.* **2006**, *11*, 177–191.

87. Tsukahara, T.; Kannagi, M.; Ohashi, T.; Kato, H.; Arai, M.; Nunez, G.; Iwanaga, Y.; Yamamoto, N.; Ohtani, K.; Nakamura, M.; *et al.* Induction of Bcl-x(L) expression by human T-cell leukemia virus type 1 Tax through NF-kappaB in apoptosis-resistant T-cell transfectants with Tax. *J. Virol.* **1999**, *73*, 7981–7987.

88. Nicot, C.; Mahieux, R.; Takemoto, S.; Franchini, G. Bcl-X(L) is up-regulated by HTLV-I and HTLV-II *in vitro* and in *ex vivo* ATLL samples. *Blood* **2000**, *96*, 275–281.

89. Swaims, A.Y.; Khani, F.; Zhang, Y.; Roberts, A.I.; Devadas, S.; Shi, Y.; Rabson, A.B. Immune activation induces immortalization of HTLV-1 LTR-Tax transgenic CD4+ T cells. *Blood* **2010**, *116*, 2994–3003.

90. Macaire, H.; Riquet, A.; Moncollin, V.; Biemont-Trescol, M.C.; Duc Dodon, M.; Hermine, O.; Debaud, A.L.; Mahieux, R.; Mesnard, J.M.; Pierre, M.; *et al.* Tax protein-induced expression of antiapoptotic Bfl-1 protein contributes to survival of human T-cell leukemia virus type 1 (HTLV-1)-infected T-cells. *J. Biol. Chem.* **2012**, *287*, 21357–21370.

91. Zhao, T.; Yasunaga, J.; Satou, Y.; Nakao, M.; Takahashi, M.; Fujii, M.; Matsuoka, M. Human T-cell leukemia virus type 1 bZIP factor selectively suppresses the classical pathway of NF-kappaB. *Blood* **2009**, *113*, 2755–2764.

92. Zhi, H.; Yang, L.; Kuo, Y.L.; Ho, Y.K.; Shih, H.M.; Giam, C.Z. NF-kappaB hyper-activation by HTLV-1 tax induces cellular senescence, but can be alleviated by the viral anti-sense protein HBZ. *PLoS Pathog.* **2011**, *7*, e1002025.

93. Ho, Y.K.; Zhi, H.; DeBiaso, D.; Philip, S.; Shih, H.M.; Giam, C.Z. HTLV-1 tax-induced rapid senescence is driven by the transcriptional activity of NF-kappaB and depends on chronically activated IKKalpha and p65/RelA. *J. Virol.* **2012**, *86*, 9474–9483.

94. Philip, S.; Zahoor, M.A.; Zhi, H.; Ho, Y.K.; Giam, C.Z. Regulation of human T-lymphotropic virus type I latency and reactivation by HBZ and Rex. *PLoS Pathog.* **2014**, *10*, e1004040.

95. Peloponese, J.M., Jr.; Jeang, K.T. Role for Akt/protein kinase B and activator protein-1 in cellular proliferation induced by the human T-cell leukemia virus type 1 tax oncoprotein. *J. Biol. Chem.* **2006**, *281*, 8927–8938.

96. Jeong, S.J.; Dasgupta, A.; Jung, K.J.; Um, J.H.; Burke, A.; Park, H.U.; Brady, J.N. PI3K/AKT inhibition induces caspase-dependent apoptosis in HTLV-1-transformed cells. *Virology* **2008**, *370*, 264–272.

97. Tomita, M.; Semenza, G.L.; Michiels, C.; Matsuda, T.; Uchihara, J.N.; Okudaira, T.; Tanaka, Y.; Taira, N.; Ohshiro, K.; Mori, N. Activation of hypoxia-inducible factor 1 in human T-cell leukaemia virus type 1-infected cell lines and primary adult T-cell leukaemia cells. *Biochem. J.* **2007**, *406*, 317–323.

98. Higuchi, M.; Takahashi, M.; Tanaka, Y.; Fujii, M. Downregulation of proapoptotic Bim augments IL-2-independent T-cell transformation by human T-cell leukemia virus type-1 Tax. *Cancer Med.* **2014**, *3*, 1605–1614.

99. Saito, K.; Saito, M.; Taniura, N.; Okuwa, T.; Ohara, Y. Activation of the PI3K-Akt pathway by human T cell leukemia virus type 1 (HTLV-1) oncoprotein Tax increases Bcl3 expression, which is associated with enhanced growth of HTLV-1-infected T cells. *Virology* **2010**, *403*, 173–180.

100. Sugata, K.; Satou, Y.; Yasunaga, J.; Hara, H.; Ohshima, K.; Utsunomiya, A.; Mitsuyama, M.; Matsuoka, M. HTLV-1 bZIP factor impairs cell-mediated immunity by suppressing production of Th1 cytokines. *Blood* **2012**, *119*, 434–444.

101. Tanaka-Nakanishi, A.; Yasunaga, J.; Takai, K.; Matsuoka, M. HTLV-1 bZIP factor suppresses apoptosis by attenuating the function of FoxO3a and altering its localization. *Cancer Res.* **2014**, *74*, 188–200.

102. Matsumoto, J.; Ohshima, T.; Isono, O.; Shimotohno, K. HTLV-1 HBZ suppresses AP-1 activity by impairing both the DNA-binding ability and the stability of c-Jun protein. *Oncogene* **2005**, *24*, 1001–1010.

103. Hivin, P.; Basbous, J.; Raymond, F.; Henaff, D.; Arpin-Andre, C.; Robert-Hebmann, V.; Barbeau, B.; Mesnard, J.M. The HBZ-SP1 isoform of human T-cell leukemia virus type I represses JunB activity by sequestration into nuclear bodies. *Retrovirology* **2007**, *4*, e14.

104. Isono, O.; Ohshima, T.; Saeki, Y.; Matsumoto, J.; Hijikata, M.; Tanaka, K.; Shimotohno, K. Human T-cell leukemia virus type 1 HBZ protein bypasses the targeting function of ubiquitination. *J. Biol. Chem.* **2008**, *283*, 34273–34282.

105. Clerc, I.; Hivin, P.; Rubbo, P.A.; Lemasson, I.; Barbeau, B.; Mesnard, J.M. Propensity for HBZ-SP1 isoform of HTLV-I to inhibit c-Jun activity correlates with sequestration of c-Jun into nuclear bodies rather than inhibition of its DNA-binding activity. *Virology* **2009**, *391*, 195–202.

106. Ohshima, T.; Mukai, R.; Nakahara, N.; Matsumoto, J.; Isono, O.; Kobayashi, Y.; Takahashi, S.; Shimotohno, K. HTLV-1 basic leucine-zipper factor, HBZ, interacts with MafB and suppresses transcription through a Maf recognition element. *J. Cell. Biochem.* **2010**, *111*, 187–194.

107. Kashanchi, F.; Brady, J.N. Transcriptional and post-transcriptional gene regulation of HTLV-1. *Oncogene* **2005**, *24*, 5938–5951.

108. Saggioro, D.; Barp, S.; Chieco-Bianchi, L. Block of a mitochondrial-mediated apoptotic pathway in Tax-expressing murine fibroblasts. *Exp. Cell Res.* **2001**, *269*, 245–255.

109. Trevisan, R.; Daprai, L.; Acquasaliente, L.; Ciminale, V.; Chieco-Bianchi, L.; Saggioro, D. Relevance of CREB phosphorylation in the anti-apoptotic function of human T-lymphotropic virus type 1 tax protein in serum-deprived murine fibroblasts. *Exp. Cell Res.* **2004**, *299*, 57–67.

110. Trevisan, R.; Daprai, L.; Paloschi, L.; Vajente, N.; Chieco-Bianchi, L.; Saggioro, D. Antiapoptotic effect of human T-cell leukemia virus type 1 tax protein correlates with its creb transcriptional activity. *Exp. Cell Res.* **2006**, *312*, 1390–1400.

111. Saggioro, D. Anti-apoptotic effect of Tax: An NF-kappaB path or a CREB way? *Viruses* **2011**, *3*, 1001–1014.

112. Saggioro, D.; Silic-Benussi, M.; Biasiotto, R.; D'Agostino, D.M.; Ciminale, V. Control of cell death pathways by HTLV-1 proteins. *Front. Biosci.* **2009**, *14*, 3338–3351.

113. Fukuda, R.I.; Tsuchiya, K.; Suzuki, K.; Itoh, K.; Fujita, J.; Utsunomiya, A.; Tsuji, T. Human T-cell leukemia virus type I tax down-regulates the expression of phosphatidylinositol 3,4,5-trisphosphate inositol phosphatases via the NF-kappaB pathway. *J. Biol. Chem.* **2009**, *284*, 2680–2689.

114. Gaudray, G.; Gachon, F.; Basbous, J.; Biard-Piechaczyk, M.; Devaux, C.; Mesnard, J.M. The complementary strand of the human T-cell leukemia virus type 1 RNA genome encodes a bZIP transcription factor that down-regulates viral transcription. *J. Virol.* **2002**, *76*, 12813–12822.

115. Lemasson, I.; Lewis, M.R.; Polakowski, N.; Hivin, P.; Cavanagh, M.H.; Thebault, S.; Barbeau, B.; Nyborg, J.K.; Mesnard, J.M. Human T-cell leukemia virus type 1 (HTLV-1) bZIP protein interacts with the cellular transcription factor CREB to inhibit HTLV-1 transcription. *J. Virol.* **2007**, *81*, 1543–1553.

116. Clerc, I.; Polakowski, N.; Andre-Arpin, C.; Cook, P.; Barbeau, B.; Mesnard, J.M.; Lemasson, I. An interaction between the human T cell leukemia virus type 1 basic leucine zipper factor (HBZ) and the KIX domain of p300/CBP contributes to the down-regulation of tax-dependent viral transcription by HBZ. *J. Biol. Chem.* **2008**, *283*, 23903–23913.

117. Cook, P.R.; Polakowski, N.; Lemasson, I. HTLV-1 HBZ protein deregulates interactions between cellular factors and the KIX domain of p300/CBP. *J. Mol. Biol.* **2011**, *409*, 384–398.

118. Ma, G.; Yasunaga, J.; Fan, J.; Yanagawa, S.; Matsuoka, M. HTLV-1 bZIP factor dysregulates the Wnt pathways to support proliferation and migration of adult T-cell leukemia cells. *Oncogene* **2013**, *32*, 4222–4230.

119. Arnulf, B.; Villemain, A.; Nicot, C.; Mordelet, E.; Charneau, P.; Kersual, J.; Zermati, Y.; Mauviel, A.; Bazarbachi, A.; Hermine, O. Human T-cell lymphotropic virus oncoprotein Tax represses TGF-beta 1 signaling in human T cells via c-Jun activation: A potential mechanism of HTLV-I leukemogenesis. *Blood* **2002**, *100*, 4129–4138.

120. Lee, D.K.; Kim, B.C.; Brady, J.N.; Jeang, K.T.; Kim, S.J. Human T-cell lymphotropic virus type 1 tax inhibits transforming growth factor-beta signaling by blocking the association of Smad proteins with Smad-binding element. *J. Biol. Chem.* **2002**, *277*, 33766–33775.

121. Mori, N.; Morishita, M.; Tsukazaki, T.; Giam, C.Z.; Kumatori, A.; Tanaka, Y.; Yamamoto, N. Human T-cell leukemia virus type I oncoprotein Tax represses Smad-dependent transforming growth factor beta signaling through interaction with CREB-binding protein/p300. *Blood* **2001**, *97*, 2137–2144.

122. Zhao, T.; Satou, Y.; Sugata, K.; Miyazato, P.; Green, P.L.; Imamura, T.; Matsuoka, M. HTLV-1 bZIP factor enhances TGF-beta signaling through p300 coactivator. *Blood* **2011**, *118*, 1865–1876.

123. Kim, Y.M.; Geiger, T.R.; Egan, D.I.; Sharma, N.; Nyborg, J.K. The HTLV-1 tax protein cooperates with phosphorylated CREB, TORC2 and p300 to activate CRE-dependent cyclin D1 transcription. *Oncogene* **2010**, *29*, 2142–2152.

124. Dellino, G.I.; Cittaro, D.; Piccioni, R.; Luzi, L.; Banfi, S.; Segalla, S.; Cesaroni, M.; Mendoza-Maldonado, R.; Giacca, M.; Pelicci, P.G. Genome-wide mapping of human DNA-replication origins: Levels of transcription at ORC1 sites regulate origin selection and replication timing. *Genome Res.* **2013**, *23*, 1–11.

125. Ma, Y.; Zheng, S.; Wang, Y.; Zang, W.; Li, M.; Wang, N.; Li, P.; Jin, J.; Dong, Z.; Zhao, G. The HTLV-1 HBZ protein inhibits cyclin D1 expression through interacting with the cellular transcription factor CREB. *Mol. Biol. Rep.* **2013**, *40*, 5967–5975.

126. Hagiya, K.; Yasunaga, J.; Satou, Y.; Ohshima, K.; Matsuoka, M. ATF3, an HTLV-1 bZip factor binding protein, promotes proliferation of adult T-cell leukemia cells. *Retrovirology* **2011**, *8*, e19.

127. Tabakin-Fix, Y.; Azran, I.; Schavinky-Khrapunsky, Y.; Levy, O.; Aboud, M. Functional inactivation of p53 by human T-cell leukemia virus type 1 Tax protein: Mechanisms and clinical implications. *Carcinogenesis* **2006**, *27*, 673–681.

128. Ariumi, Y.; Kaida, A.; Lin, J.Y.; Hirota, M.; Masui, O.; Yamaoka, S.; Taya, Y.; Shimotohno, K. HTLV-1 tax oncoprotein represses the p53-mediated trans-activation function through coactivator CBP sequestration. *Oncogene* **2000**, *19*, 1491–1499.

129. Pise-Masison, C.A.; Mahieux, R.; Jiang, H.; Ashcroft, M.; Radonovich, M.; Duvall, J.; Guillerm, C.; Brady, J.N. Inactivation of p53 by human T-cell lymphotropic virus type 1 Tax requires activation of the NF-kappaB pathway and is dependent on p53 phosphorylation. *Mol. Cell. Biol.* **2000**, *20*, 3377–3386.

130. Miyazato, A.; Sheleg, S.; Iha, H.; Li, Y.; Jeang, K.T. Evidence for NF-kappaB- and CBP-independent repression of p53's transcriptional activity by human T-cell leukemia virus type 1 Tax in mouse embryo and primary human fibroblasts. *J. Virol.* **2005**, *79*, 9346–9350.

131. Gillet, N.; Carpentier, A.; Barez, P.Y.; Willems, L. WIP1 deficiency inhibits HTLV-1 Tax oncogenesis: Novel therapeutic prospects for treatment of ATL? *Retrovirology* **2012**, *9*, e115.

132. Zane, L.; Yasunaga, J.; Mitagami, Y.; Yedavalli, V.; Tang, S.W.; Chen, C.Y.; Ratner, L.; Lu, X.; Jeang, K.T. Wip1 and p53 contribute to HTLV-1 Tax-induced tumorigenesis. *Retrovirology* **2012**, *9*, e114.

133. Kasai, T.; Iwanaga, Y.; Iha, H.; Jeang, K.T. Prevalent loss of mitotic spindle checkpoint in adult T-cell leukemia confers resistance to microtubule inhibitors. *J. Biol. Chem.* **2002**, *277*, 5187–5193.

134. Yasunaga, J.; Jeang, K.T. Viral transformation and aneuploidy. *Environ. Mol. Mutagen.* **2009**, *50*, 733–740.

135. Liu, B.; Hong, S.; Tang, Z.; Yu, H.; Giam, C.Z. HTLV-I Tax directly binds the Cdc20-associated anaphase-promoting complex and activates it ahead of schedule. *Proc. Natl. Acad. Sci. USA* **2005**, *102*, 63–68.

136. Peloponese, J.M., Jr.; Haller, K.; Miyazato, A.; Jeang, K.T. Abnormal centrosome amplification in cells through the targeting of Ran-binding protein-1 by the human T cell leukemia virus type-1 Tax oncoprotein. *Proc. Natl. Acad. Sci. USA* **2005**, *102*, 18974–18979.

137. Haoudi, A.; Semmes, O.J. The HTLV-1 tax oncoprotein attenuates DNA damage induced G1 arrest and enhances apoptosis in p53 null cells. *Virology* **2003**, *305*, 229–239.

138. Park, H.U.; Jeong, J.H.; Chung, J.H.; Brady, J.N. Human T-cell leukemia virus type 1 Tax interacts with Chk1 and attenuates DNA-damage induced G2 arrest mediated by Chk1. *Oncogene* **2004**, *23*, 4966–4974.

139. Park, H.U.; Jeong, S.J.; Jeong, J.H.; Chung, J.H.; Brady, J.N. Human T-cell leukemia virus type 1 Tax attenuates gamma-irradiation-induced apoptosis through physical interaction with Chk2. *Oncogene* **2006**, *25*, 438–447.

140. Gupta, S.K.; Guo, X.; Durkin, S.S.; Fryrear, K.F.; Ward, M.D.; Semmes, O.J. Human T-cell leukemia virus type 1 Tax oncoprotein prevents DNA damage-induced chromatin egress of hyperphosphorylated Chk2. *J. Biol. Chem.* **2007**, *282*, 29431–29440.

141. Chandhasin, C.; Ducu, R.I.; Berkovich, E.; Kastan, M.B.; Marriott, S.J. Human T-cell leukemia virus type 1 tax attenuates the ATM-mediated cellular DNA damage response. *J. Virol.* **2008**, *82*, 6952–6961.

142. Durkin, S.S.; Guo, X.; Fryrear, K.A.; Mihaylova, V.T.; Gupta, S.K.; Belgnaoui, S.M.; Haoudi, A.; Kupfer, G.M.; Semmes, O.J. HTLV-1 Tax oncoprotein subverts the cellular DNA damage response via binding to DNA-dependent protein kinase. *J. Biol. Chem.* **2008**, *283*, 36311–36320.

143. Belgnaoui, S.M.; Fryrear, K.A.; Nyalwidhe, J.O.; Guo, X.; Semmes, O.J. The viral oncoprotein tax sequesters DNA damage response factors by tethering MDC1 to chromatin. *J. Biol. Chem.* **2010**, *285*, 32897–32905.

144. Boxus, M.; Willems, L. How the DNA damage response determines the fate of HTLV-1 Tax-expressing cells. *Retrovirology* **2012**, *9*, e2.

145. Vernin, C.; Thenoz, M.; Pinatel, C.; Gessain, A.; Gout, O.; Delfau-Larue, M.H.; Nazaret, N.; Legras-Lachuer, C.; Wattel, E.; Mortreux, F. HTLV-1 bZIP Factor HBZ Promotes Cell Proliferation and Genetic Instability by Activating OncomiRs. *Cancer Res.* **2014**, *74*, 6082–6093.

146. Mukai, R.; Ohshima, T. HTLV-1 HBZ positively regulates the mTOR signaling pathway via inhibition of GADD34 activity in the cytoplasm. *Oncogene* **2014**, *33*, 2317–2328.

147. Jeang, K.T.; Widen, S.G.; Semmes, O.J.t.; Wilson, S.H. HTLV-I trans-activator protein, tax, is a trans-repressor of the human beta-polymerase gene. *Science* **1990**, *247*, 1082–1084.

148. Philpott, S.M.; Buehring, G.C. Defective DNA repair in cells with human T-cell leukemia/bovine leukemia viruses: Role of tax gene. *J. Natl. Cancer Inst.* **1999**, *91*, 933–942.

149. Kao, S.Y.; Marriott, S.J. Disruption of nucleotide excision repair by the human T-cell leukemia virus type 1 Tax protein. *J. Virol.* **1999**, *73*, 4299–4304.

150. Lemoine, F.J.; Kao, S.Y.; Marriott, S.J. Suppression of DNA repair by HTLV type 1 Tax correlates with Tax trans-activation of proliferating cell nuclear antigen gene expression. *AIDS Res. Hum. Retrovir.* **2000**, *16*, 1623–1627.

151. Ducu, R.I.; Dayaram, T.; Marriott, S.J. The HTLV-1 Tax oncoprotein represses Ku80 gene expression. *Virology* **2011**, *416*, 1–8.

152. Majone, F.; Jeang, K.T. Unstabilized DNA breaks in HTLV-1 Tax expressing cells correlate with functional targeting of Ku80, not PKcs, XRCC4, or H2AX. *Cell Biosci.* **2012**, *2*, e15.

153. Baydoun, H.H.; Bai, X.T.; Shelton, S.; Nicot, C. HTLV-I tax increases genetic instability by inducing DNA double strand breaks during DNA replication and switching repair to NHEJ. *PLoS ONE* **2012**, *7*, e42226.

154. Baydoun, H.H.; Pancewicz, J.; Nicot, C. Human T-lymphotropic type 1 virus p30 inhibits homologous recombination and favors unfaithful DNA repair. *Blood* **2011**, *117*, 5897–5906.

Chapter 3:
Interplay between Virus and
Host in Virus Replication

HIV-1 Replication and the Cellular Eukaryotic Translation Apparatus

Santiago Guerrero, Julien Batisse, Camille Libre, Serena Bernacchi,
Roland Marquet and Jean-Christophe Paillart

Abstract: Eukaryotic translation is a complex process composed of three main steps: initiation, elongation, and termination. During infections by RNA- and DNA-viruses, the eukaryotic translation machinery is used to assure optimal viral protein synthesis. Human immunodeficiency virus type I (HIV-1) uses several non-canonical pathways to translate its own proteins, such as leaky scanning, frameshifting, shunt, and cap-independent mechanisms. Moreover, HIV-1 modulates the host translation machinery by targeting key translation factors and overcomes different cellular obstacles that affect protein translation. In this review, we describe how HIV-1 proteins target several components of the eukaryotic translation machinery, which consequently improves viral translation and replication.

Reprinted from *Viruses*. Cite as: Guerrero, S.; Batisse, J.; Libre, C.; Bernacchi, S.; Marquet, R.; Paillart, J.-C. HIV-1 Replication and the Cellular Eukaryotic Translation Apparatus. *Viruses* **2015**, *7*, 199–218.

1. Introduction

Eukaryotic translation is a complex process orchestrated by a wide range of players, including several protein factors and three classes of RNA (ribosomal RNA (rRNA), transfer RNA (tRNA), and messenger RNA (mRNA)). This process is comprised of three main steps: initiation, elongation and termination. Immediately after transcription, mRNAs are matured and exported through the nuclear pores to the cytoplasm. Once in the cytoplasm, cap-dependent translation is initiated by the recruitment of the small ribosomal subunit (40S), which scans along the mRNA until it finds an initiation codon. At this point, the large ribosomal subunit (60S) is engaged to decode the mRNA and assembles amino acids to synthesize proteins. The elongation step is completed when the ribosome reaches a stop codon that triggers the termination step. Finally, the components of the translational machinery are recycled for further protein translation.

Optimal viral protein synthesis often occurs at the expense of cellular proteins and many viruses have evolved mechanisms that redirect and control the eukaryotic translational machinery. In these cases, viral factors can target the initiation, elongation and termination steps through interactions with key translation factors and mechanisms, which interfere with or disrupt the host translation machinery.

The HIV-1 proteins are mainly synthesized by a cap-dependent mechanism. Nonetheless, different pathways such as leaky scanning, frameshifting, ribosome shunting, and cap-independent mechanisms are used to complete translation of the viral proteome. Moreover, HIV-1 has evolved sophisticated strategies to overcome cellular barriers that affect viral protein translation. This is the case of ribosomal scanning inhibition due to highly structured RNA elements and cap-dependent translation inhibition triggered by host immune responses. In this review, we focus on the HIV-1 functions that are essential to control the cellular translation apparatus in order to improve translation of viral proteins at the expense of cellular factors.

2. Overview of Eukaryotic Translation

2.1. Translation Initiation

2.1.1. Pre-Initiation Complex Assembly and mRNA Activation

Prior to translation initiation, the 5'-end of nascent mRNAs are capped with a 7-methylguanosine (m^7G) and subsequently polyadenylated at their 3'-end immediately after transcription. The m^7G cap is then recognized by the cap-binding complex (CBC), which is composed by the cap-binding proteins 80 and 20 (CBP80/20) [1]. Messenger RNA export effectors (exon-exon junction complex (EJC) and SR proteins) are then loaded on mature mRNA molecules during the splicing events. These adaptors molecules allow the recruitment of the NXF1/NXT1 heterodimer, which mediates the export of the messenger ribonucleoprotein (mRNP) from the nucleus to the cytoplasm through the nuclear pores via interactions with nucleoporins [2,3] (Figure 1, steps 1–2).

Once in the cytoplasm, mRNAs undergo a CBC-mediated pioneer round of translation, which is important for the quality control of the transcript [4]. CBC is then displaced from the m^7G and mRNAs are activated by the eukaryotic initiation factor 4F complex (composed of eIF4E, eIF4G and eIF4A) and eIF4B (Figure 1, step 3). Thus, mRNA acquires a "closed-loop" structure, required for an optimal mRNA recruitment into the 43S pre-initiation complex (PIC) (Figure 1, step 6). This conformation is achieved by the simultaneous binding of PABP (Poly(A)-binding protein) and eIF4E to eIF4G [5].

In parallel, ternary complex (TC) formation occurs through the assembly of the initiator methionyl-tRNA (Met-tRNA$_i^{Met}$), eIF2 and a GTP molecule (Figure 1, step 4). The TC is then recruited to the ribosomal subunit 40S to assemble the 43S PIC (Figure 1, step 5). In this process, other eIFs (eIF1, 1A, 3 and 5) are required to promote TC binding to the 40S subunit [6].

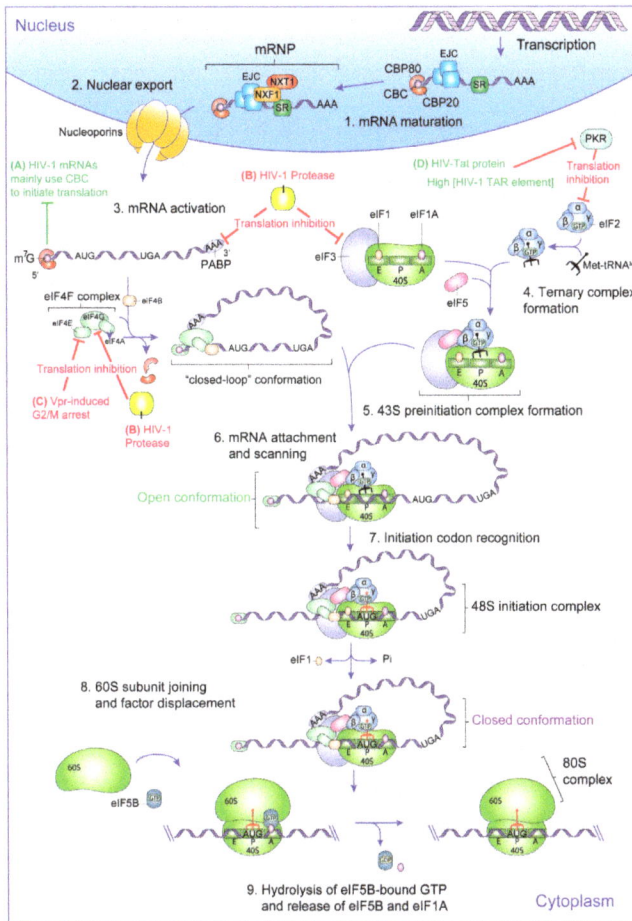

Figure 1. Eukaryotic translation initiation. mRNA maturation and nuclear export precede translation initiation (steps 1 and 2). Once in the cytoplasm, mRNA is activated (step 3) and 43S PIC formation takes place (step 4 to 5). The mRNA recruits the 43S PIC and scanning begins until an initiation codon is detected (step 6). At this point, the 48S initiation complex is formed and the 60S subunit is recruited to form the ribosome 80S complex (step 7 to 9). HIV-1 viral functions important to control the eukaryotic translation initiation are shown. (**A**) Unspliced and singly-spliced viral mRNAs could be translated due to the ability of the CBC to activate mRNA during translation initiation (green line); (**B**) HIV-1 protease partially inhibits translation initiation by targeting PABP, eIF3 and eIF4G (red lines); (**C**) Vpr-induced G2/M arrest indirectly inhibits host protein translation by targeting eIF4E activity (red line); (**D**) HIV-Tat protein and high concentration of HIV-1 TAR element indirectly promote viral translation by blocking PKR activity (red line). PKR phosphorylates eIF2α to block its recycling for ongoing translation, resulting in a potent translation inhibition of cellular and viral mRNA.

2.1.2. Initiation Codon Recognition and 80S Complex Formation

Once the mRNA has been loaded on the 43S PIC, this complex scans the 5' untranslated region (5'UTR) until it recognizes a start codon (AUG) by complementarity with the anticodon of the Met-tRNA$_i$. To prevent incorrect base pairing of the Met-tRNA$_i$ to a non-AUG codon and to promote recognition of the correct start codon, 43S PIC employs a discriminatory sequence-based mechanism. This permissive sequence (GCC(A/G)CCAUGG) surrounding the start codon is termed Kozak consensus sequence and is principally composed of a purine at position −3 and a G at position +4 (the A of the AUG codon is designated as +1) [7–9]. Once the AUG start codon has been recognized and the 48S complex formation has been accomplished, eIF1 is ejected from the scanning complex (Figure 1, step 7) [10]. This in turn triggers hydrolysis of eIF2-bound GTP and Pi release by the eIF5 GTPase activity.

These events cause the transition from an "open" to a "closed" conformation of the scanning complex, which stabilizes the binding of the Met-tRNAi with the AUG start codon [11]. At this point, the remaining factors are dissociated to allow the joining of the 60S subunit and 80S complex formation (Figure 1, step 8). This process is mediated by eIF5B which causes the dissociation of eIF3, eIF4B, eIF4F and eIF5, and eIF5B self-dissociates from the assembled 80S ribosome by its GTPase activity [12]. This reaction also triggers the release of eIF1A to finally forms an elongation competent 80S ribosome (Figure 1, step 9) [13].

2.2. Translation Elongation and Termination

After the AUG start codon has been identified and the formation of the 80S ribosome has been achieved, translation elongation begins [13]. 80S ribosome decodes the mRNA sequence and mediates the addition of amino acids to elongate the growing polypeptide chain. Elongation is accomplished by the eukaryotic elongation factors eEF1 (composed of eEF1A and eEF1B) and eEF2 (Figure 2) [14,15]. The eEF1A-GTP complex carries each aminoacylated tRNA to the 80S ribosome A site where it builds codon-anticodon interactions (Figure 2, step 1). After 80S ribosome-catalyzed peptide bond formation (Figure 2, step 2), eEF2 mediates 80S ribosome translocation through GTP hydrolysis (Figure 2, step 3) and the next round of amino acid incorporation begins (Figure 2, step 4).

Polypeptide elongation continues until a stop codon triggers the translation termination step. This process is mediated by the eukaryotic translation termination factors eRF1 and eRF3 [16]. The eRF1 factor mediates stop codon recognition, while eRF3 potently stimulates peptide release. After protein release, eRF1 remains bound to the post-termination complex (post-TC), and in conjunction with the ATP-binding cassette protein ABCE1, dissociates the post-TC into the 60S subunit, and the tRNA- and mRNA-bound 40S subunit [17].

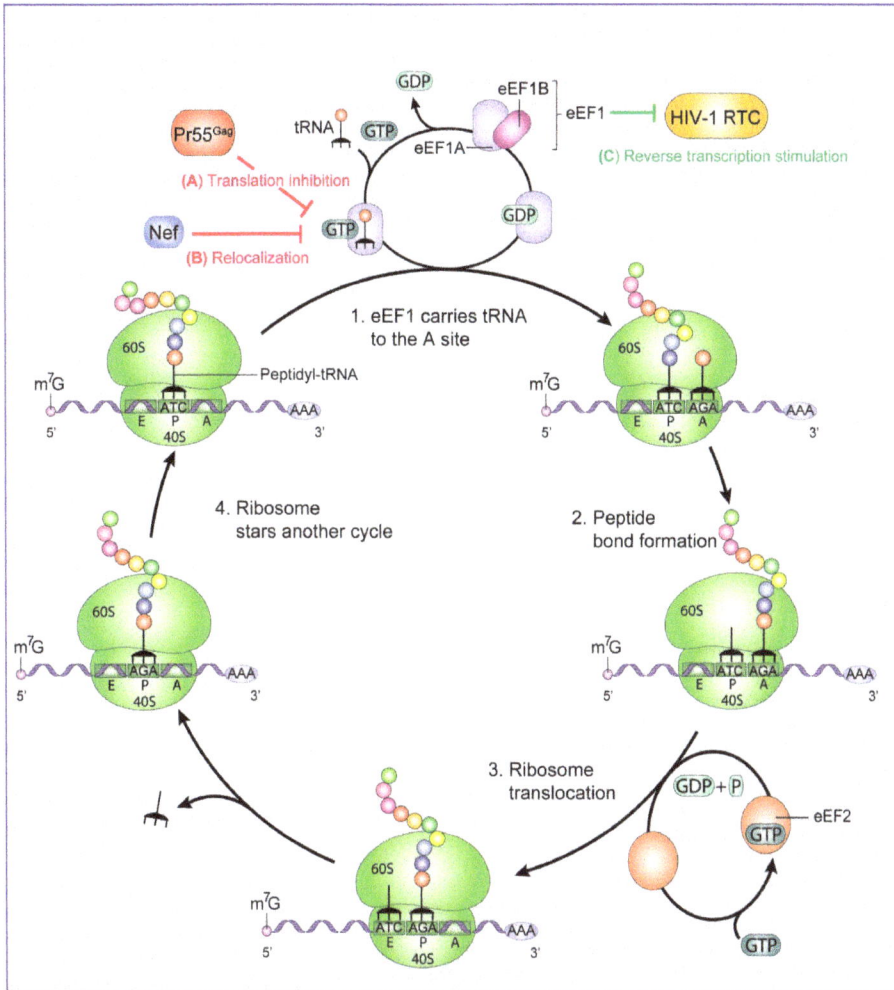

Figure 2. Eukaryotic translation elongation. eEF1A-GTP transports the aminoacylated tRNAs into the A site of the 80S ribosome and eEF1A-GDP is re-activated by eEF1B (step 1). The 80S ribosome-mediated peptide bond formation (step 2) precedes ribosome translocation mediated by eEF2 (step 3). Finally, the ribosome begins another cycle of peptide elongation (step 4). HIV-1 viral functions important to control the eukaryotic translation elongation are shown. (**A**) HIV-1 Pr55[Gag] interacts with eEF1A and induces translation inhibition (red line); (**B**) HIV-1 Nef protein also interacts with eEF1A and mediates a nucleocytoplasmic relocalization of eEF1A (red line); (**C**) The HIV-1 RTC recruits eEF1 to stimulate late steps of the HIV-1 reverse transcription process (green line).

3. HIV-1 Takes Advantage of the Host Translation Machinery

3.1. Overcoming Ribosome Scanning Barriers

All spliced and unspliced HIV-1 transcripts possess the same 289 nt long 5'UTR. Because this region presents several highly structured motifs such as the trans-activation responsive (TAR) RNA element, the unwinding step of the ribosomal scanning process is expected to be inefficient [18]. Soto-Rifo *et al.* [19] have shown that RNA helicase DDX3 is required in order to overcome this constraint. DDX3 directly binds to the HIV-1 5'UTR and interacts with eIF4G and PABP to promote translation initiation of HIV-1 genomic RNA [19], and mediates pre-initiation complex assembly in an ATP-dependent manner. In this process, the HIV-1 genomic RNA seems to be located in large cytoplasmic RNA granules alongside with DDX3, eIF4G and PABP, but not with CBP20/80 and eIF4E [20].

In vitro experiments revealed that cytoplasmic DDX3 is able to bind the m^7G cap independently of eIF4E, showing that DDX3 could promote the formation of a pre-initiation complex in the absence of eIF4E. Thus, DDX3 substitutes for eIF4E to stimulate compartmentalized translation initiation of HIV-1 unspliced mRNA [20]. Groom *et al.* [21], by using an *in vitro* transcription/translation assay from rabbit reticulocyte lysates, have demonstrated that Rev, which interacts with DDX3, stimulates translation of HIV-1 mRNAs at low concentration. This stimulation is dependent on a Rev binding site, in addition to the RRE, present in the internal loop B of stem-loop 1 (SL1) of HIV-1 RNA packaging signal [22]. On the contrary, at a high concentration, Rev inhibits mRNA translation in a non-specific manner. In this process, Rev may bind mRNAs and block ribosomal scanning by a mechanism that remains to be elucidated [21]. Lai *et al.* [23] have shown that DDX3 is recruited to the TAR region through the interaction with the Tat protein in order to promote HIV-1 mRNA translation.

Other cellular proteins also bind the TAR element and promote viral translation. RNA Helicase A (RHA) binds to the TAR region *in vitro* and *in vivo* enhancing HIV-1 LTR-directed gene expression and viral production [24]. Dorin *et al.* [25] showed that the TAR RNA-binding Protein (TRBP) promotes translation of TAR-bearing RNAs independently of its ability to inhibit protein kinase RNA-activated (PKR). Other RNA-binding proteins have been shown to stimulate HIV-1 translation. This is the case with Staufen [26] and La autoantigen [27]. Both proteins stimulate translation of TAR-containing reporter transcripts.

3.2. The Vpu-Env Bicistronic mRNA: Modulation of Ribosome Scanning Process

As many other viruses, HIV-1 optimizes its genome coding capacity by translating different proteins from a common mRNA. This is the case of the bicistronic vpu-env mRNA which encodes both Vpu and Env proteins [28,29]. The coding sequences of Vpu and Env proteins are arranged so that the Vpu Open Reading Frame (ORF) precedes the Env coding sequence (Figure 3). Schwartz *et al.* [28,29] showed that the Env protein is synthesized by a leaky scanning mechanism in which the 43S PIC passes through the Vpu start codon. According to these authors, this modulation of the 43S PIC scanning process is achieved because the Vpu initiation codon is surrounded by a weak Kozak context [28,29]. As a result, mutations of the Vpu initiation codon that improve its Kozak context inhibit Env translation from the bicistronic vpu-env mRNA [28,29].

In contrast, by studying the 5'UTR of 16 alternatively spliced Env mRNAs, some of which also including the extra upstream Rev initiation codon, Anderson *et al.* [30] demonstrated that mutations in the upstream AUG codons of the Env ORF had little effect on Env synthesis. This suggests that Env translation is achieved via a discontinuous scanning mechanism such as ribosome shunting, a process in which the ribosome bypasses parts of the 5'UTR to reach a start codon. Indeed, Krummheuer *et al.* [31] reported that translation of Env protein was inconsistent with the leaky scanning model. Instead, Env translation is stimulated by a six-nucleotide upstream ORF (uORF) (Figure 3), which is located in the vpu start codon region. uORFs are short open reading frames located within the 5'UTR of a mRNA. Mutations of the start and stop codons of this uORF reduced Env protein translation five-fold [31]. The authors suggest that this uORF acts as a ribosome pausing site supporting the ribosome shunting model [31].

3.3. The Fate of Unspliced HIV-1 mRNA

3.3.1. Cap- and IRES-Dependent Translation Initiation

The HIV-1 unspliced mRNA can initiate translation of Gag and Gag-Pol polyproteins either by a classical cap-dependent mechanism or by Internal Ribosome Entry Sites (IRES) [32–35]. Cap-dependent translation of HIV-1 unspliced mRNA has been observed *in vitro* [35] and *ex vivo* [33,36]. As described above, cap-dependent translation initiation relies on the recognition of the 5'UTR cap structure by the eIF4F complex to promote the recruitment of 43S PIC (Figure 1).

IRESs present in several viral mRNAs (mainly picornaviruses and Hepatitis C virus) overcome a global translation down-regulation established by the cell during viral infection. The HIV-1 genomic mRNA presents two IRESs: one in the viral 5'UTR, named HIV-1 IRES [32] and a second within the *gag* coding region, known as HIV-1 Gag IRES [37] (Figure 4). The HIV-1 IRES has been characterized in a proviral wild-type HIV-1 clone (pNL4.3) [32], but also in the CXCR4 (X4)-tropic primary isolate HIV-LAI [38], and in viral RNA isolated from clinical samples [39], demonstrating the importance of these elements in HIV-1 replication. The cap structure and the HIV-1 IRES both drive translation of pr55^{Gag} and pr160$^{Gag/Pol}$. The minimal active HIV-1 IRES is harbored within the region spanning nucleotides 104 to 336 [32]. This region also contains several RNA motifs involved in different functions of the HIV-1 life cycle (Figure 4).

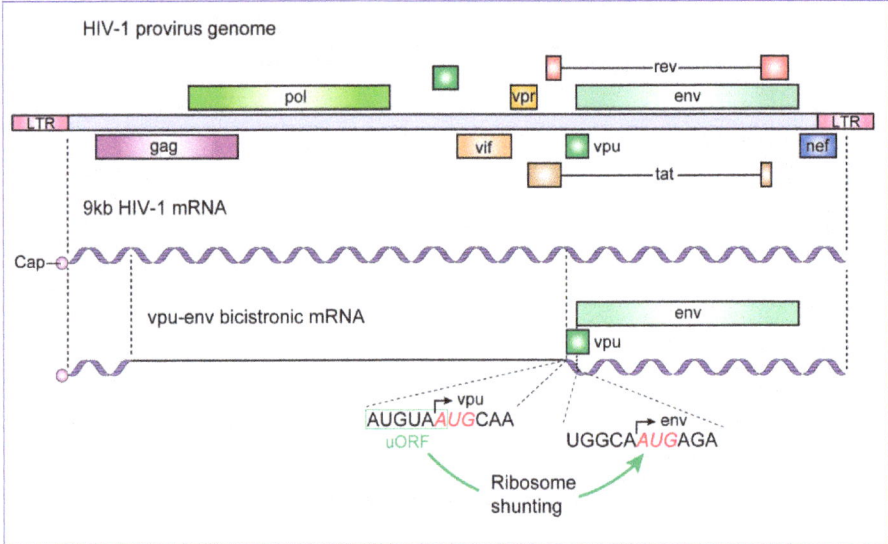

Figure 3. The HIV-1 vpu-env bicistronic mRNA. A schematic diagram of the HIV-1 provirus genome is presented along with its primary 9 kb mRNA transcript. The vpu-env bicistronic mRNA is formed after splicing of the 9 kb mRNA. Start codon sequences of both vpu and env ORFs are presented and the upstream ORF that stimulates Env translation is boxed in green.

The HIV-1 IRES has been shown to be implicated in different cellular states in which cap-dependent translation initiation is inhibited [18]. Monette *et al.* [34], by using *in vitro* artificial systems and HIV-1-expressing cells, reported that Pr55^{Gag} translation was preserved at 70% when eIF4G and PABP, two main components of the cap-dependent translation initiation, were targeted by picornavirus proteases. Using a similar system, Amorim *et al.* [40] showed that HIV-1 protein synthesis is highly dependent on cap-initiation the first 24–48 h of viral replication, while at later time points IRES-dependent translation is needed to ensure viral particles production. HIV-1 IRES also drives viral structural protein synthesis during the G2/M cell cycle transition [32,41] and is stimulated by oxidative stress [38].

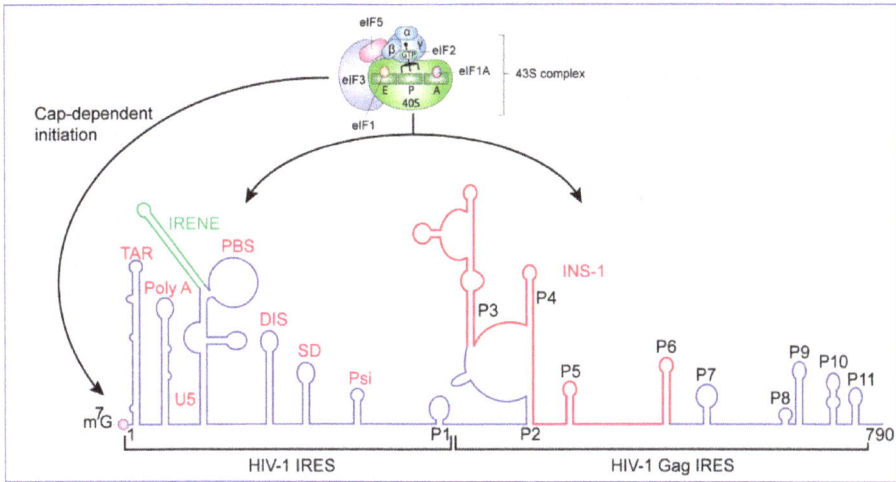

Figure 4. HIV-1 IRESs. HIV-1 IRES and HIV-1 Gag IRESs are represented. The 43S pre-initiation complex can mediate both cap-dependent and IRES-mediated translation initiation.

Although the molecular mechanism of the HIV-1 IRES-mediated translation remains largely unknown, Plank *et al.* [42] demonstrated that the HIV-1 IRES is cell type-specific. Using a plasmid encoding a dual-luciferase reporter mRNA, the authors showed that the HIV-1 IRES activity was 4-fold higher in Jurkat T-cells than in HeLa cells. Based on the fact that all spliced and unspliced HIV-1 transcripts possess the same 289 nt long 5'UTR, the authors also demonstrated that vif, vpr, vpu, and nef transcripts can initiate translation by an IRES. Interestingly, despite the fact that all HIV-1 constructs contain the same 5' leader region, the IRES activity of these transcripts differ from the IRES-containing gag transcript [42]. Based on these results, Plank *et al.* [42] proposed a model in which the structure of the 5' leader region adopts several conformations to stimulate different processes in the viral replication cycle, such as RNA dimerization prior to packaging or IRES-mediated translation. Thus, cell type-specific ITAFs (IRES Trans-Acting Factors) could promote the structural conformation of the 5' leader region required for an optimal IRES activity. In vif, vpr, vpu or nef transcripts, the specific ITAFs bind regions of the 5'UTR common to all these transcripts to form a "Core IRES" that stimulate IRES-mediated translation. However, the IRES activity of these transcripts could be regulated by specific RNA sequences present downstream of the common 5'UTR [42].

Several cellular proteins have been reported as cellular factors increasing the HIV-1 IRES activity [18]: eIF5 [43], the heterogeneous nuclear ribonucleoprotein A1 (hnRNP A1) [44] and the Rev-cofactors DDX3 and hRIP [43]. On the contrary, the human embryonic lethal abnormal vision (ELAV)-like protein (HuR) has been

231

identified as a negative regulator of the HIV-1 IRES activity [45]. *Cis*-acting elements have also been shown to negatively modulate HIV-1 IRES activity. Brasey *et al.* [32] showed that the Gag ORF impacts the HIV-1 IRES-mediated translation initiation in the context of a bi-cistronic mRNA. Gendron *et al.* [38] identified another region located upstream of the PBS, the IRES negative element (IRENE) that also negatively regulates the HIV-1 IRES activity (Figure 4). Recently, it was shown that the instability element 1 (INS-1), a *cis*-acting regulatory element present within the gag ORF, also inhibits HIV-1 IRES activity (Figure 4) [46]. Furthermore, several HIV-1 IRES ITAFs have been identified from G2/M-arrested cell extracts [41], including proteins that have already been identified to have a role in HIV-1 replication, such as the high mobility group protein HMG-I/HMG-Y (HMG-I(Y)) [47] or the activated RNA polymerase II transcriptional co-activator p15 [48].

The HIV-1 Gag IRES mediates translation of both full length Pr55^{Gag} polyprotein and a *N*-truncated 40-kDa Gag (p40) isoform lacking the matrix domain [37,49]. Despite the fact that p40 is produced at a much lower level than Pr55^{Gag}, this protein seems to be important in wild-type replication of HIV-1 in cultured cells [37]. Expression of p40 has also been observed in all HIV-1 clinical isolates from a large cohort of patients (n = 100) [18].

3.3.2. Ribosomal Frameshift

The Gag-Pol polyprotein, which is synthesized at a ~1/10 ratio compared to Gag, is translated via a −1 nucleotide ribosomal frameshift [50,51]. This process is regulated by two main factors, a slippery heptanucleotide sequence (UUUUUUA) where the frameshift takes place, and a downstream RNA element called the frameshift stimulatory signal (FSS) that controls the frameshift efficiency (Figure 5) [52]. The shift places the *gag* termination codon into an out-of-frame context and translation continues toward the downstream *pol* sequence [50,51]. The structure of the FSS and its mechanistic mode of action are not well understood. It has recently been proposed that the −1 frameshift is promoted by 4 pseudoknots (PK1-4) present in the FSS [53,54]. However, *ex virio* SHAPE (Selective 2'-Hydroxyl Acylation analyzed by Primer Extension) experiments [55] and frameshifting assays [56] do not support this model.

The most widely accepted model suggests that −1 frameshift is triggered when the FSS forces a small portion of the 80S complexes to make a pause. This pause forces the 80S complexes to shift one nucleotide backwards into the *pol* sequence [57–59]. Léger *et al.* [57] have proposed a model in which, after release of the eEF2-GDP complex from the ribosome, the two tRNAs cannot be translocated by three nucleotides (from P/A to E/P sites). Instead the two tRNAs are translocated only by two nucleotides due to the FSS. Consequently, the two tRNAs are trapped in an intermediate translocation state and frameshift is achieved [57]. This is supported by the cryo-EM structure resolved by Namy *et al.* [58], in which the ribosome is stalled by the presence of the FSS. Additionally, Mouzakis *et al.* [56] have demonstrated that the base pairs important for this process are located at an 8 nt distance from the slippery sequence, which is consistent with the paused-ribosome model.

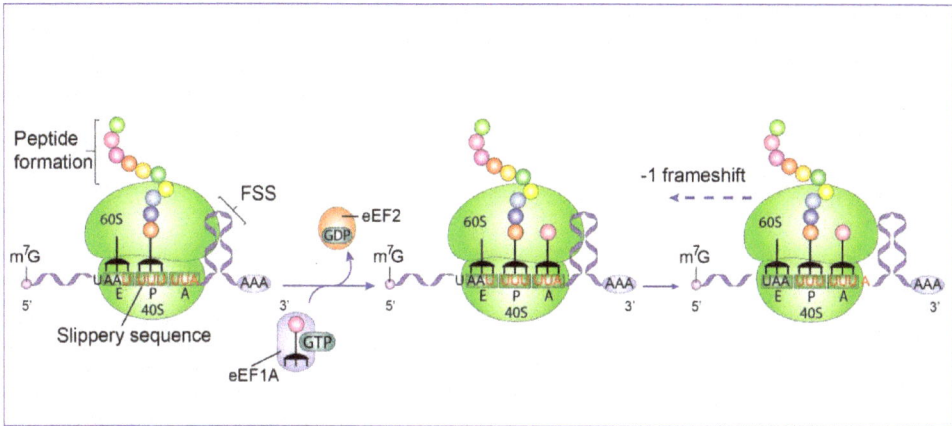

Figure 5. HIV-1 Ribosomal frameshift. A schematic diagram of the −1 nucleotide ribosomal frameshift. This process is mediated by a slippery heptanucleotide sequence (UUUUUUA) where the frameshift takes place, and an RNA element called the frameshift stimulatory signal (FSS).

Further studies have shown that the highly structured TAR element and the Tat protein modulate ribosomal frameshift [60,61]. Charbonneau *et al.* [61] demonstrated that the presence of the HIV-1 5'UTR in a reporter mRNA increases the −1 frameshift efficiency fourfold in Jurkat T-cells, compared to a control reporter with a short unstructured 5'UTR. This is associated with the presence of the TAR region within the 5'UTR [60,61]. This region slows down the rate of translation initiation during cap-dependent translation. As a consequence, the distance between ribosomes is larger and the FSS has more time to refold [60,61]. Thus, a structured FSS is required for an optimal −1 frameshift process. The increase in the −1 frameshift efficiency is antagonized by the HIV-1 Tat protein, which indirectly destabilizes TAR structure by increasing recruitment of RNA helicases, such as DDX3 [23]. Moreover, Lorgeoux *et al.* [62], using a dual-luciferase reporter assay, showed that the helicase DDX17 is required for −1 frameshift and for maintaining proper ratios of Gag *vs.* Gag-Pol proteins.

3.4. Redirecting mRNA Activation

In the G2/M phase of the cell cycle, cap-dependent translation is down-regulated by a cascade of events that lead to the disruption of the eIF4F complex, and thereby inhibition of the mRNA activation step (Figure 1, step 3) [63]. In this process, formation of the eIF4F complex is prevented by targeting of eIF4E. This factor is regulated by a family of translation inhibitor proteins, named the eIF4E-binding proteins (4E-BPs). After hypophosphorylation, 4E-BPs compete with eIF4G for the same binding site on eIF4E, preventing eIF4F complex assembly [63,64].

In HIV-1-infected cells, viral proteins alter cell function by affecting different cellular pathways. HIV-1 viral protein R (Vpr) has been identified as a viral protein capable of arresting cells in the G2/M phase. Vpr mediates G2/M arrest via a complex signaling cascade involving activation of the ataxia telangiectasia mutated and Rad3-related kinase (ATR) [65]. Vpr-induced G2/M arrest also inhibits host protein translation by a process involving the regulation of eIF4E activity (Figure 1C) [66].

Despite the global reduction of host protein synthesis by Vpr-induced G2/M arrest, translation of HIV-1 structural proteins is maintained [58]. RNA-coimmunoprecipitation experiments showed that full-length unspliced HIV-1 genomic RNA and singly spliced mRNAs are associated with CBC in contrast to multi-spliced viral and cellular mRNAs that are associated with eIF4E. Moreover, unspliced and singly-spliced viral mRNAs retain their interaction with CBC during translation and packaging. Based on these observations, Sharma *et al.* [66] hypothesized that unspliced and singly-spliced viral mRNAs are translated due to the ability of the CBC to activate mRNA during initiation (Figure 1A). Thus, CBC retention could allow viral protein synthesis while the global protein translation is inhibited by an eIF4E decrease.

3.5. Targeting Cellular Translation Factors

During the HIV-1 replication cycle, viral proteins inhibit different translation factors. This is the case of the HIV-1 protease, which partially impairs cap-dependent protein translation, in addition to its main function during virion maturation. Ventoso *et al.* [67] were the first to report that HIV-1 protease is able to cleave the initiation factor eIF4G *in cellula* (Figure 1B), by using HIV-1 C8166 target cells. Further studies using rabbit reticulocyte lysates have demonstrated that HIV-1 protease not only cleaves eIF4G but also PABP (Figure 1B) [68–70]. Degradation of eIF4G and PABP leads to an inhibition of the cap- and PABP-dependent translation initiation [67–70]. Jäger *et al.* [71] reported that the HIV-1 protease also cleaves eIF3d (Figure 1B), a subunit of eIF3. This cleavage presents a similar efficiency to the one of the Pr55[Gag]. This may also promote inhibition of cap-dependent protein synthesis.

In the early phases of HIV-1 replication, the HIV-1 TAR RNA element activates PKR, which mediates host translation inhibition [72]. PKR interacts with the

TAR element, inducing PKR dimerization, auto-phosphorylation and activation. Activated PKR phosphorylates the alpha subunit of eIF2 (eIF2α), blocking its recycling for ongoing translation, resulting in a potent translation inhibition of cellular and viral mRNA [73]. HIV-1 indirectly prevents phosphorylation by targeting PKR. Indeed, the HIV-1 Tat protein directly interacts with PKR, thus, preventing auto-phosphorylation (Figure 1D) which is essential for function [72,74]. Moreover, during the late events of HIV-1 replication, PKR activity is inhibited by the high concentration of the HIV-1 TAR element (Figure 1D). Studies with TAR and other RNAs showed that high concentration of double-stranded RNA (dsRNA) inhibits PKR dimerization and therefore its activation [75]. PKR activation is further inhibited by HIV-induced mechanisms involving the cellular factors TRBP, Adenosine deaminase ADAR1 and a change in PACT function [72,76].

Moreover, by using a yeast two-hybrid screen assay, Cimarelli and Luban [77] reported that the HIV-1 Pr55^{Gag} binds to eEF1A through its matrix (MA) and nucleocapsid (NC) domains. They demonstrated that this interaction requires RNA, and suggested that tRNA could probably mediate this interaction since both NC and eEF1A have been shown to bind tRNAs [78,79]. Thus, as cellular concentration of Pr55^{Gag} increases in the cell, Pr55^{Gag} association with eEF1A-tRNA complexes may induce translational inhibition (Figure 2A). As a consequence of this inhibition, viral genomic RNA could be released from the translation machinery. Subsequently, Pr55^{Gag} interaction with the genomic RNA may lead to a further viral translation inhibition, thus stimulating RNA packaging into virions [77]. The authors also showed that both eEF1A and a truncated form of eEF1A of 34 to 36 kDa are incorporated into virions. In addition, eEF1A can also interact with the cellular cytoskeleton, suggesting a possible role of eEF1A in virion assembly and budding [80]. Despite these reports, direct evidence for a role of eEF1A in RNA packaging, virion assembly or viral particle budding remains to be demonstrated [81].

Similar to Pr55^{Gag}, HIV-1 Nef protein also interacts with eEF1A and forms a Nef/eEF1A/tRNA complex. Thus, Nef mediates a nucleocytoplasmic relocalization of eEF1A and tRNAs to prevent stress-induced apoptosis in primary human macrophages (Figure 2B) [82]. Moreover, Warren *et al.* [83] have reported that the HIV-1 reverse transcription complex (RTC) recruits eEF1 to stimulate late steps of the HIV-1 reverse transcription process (Figure 2C). The eEF1 factor binds to the HIV-1 RTC through an interaction with reverse transcriptase (RT) and integrase (IN). This association enhances the stability of the RTC in the cytoplasm [83]. Nevertheless, further studies will be needed to establish whether eEF1 functions synergistically with the components of the RTC, or independently during reverse transcription.

3.6. Inhibiting APOBEC3G Translation

APOBEC3G (apolipoprotein *B* mRNA-editing enzyme, catalytic polypeptide-like 3G, or A3G) is a restriction factor that impairs HIV-1 replication [84]. A3G is incorporated into viral particles and during reverse transcription in the target cells, it converts cytidines into uridines in the (−) strand DNA. This will generate further reverse transcription and integration defects and potentially produce non-functional viral proteins. HIV-1 counteracts this cellular factor by at least two pathways involving the HIV-1 viral infectivity factor (Vif). First, binding of Vif to A3G allows the recruitment of an E3 ubiquitin ligase that mediates the poly-ubiquitination of A3G and its degradation through the proteasome pathway. Second, Vif impairs the translation of A3G mRNA through a putative mRNA-binding mechanism [85,86].

Mariani *et al.* [87] were the first to observe that Vif inhibits A3G translation and suggested that this repression may contribute to the reduction of A3G encapsidation. The authors reported that Vif causes a 4.6 fold reduction in A3G synthesis by pulse-chase metabolic labeling experiments. Additionally, Kao *et al.* [88] showed that Vif causes the reduction of cell-associated A3G by 20%–30% compared to up to 50-fold reduction in virus-associated protein. This result supports the idea that Vif functions at different levels to reduce the intracellular levels of A3G. A3G translational inhibition was finally confirmed by Stopak *et al.* [89]. By using kinetic analyses and *in vitro* transcription-translation experiments, the authors showed that Vif was capable of impairing A3G translation by approximately 30%–40% [89].

Based on these observations, we recently showed that Vif binds to A3G mRNA and inhibits its translation *in vitro* [86]. Indeed, filter binding assays and fluorescence titration experiments revealed that Vif tightly binds A3G mRNA. We also demonstrated that Vif is able to inhibit A3G translation in *in vitro*-coupled transcription/translation assays. In these experiments, Vif caused a two-fold reduction of A3G translation in a 5'UTR-dependent manner, most likely through mRNA binding and/or through its RNA chaperone activity [90,91]. These observations show that HIV-1 not only targets the host translational machinery to improve translation of its own proteins, but also inhibits translation of the host restriction factor A3G.

4. Concluding Remarks

HIV-1 has evolved several mechanisms that control the host translation machinery and overcome different obstacles in the cell, such as ribosomal scanning inhibition due to highly structured RNA elements and cap-dependent translation inhibition triggered by host immune responses. During this antiviral state triggered by the cell, HIV-1 proteins interact with the eukaryotic translation apparatus in different ways. Thus, these mechanisms not only promote translation of HIV-1 proteins, but also guarantee the fitness of newly produced viral particles by inhibiting

A3G incorporation. It seems that the sophisticated functions presented in this review only represent a fraction of the strategies that have evolved by HIV-1 interaction with cellular functions, leading to a control of the host translation machinery during viral replication. Indeed, Jäger *et al.* [71] reported 497 interactions between 16 HIV-1 proteins and 435 human proteins. In addition, genome-wide techniques such as ribosome profiling [92] and/or iCLIP (individual-nucleotide resolution UV cross-linking and immunoprecipitation) [93] could be used, for example to identify HIV-1 Vif mRNA targets and to understand how this protein inhibits translation of specific genes. Identifying which cellular proteins of the translational eukaryotic machinery are targeted by HIV-1 will be crucial for a global understanding of the HIV-1 replication.

Acknowledgments: We are grateful to Redmond Smyth for critical reading of the manuscript. This work was supported by grants from the French National Agency for Research on AIDS and Viral Hepatitis (ANRS) to JCP, and by post-doctoral and doctoral fellowships from ANRS to JB, SG and CL.

Author Contributions: S.G., R.M. and J.C.P. conceived the review topic. S.G. drafted the manuscript and generated the figures. J.C.P. and R.M. corrected and edited the manuscript. All authors read and approved the final manuscript.

Conflicts of Interest: The authors declare no conflict of interest.

References

1. Izaurralde, E.; Lewis, J.; McGuigan, C.; Jankowska, M.; Darzynkiewicz, E.; Mattaj, I.W. A nuclear cap binding protein complex involved in pre-mRNA splicing. *Cell* **1994**, *78*, 657–668.

2. Adams, R.L.; Wente, S.R. Uncovering nuclear pore complexity with innovation. *Cell* **2013**, *152*, 1218–1221.

3. Moore, M.J.; Proudfoot, N.J. Pre-mRNA processing reaches back to transcription and ahead to translation. *Cell* **2009**, *136*, 688–700.

4. Maquat, L.E.; Tarn, W.Y.; Isken, O. The pioneer round of translation: Features and functions. *Cell* **2010**, *142*, 368–374.

5. Uchida, N.; Hoshino, S.I.; Imataka, H.; Sonenberg, N.; Katada, T. A novel role of the mammalian GSPT/eRF3 associating with poly(A)-binding protein in Cap/Poly(A)-dependent translation. *J. Biol. Chem.* **2002**, *277*, 50286–50292.

6. Majumdar, R.; Bandyopadhyay, A.; Maitra, U. Mammalian translation initiation factor eIF1 functions with eIF1A and eIF3 in the formation of a stable 40S preinitiation complex. *J. Biol. Chem.* **2003**, *278*, 6580–6587.

7. Kozak, M. Point mutations close to the AUG initiator codon affect the efficiency of translation of rat preproinsulin *in vivo*. *Nature* **1984**, *308*, 241–246.

8. Kozak, M. Point mutations define a sequence flanking the AUG initiator codon that modulates translation by eukaryotic ribosomes. *Cell* **1986**, *44*, 283–292.

9.   Kozak, C.A.; Chakraborti, A. Single amino acid changes in the murine leukemia virus capsid protein gene define the target of Fv1 resistance. *Virology* **1996**, *225*, 300–305.

10. Maag, D.; Fekete, C.A.; Gryczynski, Z.; Lorsch, J.R. A conformational change in the eukaryotic translation preinitiation complex and release of eIF1 signal recognition of the start codon. *Mol. Cell* **2005**, *17*, 265–275.

11. Nanda, J.S.; Cheung, Y.N.; Takacs, J.E.; Martin-Marcos, P.; Saini, A.K.; Hinnebusch, A.G.; Lorsch, J.R. eIF1 controls multiple steps in start codon recognition during eukaryotic translation initiation. *J. Mol. Biol.* **2009**, *394*, 268–285.

12. Pestova, T.V.; Lomakin, I.B.; Lee, J.H.; Choi, S.K.; Dever, T.E.; Hellen, C.U. The joining of ribosomal subunits in eukaryotes requires eIF5B. *Nature* **2000**, *403*, 332–335.

13. Jackson, R.J.; Hellen, C.U.T.; Pestova, T.V. The mechanism of eukaryotic translation initiation and principles of its regulation. *Nat. Rev. Mol. Cell Biol.* **2010**, *11*, 113–127.

14. Groppo, R.; Richter, J.D. Translational control from head to tail. *Curr. Opin. Cell Biol.* **2009**, *21*, 444–451.

15. Kapp, L.D.; Lorsch, J.R. The molecular mechanics of eukaryotic translation. *Annu. Rev. Biochem.* **2004**, *73*, 657–704.

16. Jackson, R.J.; Hellen, C.U.T.; Pestova, T.V. Termination and post-termination events in eukaryotic translation. *Adv. Protein Chem. Struct. Biol.* **2012**, *86*, 45–93.

17. Pisarev, A.V.; Skabkin, M.A.; Pisareva, V.P.; Skabkina, O.V.; Rakotondrafara, A.M.; Hentze, M.W.; Hellen, C.U.T.; Pestova, T.V. The role of ABCE1 in eukaryotic posttermination ribosomal recycling. *Mol. Cell* **2010**, *37*, 196–210.

18. De Breyne, S.; Soto-Rifo, R.; López-Lastra, M.; Ohlmann, T. Translation initiation is driven by different mechanisms on the HIV-1 and HIV-2 genomic RNAs. *Virus Res.* **2013**, *171*, 366–381.

19. Soto-Rifo, R.; Rubilar, P.S.; Limousin, T.; de Breyne, S.; Décimo, D.; Ohlmann, T. DEAD-box protein DDX3 associates with eIF4F to promote translation of selected mRNAs. *EMBO J.* **2012**, *31*, 3745–3756.

20. Soto-Rifo, R.; Rubilar, P.S.; Ohlmann, T. The DEAD-box helicase DDX3 substitutes for the cap-binding protein eIF4E to promote compartmentalized translation initiation of the HIV-1 genomic RNA. *Nucleic Acids Res.* **2013**, *41*, 6286–6299.

21. Groom, H.C.T.; Anderson, E.C.; Dangerfield, J.A.; Lever, A.M.L. Rev regulates translation of human immunodeficiency virus type 1 RNAs. *J. Gen. Virol.* **2009**, *90*, 1141–1147.

22. Greatorex, J.; Gallego, J.; Varani, G.; Lever, A. Structure and stability of wild-type and mutant RNA internal loops from the SL-1 domain of the HIV-1 packaging signal. *J. Mol. Biol.* **2002**, *322*, 543–557.

23. Lai, M.C.; Wang, S.W.; Cheng, L.; Tarn, W.Y.; Tsai, S.J.; Sun, H.S. Human DDX3 Interacts with the HIV-1 Tat Protein to Facilitate Viral mRNA Translation. *PLoS One* **2013**, *8*, e68665.

24. Fujii, R.; Okamoto, M.; Aratani, S.; Oishi, T.; Ohshima, T.; Taira, K.; Baba, M.; Fukamizu, A.; Nakajima, T. A Role of RNA Helicase A in cis-Acting Transactivation Response Element-mediated Transcriptional Regulation of Human Immunodeficiency Virus Type 1. *J. Biol. Chem.* **2001**, *276*, 5445–5451.

238

25. Dorin, D.; Bonnet, M.C.; Bannwarth, S.; Gatignol, A.; Meurs, E.F.; Vaquero, C. The TAR RNA-binding protein, TRBP, stimulates the expression of TAR-containing RNAs *in vitro* and *in vivo* independently of its ability to inhibit the dsRNA-dependent kinase PKR. *J. Biol. Chem.* **2003**, *278*, 4440–4448.

26. Dugré-Brisson, S.; Elvira, G.; Boulay, K.; Chatel-Chaix, L.; Mouland, A.J.; DesGroseillers, L. Interaction of Staufen1 with the 5' end of mRNA facilitates translation of these RNAs. *Nucleic Acids Res.* **2005**, *33*, 4797–4812.

27. Svitkin, Y.V.; Pause, A.; Sonenberg, N. La autoantigen alleviates translational repression by the 5' leader sequence of the human immunodeficiency virus type 1 mRNA. *J. Virol.* **1994**, *68*, 7001–7007.

28. Schwartz, S.; Felber, B.K.; Pavlakis, G.N. Mechanism of translation of monocistronic and multicistronic human immunodeficiency virus type 1 mRNAs. *Mol. Cell. Biol.* **1992**, *12*, 207–219.

29. Schwartz, S.; Felber, B.K.; Fenyö, E.M.; Pavlakis, G.N. Env and Vpu proteins of human immunodeficiency virus type 1 are produced from multiple bicistronic mRNAs. *J. Virol.* **1990**, *64*, 5448–5456.

30. Anderson, J.L.; Johnson, A.T.; Howard, J.L.; Purcell, D.F.J. Both linear and discontinuous ribosome scanning are used for translation initiation from bicistronic human immunodeficiency virus type 1 env mRNAs. *J. Virol.* **2007**, *81*, 4664–4676.

31. Krummheuer, J.; Johnson, A.T.; Hauber, I.; Kammler, S.; Anderson, J.L.; Hauber, J.; Purcell, D.F.J.; Schaal, H. A minimal uORF within the HIV-1 vpu leader allows efficient translation initiation at the downstream env AUG. *Virology* **2007**, *363*, 261–271.

32. Brasey, A.; Lopez-Lastra, M.; Ohlmann, T.; Beerens, N.; Berkhout, B.; Darlix, J.L.; Sonenberg, N. The leader of human immunodeficiency virus type 1 genomic RNA harbors an internal ribosome entry segment that is active during the G2/M phase of the cell cycle. *J. Virol.* **2003**, *77*, 3939–3949.

33. Berkhout, B.; Arts, K.; Abbink, T.E.M. Ribosomal scanning on the 5'-untranslated region of the human immunodeficiency virus RNA genome. *Nucleic Acids Res.* **2011**, *39*, 5232–5244.

34. Monette, A.; Valiente-Echeverría, F.; Rivero, M.; Cohen, E.A.; Lopez-Lastra, M.; Mouland, A.J. Dual Mechanisms of Translation Initiation of the Full-Length HIV-1 mRNA Contribute to Gag Synthesis. *PloS One* **2013**, *8*, e68108.

35. Ricci, E.P.; Soto Rifo, R.; Herbreteau, C.H.; Decimo, D.; Ohlmann, T. Lentiviral RNAs can use different mechanisms for translation initiation. *Biochem. Soc. Trans.* **2008**, *36*, 690–693.

36. Miele, G.; Mouland, A.; Harrison, G.P.; Cohen, E.; Lever, A.M. The human immunodeficiency virus type 1 5' packaging signal structure affects translation but does not function as an internal ribosome entry site structure. *J. Virol.* **1996**, *70*, 944–951.

37. Buck, C.B.; Shen, X.; Egan, M.A.; Pierson, T.C.; Walker, C.M.; Siliciano, R.F. The human immunodeficiency virus type 1 gag gene encodes an internal ribosome entry site. *J. Virol.* **2001**, *75*, 181–191.

38. Gendron, K.; Ferbeyre, G.; Heveker, N.; Brakier-Gingras, L. The activity of the HIV-1 IRES is stimulated by oxidative stress and controlled by a negative regulatory element. *Nucleic Acids Res.* **2011**, *39*, 902–912.

39. Vallejos, M.; Carvajal, F.; Pino, K.; Navarrete, C.; Ferres, M.; Huidobro-Toro, J.P.; Sargueil, B.; López-Lastra, M. Functional and Structural Analysis of the Internal Ribosome Entry Site Present in the mRNA of Natural Variants of the HIV-1. *PLoS One* **2012**, *7*, e35031.

40. Amorim, R.; Costa, S.M.; Cavaleiro, N.P.; da Silva, E.E.; da Costa, L.J. HIV-1 transcripts use IRES-initiation under conditions where Cap-dependent translation is restricted by poliovirus 2A protease. *PloS One* **2014**, *9*, e88619.

41. Vallejos, M.; Deforges, J.; Plank, T.D.M.; Letelier, A.; Ramdohr, P.; Abraham, C.G.; Valiente-Echeverría, F.; Kieft, J.S.; Sargueil, B.; López-Lastra, M. Activity of the human immunodeficiency virus type 1 cell cycle-dependent internal ribosomal entry site is modulated by IRES trans-acting factors. *Nucleic Acids Res.* **2011**, *39*, 6186–6200.

42. Plank, T.D.M.; Whitehurst, J.T.; Kieft, J.S. Cell type specificity and structural determinants of IRES activity from the 5' leaders of different HIV-1 transcripts. *Nucleic Acids Res.* **2013**, *41*, 6698–6714.

43. Liu, J.; Henao-Mejia, J.; Liu, H.; Zhao, Y.; He, J.J. Translational Regulation of HIV-1 Replication by HIV-1 Rev Cellular Cofactors Sam68, eIF5A, hRIP, and DDX3. *J. Neuroimmune Pharmacol.* **2011**, *6*, 308–321.

44. Monette, A.; Ajamian, L.; López-Lastra, M.; Mouland, A.J. Human immunodeficiency virus type 1 (HIV-1) induces the cytoplasmic retention of heterogeneous nuclear ribonucleoprotein A1 by disrupting nuclear import: Implications for HIV-1 gene expression. *J. Biol. Chem.* **2009**, *284*, 31350–31362.

45. Rivas-Aravena, A.; Ramdohr, P.; Vallejos, M.; Valiente-Echeverría, F.; Dormoy-Raclet, V.; Rodríguez, F.; Pino, K.; Holzmann, C.; Huidobro-Toro, J.P.; Gallouzi, I.E.; *et al.* The Elav-like protein HuR exerts translational control of viral internal ribosome entry sites. *Virology* **2009**, *392*, 178–185.

46. Valiente-Echeverría, F.; Vallejos, M.; Monette, A.; Pino, K.; Letelier, A.; Huidobro-Toro, J.P.; Mouland, A.J.; López-Lastra, M. A cis-acting element present within the Gag open reading frame negatively impacts on the activity of the HIV-1 IRES. *PloS One* **2013**, *8*, e56962.

47. Li, L.; Yoder, K.; Hansen, M.S.; Olvera, J.; Miller, M.D.; Bushman, F.D. Retroviral cDNA integration: Stimulation by HMG I family proteins. *J. Virol.* **2000**, *74*, 10965–10974.

48. Holloway, A.F.; Occhiodoro, F.; Mittler, G.; Meisterernst, M.; Shannon, M.F. Functional interaction between the HIV transactivator Tat and the transcriptional coactivator PC4 in T cells. *J. Biol. Chem.* **2000**, *275*, 21668–21677.

49. Weill, L.; James, L.; Ulryck, N.; Chamond, N.; Herbreteau, C.H.; Ohlmann, T.; Sargueil, B. A new type of IRES within gag coding region recruits three initiation complexes on HIV-2 genomic RNA. *Nucleic Acids Res.* **2010**, *38*, 1367–1381.

50. Brierley, I.; Dos Ramos, F.J. Programmed ribosomal frameshifting in HIV-1 and the SARS–CoV. *Virus Res.* **2006**, *119*, 29–42.

51. Giedroc, D.P.; Cornish, P.V. Frameshifting RNA pseudoknots: Structure and mechanism. *Virus Res.* **2009**, *139*, 193–208.

52. Staple, D.W.; Butcher, S.E. Solution structure and thermodynamic investigation of the HIV-1 frameshift inducing element. *J. Mol. Biol.* **2005**, *349*, 1011–1023.

53. Huang, X.; Yang, Y.; Wang, G.; Cheng, Q.; Du, Z. Highly conserved RNA pseudoknots at the Gag-Pol junction of HIV-1 suggest a novel mechanism of −1 ribosomal frameshifting. *RNA N. Y. N* **2014**, *20*, 587–593.

54. Wang, G.; Yang, Y.; Huang, X.; Du, Z. Possible involvement of coaxially stacked double pseudoknots in the regulation of −1 programmed ribosomal frameshifting in RNA viruses. *J. Biomol. Struct. Dyn.* **2014**, 1–11.

55. Low, J.T.; Garcia-Miranda, P.; Mouzakis, K.D.; Gorelick, R.J.; Butcher, S.E.; Weeks, K.M. Structure and Dynamics of the HIV-1 Frameshift Element RNA. *Biochemistry (Mosc.)* **2014**, *53*, 4282–4291.

56. Mouzakis, K.D.; Lang, A.L.; Meulen, K.A.V.; Easterday, P.D.; Butcher, S.E. HIV-1 frameshift efficiency is primarily determined by the stability of base pairs positioned at the mRNA entrance channel of the ribosome. *Nucleic Acids Res.* **2012**.

57. Léger, M.; Dulude, D.; Steinberg, S.V.; Brakier-Gingras, L. The three transfer RNAs occupying the A, P and E sites on the ribosome are involved in viral programmed −1 ribosomal frameshift. *Nucleic Acids Res.* **2007**, *35*, 5581–5592.

58. Namy, O.; Moran, S.J.; Stuart, D.I.; Gilbert, R.J.C.; Brierley, I. A mechanical explanation of RNA pseudoknot function in programmed ribosomal frameshifting. *Nature* **2006**, *441*, 244–247.

59. Liao, P.Y.; Choi, Y.S.; Dinman, J.D.; Lee, K.H. The many paths to frameshifting: Kinetic modelling and analysis of the effects of different elongation steps on programmed −1 ribosomal frameshifting. *Nucleic Acids Res.* **2011**, *39*, 300–312.

60. Gendron, K.; Charbonneau, J.; Dulude, D.; Heveker, N.; Ferbeyre, G.; Brakier-Gingras, L. The presence of the TAR RNA structure alters the programmed −1 ribosomal frameshift efficiency of the human immunodeficiency virus type 1 (HIV-1) by modifying the rate of translation initiation. *Nucleic Acids Res.* **2008**, *36*, 30–40.

61. Charbonneau, J.; Gendron, K.; Ferbeyre, G.; Brakier-Gingras, L. The 5'UTR of HIV-1 full-length mRNA and the Tat viral protein modulate the programmed −1 ribosomal frameshift that generates HIV-1 enzymes. *RNA N. Y. N* **2012**, *18*, 519–529.

62. Lorgeoux, R.P.; Pan, Q.; Le Duff, Y.; Liang, C. DDX17 promotes the production of infectious HIV-1 particles through modulating viral RNA packaging and translation frameshift. *Virology* **2013**, *443*, 384–392.

63. Kronja, I.; Orr-Weaver, T.L. Translational regulation of the cell cycle: When, where, how and why? *Philos. Trans. R. Soc. Lond. B. Biol. Sci.* **2011**, *366*, 3638–3652.

64. Richter, J.D.; Sonenberg, N. Regulation of cap-dependent translation by eIF4E inhibitory proteins. *Nature* **2005**, *433*, 477–480.

65. Planelles, V.; Barker, E. Roles of Vpr and Vpx in modulating the virus-host cell relationship. *Mol. Asp. Med.* **2010**, *31*, 398–406.

66. Sharma, A.; Yilmaz, A.; Marsh, K.; Cochrane, A.; Boris-Lawrie, K. Thriving under stress: Selective translation of HIV-1 structural protein mRNA during Vpr-mediated impairment of eIF4E translation activity. *PLoS Pathog.* **2012**, *8*, e1002612.

67. Ventoso, I.; Blanco, R.; Perales, C.; Carrasco, L. HIV-1 protease cleaves eukaryotic initiation factor 4G and inhibits cap-dependent translation. *Proc. Natl. Acad. Sci. USA* **2001**, *98*, 12966–12971.

68. Alvarez, E.; Castelló, A.; Menéndez-Arias, L.; Carrasco, L. HIV protease cleaves poly(A)-binding protein. *Biochem. J.* **2006**, *396*, 219–226.

69. Castelló, A.; Franco, D.; Moral-López, P.; Berlanga, J.J.; Alvarez, E.; Wimmer, E.; Carrasco, L. HIV-1 protease inhibits Cap- and poly(A)-dependent translation upon eIF4GI and PABP cleavage. *PloS One* **2009**, *4*, e7997.

70. Ohlmann, T.; Prévôt, D.; Décimo, D.; Roux, F.; Garin, J.; Morley, S.J.; Darlix, J.L. In vitro cleavage of eIF4GI but not eIF4GII by HIV-1 protease and its effects on translation in the rabbit reticulocyte lysate system. *J. Mol. Biol.* **2002**, *318*, 9–20.

71. Jäger, S.; Cimermancic, P.; Gulbahce, N.; Johnson, J.R.; McGovern, K.E.; Clarke, S.C.; Shales, M.; Mercenne, G.; Pache, L.; Li, K.; *et al.* Global landscape of HIV-human protein complexes. *Nature* **2012**, *481*, 365–370.

72. Clerzius, G.; Gélinas, J.F.; Gatignol, A. Multiple levels of PKR inhibition during HIV-1 replication. *Rev. Med. Virol.* **2011**, *21*, 42–53.

73. Sadler, A.J.; Williams, B.R.G. Structure and function of the protein kinase R. *Curr. Top. Microbiol. Immunol.* **2007**, *316*, 253–292.

74. Mcmillan, N.A.J.; Chun, R.F.; Siderovski, D.P.; Galabru, J.; Toone, W.M.; Samuel, C.E.; Mak, T.W.; Hovanessian, A.G.; Jeang, K.T.; Williams, B.R.G. HIV-1 Tat Directly Interacts with the Interferon-Induced, Double-Stranded RNA-Dependent Kinase, PKR. *Virology* **1995**, *213*, 413–424.

75. Lemaire, P.A.; Anderson, E.; Lary, J.; Cole, J.L. Mechanism of PKR Activation by dsRNA. *J. Mol. Biol.* **2008**, *381*, 351–360.

76. Clerzius, G.; Shaw, E.; Daher, A.; Burugu, S.; Gélinas, J.F.; Ear, T.; Sinck, L.; Routy, J.P.; Mouland, A.J.; Patel, R.C.; Gatignol, A. The PKR activator, PACT, becomes a PKR inhibitor during HIV-1 replication. *Retrovirology* **2013**, *10*, 96.

77. Cimarelli, A.; Luban, J. Translation elongation factor 1-alpha interacts specifically with the human immunodeficiency virus type 1 Gag polyprotein. *J. Virol.* **1999**, *73*, 5388–5401.

78. Khan, R.; Giedroc, D.P. Recombinant human immunodeficiency virus type 1 nucleocapsid (NCp7) protein unwinds tRNA. *J. Biol. Chem.* **1992**, *267*, 6689–6695.

79. Merrick, W.C. Eukaryotic Protein Synthesis: Still a Mystery. *J. Biol. Chem.* **2010**, *285*, 21197–21201.

80. Ott, D.E.; Coren, L.V.; Johnson, D.G.; Kane, B.P.; Sowder, R.C., 2nd; Kim, Y.D.; Fisher, R.J.; Zhou, X.Z.; Lu, K.P.; Henderson, L.E. Actin-binding cellular proteins inside human immunodeficiency virus type 1. *Virology* **2000**, *266*, 42–51.

81. Li, D.; Wei, T.; Abbott, C.M.; Harrich, D. The unexpected roles of eukaryotic translation elongation factors in RNA virus replication and pathogenesis. *Microbiol. Mol. Biol. Rev. MMBR* **2013**, *77*, 253–266.

82. Abbas, W.; Khan, K.A.; Tripathy, M.K.; Dichamp, I.; Keita, M.; Rohr, O.; Herbein, G. Inhibition of ER stress-mediated apoptosis in macrophages by nuclear-cytoplasmic relocalization of eEF1A by the HIV-1 Nef protein. *Cell Death Dis.* **2012**, *3*, e292.

83. Warren, K.; Wei, T.; Li, D.; Qin, F.; Warrilow, D.; Lin, M.H.; Sivakumaran, H.; Apolloni, A.; Abbott, C.M.; Jones, A.; Anderson, J.L.; Harrich, D. Eukaryotic elongation factor 1 complex subunits are critical HIV-1 reverse transcription cofactors. *Proc. Natl. Acad. Sci. USA* **2012**, *109*, 9587–9592.

84. Sheehy, A.M.; Gaddis, N.C.; Choi, J.D.; Malim, M.H. Isolation of a human gene that inhibits HIV-1 infection and is suppressed by the viral Vif protein. *Nature* **2002**, *418*, 646–650.

85. Henriet, S.; Mercenne, G.; Bernacchi, S.; Paillart, J.C.; Marquet, R. Tumultuous Relationship between the Human Immunodeficiency Virus Type 1 Viral Infectivity Factor (Vif) and the Human APOBEC-3G and APOBEC-3F Restriction Factors. *Microbiol. Mol. Biol. Rev.* **2009**, *73*, 211–232.

86. Mercenne, G.; Bernacchi, S.; Richer, D.; Bec, G.; Henriet, S.; Paillart, J.C.; Marquet, R. HIV-1 Vif binds to APOBEC3G mRNA and inhibits its translation. *Nucleic Acids Res.* **2010**, *38*, 633–646.

87. Mariani, R.; Chen, D.; Schröfelbauer, B.; Navarro, F.; König, R.; Bollman, B.; Münk, C.; Nymark-McMahon, H.; Landau, N.R. Species-specific exclusion of APOBEC3G from HIV-1 virions by Vif. *Cell* **2003**, *114*, 21–31.

88. Kao, S.; Khan, M.A.; Miyagi, E.; Plishka, R.; Buckler-White, A.; Strebel, K. The human immunodeficiency virus type 1 Vif protein reduces intracellular expression and inhibits packaging of APOBEC3G (CEM15), a cellular inhibitor of virus infectivity. *J. Virol.* **2003**, *77*, 11398–11407.

89. Stopak, K.; de Noronha, C.; Yonemoto, W.; Greene, W.C. HIV-1 Vif blocks the antiviral activity of APOBEC3G by impairing both its translation and intracellular stability. *Mol. Cell* **2003**, *12*, 591–601.

90. Batisse, J.; Guerrero, S.; Bernacchi, S.; Sleiman, D.; Gabus, C.; Darlix, J.L.; Marquet, R.; Tisné, C.; Paillart, J.C. The role of Vif oligomerization and RNA chaperone activity in HIV-1 replication. *Virus Res.* **2012**, *169*, 361–376.

91. Sleiman, D.; Bernacchi, S.; Xavier Guerrero, S.; Brachet, F.; Larue, V.; Paillart, J.C.; Tisne, C. Characterization of RNA binding and chaperoning activities of HIV-1 Vif protein: Importance of the C-terminal unstructured tail. *RNA Biol.* **2014**, *11*, 906–920.

92. Ingolia, N.T. Ribosome profiling: New views of translation, from single codons to genome scale. *Nat. Rev. Genet.* **2014**, *15*, 205–213.

93. König, J.; Zarnack, K.; Rot, G.; Curk, T.; Kayikci, M.; Zupan, B.; Turner, D.J.; Luscombe, N.M.; Ule, J. iCLIP reveals the function of hnRNP particles in splicing at individual nucleotide resolution. *Nat. Struct. Mol. Biol.* **2010**, *17*, 909–915.

APOBEC3 Interference during Replication of Viral Genomes

Luc Willems and Nicolas Albert Gillet

Abstract: Co-evolution of viruses and their hosts has reached a fragile and dynamic equilibrium that allows viral persistence, replication and transmission. In response, infected hosts have developed strategies of defense that counteract the deleterious effects of viral infections. In particular, single-strand DNA editing by Apolipoprotein B Editing Catalytic subunits proteins 3 (APOBEC3s) is a well-conserved mechanism of mammalian innate immunity that mutates and inactivates viral genomes. In this review, we describe the mechanisms of APOBEC3 editing during viral replication, the viral strategies that prevent APOBEC3 activity and the consequences of APOBEC3 modulation on viral fitness and host genome integrity. Understanding the mechanisms involved reveals new prospects for therapeutic intervention.

Reprinted from *Viruses*. Cite as: Willems, L.; Gillet, N.A. APOBEC3 Interference during Replication of Viral Genomes. *Viruses* **2015**, *7*, 2999–3018.

1. APOBEC3s Edit Single-Stranded DNA

The APOBEC3 enzymes are deaminases that edit single-stranded DNA (ssDNA) sequences by transforming deoxycytidine into deoxyuridine [1–3]. APOBEC3s are involved in the mechanisms of innate defense against exogenous viruses and endogenous retroelements [3]. The human genome codes for seven APOBEC3 genes clustered in tandem on chromosome 22 (namely A3A, A3B, A3C, A3DE, A3F, A3G, and A3H) and surrounded by the CBX6 and CBX7 genes. All APOBEC3 genes encode a single- or a double-zinc-coordinating-domain protein. Each zinc-domain belongs to one of the three distinct phylogenic clusters termed Z1, Z2 and Z3. The seven APOBEC3 genes arose via gene duplications and fusions of a key mammalian ancestor with a CBX6-Z1-Z2-Z3-CBX7 locus organization. Aside from mice and pigs, duplications of APOBEC3 genes have occurred independently in different lineages: humans and chimpanzees ($n = 7$), horses ($n = 6$), cats ($n = 4$), and sheep and cattle ($n = 3$) [4,5]. Read-through transcription, alternative splicing and internal transcription initiation may further extend the diversity of APOBEC3 proteins.

APOBEC3s are interferon-inducible genes [6] that are highly expressed in immune cells despite being present in almost all cell types [7,8]. The sub-cellular localization differs between the APOBEC3s isoforms: A3DE/A3F/A3G are excluded from chromatin throughout mitosis and become cytoplasmic during interphase, A3B is nuclear and A3A/A3C/A3H are cell-wide during interphase [9].

APOBEC3s exert an antiviral effect either dependently or independently of their deaminase activity. The deaminase activity involves the removal of the exocyclic amine group from deoxycytidine to form deoxyuridine. This process can generate different types of substitutions. First, DNA replication through deoxyuridine leads to the insertion of a deoxyadenosine, therefore causing a C to T transition. Alternatively, Rev1 translesion synthesis DNA polymerase can insert a C in front of an abasic site that is produced through uracil excision by uracil-DNA glycosylase (UNG2) leading to a C-to-G transversion [10]. In addition to inducing deleterious mutations in the viral genome, deamination of deoxycytidine can also initiate degradation of uracilated viral DNA via a UNG2-dependent pathway [11,12]. On the other hand, deaminase-independent inhibition requires binding of APOBEC3s to single-stranded DNA or RNA viral sequences at various steps of the replication cycle [13–23].

2. APOBEC3 Edition during Viral Replication Cycles

The mechanism of APOBEC3s inactivation is dependent on the type of virus and its mode of replication.

2.1. Retroviruses

Retroviruses are plus-strand single-stranded RNA viruses replicating via a DNA intermediate generated in the cytoplasm by reverse transcription. Human retroviruses notably include HIV (human immunodeficiency virus) and HTLV (human T-lymphotropic virus).

2.1.1. HIV-1

Historically, the first member of the APOBEC3 family was discovered in a groundbreaking study on HIV-1 [24]. A3G has indeed been shown to inhibit HIV infection and to be repressed by the viral Vif protein. Later on, a similar function was also attributed to other APOBEC3 proteins, namely A3DE, A3F and A3H [25–27]. Figure 1 illustrates the different mechanisms of HIV-1 inhibition by APOBEC3s. After binding of the HIV virion to the host cell membrane, the viral single-stranded RNA (ssRNA) genome is released into the cytoplasm and converted into double-stranded DNA (dsDNA) by reverse transcription. This dsDNA is then inserted into the host genome as an integrated provirus.

A3DE, A3F, A3G and A3H are expressed by CD4+ T cells upon HIV infection, are packaged into virions and lead to proviral DNA mutations [27]. A3G and A3F notably concentrate in cytoplasmic microdomains (non-membrane structures) called mRNA-processing bodies or P-bodies [28–30]. P-bodies are sites of RNA storage, translational repression and decay [31]. A3G exerts its anti-HIV effect mainly via its deaminase function inducing abundant and deleterious mutations

245

within the HIV provirus, whereas A3F acts more preferentially through its deaminase-independent activity [32]. This deaminase-independent effect involves inhibition of reverse transcription priming and extension [14–17] and interference with proviral integration [21–23].

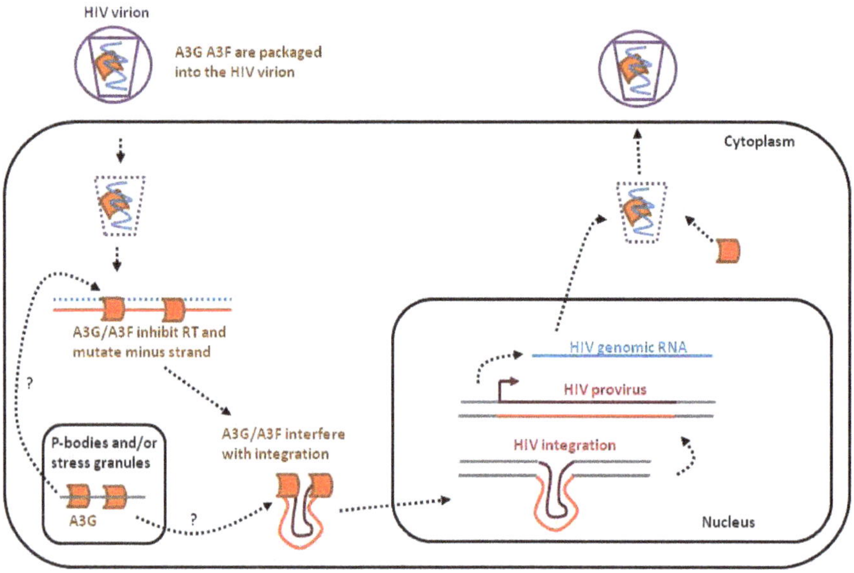

Figure 1. APOBEC3s interfere with several key steps of the HIV infectious cycle. After binding of the HIV virion to the cell membrane, the single-stranded RNA genome (**in blue**) is released into the cytoplasm together with APOBEC3G and 3F (**orange**). APOBEC3 proteins expressed by the host cell concentrate in P-bodies and stress granules. A3G and A3F inhibit reverse transcription, mutate viral DNA and perturb proviral integration into the host genome. In the absence of HIV Vif, A3G and A3F will be incorporated into the budding virions.

APOBEC3-induced mutations are almost always G-to-A transitions of the plus-strand genetic code. Moreover, the mutation load is not homogeneous along the HIV provirus but presents two highly polarized gradients, each peaking just 5′ to the central polypurine tract (cPPT) and 5′ to the LTR (long terminal repeat) proximal polypurine tract (3′PPT) [33]. As illustrated in Figure 2, this mutational signature is due to the mechanism of HIV reverse transcription. Binding of the human tRNALys3 to the primer binding sequence (PBS) initiates the minus strand DNA synthesis by the virus-encoded reverse transcriptase protein (RT). The RT-associated ribonuclease H activity (RNAse H) selectively degrades the RNA strand of the RNA:DNA hybrid leaving the nascent minus-strand DNA free to hybridize with the complementary sequence at the 3′ end of the viral genomic ssRNA. After minus strand transfer, the viral RNA is reverse-transcribed into DNA. Whilst DNA synthesis proceeds, the RNAse H function cleaves the RNA strand of the RNA:DNA. Two specific purine-rich sequences (polypurine tracts cPPT and 3'PPT) that are resistant to RNAse H remain annealed with the nascent minus strand DNA. The reverse transcriptase uses the PPTs as primers to synthesize the plus-strand DNA. Finally, another strand transfer allows the production of the 5' end of the plus-strand DNA (reviewed in [34]). From this complex multistep process, it appears that only the minus strand can be single-stranded (light red in Figure 2). Thus, G to A mutations observed on the plus strand (dark red in Figure 2) originate from C-to-T mutations on the minus strand. The gradient of mutational load actually correlates with the time that the minus strand remains single-chain [33].

To counteract inactivation, HIV-1 encodes the Vif protein that inhibits APOBEC3s. Vif prevents A3G, A3F and A3H from being packaged into the virion by recruitment to a cullin5-elonginB/C-Rbx2-CBFβ E3 ubiquitin ligase complex, resulting in their polyubiquitination and subsequent proteasomal degradation [35–37]. Other mechanisms can limit APOBEC3 access to the single-chain minus-strand DNA generated during reverse transcription. By stabilizing the viral core, the glycosylated Gag protein of the murine leukemia virus renders the reverse transcription complex resistant to APOBEC3 and to other cytosolic viral sensors [38]. The HIV nucleocapsid protein (NCp) is able to bind ssDNA in a sequence aspecific manner and prevents A3A from mutating genomic DNA during transient strand separation [39]. Degradation of the RNA strand from the RNA:DNA hybrid by the RNAse H activity of the reverse transcriptase contributes to expose the minus strand as a single-chain nucleic acid. Interestingly, the host factor SAMHD1 (sterile alpha motif and histidine-aspartic acid domain containing protein 1) restricts HIV via its RNAse H function, activity that may facilitate the access of the APOBEC3s to the transiently single-stranded minus strand [40,41].

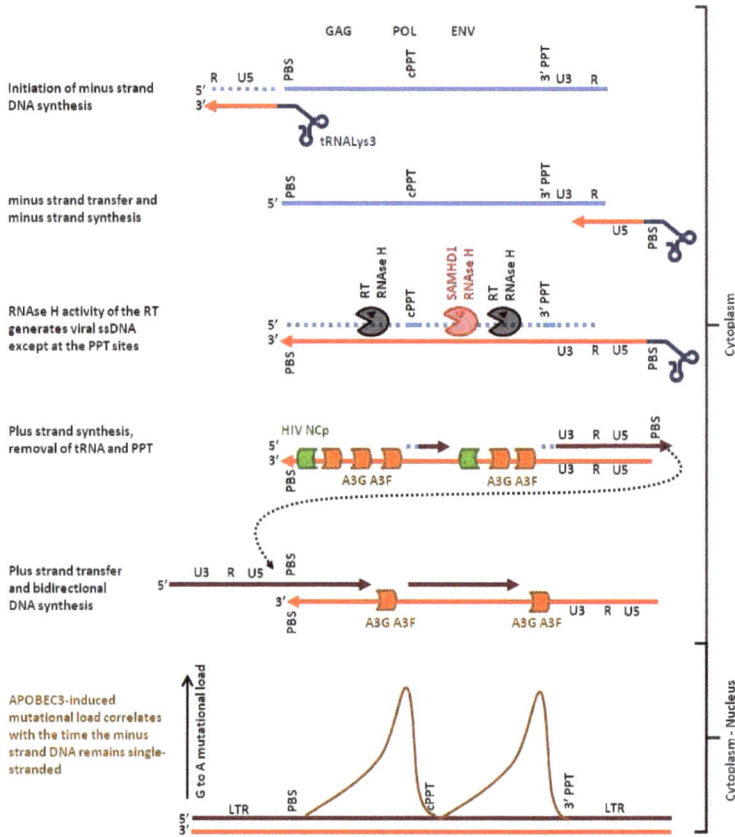

Figure 2. Hotspots of APOBEC3 editing in the HIV genome. Host cell tRNALys3 (**dark blue**) hybridizes to the primer binding sequence (PBS) of the single stranded plus-strand RNA genome (**light blue**) and initiates minus strand DNA synthesis (**light red**). After strand transfer, reverse transcription proceeds up to the PBS yielding minus-strand DNA. RNAse H then hydrolyses the RNA (**dotted light blue**) of the RNA:DNA hybrid leaving the minus-strand DNA single-stranded. APOBEC3 G and F (**orange**) have now access to the ssDNA genome, deaminate deoxycytidine and inhibit plus strand DNA synthesis (**dark red**). RNAse H activity of SAMHD1 promotes exposure of the minus-strand DNA (red pacman) whereas HIV nucleocapsid (**green**) limits APOBEC3-edition. Deoxycytidine deamination of the minus strand generates G-to-A mutations on the plus strand. Since plus-strand DNA synthesis starts from the PolyPurine Tracts (cPPT, 3'PPT), ssDNA located distant to these sites will be accessible to APOBEC3 edition over a longer period of time. Therefore, the APOBEC3-related mutational load will also be higher (**brown curve**, schematic representation of the data from [33]).

248

2.1.2. HTLV-1

Another human retrovirus, human T-lymphotropic virus 1 (HTLV-1), is also a target of A3G [42,43]. As in HIV-1 infection, A3G induces G-to-A transitions on the plus strand via deamination of deoxycytidines on the minus strand. HTLV-1 proviruses contain A3G-related base substitutions, including non-sense mutations [43]. Because HTLV-1 proviral loads mainly result from clonal expansion of infected cells, non-sense mutations are stabilized and amplified by mitosis, provided that viral factors stimulating proliferation are functional [44–47]. Although HTLV-1 does not seem to encode for a Vif-like protein, the frequencies of G-to-A changes in HTLV-1 proviruses are low, likely due to the mode of replication of HTLV-1 by clonal expansion [43,48]. This phenotype has also been associated with the ability of the viral nucleocapsid to limit A3G encapsidation [49].

2.1.3. HERVs

Human endogenous retroviruses (HERV) are transposable elements which were evolutionary integrated into human lineage after infection of germline cells. HERVs are abundant in the human genome (about 8%) and exert important regulatory functions such as control of cellular gene transcription [50]. HERVs contain canonical retroviral *gag*, *pol* and *env* genes surrounded by two LTRs. Nevertheless, most HERVs are defective for replication because of inactivating mutations or deletions [51]. These mutations are likely associated with A3G activity because of a particular signature with a mutated C present in a 5'GC context instead of 5'TC for other APOBEC3s [52–54]. Interestingly, A3G is still able to inhibit a reconstituted functional form of HERV-K in cell culture [52].

2.1.4. Simian Foamy Virus

SFV (simian foamy virus) is a retrovirus that is widespread among non-human primates and can be transmitted to humans [55]. A3F and A3G target SFV genome *in vitro*, leading to G-to-A transitions on the plus strand [56]. SFV genomes found in humans also display G-to-A mutations [57–59]. SFV codes for the accessory protein Bet, limiting APOBEC3 action [60–63].

2.2. Retroelements

About half of the human genome is constituted by repetitive elements. Among them, non-LTR retroelements LINE-1 (long interspersed nuclear element-1), SINE (short interspersed nuclear elements) and Alu are capable of retrotransposition, *i.e.*, inserting a copy of themselves elsewhere in the genome. Since retrotranpositions can be harmful for genome integrity, these events are tightly controlled. In fact, only a small proportion of endogenous retroelements remains active in the germline cells because APOBEC3s protect the host genome from unscheduled retrotransposition (Figure 3). LINE-1 retrotransposition is initiated by transcription of a full-length LINE-1 RNA and translation of ORF1p and ORF2p. These two proteins associate with LINE-1 RNA to form the LINE-1 RiboNucleoProtein (L1 RNP) complex. Upon translocation of L1 RNP into the nucleus, LINE-1 is reverse transcribed and integrated into a new site of the host genome. A3C restricts LINE-1 retrotransposition in a deaminase-independent manner by redirecting and degrading the L1 RNP complex in P-bodies [20]. Within the nucleus, A3C also impairs LINE-1 minus strand DNA synthesis [20]. A3A prevents LINE-1 retrotransposition by deaminating the LINE-1 minus strand DNA [64]. Consistently, RNAse H treatment increases deamination of the LINE-1 minus strand [64].

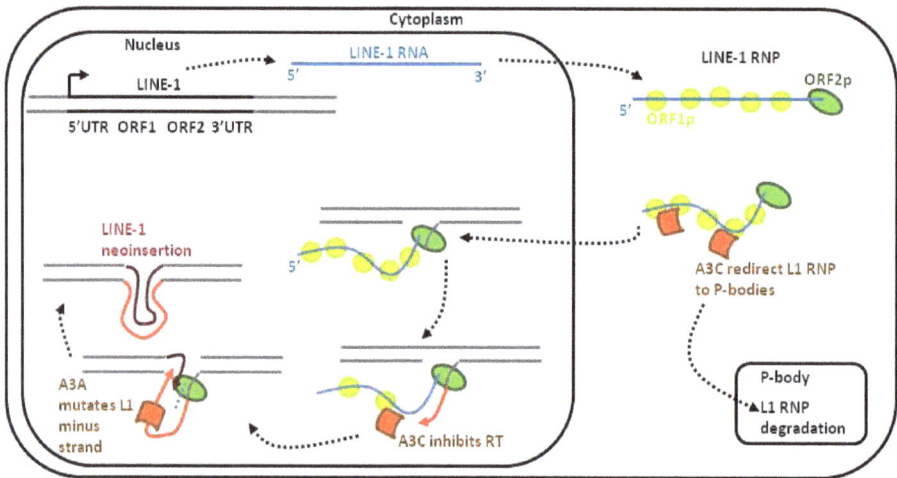

Figure 3. LINE-1 retrotransposons are targeted by APOBEC3s. After transcription, the LINE-1 mRNA is transported into the cytoplasm. After translation, the ORF1- and ORF2-encoded proteins associate with the LINE-1 RNA and form a ribonucleoprotein (RNP) complex. The LINE-1 RNP enters the nucleus, where the ORF2p endonuclease domain cleaves the chromosomal DNA. After cleavage, the 3′-hydroxyl is used by the LINE-1 reverse transcriptase to synthesize a cDNA of LINE-1. This target-site-primed reverse transcription typically results in the insertion of a 5′-truncated LINE-1 element into a new genomic location. Different APOBEC3s-dependent mechanisms control LINE-1 retrotransposition: (1) in the cytoplasm, A3C interacts with and redirects the L1-RNP into P-bodies for degradation; (2) in the nucleus, A3C inhibits reverse transcriptase processing while A3A mutates the minus strand LINE-1 DNA.

2.3. Hepadnaviruses

Since their genome is partially single-strand, hepadnaviruses, such as human hepatitis B virus (HBV), are susceptible to APOBEC3 editing. Except A3DE, all APOBEC3s are able to edit the HBV genome *in vitro*, A3A being the most efficient [65,66]. APOBEC3 editing of HBV DNA has also been validated *in vivo* [67,68]. Since both minus and plus strands are susceptible to APOBEC3 editing, the mutational signature is more complex than in retroviruses [65–67]. HBV viral particles contain a partially double-stranded circular DNA genome (relaxed circular DNA or rcDNA; Figure 4). After uncoating of the viral particle, the rcDNA migrates into the nucleus, where minus-strand DNA synthesis is completed to generate the covalently closed circular double-stranded DNA genome (cccDNA).

In the nucleus, A3A and A3B deaminate HBV cccDNA (Figure 4). Since APOBEC3s require a ssDNA substrate, it is predicted that cccDNA melts during transcription. APOBEC3 deamination of deoxycytidine introduces deleterious mutations in the viral genome and initiates its catabolism via the uracil DNA glycosylase dependent pathway [12].

After transcription of the cccDNA, the pregenomic RNA (pgRNA) translocates into the cytoplasm and is reverse-transcribed into circular partially double-stranded DNA. This mechanism involves priming by the viral P protein, a strand transfer directed by DR1 annealing and degradation of the RNA template by the RNAse H activity of the reverse transcriptase (Figure 5, dotted light blue). The $5'$ end of the pgRNA anneals with DR2, directs a second strand transfer and primes plus-strand DNA synthesis, yielding rcDNA. The minus strand DNA is deaminated proportionally to its exposure to APOBEC3s (Figure 5, orange curve) [68]. Since different subcellular compartments are involved (cytoplasm, nucleus, extracellular viral particles), multiple nuclear and cytoplasmic APOBEC3s (*i.e.*, A3A, A3B, A3C, A3F and A3G) edit the HBV genome [65].

Compared to HIV, additional APOBEC3 proteins (A3A, A3B and A3C) target the HBV genome in the nucleus. Incorporation of HIV into chromatin instead of an episome for HBV may protect the provirus from APOBEC3s editing by a mechanism involving Tribbles 3 proteins [69].

2.4. Herpesviruses

Herpesviruses such as herpes simplex virus-1 (HSV-1) and Epstein-Barr virus (EBV) have a linear double-stranded DNA genome that is edited by APOBEC3 on both strands [70]. After infection, the HSV-1 capsid is transported to the nuclear pores and delivers the double-stranded linear DNA into the nucleus. After circularization of the viral genome, bidirectional DNA synthesis is initiated at the origins of replication [71,72]. This process requires DNA denaturation by the origin binding protein (UL9). The helicase/primase (UL5/UL8/UL52) and single-stranded DNA binding proteins (ICP8 coded by the UL29 gene) then associate with the origin of replication and recruit the DNA polymerase/UL42 complex (Figure 6). During DNA synthesis and transcription, nuclear APOBEC3s have access to single-stranded viral DNA. APOBEC3-edition of HSV-1 and EBV genomes is higher in the minus strand (G to A as opposed to C to T) [70]. It is hypothesized that, due to discontinued replication, the lagging strand exposes more viral ssDNA than the leading strand. HSV-1 and EBV encode orthologs of uracil-DNA glycosylases (UDG) that excise uridine at the replication fork. The HSV-1 UDG (UL2) binds to UL30, associates with the viral replisome and directs replication-coupled BER (base excision repair) to ensure genome integrity [73]. The viral UDG might therefore protect against APOBEC3 editing.

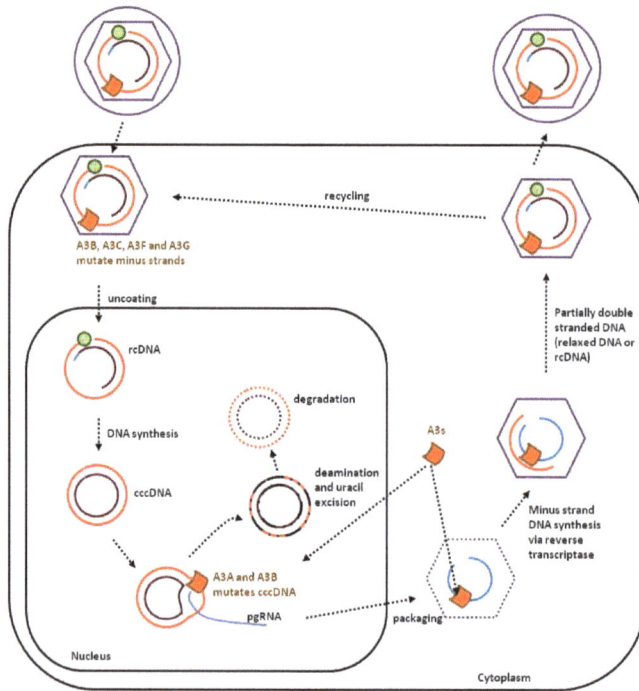

Figure 4. APOBEC3s interfere with several steps of the HBV replication cycle. The HBV viral particle contains a partially double-stranded DNA genome (relaxed circular DNA or rcDNA) that can be edited by A3G and A3F. Unlike HIV, HBV does not appear to encode Vif-like protein. Upon transfer into the nucleus, the plus strand of the rcDNA is replicated to form the covalently closed circular DNA genome (cccDNA). A3A and A3B deaminate the cccDNA genome leading to uracil excision and subsequent degradation.

252

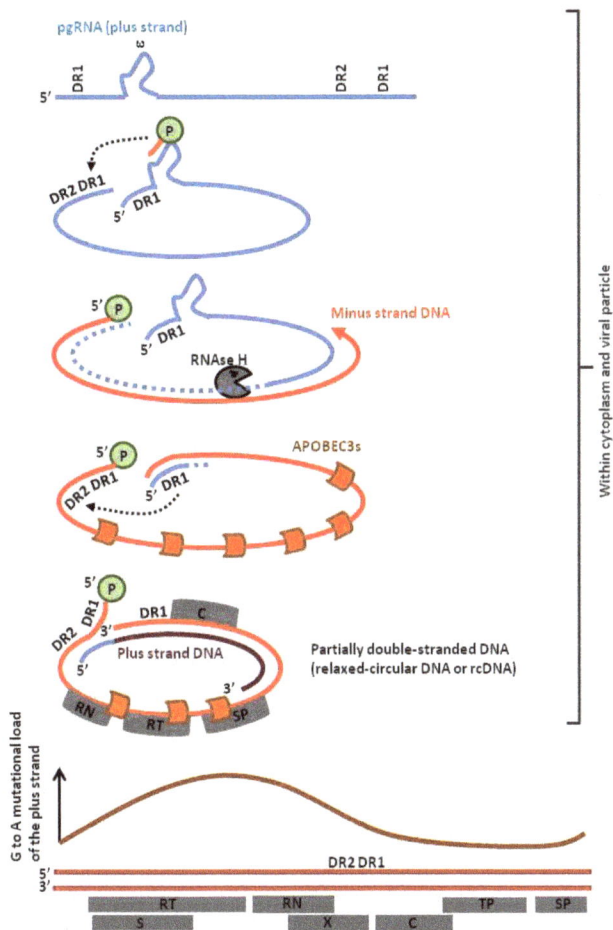

Figure 5. Profile of APOBEC3 editing of the HBV genome. The viral P protein initiates reverse transcription at the stem loop structure ε. The pregenomic RNA (pgRNA) contains two direct repeat sequences (DR1) at the 5′ and 3′ end of the viral genome, allowing strand transfer to the 5′ end of the viral genome. While synthesis of the minus-strand DNA proceeds, the RNAse H activity of the reverse transcriptase degrades the pgRNA except at the 5′ end. After a second strand transfer, the undigested pgRNA anneals with the direct repeat sequence DR2 and primes plus-strand gDNA synthesis, yielding relaxed circular DNA (rcDNA). Mutational load correlates with the time of exposure of ssDNA (**orange curve**, schematic representation of data extracted from reference [68]). Abbreviations within grey boxes read as follow: RT, Reverse Transcriptase; RN, RNAse; TP, Terminal Protein; SP, Spacer Domain; S, short surface gene; X, X gene, C, Core gene.

253

2.5. Papillomavirus

Human papillomaviruses (HPVs) are circular double-stranded DNA viruses. A3A, A3C and A3H are able to deaminate both strands of the 8Kb viral genome [74–76]. APOBEC3-edited HPV DNA is found in benign and precancerous cervical lesions [74]. Replication of the HPV genome occurs in the nucleus and is primarily based on the host replication machinery. The HPV protein E1 recruits ssDNA-binding protein RPA (replication protein A) during replication to cover the transiently exposed viral ssDNA [77].

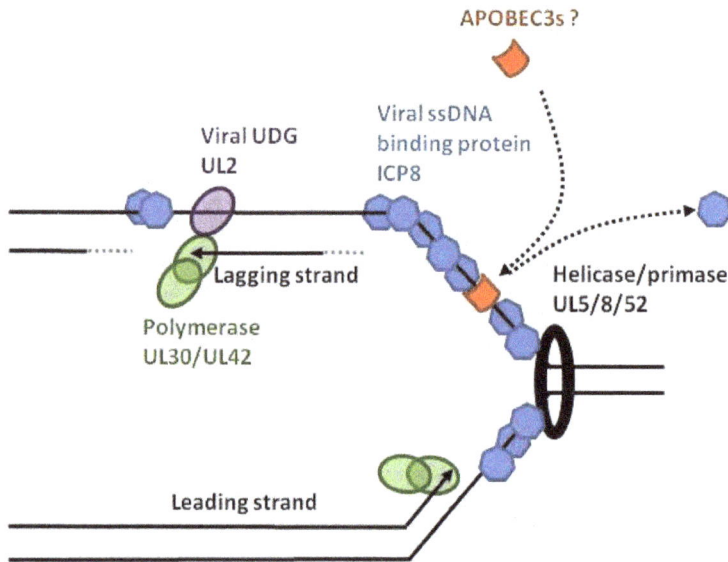

Figure 6. HSV-1 replication fork and hypothetical model of APOBEC3 editing. In the nucleus, replication of HSV-1 is initiated by the origin binding protein (UL9) that melts double-stranded DNA. The helicase/primase complex (UL5/UL8/UL52) unwinds and anneals RNA primers, allowing DNA replication by the UL30/UL42 complex. The viral protein ICP8 covers the transiently exposed single-stranded DNA and competes with the APOBEC3 binding. The viral UL2 is a uracil-DNA glycosylase (UDG) that favors replication-coupled base excision DNA repair.

2.6. TT Virus

Transfusion-transmitted virus (TTV) is a non-enveloped virus causing a persistent and asymptomatic infection. Having a circular single-stranded DNA genome, TTV is a prototypical substrate of APOBEC3s and shows APOBEC3-related mutations [78].

Together, these data show that viruses are targeted by particular isoforms of APOBEC3 depending on their modes of replication (Table 1) and have developed strategies to dampen ssDNA edition. A3G and A3F are restricted to the cytoplasm whereas A3A, A3B and A3C preferentially act in the nucleus. Importantly, the mutational load is proportional to the duration of single-stranded DNA exposure to APOBEC3s.

Table 1. Summary of the anti-viral activity of the different APOBEC3 isoforms. * It has been recently shown that A3A can also edit RNA transcripts [79].

	Sub-cellular localization	Substrate edited	Retro viruses				Retro elements	Hepadna viruses	Herpes viruses
			HIV-1	HTLV-1	HERVs	SFV			
A3A	cell wide	single stranded DNA, RNA *					+	+	+
A3B	nuclear	single stranded DNA						+	
A3C	cell wide	single stranded DNA					+	+	+
A3D	cytoplasmic	single stranded DNA	+						
A3F	cytoplasmic	single stranded DNA	+		+			+	
A3G	cytoplasmic	single stranded DNA	+	+	+	+		+	
A3H	cell wide	single stranded DNA	+					+	+

3. Therapeutic Strategies by Perturbation of the Viral Mutation Rate

Viral quasi-species refer to a population of distinct but closely related viral genomes that differ only by a limited number of mutations. The distribution of variants is dominated by a master sequence that displays the highest fitness within a given environment (Figure 7A). High mutation rates during viral replication are the driving force for quasi-species generation. Lethal mutations or inappropriate adaptation to the environmental conditions (e.g., anti-viral therapy, immune pressure) will clear unfit genomes. When conditions change, the fittest quasi-species may differ from the master sequence. Providing that the distribution contains an adequate variant, a new population will grow (Figure 7B). The rate of mutation and the selection pressure will dictate the wideness of the distribution. If the environmental changes are too drastic or the quasi-species distribution too narrow, the viral population will be unable to recover [80].

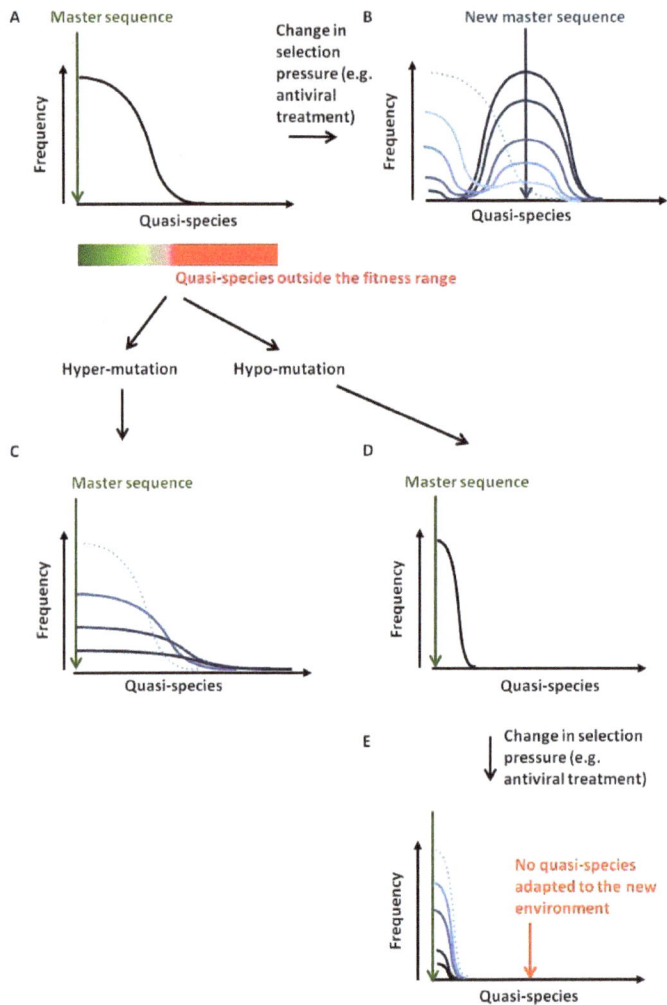

Figure 7. Antiviral strategy by hypo- or hyper-mutation. (**A**) Viral quasi-species refer to as a population of distinct but closely related viral genomes that only differ by a limited number of mutations. The frequency of these quasi-species spreads around a master sequence. The boundary of this population is dictated by the selection forces acting against the viral diversification. At equilibrium, quasi-species generated outside the fitness range will not persist. (**B**) If selection criteria are modified, the fittest sequence will change. If the original distribution contained this sequence, the population will first shrink and then re-grow around a new master sequence (from light to dark blue corresponding to the time evolution). (**C**) Excess of APOBEC3-directed mutations will affect fitness of the newly created quasi-species up to complete disappearance. (**D,E**) Hypo-mutation will restrict the range of quasi-species and limit adaptability to new environmental conditions.

Emergence of quasi-species is thus a major issue that limits antiviral therapy. Viral populations can indeed accommodate environmental changes due to improved immunity (vaccination) or pharmacological inhibition. It is possible to affect quasi-species adaptability by modulating the frequencies of mutation [81,82].

The first approach, referred to as lethal mutagenesis or the hyper-mutation strategy, aims to introduce an excess of mutations in the viral genomes. If the mutational load per viral genome is too high, a substantial proportion of the new viruses will be defective or inadequately adapted to their environment. Introducing mutations in viruses would therefore decrease viral load (Figure 7C). In principle, exogenous induction of APOBEC3 expression could achieve this goal. This strategy has recently been exemplified for HBV, where forced expression of A3A and A3B induced HBV cccDNA hypermutation with no detectable effect on genomic DNA [12]. Nevertheless, this approach raises serious safety issues because APOBEC3 mutations could also drive cancer development [83–85]. Indeed, A3A and A3B over-expression in yeast creates mutational clusters and genomic rearrangements similar to those observed in human cancers, the mutational burden being magnified by DNA strand breaks [86–89]. Because the processing of double-strand break repair transiently exposes single-stranded nucleic acids, DNA repair could provide a substrate for nuclear deaminases. What would, for example, happen if an HBV-infected liver cell is being forced to express deaminases and at the same time has to repair DNA strand breaks generated by reactive oxygen species produced during alcohol catabolism [90]?

Therefore, it would be safer to promote hyper-mutation by targeting the viral factors that inhibit APOBEC3s. In that respect, Vif inhibitors are being developed [91,92]. Inhibition of viral ssDNA-binding proteins (like HSV-1 ICP8) might lead to increased access for endogenously expressed APOBEC3s to the viral ssDNA (Figure 6). Promotion of RNAse H activity during retrotranscription might facilitate the binding of the APOBEC3s to the viral ssDNA (Figure 2). Because reverse transcription is thought to start within the virion, promotion of APOBEC3s loading in to the viral particle will increase editing (MLV glyco-Gag shields the reverse transcription complex from APOBEC3 and cytosolic sensors [38]). Alternatively, it would be possible to target viral DNA repair mechanisms (e.g., via inhibitors against the viral UDG UL2 of HSV-1, Figure 6). In these cases, safety issues are related to the emergence of sub-lethal APOBEC3-mutations and promotion of drug resistant quasi-species.

The reverse strategy would be to reduce mutation rate by inhibiting APOBEC3, thereby narrowing the quasi-species spectrum and limiting viral adaptability to new environmental conditions (Figure 7D,E). APOBEC3s inhibitors are currently being developed and evaluated [93,94]. This approach, which paradoxically targets a well-conserved mechanism of mammalian innate immunity, would preserve host genome integrity. Potential risks of this therapy pertain to adequate control of endogenous retroelements and opportunistic infections.

4. Conclusions

Single-strand DNA editing by APOBEC3 proteins is a very powerful mechanism of mammalian innate immunity that mutates and inactivates viral genomes. The outcome of infection is the result of a finely tuned balance between onset of mutations, generation of quasi-species and APOBEC inhibition by viral factors. Understanding the mechanisms involved reveals new prospects for therapeutic strategies that interfere with APOBEC3 deamination of cytosine residues in nascent viral DNA.

Acknowledgments: This work was supported by the "Fonds National de la Recherche Scientifique" (FNRS), the Télévie, the Interuniversity Attraction Poles (IAP) Program "Virus-host interplay at the early phases of infection" BELVIR initiated by the Belgian Science Policy Office, the Belgian Foundation against Cancer (FBC), the Sixth Research Framework Programme of the European Union (project "The role of infections in cancer" INCA LSHC-CT-2005-018704), the "Neoangio" excellence program and the "Partenariat Public Privé", PPP INCA, of the "Direction générale des Technologies, de la Recherche et de L'Energie/DG06" of the Walloon government, the "Action de Recherche Concertée Glyvir" (ARC) of the "Communauté française de Belgique", the "Centre anticancéreux près ULg" (CAC), the "Subside Fédéral de Soutien à la Recherche Synbiofor and Agricultureislife" projects of Gembloux Agrobiotech (GxABT), the "ULg Fonds Spéciaux pour la Recherche" and the "Plan Cancer" of the "Service Public Fédéral". We thank Sathya Neelature Sriramareddy, Srikanth Perike, Hélène Gazon, Alix de Brogniez, Alexandre Carpentier, Pierre-Yves Barez, Bernard Staumont and Malik Hamaidia for their careful reading and helpful comments.

Author Contributions: L.W. and N.A.G. wrote the paper.

Conflicts of Interest: The authors declare no conflict of interest.

References

1. Smith, H.C.; Bennett, R.P.; Kizilyer, A.; McDougall, W.M.; Prohaska, K.M. Functions and regulation of the APOBEC family of proteins. *Semin. Cell Dev. Biol.* **2012**, *23*, 258–268.
2. Refsland, E.W.; Harris, R.S. The APOBEC3 family of retroelement restriction factors. *Curr. Top. Microbiol. Immunol.* **2013**, *371*, 1–27.
3. Vieira, V.C.; Soares, M.A. The role of cytidine deaminases on innate immune responses against human viral infections. *Biomed. Res. Int.* **2013**, *2013*, e683095.
4. LaRue, R.S.; Andresdottir, V.; Blanchard, Y.; Conticello, S.G.; Derse, D.; Emerman, M.; Greene, W.C.; Jonsson, S.R.; Landau, N.R.; Lochelt, M.; *et al.* Guidelines for naming nonprimate APOBEC3 genes and proteins. *J. Virol.* **2009**, *83*, 494–497.
5. Munk, C.; Willemsen, A.; Bravo, I.G. An ancient history of gene duplications, fusions and losses in the evolution of APOBEC3 mutators in mammals. *BMC Evol. Biol.* **2012**, *12*, e71.
6. Mehta, H.V.; Jones, P.H.; Weiss, J.P.; Okeoma, C.M. IFN-alpha and lipopolysaccharide upregulate APOBEC3 mRNA through different signaling pathways. *J. Immunol.* **2012**, *189*, 4088–4103.

7. Refsland, E.W.; Stenglein, M.D.; Shindo, K.; Albin, J.S.; Brown, W.L.; Harris, R.S. Quantitative profiling of the full APOBEC3 mRNA repertoire in lymphocytes and tissues: Implications for HIV-1 restriction. *Nucleic Acids Res.* **2010**, *38*, 4274–4284.

8. Koning, F.A.; Newman, E.N.; Kim, E.Y.; Kunstman, K.J.; Wolinsky, S.M.; Malim, M.H. Defining APOBEC3 expression patterns in human tissues and hematopoietic cell subsets. *J. Virol.* **2009**, *83*, 9474–9485.

9. Lackey, L.; Law, E.K.; Brown, W.L.; Harris, R.S. Subcellular localization of the APOBEC3 proteins during mitosis and implications for genomic DNA deamination. *Cell Cycle* **2013**, *12*, 762–772.

10. Krokan, H.E.; Saetrom, P.; Aas, P.A.; Pettersen, H.S.; Kavli, B.; Slupphaug, G. Error-free *versus* mutagenic processing of genomic uracil—Relevance to cancer. *DNA Repair* **2014**, *19*, 38–47.

11. Weil, A.F.; Ghosh, D.; Zhou, Y.; Seiple, L.; McMahon, M.A.; Spivak, A.M.; Siliciano, R.F.; Stivers, J.T. Uracil DNA glycosylase initiates degradation of HIV-1 cDNA containing misincorporated dUTP and prevents viral integration. *Proc. Natl. Acad Sci. USA* **2013**, *110*, E448–E457.

12. Lucifora, J.; Xia, Y.; Reisinger, F.; Zhang, K.; Stadler, D.; Cheng, X.; Sprinzl, M.F.; Koppensteiner, H.; Makowska, Z.; Volz, T.; *et al.* Specific and nonhepatotoxic degradation of nuclear hepatitis B virus cccDNA. *Science* **2014**, *343*, 1221–1228.

13. Huthoff, H.; Autore, F.; Gallois-Montbrun, S.; Fraternali, F.; Malim, M.H. RNA-dependent oligomerization of APOBEC3G is required for restriction of HIV-1. *PLoS Pathog.* **2009**, *5*, e1000330.

14. Guo, F.; Cen, S.; Niu, M.; Saadatmand, J.; Kleiman, L. Inhibition of tRNA(3)(Lys)—Primed reverse transcription by human APOBEC3G during human immunodeficiency virus type 1 replication. *J. Virol.* **2006**, *80*, 11710–11722.

15. Anderson, J.L.; Hope, T.J. APOBEC3G restricts early HIV-1 replication in the cytoplasm of target cells. *Virology* **2008**, *375*, 1–12.

16. Wang, X.; Ao, Z.; Chen, L.; Kobinger, G.; Peng, J.; Yao, X. The cellular antiviral protein APOBEC3G interacts with HIV-1 reverse transcriptase and inhibits its function during viral replication. *J. Virol.* **2012**, *86*, 3777–3786.

17. Gillick, K.; Pollpeter, D.; Phalora, P.; Kim, E.Y.; Wolinsky, S.M.; Malim, M.H. Suppression of HIV-1 infection by APOBEC3 proteins in primary human CD4(+) T cells is associated with inhibition of processive reverse transcription as well as excessive cytidine deamination. *J. Virol.* **2013**, *87*, 1508–1517.

18. Nguyen, D.H.; Gummuluru, S.; Hu, J. Deamination-independent inhibition of hepatitis B virus reverse transcription by APOBEC3G. *J. Virol.* **2007**, *81*, 4465–4472.

19. Narvaiza, I.; Linfesty, D.C.; Greener, B.N.; Hakata, Y.; Pintel, D.J.; Logue, E.; Landau, N.R.; Weitzman, M.D. Deaminase-independent inhibition of parvoviruses by the APOBEC3A cytidine deaminase. *PLoS Pathog.* **2009**, *5*, e1000439.

20. Horn, A.V.; Klawitter, S.; Held, U.; Berger, A.; Vasudevan, A.A.; Bock, A.; Hofmann, H.; Hanschmann, K.M.; Trosemeier, J.H.; Flory, E.; *et al.* Human LINE-1 restriction by APOBEC3C is deaminase independent and mediated by an ORF1p interaction that affects LINE reverse transcriptase activity. *Nucleic Acids Res.* **2014**, *42*, 396–416.

21. Luo, K.; Wang, T.; Liu, B.; Tian, C.; Xiao, Z.; Kappes, J.; Yu, X.F. Cytidine deaminases APOBEC3G and APOBEC3F interact with human immunodeficiency virus type 1 integrase and inhibit proviral DNA formation. *J. Virol.* **2007**, *81*, 7238–7248.

22. Mbisa, J.L.; Barr, R.; Thomas, J.A.; Vandegraaff, N.; Dorweiler, I.J.; Svarovskaia, E.S.; Brown, W.L.; Mansky, L.M.; Gorelick, R.J.; Harris, R.S.; *et al.* Human immunodeficiency virus type 1 cDNAs produced in the presence of APOBEC3G exhibit defects in plus-strand DNA transfer and integration. *J. Virol.* **2007**, *81*, 7099–7110.

23. Vetter, M.L.; D'Aquila, R.T. Cytoplasmic APOBEC3G restricts incoming Vif-positive human immunodeficiency virus type 1 and increases two-long terminal repeat circle formation in activated T-helper-subtype cells. *J. Virol.* **2009**, *83*, 8646–8654.

24. Sheehy, A.M.; Gaddis, N.C.; Choi, J.D.; Malim, M.H. Isolation of a human gene that inhibits HIV-1 infection and is suppressed by the viral Vif protein. *Nature* **2002**, *418*, 646–650.

25. Liddament, M.T.; Brown, W.L.; Schumacher, A.J.; Harris, R.S. APOBEC3F properties and hypermutation preferences indicate activity against HIV-1 *in vivo*. *Curr. Biol.* **2004**, *14*, 1385–1391.

26. Wiegand, H.L.; Doehle, B.P.; Bogerd, H.P.; Cullen, B.R. A second human antiretroviral factor, APOBEC3F, is suppressed by the HIV-1 and HIV-2 Vif proteins. *EMBO J.* **2004**, *23*, 2451–2458.

27. Hultquist, J.F.; Lengyel, J.A.; Refsland, E.W.; LaRue, R.S.; Lackey, L.; Brown, W.L.; Harris, R.S. Human and rhesus APOBEC3D, APOBEC3F, APOBEC3G, and APOBEC3H demonstrate a conserved capacity to restrict Vif-deficient HIV-1. *J. Virol.* **2011**, *85*, 11220–11234.

28. Chiu, Y.L.; Witkowska, H.E.; Hall, S.C.; Santiago, M.; Soros, V.B.; Esnault, C.; Heidmann, T.; Greene, W.C. High-molecular-mass APOBEC3G complexes restrict Alu retrotransposition. *Proc. Natl. Acad. Sci. USA* **2006**, *103*, 15588–15593.

29. Gallois-Montbrun, S.; Kramer, B.; Swanson, C.M.; Byers, H.; Lynham, S.; Ward, M.; Malim, M.H. Antiviral protein APOBEC3G localizes to ribonucleoprotein complexes found in P bodies and stress granules. *J. Virol.* **2007**, *81*, 2165–2178.

30. Wichroski, M.J.; Robb, G.B.; Rana, T.M. Human retroviral host restriction factors APOBEC3G and APOBEC3F localize to mRNA processing bodies. *PLoS Pathog.* **2006**, *2*, e41.

31. Phalora, P.K.; Sherer, N.M.; Wolinsky, S.M.; Swanson, C.M.; Malim, M.H. HIV-1 replication and APOBEC3 antiviral activity are not regulated by P bodies. *J. Virol.* **2012**, *86*, 11712–11724.

32. Kobayashi, T.; Koizumi, Y.; Takeuchi, J.S.; Misawa, N.; Kimura, Y.; Morita, S.; Aihara, K.; Koyanagi, Y.; Iwami, S.; Sato, K. Quantification of deaminase activity-dependent and -independent restriction of HIV-1 replication mediated by APOBEC3F and APOBEC3G through experimental-mathematical investigation. *J. Virol.* **2014**, *88*, 5881–5887.

33. Suspene, R.; Rusniok, C.; Vartanian, J.P.; Wain-Hobson, S. Twin gradients in APOBEC3 edited HIV-1 DNA reflect the dynamics of lentiviral replication. *Nucleic Acids Res.* **2006**, *34*, 4677–4684.

34. Esposito, F.; Corona, A.; Tramontano, E. HIV-1 Reverse Transcriptase Still Remains a New Drug Target: Structure, Function, Classical Inhibitors, and New Inhibitors with Innovative Mechanisms of Actions. *Mol. Biol. Int.* **2012**, *2012*, e586401.

35. Mariani, R.; Chen, D.; Schrofelbauer, B.; Navarro, F.; Konig, R.; Bollman, B.; Munk, C.; Nymark-McMahon, H.; Landau, N.R. Species-specific exclusion of APOBEC3G from HIV-1 virions by Vif. *Cell* **2003**, *114*, 21–31.

36. Jager, S.; Kim, D.Y.; Hultquist, J.F.; Shindo, K.; LaRue, R.S.; Kwon, E.; Li, M.; Anderson, B.D.; Yen, L.; Stanley, D.; *et al.* Vif hijacks CBF-beta to degrade APOBEC3G and promote HIV-1 infection. *Nature* **2012**, *481*, 371–375.

37. Zhao, K.; Du, J.; Rui, Y.; Zheng, W.; Kang, J.; Hou, J.; Wang, K.; Zhang, W.; Simon, V.A.; Yu, X.F. Evolutionarily conserved pressure for the existence of distinct G2/M cell cycle arrest and A3H inactivation functions in HIV-1 Vif. *Cell Cycle* **2015**, *14*, 838–847.

38. Stavrou, S.; Nitta, T.; Kotla, S.; Ha, D.; Nagashima, K.; Rein, A.R.; Fan, H.; Ross, S.R. Murine leukemia virus glycosylated Gag blocks apolipoprotein B editing complex 3 and cytosolic sensor access to the reverse transcription complex. *Proc. Natl. Acad. Sci. USA* **2013**, *110*, 9078–9083.

39. Mitra, M.; Hercik, K.; Byeon, I.J.; Ahn, J.; Hill, S.; Hinchee-Rodriguez, K.; Singer, D.; Byeon, C.H.; Charlton, L.M.; Nam, G.; *et al.* Structural determinants of human APOBEC3A enzymatic and nucleic acid binding properties. *Nucleic Acids Res.* **2014**, *42*, 1095–1110.

40. Beloglazova, N.; Flick, R.; Tchigvintsev, A.; Brown, G.; Popovic, A.; Nocek, B.; Yakunin, A.F. Nuclease activity of the human SAMHD1 protein implicated in the Aicardi-Goutieres syndrome and HIV-1 restriction. *J. Biol. Chem.* **2013**, *288*, 8101–8110.

41. Ryoo, J.; Choi, J.; Oh, C.; Kim, S.; Seo, M.; Kim, S.Y.; Seo, D.; Kim, J.; White, T.E.; Brandariz-Nunez, A.; *et al.* The ribonuclease activity of SAMHD1 is required for HIV-1 restriction. *Nat. Med.* **2014**, *20*, 936–941.

42. Mahieux, R.; Suspene, R.; Delebecque, F.; Henry, M.; Schwartz, O.; Wain-Hobson, S.; Vartanian, J.P. Extensive editing of a small fraction of human T-cell leukemia virus type 1 genomes by four APOBEC3 cytidine deaminases. *J. Gen. Virol.* **2005**, *86*, 2489–2494.

43. Fan, J.; Ma, G.; Nosaka, K.; Tanabe, J.; Satou, Y.; Koito, A.; Wain-Hobson, S.; Vartanian, J.P.; Matsuoka, M. APOBEC3G generates nonsense mutations in human T-cell leukemia virus type 1 proviral genomes *in vivo*. *J. Virol.* **2010**, *84*, 7278–7287.

44. Cavrois, M.; Wain-Hobson, S.; Gessain, A.; Plumelle, Y.; Wattel, E. Adult T-cell leukemia/lymphoma on a background of clonally expanding human T-cell leukemia virus type-1-positive cells. *Blood* **1996**, *88*, 4646–4650.

45. Etoh, K.; Tamiya, S.; Yamaguchi, K.; Okayama, A.; Tsubouchi, H.; Ideta, T.; Mueller, N.; Takatsuki, K.; Matsuoka, M. Persistent clonal proliferation of human T-lymphotropic virus type I-infected cells *in vivo*. *Cancer Res.* **1997**, *57*, 4862–4867.

46. Mortreux, F.; Gabet, A.S.; Wattel, E. Molecular and cellular aspects of HTLV-1 associated leukemogenesis *in vivo*. *Leukemia* **2003**, *17*, 26–38.

47. Gillet, N.A.; Malani, N.; Melamed, A.; Gormley, N.; Carter, R.; Bentley, D.; Berry, C.; Bushman, F.D.; Taylor, G.P.; Bangham, C.R. The host genomic environment of the provirus determines the abundance of HTLV-1-infected T-cell clones. *Blood* **2011**, *117*, 3113–3122.

48. Yu, Q.; Konig, R.; Pillai, S.; Chiles, K.; Kearney, M.; Palmer, S.; Richman, D.; Coffin, J.M.; Landau, N.R. Single-strand specificity of APOBEC3G accounts for minus-strand deamination of the HIV genome. *Nat. Struct. Mol. Biol.* **2004**, *11*, 435–442.

49. Derse, D.; Hill, S.A.; Princler, G.; Lloyd, P.; Heidecker, G. Resistance of human T cell leukemia virus type 1 to APOBEC3G restriction is mediated by elements in nucleocapsid. *Proc. Natl. Acad. Sci. USA* **2007**, *104*, 2915–2920.

50. Medstrand, P.; van de Lagemaat, L.N.; Dunn, C.A.; Landry, J.R.; Svenback, D.; Mager, D.L. Impact of transposable elements on the evolution of mammalian gene regulation. *Cytogenet. Genome Res.* **2005**, *110*, 342–352.

51. Stoye, J.P. Studies of endogenous retroviruses reveal a continuing evolutionary saga. *Nat. Rev. Microbiol.* **2012**, *10*, 395–406.

52. Lee, Y.N.; Malim, M.H.; Bieniasz, P.D. Hypermutation of an ancient human retrovirus by APOBEC3G. *J. Virol.* **2008**, *82*, 8762–8770.

53. Esnault, C.; Priet, S.; Ribet, D.; Heidmann, O.; Heidmann, T. Restriction by APOBEC3 proteins of endogenous retroviruses with an extracellular life cycle: *Ex vivo* effects and *in vivo* "traces" on the murine IAPE and human HERV-K elements. *Retrovirology* **2008**, *5*, e75.

54. Anwar, F.; Davenport, M.P.; Ebrahimi, D. Footprint of APOBEC3 on the genome of human retroelements. *J. Virol.* **2013**, *87*, 8195–8204.

55. Rua, R.; Gessain, A. Origin, evolution and innate immune control of simian foamy viruses in humans. *Curr. Opin. Virol.* **2015**, *10*, 47–55.

56. Gartner, K.; Wiktorowicz, T.; Park, J.; Mergia, A.; Rethwilm, A.; Scheller, C. Accuracy estimation of foamy virus genome copying. *Retrovirology* **2009**, *6*, e32.

57. Delebecque, F.; Suspene, R.; Calattini, S.; Casartelli, N.; Saib, A.; Froment, A.; Wain-Hobson, S.; Gessain, A.; Vartanian, J.P.; Schwartz, O. Restriction of foamy viruses by APOBEC cytidine deaminases. *J. Virol.* **2006**, *80*, 605–614.

58. Rua, R.; Betsem, E.; Gessain, A. Viral latency in blood and saliva of simian foamy virus-infected humans. *PLoS ONE* **2013**, *8*, e77072.

59. Matsen, F.A.T.; Small, C.T.; Soliven, K.; Engel, G.A.; Feeroz, M.M.; Wang, X.; Craig, K.L.; Hasan, M.K.; Emerman, M.; Linial, M.L.; *et al.* A novel bayesian method for detection of APOBEC3-mediated hypermutation and its application to zoonotic transmission of simian foamy viruses. *PLoS Comput. Biol.* **2014**, *10*, e1003493.

60. Russell, R.A.; Wiegand, H.L.; Moore, M.D.; Schafer, A.; McClure, M.O.; Cullen, B.R. Foamy virus Bet proteins function as novel inhibitors of the APOBEC3 family of innate antiretroviral defense factors. *J. Virol.* **2005**, *79*, 8724–8731.

61. Lochelt, M.; Romen, F.; Bastone, P.; Muckenfuss, H.; Kirchner, N.; Kim, Y.B.; Truyen, U.; Rosler, U.; Battenberg, M.; Saib, A.; *et al.* The antiretroviral activity of APOBEC3 is inhibited by the foamy virus accessory Bet protein. *Proc. Natl. Acad. Sci. USA* **2005**, *102*, 7982–7987.

62. Perkovic, M.; Schmidt, S.; Marino, D.; Russell, R.A.; Stauch, B.; Hofmann, H.; Kopietz, F.; Kloke, B.P.; Zielonka, J.; Strover, H.; *et al.* Species-specific inhibition of APOBEC3C by the prototype foamy virus protein bet. *J. Biol. Chem.* **2009**, *284*, 5819–5826.

63. Jaguva Vasudevan, A.A.; Perkovic, M.; Bulliard, Y.; Cichutek, K.; Trono, D.; Haussinger, D.; Munk, C. Prototype foamy virus Bet impairs the dimerization and cytosolic solubility of human APOBEC3G. *J. Virol.* **2013**, *87*, 9030–9040.

64. Richardson, S.R.; Narvaiza, I.; Planegger, R.A.; Weitzman, M.D.; Moran, J.V. APOBEC3A deaminates transiently exposed single-strand DNA during LINE-1 retrotransposition. *eLife* **2014**, *3*, e02008.

65. Suspene, R.; Guetard, D.; Henry, M.; Sommer, P.; Wain-Hobson, S.; Vartanian, J.P. Extensive editing of both hepatitis B virus DNA strands by APOBEC3 cytidine deaminases *in vitro* and *in vivo*. *Proc. Natl. Acad. Sci. USA* **2005**, *102*, 8321–8326.

66. Henry, M.; Guetard, D.; Suspene, R.; Rusniok, C.; Wain-Hobson, S.; Vartanian, J.P. Genetic editing of HBV DNA by monodomain human APOBEC3 cytidine deaminases and the recombinant nature of APOBEC3G. *PLoS ONE* **2009**, *4*, e4277.

67. Vartanian, J.P.; Henry, M.; Marchio, A.; Suspene, R.; Aynaud, M.M.; Guetard, D.; Cervantes-Gonzalez, M.; Battiston, C.; Mazzaferro, V.; Pineau, P.; *et al.* Massive APOBEC3 editing of hepatitis B viral DNA in cirrhosis. *PLoS Pathog.* **2010**, *6*, e1000928.

68. Beggel, B.; Munk, C.; Daumer, M.; Hauck, K.; Haussinger, D.; Lengauer, T.; Erhardt, A. Full genome ultra-deep pyrosequencing associates G-to-A hypermutation of the hepatitis B virus genome with the natural progression of hepatitis B. *J. Viral. Hepat.* **2013**, *20*, 882–889.

69. Aynaud, M.M.; Suspene, R.; Vidalain, P.O.; Mussil, B.; Guetard, D.; Tangy, F.; Wain-Hobson, S.; Vartanian, J.P. Human Tribbles 3 protects nuclear DNA from cytidine deamination by APOBEC3A. *J. Biol. Chem.* **2012**, *287*, 39182–39192.

70. Suspene, R.; Aynaud, M.M.; Koch, S.; Pasdeloup, D.; Labetoulle, M.; Gaertner, B.; Vartanian, J.P.; Meyerhans, A.; Wain-Hobson, S. Genetic editing of herpes simplex virus 1 and Epstein-Barr herpesvirus genomes by human APOBEC3 cytidine deaminases in culture and *in vivo*. *J. Virol.* **2011**, *85*, 7594–7602.

71. Muylaert, I.; Tang, K.W.; Elias, P. Replication and recombination of herpes simplex virus DNA. *J. Biol. Chem.* **2011**, *286*, 15619–15624.

72. Weller, S.K.; Coen, D.M. Herpes simplex viruses: Mechanisms of DNA replication. *Cold Spring Harb. Perspect. Biol.* **2012**, *4*, a013011.

73. Bogani, F.; Corredeira, I.; Fernandez, V.; Sattler, U.; Rutvisuttinunt, W.; Defais, M.; Boehmer, P.E. Association between the herpes simplex virus-1 DNA polymerase and uracil DNA glycosylase. *J. Biol. Chem.* **2010**, *285*, 27664–27672.

74. Vartanian, J.P.; Guetard, D.; Henry, M.; Wain-Hobson, S. Evidence for editing of human papillomavirus DNA by APOBEC3 in benign and precancerous lesions. *Science* **2008**, *320*, 230–233.

75. Wang, Z.; Wakae, K.; Kitamura, K.; Aoyama, S.; Liu, G.; Koura, M.; Monjurul, A.M.; Kukimoto, I.; Muramatsu, M. APOBEC3 deaminases induce hypermutation in human papillomavirus 16 DNA upon beta interferon stimulation. *J. Virol.* **2014**, *88*, 1308–1317.

76. Warren, C.J.; Xu, T.; Guo, K.; Griffin, L.M.; Westrich, J.A.; Lee, D.; Lambert, P.F.; Santiago, M.L.; Pyeon, D. APOBEC3A functions as a restriction factor of human papillomavirus. *J. Virol.* **2015**, *89*, 688–702.

77. Loo, Y.M.; Melendy, T. Recruitment of replication protein A by the papillomavirus E1 protein and modulation by single-stranded DNA. *J. Virol.* **2004**, *78*, 1605–1615.

78. Tsuge, M.; Noguchi, C.; Akiyama, R.; Matsushita, M.; Kunihiro, K.; Tanaka, S.; Abe, H.; Mitsui, F.; Kitamura, S.; Hatakeyama, T.; *et al.* G to A hypermutation of TT virus. *Virus Res.* **2010**, *149*, 211–216.

79. Sharma, S.; Patnaik, S.K.; Thomas Taggart, R.; Kannisto, E.D.; Enriquez, S.M.; Gollnick, P.; Baysal, B.E. APOBEC3A cytidine deaminase induces RNA editing in monocytes and macrophages. *Nat. Commun.* **2015**, *6*, 6881.

80. Domingo, E.; Sheldon, J.; Perales, C. Viral quasispecies evolution. *Microbiol. Mol. Biol. Rev.* **2012**, *76*, 159–216.

81. Hultquist, J.F.; Harris, R.S. Leveraging APOBEC3 proteins to alter the HIV mutation rate and combat AIDS. *Future Virol.* **2009**, *4*, 605.

82. Harris, R.S. Enhancing immunity to HIV through APOBEC. *Nat. Biotechnol.* **2008**, *26*, 1089–1090.

83. Alexandrov, L.B.; Nik-Zainal, S.; Wedge, D.C.; Aparicio, S.A.; Behjati, S.; Biankin, A.V.; Bignell, G.R.; Bolli, N.; Borg, A.; Borresen-Dale, A.L.; *et al.* Signatures of mutational processes in human cancer. *Nature* **2013**, *500*, 415–421.

84. Burns, M.B.; Temiz, N.A.; Harris, R.S. Evidence for APOBEC3B mutagenesis in multiple human cancers. *Nat. Genet.* **2013**, *45*, 977–983.

85. Roberts, S.A.; Lawrence, M.S.; Klimczak, L.J.; Grimm, S.A.; Fargo, D.; Stojanov, P.; Kiezun, A.; Kryukov, G.V.; Carter, S.L.; Saksena, G.; *et al.* An APOBEC cytidine deaminase mutagenesis pattern is widespread in human cancers. *Nat. Genet.* **2013**, *45*, 970–976.

86. Taylor, B.J.; Nik-Zainal, S.; Wu, Y.L.; Stebbings, L.A.; Raine, K.; Campbell, P.J.; Rada, C.; Stratton, M.R.; Neuberger, M.S. DNA deaminases induce break-associated mutation showers with implication of APOBEC3B and 3A in breast cancer kataegis. *eLife* **2013**, *2*, e00534.

87. Lada, A.G.; Dhar, A.; Boissy, R.J.; Hirano, M.; Rubel, A.A.; Rogozin, I.B.; Pavlov, Y.I. AID/APOBEC cytosine deaminase induces genome-wide kataegis. *Biol. Direct.* **2012**, *7*, e47.

88. Lada, A.G.; Stepchenkova, E.I.; Waisertreiger, I.S.; Noskov, V.N.; Dhar, A.; Eudy, J.D.; Boissy, R.J.; Hirano, M.; Rogozin, I.B.; Pavlov, Y.I. Genome-wide mutation avalanches induced in diploid yeast cells by a base analog or an APOBEC deaminase. *PLoS Genet.* **2013**, *9*, e1003736.

89. Roberts, S.A.; Sterling, J.; Thompson, C.; Harris, S.; Mav, D.; Shah, R.; Klimczak, L.J.; Kryukov, G.V.; Malc, E.; Mieczkowski, P.A.; *et al.* Clustered mutations in yeast and in human cancers can arise from damaged long single-strand DNA regions. *Mol. Cell* **2012**, *46*, 424–435.

90. Seitz, H.K.; Stickel, F. Molecular mechanisms of alcohol-mediated carcinogenesis. *Nat. Rev. Cancer* **2007**, *7*, 599–612.

91. Zuo, T.; Liu, D.; Lv, W.; Wang, X.; Wang, J.; Lv, M.; Huang, W.; Wu, J.; Zhang, H.; Jin, H.; *et al.* Small-molecule inhibition of human immunodeficiency virus type 1 replication by targeting the interaction between Vif and ElonginC. *J. Virol.* **2012**, *86*, 5497–5507.

92. Pery, E.; Sheehy, A.; Nebane, N.M.; Brazier, A.J.; Misra, V.; Rajendran, K.S.; Buhrlage, S.J.; Mankowski, M.K.; Rasmussen, L.; White, E.L.; *et al.* Identification of a Novel HIV-1 Inhibitor Targeting Vif-dependent Degradation of Human APOBEC3G. *J. Biol. Chem.* **2015**, *290*, 10504–10517.

93. Li, M.; Shandilya, S.M.; Carpenter, M.A.; Rathore, A.; Brown, W.L.; Perkins, A.L.; Harki, D.A.; Solberg, J.; Hook, D.J.; Pandey, K.K.; *et al.* First-in-class small molecule inhibitors of the single-strand DNA cytosine deaminase APOBEC3G. *ACS Chem. Biol.* **2012**, *7*, 506–517.

94. Olson, M.E.; Li, M.; Harris, R.S.; Harki, D.A. Small-molecule APOBEC3G DNA cytosine deaminase inhibitors based on a 4-amino-1,2,4-triazole-3-thiol scaffold. *ChemMedChem* **2013**, *8*, 112–117.

Hydrogen Peroxide Induce Human Cytomegalovirus Replication through the Activation of p38-MAPK Signaling Pathway

Jun Xiao, Jiang Deng, Liping Lv, Qiong Kang, Ping Ma, Fan Yan, Xin Song, Bo Gao, Yanyu Zhang and Jinbo Xu

Abstract: Human cytomegalovirus (HCMV) is a major risk factor in transplantation and AIDS patients, which induces high morbidity and mortality. These patients infected with HCMV experience an imbalance of redox homeostasis that cause accumulation of reactive oxygen species (ROS) at the cellular level. H_2O_2, the most common reactive oxygen species, is the main byproduct of oxidative metabolism. However, the function of H_2O_2 on HCMV infection is not yet fully understood and the effect and mechanism of N-acetylcysteine (NAC) on H_2O_2-stimulated HCMV replication is unclear. We, therefore, examined the effect of NAC on H_2O_2-induced HCMV production in human foreskin fibroblast cells. In the present study, we found that H_2O_2 enhanced HCMV lytic replication through promoting major immediate early (MIE) promoter activity and immediate early (IE) gene transcription. Conversely, NAC inhibited H_2O_2-upregulated viral IE gene expression and viral replication. The suppressive effect of NAC on CMV in an acute CMV-infected mouse model also showed a relationship between antioxidants and viral lytic replication. Intriguingly, the enhancement of HCMV replication via supplementation with H_2O_2 was accompanied with the activation of the p38 mitogen-activated protein kinase pathway. Similar to NAC, the p38 inhibitor SB203580 inhibited H_2O_2-induced p38 phosphorylation and HCMV upregulation, while upregulation of inducible ROS was unaffected. These results directly relate HCMV replication to H_2O_2, suggesting that treatment with antioxidants may be an attractive preventive and therapeutic strategy for HCMV.

Reprinted from *Viruses*. Cite as: Xiao, J.; Deng, J.; Lv, L.; Kang, Q.; Ma, P.; Yan, F.; Song, X.; Gao, B.; Zhang, Y.; Xu, J. Hydrogen Peroxide Induce Human Cytomegalovirus Replication through the Activation of p38-MAPK Signaling Pathway. *Viruses* **2015**, *7*, 2816–2833.

1. Introduction

Human cytomegalovirus (HCMV), a β herpesvirus, is an enveloped, large and double-stranded DNA virus. Like most herpesviruses, HCMV is able to establish a latent state in its hosts after primary infection, which can result in serious health conditions when the virus reactivates and performs lytic replication [1].

HMCV infection can be asymptomatic among immunocompetent people, but it can become an important and common cause of morbidity and mortality in immunocompromised patients, such as those with AIDS, solid organ transplantation, and hematopoietic stem cell transplantation [1–5]. Previous studies have shown that HCMV infection is highly associated with atherosclerosis, cardiovascular diseases [6–8], and inflammatory bowel disease [9].

Several mechanisms of the regulation of HCMV and mouse cytomegalovirus (MCMV) latency, reactivation, and lytic replication, such as chromatin remodeling and mitogen-activated protein kinase (MAPK) pathways, have been reported by previous studies [10–14]. However, none of these proposed triggers can be the cause of all clinical cases of HCMV reactivation and replication.

Although the common physiological trigger that stimulates HCMV replication remains unclear, many clinical diseases are characterized by high levels of oxidative stress. Patients who undergo solid organ transplantation generally suffer from oxidative stress and inflammation associated with ischemia/reperfusion, organ rejection, and as a side effect of immunosuppressive therapy [15–17]. AIDS patients have high levels of oxidative stress and inflammation as a result of the defensive mechanism of the immune system in the response to HIV infection [18]. In cardiovascular diseases, oxidative stress and inflammation can also be found and are believed to contribute to the development of atherosclerosis [19,20]. Thus, one potential key to determining the trigger of HCMV replication may be oxidative stress.

As a byproduct of oxidative metabolism [21], H_2O_2 was produced and released to impair redox homeostasis during oxidative stress. Mechanically, transcription of the major HCMV immediate early (IE) gene is driven by the complex major IE (MIE) promoter/enhancer. Within the region, there are several binding sites for known cellular transcription factors [22], such as NF-κB, CREB, ATF, and YY1. Studies have shown that high levels of oxidative stress result in activation of these transcription factors through mitogen- and stress-activated protein kinase (MSK), which is activated by the p38-MAPK pathway [23,24].

Thus, we hypothesize that the activation of p38-MAPK signaling plays a significant role in the HCMV replication caused by hydrogen peroxide, and this could be decreased by antioxidant treatment.

2. Materials and Methods

2.1. Cell Culture, Chemical Reagents and Antibodies

Human foreskin fibroblast (HFF) cells of no more than 15 passages, HEK 293 cells, mouse embryonic fibroblast (MEF) cells, and MRC-5 cells were cultured in Dulbecco's modified Eagle's Medium (DMEM) supplemented with 10% fetal bovine serum (FBS) at 37 °C under a 5% CO_2 atmosphere. Confluent cell monolayers were starved from serum for 24 h before infection. Serum free DMEM was used during drug treatment, virus incubation, and infection until the cells were harvested.

H_2O_2 solution, 3-amino-1,2,4-triazole (ATA), N-acetylcysteine (NAC), bovine liver catalase, reduced L-glutathione, 2′,7′-dichlorodihydrofluorescein diacetate (H_2DCF-DA) and the p38 inhibitor SB203580 were purchased from Sigma Life Science (St. Louis, MO, USA).

The rabbit polyclonal antibodies used in this study included phospho-p38 (T180/Y182), p38 and β-actin (all from ABclonal technology, Cambridge, MA, USA) and the mouse monoclonal antibodies to HCMV, pp72 and pp65, were purchased from Santa Cruz Biotechnology (Santa Cruz, CA, USA).

2.2. Plasmids

The MIEP-pGL3 and pRL-TK plasmids were kindly provided by Dr. Dongqing Wen [25]. The plasmid expression specific siRNA used to target human catalase (5′-GCCTGGGACCCAATTATCTTCATAT-3′) and the scrambled control siRNA were purchased from Hanbio (Shanghai, China). The transfection of HEK 293 and HFF cells was performed using Lipofectamine 3000 reagent from Invitrogen (Carlsbad, CA, USA).

2.3. Virus Preparation, Titration and Infection

HCMV (AD169) and MCMV (Smith strain) stocks were prepared in MRC-5 cells and MEF cells, and aliquots were stored at -80 °C. Viral titers were determined using the 50% tissue culture infective dose ($TCID_{50}$) method, as previously described [26]. Briefly, HFF or MEF cells were incubated with mock, UV-inactivated HCMV (UV-HCMV) or HCMV for 1 h under serum free DMEM at a multiplicity of infection (MOI) of 0.5. Then, the medium was removed, the cells were washed with PBS and fresh serum free medium was added. All experiments were examined at least three times using Reed and Muench's method.

2.4. DCF Staining

The measurement of reactive oxygen species (ROS) production in response to H_2O_2 was performed, as previously described [27]. After treatment with H_2O_2 (0, 100, or 200 µM) for 24 h, cells were washed with PBS and incubated with 10 µM H_2DCF-DA in warmed, serum free DMEM for 30 min in a CO_2 incubator at 37 °C. Cells were then washed three times with PBS and images were taken using a Leica microscope.

2.5. Detection of Cellular Catalase Activity and Intracellular H_2O_2 Level

HFF cells cultured with H_2O_2 (0, 100 or 200 µM) or ATA (0, 1.0, 2.0 or 4.0 mM) for 24 h were harvested by centrifugation. The cell pellets were sonicated in 100 µL cold PBS. After centrifugation at $10,000 \times g$ and 4 °C for 5 min, the supernatants were used for the detection of catalase activity and intracellular H_2O_2 level. The assay kits were all purchased from Jiancheng Bioengineering Institute (Nanjing, China).

2.6. Luciferase Assays

HEK 293 cells were transiently transfected with the MIEP-pGL3 luciferase reporter plasmid and the pRL-TK vector. At 12 h after transfection, the cells were treated with or without the antioxidants NAC (5 mM), L-glutathione (5 mM), and catalase (800 U/mL) for 2 h and then stimulated for 24 h with H_2O_2 (200 µM) or ATA (4 mM), or were treated with H_2O_2 (0, 50, 100, or 200 µM) or ATA (0, 1, 2, or 4 mM) alone for 24 h. Luciferase activity was determined as previously described [25] using the Dual-Luciferase® Reporter Assay System (Promega, Madison, WI, USA).

2.7. Real-Time PCR

Total DNA was isolated from the supernatants of infected cells using a cell and tissue genomic DNA extraction kit (BioTeke Corporation, Beijing, China). Changes in viral DNA loads were monitored using absolute quantitative real-time PCR. Viral DNA levels were detected using primers against the HCMV IE1 gene (forward primer, 5′-ATGTACGGGGGCATCTCTCT-3′ and reverse primer, 5′-GGCTTGGTTATCAGA GGCCG-3′) or the MCMV IE1 gene (forward primer, 5′-GTGGGCATGAAGTG TGGGTA-3′ and reverse primer, 5′-CGCATCGAAAGACAACGCAA-3′).

2.8. qRT-PCR

Total RNA was extracted using TRIzol reagent (Invitrogen) at 24 h after HCMV infection (MOI = 0.5). cDNA was prepared using ReverTra Ace® qRCR RT Master Mix with gDNA Remover (TOYOBO, Osaka, Japan). Each sample was measured in triplicate. The expression level of the IE1 gene transcript (forward primer, 5′-GTTGGCCGAAGAATCCCTCA-3′ and reverse primer, 5′-CACCATG TCCACTCGAACCT-3′) was normalized to GAPDH mRNA (forward primer, 5′-CA TGAGAAGTATGACAACAGCCT-3′ and reverse primer, 5′-AGTCCTTCCAC GATACCAAAGT-3′). Compared to the untreated cells, the relative expression levels in treated cells were calculated as fold changes.

2.9. Western Blot Analysis

Cells pellets were lysed in lysis buffer (Promega) with a cocktail of protein inhibitors (Roche, Mannheim, Germany) and then centrifuged at $13,000 \times g$ and 4 °C for 10 min. In brief, 30 μg whole cell extract was heated for 5 min at 98 °C with Laemmli buffer, and then, samples were separated by 12% sodium dodecyl sulfate polymerase gel electrophoresis (SDS-PAGE) and transferred to polyvinylidene difluoride (PVDF) membranes (Millipore, Billerica, MA, USA). After blocking in 5% (*w/v*) skim milk (Applygen, Beijing, China) or 5% (*w/v*) bovine serum albumin (BSA) (MP, Auckland, New Zealand), the blots were probed with primary antibodies overnight at 4 °C. Protein bands were detected using Western blotting luminol reagent (Santa Cruz Biotechnology). The membranes were incubated with Western Blot stripping buffer (CWBio) to re-probe for other proteins in the same membrane.

2.10. Animal Studies

BALB/c mice (male, 6–8 weeks old, 20–25 g body weight) were purchased from Vital River (Beijing, China). The animal study was performed according to the protocols approved by the Ethics Committee at the Beijing Institute of Transfusion Medicine and in accordance with Institutional Animal Care and Use Committee (IACUC) guidelines.

Mice were treated intragastrically with 400 μL of 40 mM NAC in water every day, from 3 days before intraperitoneal inoculation with MCMV (Smith strain, 5×10^3 p.f.u). At the proper time (day 7, 14, 21 and 28 post infection), DNA was extracted from 100 μL whole blood and used to determine the viral DNA load. To detect infectious virions in mice organs, the salivary glands (50 mg) and the lung (50 mg) were collected and homogenized on day 14 and 28, and viral titer was calculated with $TCID_{50}$ assays in MEF monolayers.

2.11. Statistical Analysis

All values are expressed as the means ± standard deviations. Statistical analyses were performed using SPSS statistical software V.17 (SPSS Inc., Chicago, IL, USA). Significant differences were evaluated by the two-tailed Student's *t*-test when two groups were compared, one-way analysis of variance (ANOVA) followed by the Dunnett's test when multiple groups were tested against a control group and the Bonferroni *post hoc* test when performing multiple comparisons between groups. A *p*-value lower than 0.05 was considered to indicate a statistically significant difference.

3. Results

3.1. ROS Enhance HCMV Replication through Paracrine and Autocrine Mechanisms

After treatment with exogenous hydrogen peroxide for 24 h, HFF cells dose-dependently formed an increasing amount of ROS (Figure 1A,B) and H_2O_2 content (Figure 1C).

To investigate the role of H_2O_2 in HCMV lytic replication, we observed whether exogenous H_2O_2 is sufficient to enhance HCMV replication at both the mRNA and protein level. HCMV MIE promoter activities were induced by H_2O_2 (Figure 2A), and H_2O_2 increased the expression of the IE gene in a dose-dependent manner (Figure 2B). Furthermore, H_2O_2 increased the levels of viral proteins, including pp72 and pp65 (Figure 2C,D), as well as the production of HCMV DNA in the culture supernatant (Figure 2E) and infectious virions (Figure 2F).

Next, we verified whether an increase in intracellular H_2O_2 is sufficient to induce HCMV replication. Treatment of HFF cells with ATA, an inhibitor of the H_2O_2-scavenging enzyme catalase, reduced the activity of cellular catalase (Figure 3A left panel) and increased the intracellular H_2O_2 level (Figure 3A right panel). ATA increased the MIE promoter activities, the IE1 gene transcripts, the expression of HCMV pp72 and pp65, and production of infectious virions (Figure 3B–F).

To confirm that the effect of ATA on HCMV replication was the result of an increase in the intracellular level of H_2O_2, we transiently expressed a catalase-specific siRNA in HFF cells. Compared to cells expressing a control siRNA, those transfected with a catalase-specific specific siRNA showed greatly lower catalase protein expression. Similar to ATA treatment, knockdown of catalase enhanced HCMV lytic pp72 and pp65 protein levels (Figure 3G).

3.2. H_2O_2 Scavengers Inhibit H_2O_2-Upregulated HCMV Lytic Replication

To detect whether H_2O_2 is required for HCMV lytic replication, we used a H_2O_2 scavenger to decrease the intracellular H_2O_2 level. Treatment with NAC, a common H_2O_2 scavenger, decreased the cellular H_2O_2 level, as indicated by a reduction in the median fluorescent level, in H_2-DCFH-treated HFF cells (Figure 4A). As expected, NAC inhibited the upregulation of MIE promoter activities (Figure 4B) and IE1 transcription (Figure 4C) by supplementation with ATA and H_2O_2. The effects of H_2O_2 on upregulation of HCMV MIE promoter activity and viral gene expression were also inhibited by another two scavengers, catalase and reduced glutathione (Figure 4D). Furthermore, NAC impaired the H_2O_2-upregulated HCMV lytic protein and infectious virions (Figure 4E–H). These results indicate that H_2O_2 scavengers, such as NAC and catalase, can suppress the stimulation of HCMV lytic replication by H_2O_2.

3.3. H_2O_2 Scavenger NAC Inhibits MCMV Lytic Replication in Vivo

Because of the strict species specificity of CMV, it is difficult to establish an animal model of HCMV infection. MCMV, which is similar to HCMV biological characteristics [28,29], has been regularly used to mimic HCMV infection *in vitro* and *in vivo*. We sought to inhibit MCMV lytic replication in mice using H_2O_2 scavenger NAC. We found that oxidative stress production was induced during primary infection of MEF cells with MCMV (Figure 5A,B). Conversely, supplement with NAC strongly inhibited MCMV infection of MEF (Figure 5C).

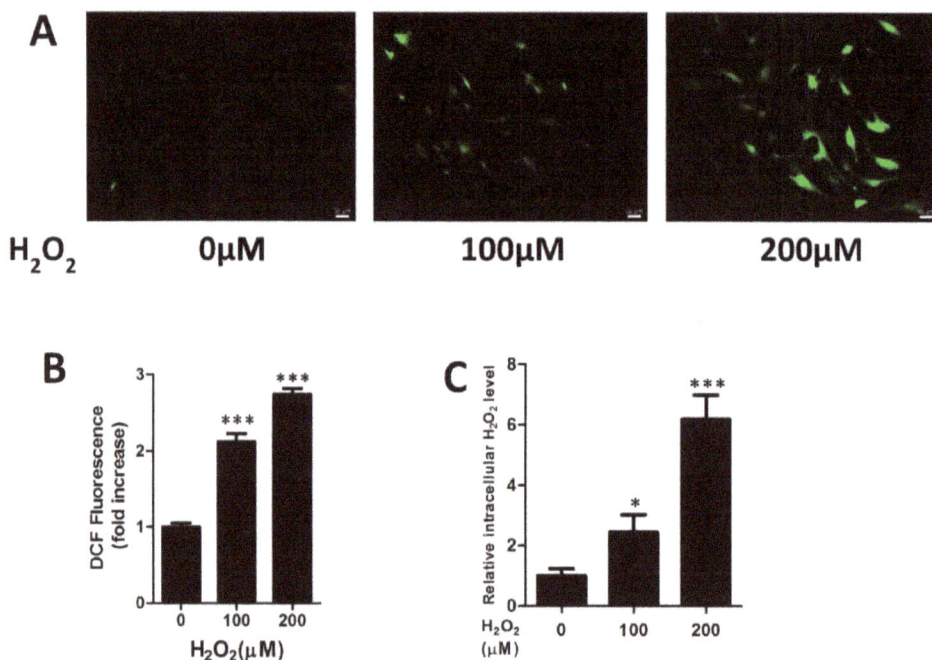

Figure 1. Oxidative stress was induced in human fibroblast cells by treatment with H_2O_2. After supplementation with 0, 100, 200 μM H_2O_2 for 24 h, increasing reactive oxygen species (ROS) production in human foreskin fibroblast (HFF) cells was determined by staining (**A**) or by measuring (**B**) the fluorescence produced after a 30 min incubation at 37 °C with 10 μM 2′,7′-dichlorodihydrofluorescein diacetate (H_2DCFH-DA) (for total ROS). (**C**) Treatment of HFF cells with 0, 100, 200 μM H_2O_2 increased the intracellular H_2O_2 concentration. The data are expressed as the means ± SD. * $p < 0.05$ or *** $p < 0.001$ for treated cells *versus* untreated cells by Dunnett's test.

To examine CMV lytic replication and verify the inhibitory effect of antioxidant NAC *in vivo*, we treated BALB/c mice intragastrically with 400 μL of 40 mM

NAC. The viral DNA load in whole blood was measured on day 7, 14, 21, and 28 post-infection. We found that mice fed NAC had a lower viral load than those fed drinking water alone (Figure 5D). To determine the production of infectious virions in mice organs, we used cell-free supernatants from ultrasonic homogenates of the salivary glands and the lung to infect MEF. We observed a high viral titer in cells infected with supernatants from the control group, while those infected with supernatants from NAC-treated mice had lower ones on days 14 and 28, respectively (Figure 5E). Collectively, the results of these *in vivo* experiments indicate that the antioxidant NAC effectively decreased MCMV replication in these mice.

3.4. H_2O_2 Upregulates HCMV Replication by Activating the p38 MAPK Pathway

As previously presented (Figure 2), H_2O_2 stimulates the upregulation of HCMV replication, but the mechanism is unclear. Here, the results revealed that p38-MAPK was rapidly and strongly activated by H_2O_2 treatment, following a time- and dose-dependent pattern (Figure 6A,B). In particular, p38-MAPK activation displayed a rapid onset within 1 h of treatment, followed by a progressive increase, returning to basal levels within 48 h, while a second peak was observed at 72 h after treatment (Figure 6A). Increasing H_2O_2 concentration led to an increase of p38-MAPK phosphorylation (Figure 6B) and the minimal concentration of H_2O_2 was 25 μM. Next, treatment with 10 μM SB203580 reduced both H_2O_2- and ATA-activated p38 phosphorylation (Figure 6C). Since supplementation with NAC inhibited H_2O_2- and ATA-induced HCMV replication (Figure 5), we suppose that NAC could inhibit H_2O_2-induced p38-MAPK activation. As expected, pretreatment with NAC (5 mM) also strongly decreased ATA- and H_2O_2-induced activation of p38-MAPK (Figure 6D). This effect was consistent with a decline of H_2O_2-induced oxidative stress in cells (Figure 6E,F), while SB203580 inhibited the H_2O_2-induced activation of p38 without affecting the redox status. At the same time, the upregulation of IE1 gene transcription, the expression of viral pp72 and pp65 proteins and the production of infectious virions were inhibited by treatment with SB203580 (Figure 6G–I). These results indicated that H_2O_2 upregulation of HCMV replication was mediated by the p38 MAPK pathway and that the inhibitory effect of NAC on H_2O_2-induced HCMV replication was also involved in this pathway.

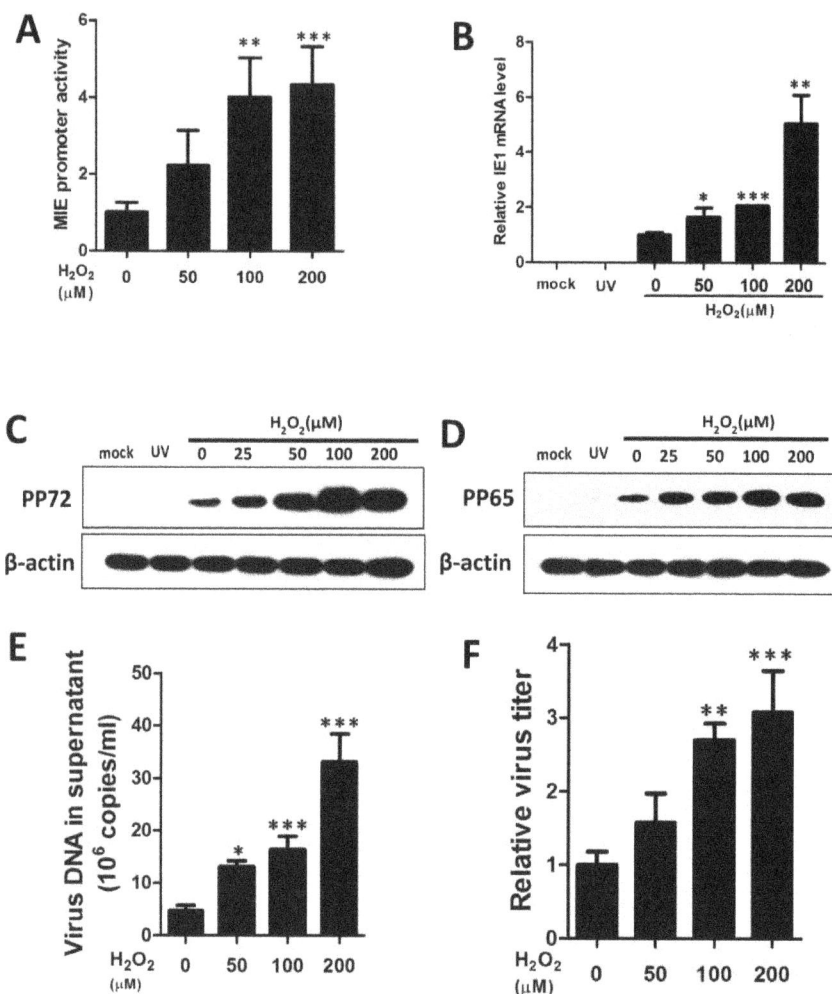

Figure 2. Exogenous H_2O_2 induces HCMV replication in HFF cells. Luciferase activities were measured for 293 cells transfected with promoter reporter plasmids for 24 h and treated with H_2O_2 (0, 50, 100, 200 µM) for 12 h (**A**). Cells were infected with UV-HCMV (UV) or HCMV at a multiplicity of infection (MOI) of 0.5. Incubation with exogenous H_2O_2 (0, 50, 100, or 200 µM) for 24 h induced the expression of immediately early (IE1) transcript in HCMV infected cells (**B**). Proteins were collected from infected cells for treatment with 0, 25, 50, 100, 200 µM H_2O_2 for 72 h. Viral proteins pp72 (**C**) and pp65 (**D**) were detected by Western blotting using β-actin for calibration of sample loading. Viral DNA load and viral titer and were measured after a 72 h treatment with 0, 50, 100, or 200 µM H_2O_2 by real-time quantitative PCR (**E**) and $TCID_{50}$ assay (**F**). * $p < 0.05$; ** $p < 0.01$ or *** $p < 0.001$ for treated *versus* untreated cells.

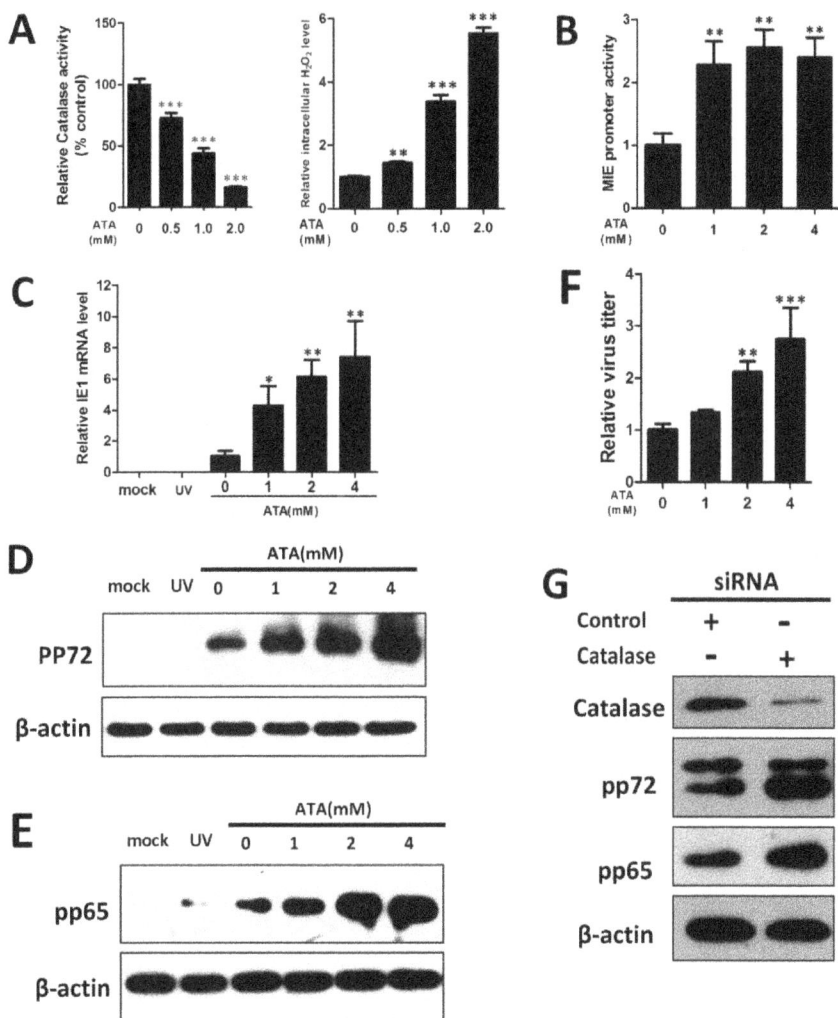

Figure 3. ATA-induced intracellular H_2O_2, enhancing viral replication in HFF cells. Treatment of HFF cells with the catalase inhibitor 3-amino-1,2,4-triazole (ATA) (0, 1, 2, 4 mM) for 24 h reduced catalase activity and increased intracellular H_2O_2 level in a dose-dependent manner (**A**). Cells were cultured with ATA for 24 h. ATA-induced intracellular H_2O_2 increased MIE promoter activity (**B**) and HCMV IE1 transcription (**C**). Cells were infected with UV-HCMV (UV) or HCMV at an MOI of 0.5. An increase in pp72 (**D**) and pp65 (**E**) protein levels were detected by Western blotting under 0, 1, 2, 4 mM ATA treatments for 72 h, and β-actin was used to calibrate sample loading. (**F**) Relative virus titers were measured using $TCID_{50}$ assay within five days. (**G**) Catalase, HCMV lytic protein pp72, pp65 and β-actin levels were determined by Western blotting after treatment with siRNA for five days. * $p < 0.05$; ** $p < 0.01$ or *** $p < 0.001$ for ATA-treated *versus* untreated cells.

275

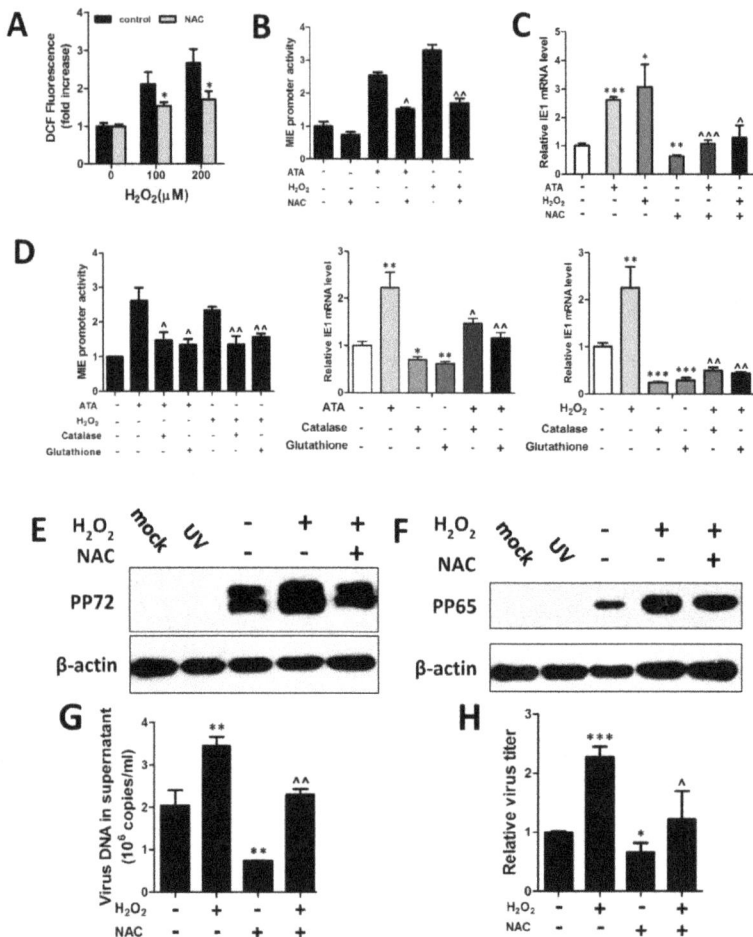

Figure 4. H_2O_2 scavengers inhibit H_2O_2-induced HCMV lytic replication *in vitro*. Treatment with the H_2O_2 scavenger *N*-acetylcysteine (NAC) (5 mM) was shown to decrease ROS production in HFF cells by measurement of fluorescence (**A**). Cells were treated with 200 µM H_2O_2 without any scavengers, or with the scavenger NAC at 5 mM, catalase at 800 U/mL or reduced glutathione at 5 mM for 24 h and then HCMV infection (MOI = 0.5). H_2O_2 scavengers reduced H_2O_2 (200 µM) and ATA (4 mM)-induction of MIE promoter activity and IE transcription in HFF cells (**B,C**). Treated with 200 µM H_2O_2 upregulated HCMV lytic replication, but inhibited by treatment with 5 mM NAC. pp72 and pp65 viral proteins in UV-HCMV (UV) or HCMV infected HFF cells treated with NAC (5 mM) for 72 h were determined by Western blotting with β-actin to calibrate sample loading (**E,F**). Treatment with 5 mM NAC downregulated viral DNA load (**G**) in culture supernatants and viral titer (**H**) in the presence or absence of 200 µM H_2O_2. * $p < 0.05$; ** $p < 0.01$ or *** $p < 0.001$ for treated *versus* untreated cells. ^ $p < 0.05$; ^^ $p < 0.01$ or ^^^ $p < 0.001$ for H_2O_2 scavenger-treated cells *versus* H_2O_2- or ATA-treated cells.

276

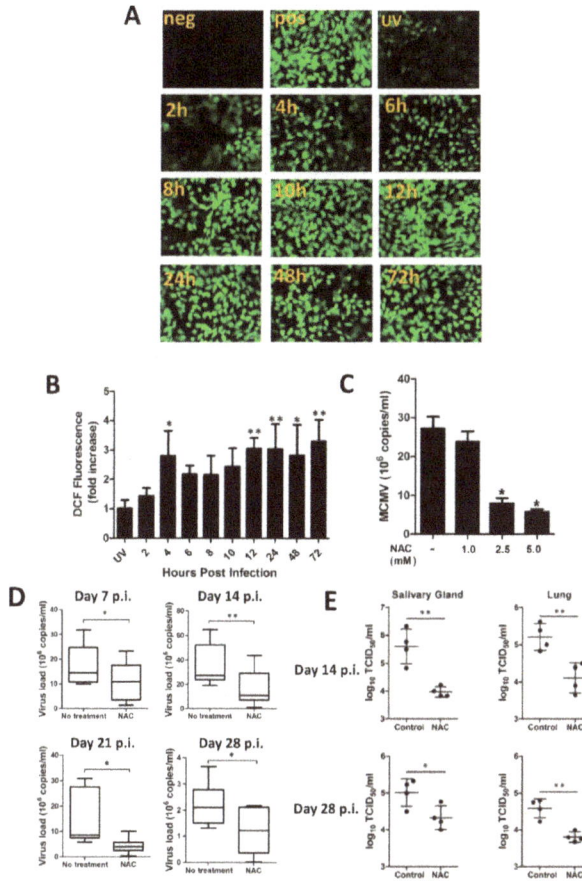

Figure 5. The H_2O_2 scavenger NAC inhibits MCMV lytic replication *in vivo*. ROS production upon primary infection of MEF cells with MCMV. Confluent MEF cells in 24-well plates were serum starved for 2 h, incubated with DMEM containing 10 μM H_2-DCFDA for 30 min at 37 °C, and infected with either UV-inactivated MCMV or MCMV (MOI = 0.5). H_2-DCFDA fluorescence were stained between infected and uninfected cells at indicated times (**A**). Fold induction of ROS production in infected cells relative to UV-HCMV (UV) infected cells (**B**). The culture supernatant collected at 72 h post infection with MCMV at an MOI of 0.5 detected the viral DNA (**C**). Mice were treated intragastrically with 400 μL of 40 mM NAC in water every day, from three days before intraperitoneal inoculation with MCMV (Smith strain, 5×10^3 p.f.u). One hundred microliters of whole blood from each mouse were examined. Viral loads in blood samples of control ($n = 14$) and NAC-treated ($n = 14$) mice at indicated days post infection (**D**). Related infectious viral titer in the salivary glands and in the lung was detected at 14 and 28 days post infection by $TCID_{50}$ assay (**E**). * $p < 0.05$ or ** $p < 0.01$ for treated *versus* untreated cells and mice.

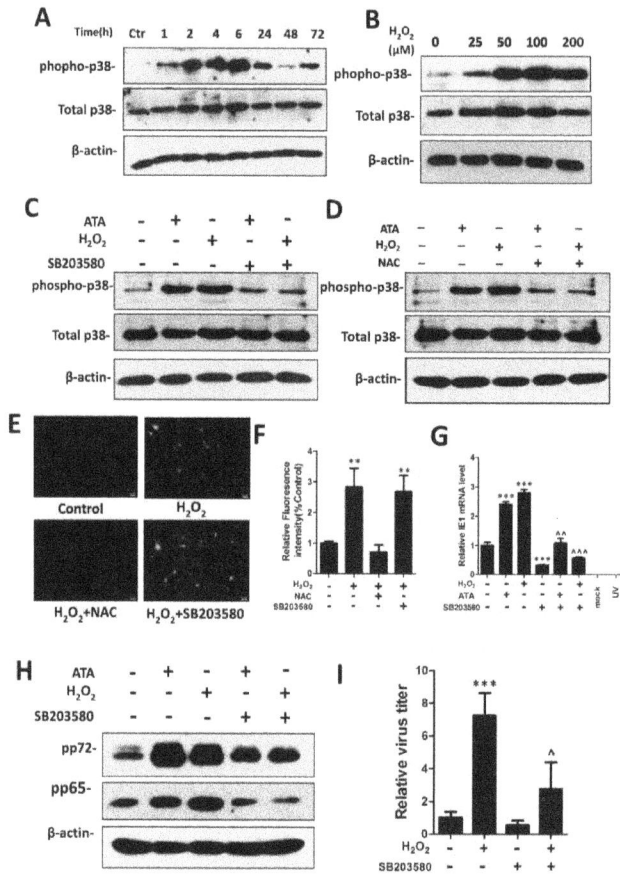

Figure 6. H_2O_2 facilitates HCMV replication by activating the p38-MAPK pathway. HFF cells were left untreated or were treated with 200 μM H_2O_2 for the indicated times (**A**) or with various H_2O_2 concentrations for 6 h (**B**). The kinases were detected by Western blotting, using specific primary antibodies against phospho-p38 and p38 and β-actin to calibrate sample loading. HFF cells were treated with SB203580 (10 μM) or NAC (5 mM) 1 h prior to H_2O_2 (200 μM) and ATA (5 mM) stimulation. Cells were harvested at 6 h post H_2O_2 and ATA treatment (**C,D**). Cells were determined by staining (**E**) or by measuring (**F**) the fluorescence produced after a 30 min incubation at 37 °C with 10 μM H_2DCFH-DA. Cells were infected with UV-HCMV (UV) or HCMV at an MOI of 0.5. Real-time PCR analysis of IE1 mRNA levels in cells allowed comparisons to untreated cells. Total mRNA was extracted from HFF at 24 h post infection. At 72 h, pp72 and pp65 protein were detected by Western blotting (**H**). Viral titer was detected in the presence or absence of H_2O_2 under a treatment with 10 μM SB203580 (**I**). ** $p < 0.01$ or *** $p < 0.001$ for treated *versus* untreated cells. ˆ $p < 0.05$; ˆˆ $p < 0.01$ or ˆˆˆ $p < 0.001$ for p38 inhibitor SB203580-treated cells *versus* H_2O_2 or ATA-treated cells.

278

4. Discussion

In this study, we investigated the role of H_2O_2 in the regulation of viral lytic replication in HFF. We demonstrate that hydrogen peroxide upregulates HCMV lytic replication through extracellular and intracellular mechanisms in fibroblasts. In addition, pretreatment with antioxidants inhibits HCMV replication *in vitro* and *in vivo*. Mechanistically, the p38-MAPK pathway contributes to the stimulation of HCMV replication by H_2O_2.

Since AIDS, organ transplantation, and atherosclerosis are characterized by oxidative stress, and HCMV infection is of high morbidity and mortality among these patients, oxidative stress might be a crucial physiological factor that upregulates HCMV lytic replication in these cases. Regularly, the reactive oxygen species H_2O_2 is often used to induce intracellular oxidative stress *in vitro* [30] and our results demonstrated that treatment with exogenous H_2O_2 for 24 h impairs cellular redox homeostasis and acted as an active ROS in cells (Figure 1). ATA, a small molecule irreversible inhibitor of the H_2O_2-scavenging enzyme catalase, was utilized to induce intracellular H_2O_2 in cells (Figure 3A) and thereby performed the same role as exogenous H_2O_2 (Figure 3B–F).

HCMV lytic replication is initiated by IE1 transcription, which is activated by MIE promoter/enhancer activity. Studies have shown that ROS can enhance IE transcription products in human endothelial cells and smooth muscle cells [31,32], but this has rarely been declared in fibroblasts. Interestingly, it has been shown that oxidative stress can lead to the reactivation and replication of KSHV, another member of the herpesvirus family, in PEL cells and endothelial cells [33,34]. The luciferase reporter assay showed that treatment with H_2O_2 enhances the activity of HCMV MIE promoter in a dose-dependent manner. Consistent with this result, we detected an increasing expression of viral IE1 gene and production of virions under the treatment of both H_2O_2- and ATA-induced oxidative stress. These results indicated that viral gene transcription and viral replication in the permissible cells, HFF, was initiated by H_2O_2-upregulated HCMV major immediately promoter activity.

Cellular antioxidants, such as superoxide dismutase (SOD) and catalase (CAT), protect cells from oxidative stress. SOD catalyzes the transition of superoxide into H_2O_2, which can be further converted into H_2O and O_2 by catalase. Antioxidants used to inhibit oxidative stress have been shown to block the replication of RNA viruses, including influenza virus, EV71, and HIV-1 [35,36]. However, it remains unclear whether H_2O_2 scavengers can inhibit H_2O_2-induced HCMV replication. In this study, NAC treatment resulted in inhibition of H_2O_2-induced oxidative stress and H_2O_2-upregulated HCMV replication. Similar to NAC, catalase and reduced glutathione also inhibited H_2O_2 induced MIE promoter transcription and IE gene expression. In addition, we illustrated that ROS were required for upregulation of HCMV replication induced by CAT inhibition and depletion. These results support the hypothesis that the reactive oxygen species hydrogen peroxide is a key factor in the enhancement of HCMV gene expression and replication and that this effect could be inhibited by treatment with H_2O_2 scavengers *in vitro*. Therefore, our results exhibited a critical role for H_2O_2 and cellular antioxidants in regulating HCMV replication.

Accumulating evidence has suggested an essential role for oxidative stress during viral infection [37–41]. Oxidative stress can be considered a protective means of the cell, which can contribute to apoptosis [42], and thus prevent the virus from replicating and infecting other cells. Interestingly, CMV appears to utilize virus-specific mechanisms to protect the cell from the effects of ROS and maintain a redox homeostasis [43]. However, the results showed that the levels of ROS increased remarkably upon MCMV infection, with the increase first appearing at approximately 2 h after infection (Figure 5A) and sustained until 72 h post-infection. (Figure 5B). Thus, it seems antioxidant therapy could be a potential treatment method for primary MCMV infection. In support of the results of previous studies we conducted, NAC was shown to prevent MCMV replication and production *in vitro* (Figure 5C). Significantly, NAC strongly reduced MCMV DNA load in whole blood and the production of infectious virions in the salivary gland and the lung.

It is widely accepted that the HCMV major immediate early promoter contains several types of transcription factor binding sites [44], such as NF-κB, that can be induced by H_2O_2 [45]. Furthermore, previous studies have shown that H_2O_2 can induce NF-κB transcription through multiple signaling pathways [23,46], including the JNK and p38 MAPK pathways. Furthermore, studies have shown that p38 MAPK, which is mediated by MSK1, is involved in NF-κB transactivation by H_2O_2 stimulus [23,24], but very little is known regarding the possible linkage between this pathway under H_2O_2-upregulated HCMV replication in fibroblast cells. Here, consistent with the results of previous studies, we showed that p38 was rapidly and strongly activated by H_2O_2 treatment. Additionally, co-culturing fibroblasts with HCMV and the p38 specific inhibitor SB203580 decreased the phosphorylation of p38 and HCMV transcription and production. Similar to SB203580, NAC also hampered the activation of p38 by H_2O_2 and inhibited the viral replication. However, NAC inhibited the H_2O_2-induced phosphorylation of p38 through inhibition of H_2O_2-induced oxidative stress, while SB203580 directly inhibited the p38 activation without affecting the production of ROS. Thus, we first demonstrated that H_2O_2 induced HCMV replication through the ROS/P38 MAPK signaling pathway.

Conclusively, our findings suggest that further studies of the antiviral and immune-modulatory effects of antioxidants are warranted. Furthermore, targeting of hydrogen peroxide and H_2O_2-mediated signaling is a potential therapeutic or preventive approach in HCMV infection.

Acknowledgments: This work was supported by Beijing Key Laboratory of Science and Technology Fund (Z141102004414034). We are grateful to Dongqing Wen for providing us with the plasmids used in this study and her expert technical assistance.

Author Contributions: J.X. designed, performed experiments and wrote the paper; J.D., L.P.L., Q.K., F.Y., and S.X. performed experiments; P.M. and B.G. performed virus titer experiment; J.X. and J.D. analyzed the data; Y.Y.Z. and J.B.X. gave scientific advices and contributed to a deep manuscript revision. All authors contributed substantially to the present work, then read and approved the final manuscript.

Conflicts of Interest: The authors declare no conflict of interest.

References

1. Gerna, G.; Baldanti, F.; Revello, M.G. Pathogenesis of human cytomegalovirus infection and cellular targets. *Hum. Immunol.* **2004**, *65*, 381–386.
2. Rubin, R.H. Impact of cytomegalovirus infection on organ transplant recipients. *Rev. Infect. Dis.* **1990**, *12* (Suppl. 7), S754–S766.
3. Patel, R.; Snydman, D.R.; Rubin, R.H.; Ho, M.; Pescovitz, M.; Martin, M.; Paya, C.V. Cytomegalovirus prophylaxis in solid organ transplant recipients. *Transplantation* **1996**, *61*, 1279–1289.
4. Castro-Malaspina, H.; Harris, R.E.; Gajewski, J.; Ramsay, N.; Collins, R.; Dharan, B.; King, R.; Deeg, H.J. Unrelated donor marrow transplantation for myelodysplastic syndromes: Outcome analysis in 510 transplants facilitated by the National Marrow Donor Program. *Blood* **2002**, *99*, 1943–1951.
5. Steininger, C.; Puchhammer-Stockl, E.; Popow-Kraupp, T. Cytomegalovirus disease in the era of highly active antiretroviral therapy (HAART). *J. Clin. Virol.* **2006**, *37*, 1–9.
6. Weis, M.; Kledal, T.N.; Lin, K.Y.; Panchal, S.N.; Gao, S.Z.; Valantine, H.A.; Mocarski, E.S.; Cooke, J.P. Cytomegalovirus infection impairs the nitric oxide synthase pathway: Role of asymmetric dimethylarginine in transplant arteriosclerosis. *Circulation* **2004**, *109*, 500–505.
7. Simmonds, J.; Fenton, M.; Dewar, C.; Ellins, E.; Storry, C.; Cubitt, D.; Deanfield, J.; Klein, N.; Halcox, J.; Burch, M. Endothelial dysfunction and cytomegalovirus replication in pediatric heart transplantation. *Circulation* **2008**, *117*, 2657–2661.
8. Arasaratnam, R.J. Cytomegalovirus and cardiovascular disease–the importance of covariates. *J. Infect. Dis.* **2013**, *208*, 1349.
9. Rahbar, A.; Bostrom, L.; Lagerstedt, U.; Magnusson, I.; Soderberg-Naucler, C.; Sundqvist, V.A. Evidence of active cytomegalovirus infection and increased production of IL-6 in tissue specimens obtained from patients with inflammatory bowel diseases. *Inflamm. Bowel Dis.* **2003**, *9*, 154–161.
10. Reeves, M.B.; MacAry, P.A.; Lehner, P.J.; Sissons, J.G.; Sinclair, J.H. Latency, chromatin remodeling, and reactivation of human cytomegalovirus in the dendritic cells of healthy carriers. *Proc. Natl. Acad. Sci. USA* **2005**, *102*, 4140–4145.

11. Reeves, M.B.; Compton, T. Inhibition of inflammatory interleukin-6 activity via extracellular signal-regulated kinase-mitogen-activated protein kinase signaling antagonizes human cytomegalovirus reactivation from dendritic cells. *J. Virol.* **2011**, *85*, 12750–12758.

12. Reeves, M.B.; Breidenstein, A.; Compton, T. Human cytomegalovirus activation of ERK and myeloid cell leukemia-1 protein correlates with survival of latently infected cells. *Proc. Natl. Acad. Sci. USA* **2012**, *109*, 588–593.

13. Rodems, S.M.; Spector, D.H. Extracellular signal-regulated kinase activity is sustained early during human cytomegalovirus infection. *J. Virol.* **1998**, *72*, 9173–9180.

14. Johnson, R.A.; Ma, X.L.; Yurochko, A.D.; Huang, E.S. The role of MKK1/2 kinase activity in human cytomegalovirus infection. *J. Gen. Virol.* **2001**, *82*, 493–497.

15. Jassem, W.; Fuggle, S.V.; Rela, M.; Koo, D.D.; Heaton, N.D. The role of mitochondria in ischemia/reperfusion injury. *Transplantation* **2002**, *73*, 493–499.

16. Kedzierska, K.; Sporniak-Tutak, K.; Bober, J.; Safranow, K.; Olszewska, M.; Jakubowska, K.; Domanski, L.; Golembiewska, E.; Kwiatkowska, E.; Laszczynska, M.; *et al.* Oxidative stress indices in rats under immunosuppression. *Transplant. Proc.* **2011**, *43*, 3939–3945.

17. Lamoureux, F.; Mestre, E.; Essig, M.; Sauvage, F.L.; Marquet, P.; Gastinel, L.N. Quantitative proteomic analysis of cyclosporine-induced toxicity in a human kidney cell line and comparison with tacrolimus. *J. Proteomics* **2011**, *75*, 677–694.

18. Sharma, B. Oxidative stress in HIV patients receiving antiretroviral therapy. *Curr. HIV Res.* **2014**, *12*, 13–21.

19. Pastori, D.; Carnevale, R.; Pignatelli, P. Is there a clinical role for oxidative stress biomarkers in atherosclerotic diseases? *Intern. Emerg. Med.* **2014**, *9*, 123–131.

20. Li, H.; Horke, S.; Forstermann, U. Vascular oxidative stress, nitric oxide and atherosclerosis. *Atherosclerosis* **2014**, *237*, 208–219.

21. Pan, J.S.; Hong, M.Z.; Ren, J.L. Reactive oxygen species: A double-edged sword in oncogenesis. *World J. Gastroenterol.* **2009**, *15*, 1702–1707.

22. Sinclair, J. Chromatin structure regulates human cytomegalovirus gene expression during latency, reactivation and lytic infection. *Biochim. Biophys. Acta* **2010**, *1799*, 286–295.

23. Aggeli, I.K.; Gaitanaki, C.; Beis, I. Involvement of JNKs and p38-MAPK/MSK1 pathways in H2O2-induced upregulation of heme oxygenase-1 mRNA in H9c2 cells. *Cell. Signal.* **2006**, *18*, 1801–1812.

24. Kefaloyianni, E.; Gaitanaki, C.; Beis, I. ERK1/2 and p38-MAPK signalling pathways, through MSK1, are involved in NF-kappaB transactivation during oxidative stress in skeletal myoblasts. *Cell. Signal.* **2006**, *18*, 2238–2251.

25. Wen, D.Q.; Zhang, Y.Y.; Lv, L.P.; Zhou, X.P.; Yan, F.; Ma, P.; Xu, J.B. Human cytomegalovirus-encoded chemokine receptor homolog US28 stimulates the major immediate early gene promoter/enhancer via the induction of CREB. *J. Recept. Signal Transduct. Res.* **2009**, *29*, 266–273.

26. Keyes, L.R.; Bego, M.G.; Soland, M.; St Jeor, S. Cyclophilin A is required for efficient human cytomegalovirus DNA replication and reactivation. *J. Gen. Virol.* **2012**, *93*, 722–732.

27. Satoh, K.; Nigro, P.; Matoba, T.; O'Dell, M.R.; Cui, Z.; Shi, X.; Mohan, A.; Yan, C.; Abe, J.; Illig, K.A.; *et al.* Cyclophilin A enhances vascular oxidative stress and the development of angiotensin II-induced aortic aneurysms. *Nat. Med.* **2009**, *15*, 649–656.

28. Rawlinson, W.D.; Farrell, H.E.; Barrell, B.G. Analysis of the complete DNA sequence of murine cytomegalovirus. *J. Virol.* **1996**, *70*, 8833–8849.

29. Krmpotic, A.; Bubic, I.; Polic, B.; Lucin, P.; Jonjic, S. Pathogenesis of murine cytomegalovirus infection. *Microbes Infect./Inst. Pasteur* **2003**, *5*, 1263–1277.

30. Bak, M.J.; Jeong, W.S.; Kim, K.B. Detoxifying effect of fermented black ginseng on H2O2-induced oxidative stress in HepG2 cells. *Int. J. Mol. Med.* **2014**, *34*, 1516–1522.

31. Scholz, M.; Cinatl, J.; Gross, V.; Vogel, J.U.; Blaheta, R.A.; Freisleben, H.J.; Markus, B.H.; Doerr, H.W. Impact of oxidative stress on human cytomegalovirus replication and on cytokine-mediated stimulation of endothelial cells. *Transplantation* **1996**, *61*, 1763–1770.

32. Speir, E.; Shibutani, T.; Yu, Z.X.; Ferrans, V.; Epstein, S.E. Role of reactive oxygen intermediates in cytomegalovirus gene expression and in the response of human smooth muscle cells to viral infection. *Circ. Res.* **1996**, *79*, 1143–1152.

33. Li, X.; Feng, J.; Sun, R. Oxidative stress induces reactivation of Kaposi's sarcoma-associated herpesvirus and death of primary effusion lymphoma cells. *J. Virol.* **2011**, *85*, 715–724.

34. Ye, F.; Zhou, F.; Bedolla, R.G.; Jones, T.; Lei, X.; Kang, T.; Guadalupe, M.; Gao, S.J. Reactive oxygen species hydrogen peroxide mediates Kaposi's sarcoma-associated herpesvirus reactivation from latency. *PLoS Pathog.* **2011**, *7*, e1002054.

35. Cai, J.; Chen, Y.; Seth, S.; Furukawa, S.; Compans, R.W.; Jones, D.P. Inhibition of influenza infection by glutathione. *Free Radic. Biol. Med.* **2003**, *34*, 928–936.

36. Staal, F.J.; Roederer, M.; Herzenberg, L.A.; Herzenberg, L.A. Intracellular thiols regulate activation of nuclear factor kappa B and transcription of human immunodeficiency virus. *Proc. Natl. Acad. Sci. USA* **1990**, *87*, 9943–9947.

37. McGuire, K.A.; Barlan, A.U.; Griffin, T.M.; Wiethoff, C.M. Adenovirus type 5 rupture of lysosomes leads to cathepsin B-dependent mitochondrial stress and production of reactive oxygen species. *J. Virol.* **2011**, *85*, 10806–10813.

38. Barlan, A.U.; Griffin, T.M.; McGuire, K.A.; Wiethoff, C.M. Adenovirus membrane penetration activates the NLRP3 inflammasome. *J. Virol.* **2011**, *85*, 146–155.

39. Tung, W.H.; Hsieh, H.L.; Lee, I.T.; Yang, C.M. Enterovirus 71 induces integrin beta1/EGFR-Rac1-dependent oxidative stress in SK-N-SH cells: Role of HO-1/CO in viral replication. *J. Cell. Physiol.* **2011**, *226*, 3316–3329.

40. Kavouras, J.H.; Prandovszky, E.; Valyi-Nagy, K.; Kovacs, S.K.; Tiwari, V.; Kovacs, M.; Shukla, D.; Valyi-Nagy, T. Herpes simplex virus type 1 infection induces oxidative stress and the release of bioactive lipid peroxidation by-products in mouse P19N neural cell cultures. *J. Neurovirol.* **2007**, *13*, 416–425.

41. Aubert, M.; Chen, Z.; Lang, R.; Dang, C.H.; Fowler, C.; Sloan, D.D.; Jerome, K.R. The antiapoptotic herpes simplex virus glycoprotein J localizes to multiple cellular organelles and induces reactive oxygen species formation. *J. Virol.* **2008**, *82*, 617–629.

42. Wang, J.; Shen, Y.H.; Utama, B.; Wang, J.; LeMaire, S.A.; Coselli, J.S.; Vercellotti, G.M.; Wang, X.L. HCMV infection attenuates hydrogen peroxide induced endothelial apoptosis-involvement of ERK pathway. *FEBS Lett.* **2006**, *580*, 2779–2787.

43. Tilton, C.; Clippinger, A.J.; Maguire, T.; Alwine, J.C. Human cytomegalovirus induces multiple means to combat reactive oxygen species. *J. Virol.* **2011**, *85*, 12585–12593.

44. Sinclair, J.; Sissons, P. Latency and reactivation of human cytomegalovirus. *J. Gen. Virol.* **2006**, *87*, 1763–1779.

45. Korbecki, J.; Baranowska-Bosiacka, I.; Gutowska, I.; Chlubek, D. The effect of reactive oxygen species on the synthesis of prostanoids from arachidonic acid. *J. Physiol. Pharmacol.* **2013**, *64*, 409–421.

46. Chen, K.; Vita, J.A.; Berk, B.C.; Keaney, J.F., Jr. c-Jun N-terminal kinase activation by hydrogen peroxide in endothelial cells involves SRC-dependent epidermal growth factor receptor transactivation. *J. Biol. Chem.* **2001**, *276*, 16045–16050.

Both ERK1 and ERK2 Are Required for Enterovirus 71 (EV71) Efficient Replication

Meng Zhu, Hao Duan, Meng Gao, Hao Zhang and Yihong Peng

Abstract: It has been demonstrated that MEK1, one of the two MEK isoforms in Raf-MEK-ERK1/2 pathway, is essential for successful EV71 propagation. However, the distinct function of ERK1 and ERK2 isoforms, the downstream kinases of MEKs, remains unclear in EV71 replication. In this study, specific ERK siRNAs and selective inhibitor U0126 were applied. Silencing specific ERK did not significantly impact on the EV71-caused biphasic activation of the other ERK isoform, suggesting the EV71-induced activations of ERK1 and ERK2 were non-discriminative and independent to one another. Knockdown of either ERK1 or ERK2 markedly impaired progeny EV71 propagation (both by more than 90%), progeny viral RNA amplification (either by about 30% to 40%) and protein synthesis (both by around 70%), indicating both ERK1 and ERK2 were critical and not interchangeable to EV71 propagation. Moreover, suppression of EV71 replication by inhibiting both early and late phases of ERK1/2 activation showed no significant difference from that of only blocking the late phase, supporting the late phase activation was more importantly responsible for EV71 life cycle. Taken together, this study for the first time identified both ERK1 and ERK2 were required for EV71 efficient replication and further verified the important role of MEK1-ERK1/2 in EV71 replication.

Reprinted from *Viruses*. Cite as: Zhu, M.; Duan, H.; Gao, M.; Zhang, H.; Peng, Y. Both ERK1 and ERK2 Are Required for Enterovirus 71 (EV71) Efficient Replication. *Viruses* **2015**, *7*, 1344–1356.

1. Introduction

Enterovirus 71 (EV71) is a non-enveloped, positive single-stranded RNA virus belonging to the enterovirus A species of the genus *Enterovirus*, family *Picornaviridae* [1]. The EV71 genome, about 7.5 kb in length, consists of a single open reading frame encoding seven nonstructural proteins (2A, 2B, 2C, 3A, 3B, 3C, and 3D) and four structural proteins (VP1, VP2, VP3, and VP4) [2]. As one of the major causative agents leading to large outbreaks of hand, foot and mouth disease (HFMD) worldwide, especially in the Asia-Pacific area recently, EV71 has become the most dangerous neurotropic enterovirus after the control of poliovirus [3,4]. There is currently no effective vaccine or specific therapy that has been applied to prevent or treat EV71-caused severe HFMD, due to insufficient understanding of the molecular mechanisms of EV71 replication and host response to EV71 infection.

It is universally acknowledged that successful viral replication is reliant on many functioning components of cellular metabolism and a prerequisite for all the following pathogenic consequences in host cells. Accumulated data show that virus takes advantage of various signaling cascades for its life cycle, among which studies done by us and others have demonstrated that the extracellular signal regulated kinase (ERK) pathway is essential for EV71 and other viruses replication, inhibition of this signaling pathway has been found to severely impair EV71 and other variety of viruses production [5–11].

Cellular ERK signaling pathway, one of the three major mitogen-activated protein kinase (MAPK) cascades, which consists of three tiered serine/threonine kinases of Raf, MEK and ERK, plays an important role in regulating cell physiological functions [12–14] as well as many pathologic processes, including brain injury, cancer, diabetes, infectious diseases and inflammation *etc.* [15–18]. The two isoforms of ERKs, ERK1 and ERK2 (also referring to ERK1/2), are considered to be the only downstream substrates of MEK (including MEK1 and MEK2, also referring to MEK1/2) to date [19]. Therefore, ERK1 and ERK2, undertaking the upstream signals from MEK1/2 and in turn activating variety of their downstream substrates, are key players in ERK pathway. They share 85% similarity at the amino acid level [20] and yet it is still controversial whether the individual ERK isoform plays a distinctive role(s). Some studies suggest that ERK1 and ERK2 are interchangeable [21–23]. However, considerable evidences indicate that they might act differentially [24–27]. Thus, it is still an open question, which needs to be further explored as to whether roles are unique or preferred to one or the other ERK isoform in the physiological and/or pathological processes.

Our previous work have proved that MEK1 and MEK2 play differential roles, and MEK1, rather than MEK2, is critical to promote EV71 efficient replication [6], highlighting that MEK1 and MEK2 could exert distinct effects on the replication of EV71. However, as the downstream kinases of MEK1, the specific contributions of ERK1 and ERK2 to EV71 replication have not been addressed yet. The objective of the present study is to determine the role(s) of individual ERK isoform on the life cycle of EV71. In addition, here we showed that either ERK1 or ERK2 were both required and not functionally redundant for EV71 efficient replication.

2. Materials and Methods

2.1. Cells Culture and Virus Preparation

Rhabdomyosarcoma (RD) cell line was obtained from The National Institute for the Control of Pharmaceutical and Biological Products. Cells were cultured in Dulbecco's modified Eagle's medium (DMEM, GIBCO) supplemented with 10% fetal bovine serum (FBS, Gibco) at 37 °C in an atmosphere of 5% CO_2.

The titer of Enterovirus 71 (EV71-BC08 stain) was determined by titration in RD cells and stored at −80 °C until use [28].

2.2. Inhibitor against ERK Pathway and Antibodies

U0126 (Pierce, Thermo Scientific, Waltham, MA, USA), the inhibitor of ERK pathway, was dissolved in DMSO at the stocking concentration of 2 mM. Antibodies were purchased from Cell Signaling Technology (Danvers, MA, USA) (CST, anti-ERK1/2, anti-phospho-ERK1/2), Abcam (anti-EV71 VP1 and anti-VP3/4), Santa Cruz (anti-β-actin).

2.3. siRNAs and Transfection

siRNAs targeting human ERK1 (siERK1) and ERK2 (siERK2) were synthesized from Genepharma Co., Ltd. (Shanghai, China). The sequences, coming from Christopher A. Dimitri's paper [29], are showing as follows:

ERK1 siRNA (siERK1): 5'-CCCUGACCCGUCUAAUAUAdTdT-3' (sense),
5'-UAUAUUAGACGGGUCAGGGdAdG-3' (antisense);
ERK2 siRNA (siERK2): 5'-CAUGGUAGUCACUAACAUAdTdT-3'(sense),
5'-UAUGUUAGUGACUACCAUGdAdT-3' (antisense).

In addition, the negative control siRNAs (siNC) were purchased from Genepharma Co., Ltd. siERK1 and siERK2 (siERK1+2) were used together to knock down both ERK1 and ERK2.

Lipofectamine 2000 (Invitrogen) was used according to the manufacturer's instructions for siRNA transfection. RD cells were grown to 60% confluency in 6 or 12 well plates before transfection. RD cells were then transfected with siRNAs at the indicated concentrations.

2.4. Morphological Analysis

RD cells infected with EV71 were examined at every 8 h intervals post infection (p.i.) for the cytopathic effect (CPE) with phase-contrast microscopy.

2.5. Real Time Quantitative PCR (qPCR)

Total and intracellular viral RNAs were prepared for relative qPCR by Trizol reagent (Invitrogen). Then, according to the manufacturer's instructions of the ReverAid First strand cDNA synthesis kit (Thermo Scientific), 11 μL of viral RNAs and 1.5 μg of total RNAs were reversed transcribed into cDNA. Then 1 μL of cDNA was amplified with forward and reverse primers for EV71 VP1 gene and GAPDH control using LightCycler DNA Master SYBR Green I kit (Roche Diagnostics Corporation, Basel, Switzerland). The forward and reverse EV71 VP1 gene primers were: 5'- GCA GCC CAA AAG AAC TTC AC-3' and 5'- ATT TCA GCA GCT TGG AGT GC-3', respectively. The forward and reverse GAPDH primers were:

5'-TGTTCCAATATGATTCCACCC-3' and 5'- CTTCTCCATGGTGCGTGAAGA-3',
respectively. The reactions were performed with the Roche Light Cycler 480 system
under the following conditions: Initial denaturation step at 95 °C for 10 min, followed
by 40 cycles of 30 sat 94 °C, at 55 °C for 30 s and at 72 °C for 30 s. The CT value was
normalized to that of GAPDH. All samples were run in triplicate.

Intracellular EV71 virions for absolute qPCR were prepared as described in
Dr. Mingliang He's paper [30]. A quantitative standard curve was achieved as
described in our previous paper [6]. Quantified results were extrapolated from the
standard curve with all samples being run in triplicate.

2.6. Western Blot Analysis

Western blots were performed as described in our previous study [9]. Cells
were harvested at indicated time points and lysed for 1 h in lysis buffer (Santa
Cruz) containing complete protease inhibitors (Roche Applied Science). Total protein
concentration was determined by the Bicinchoninic Acid Protein Assay Kit (Pierce,
Thermo Scientific) after obtaining cell extracts by centrifugation at 13,000 rpm and
4 °C. Before transferred to PVDF membranes (Millipore), proteins were resolved on
the sodium dodecyl sulfatesulfate polyacrylamide gel electrophoresis (SDS-PAGE).
Then the membranes were blocked in 5% non-fat-dry-milk solution for 1 h at
room temperature and then blotted with specific primary antibodies over night
at 4 °C, following incubated with horseradish peroxidase antibodies for 1 h at
room temperature. The immunoreactive bands were developed with SuperSignal
West Femto Maximum Sensitivity Substrate (Pierce, Thermo Scientific) or enhanced
chemiluminescent substrate (ECL), followed by autoradiography.

2.7. Statistics

All the curves and diagrams were made by using the Graph Pad Prism 5
Program (GraphPad). Data were shown as the mean $\pm$ standard deviation (SD)
and analyzed by Student's t-test. $p < 0.05$ was considered statistically significant.

3. Results

*3.1. Specific Knockdown of ERK Isoform Did Not Affect the Activation of the Other Isoform
Induced by EV71*

Biphasic activation of ERK1/2 caused by EV71 infection, including early
transient and late sustained phases, has been demonstrated in our previous study [6].
To further identify the activation status of specific ERK isoform (ERK1 or ERK2) in
viral infection, the phosph-ERK1 (pERK1) and phosph-ERK2 (pERK2) under EV71
infection were determined separately. As shown in Figures 1A and 1B, a biphasic
activation of ERK2 induced by EV71 was observed in RD cells pre-treated with ERK1

siRNA (designated as siERK1) compared with that of respective uninfected groups pre-transfected with negative control siRNA (designated as siNC).

No compensatory activation of ERK2 was found when specific knockdown of ERK1. Similarly, the activation of ERK1 was not impacted when specifically silencing of ERK2 with ERK2 siRNA (designated as siERK2) in RD cells infected with EV71 (Figure 1C, D). These data indicated that the activation of ERK1 and ERK2 did not affect one another in the presence of EV71 infection in RD cells.

Notably, siERK1 and siERK2 were confirmed very efficient in knocking down corresponding ERK protein in quiescent RD cells collected 36 h post transfection and detected by Western blot analysis, which correlated with a strong decrease in ERK phosphorylation (Figure 1E). Moreover, treatment of U0126, a specific inhibitor of ERK activation, decreased more than 99% of ERK1/2 activation at 12 h after addition. No significant cytotoxicity of siERK1, siERK2, siERK1+2 and U0126 to the proliferation and survival of RD cells were observed in the current study (Figure S1).

3.2. Depletion of Individual ERK Isoform Resulted in a Similar Reduction of EV71 Proliferation

It has been demonstrated that MEK1 and MEK2 play a different role in EV71 replication [6]. To further elucidate the roles of downstream kinases of MEKs, ERK1 and ERK2 in EV71 replication, the effects of distinct knockdown of ERK1 or ERK2, or both on progeny viral titers were investigated. A clear reduction of progeny EV71 titers by about 90% were obtained in cells pre-treated with siERK1, siERK2 and siERK1+2, respectively, when compared with that of siNC control (Figure 2A).

No obvious difference of progeny viral titers was observed either knocking down of both ERK1 and ERK2, or each, whereas a stronger reduction of progeny viral titers was found when treating cells with U0126. In addition, EV71-induced cytopathic effect (CPE) was examined under phase-contrast microscopy. At 24 h p.i., pre-treatment with siERK1, siERK2, siERK1+2 or U0126 remarkably suppressed the morphological changes caused by EV71 infection (Figure 2B), as compared to corresponding controls, which was consistent with the viral titers in corresponding groups. Taken together, these results indicated ERK1 and ERK2 were required, but not functionally redundant for EV71 proliferation.

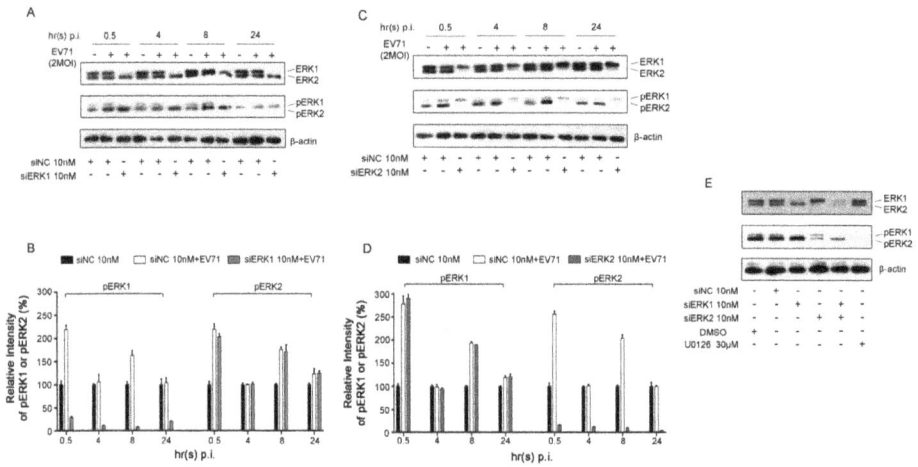

Figure 1. EV71-induced activation of ERK1/2 in Rhabdomyosarcoma (RD) cells. (**A**) RD cells were infected with EV71 (MOI = 2) at 36 h post-transfection with ERK1 siRNA (siERK1). Cells transfected with negative control siRNA (siNC) were used as controls. Cell lysates were collected at the indicated time points post infection (p.i.). ERK1/2, pERK1/2 and β-actin were detected by Western blot analysis, respectively; (**B**) The intensity of pERK1 or pERK2 normalized to that of corresponding ERK was determined by densitometric scanning based on the results from panel A; (**C**) The experiment was performed as described in panel A except that RD cells were transfected with ERK2 siRNA (siERK2); (**D**) Based on the results from panel C, the intensity was determined as described in panel B; (**E**) Cell lysates were collected from RD cells treated with siERK1, siERK2 and both (siERK1+2) at 36 h after transfection and cells treated with U0126 at 12 h after addition. ERK1/2 and pERK1/2 were blotted with specific antibodies. Experiments were repeated three times.

3.3. Distinct Knockdown of ERK Isoform Reduced Viral Genomic RNAs

To further identify the effects of ERK1 and ERK2 on EV71 replication cycle, VP1 gene among total viral RNAs extracted from siERK(s)- and U0126-treated RD cells was quantified by relative quantitative PCR (qPCR). Viral VP1 gene was significantly suppressed by about 30% to 40% at 4 h, 8 h and 12 h p.i. by siERK1, siERK2, siERK1+2 and U0126 when compared with corresponding controls in RD cells infected with EV71 (Figure 3).

290

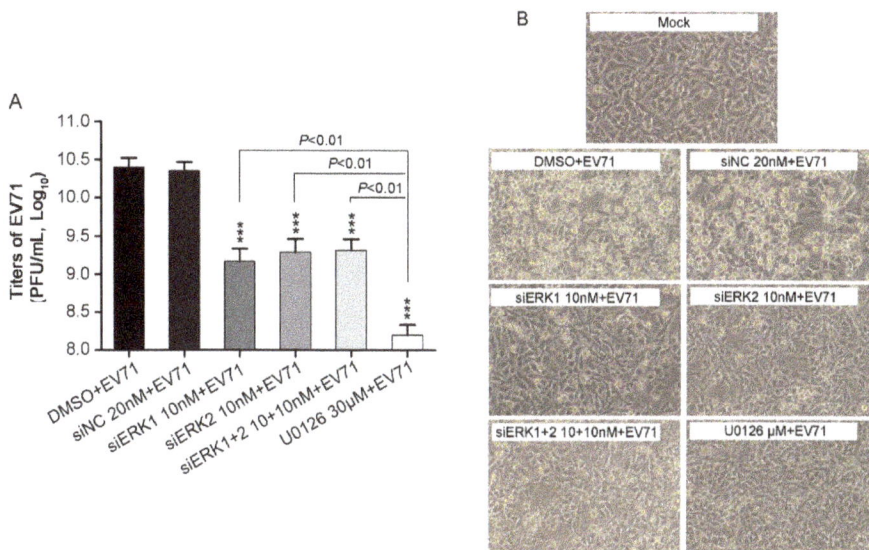

Figure 2. EV71 titers and CPE in RD cells treatment with distinct siERK (s) or U0126. (**A**) RD cells, pre-transfected with siERK1, siERK2 or siERK1+2 for 36 h, respectively, or treated with U0126 1 h prior to infection, were infected with EV71 at an MOI of 2. Cells pre-treated with DMSO or siNC were done as parallels. At 24 h p.i., both supernatants and cell lysates were collected and applied for determining the total titers of EV71 by titration; (**B**) Experiments were performed as described in panel A except CPE was examined every 8 h and images were taken at 24 h p.i.. Each result represents the average of three independent experiments and is shown as the means $\pm$ standard deviations (SD). *** $p < 0.001$, *versus* corresponding controls by Student's *t*-test.

In addition, U0126 caused a stronger reduction of viral genomic RNA than those of siERK1 and siERK1+2 at 4 and 8 h p.i., when RD cells was infected with 0.4 MOI of EV71 (Figure 3A). It should be noted that amounts of VP1 gene remained almost unchanged in each group from 8 h p.i. to 12 h p.i., indicating that the life cycle of EV71 might be less than 8 h. All these results revealed that both ERK1 and ERK2 might play important roles in viral RNA synthesis.

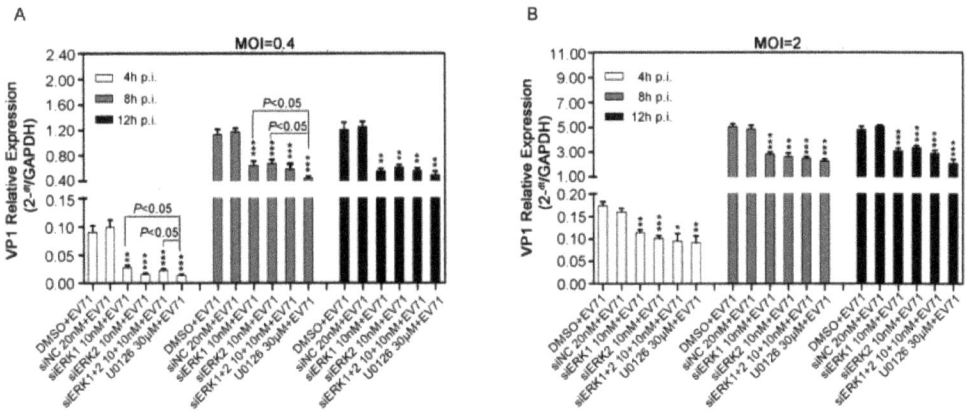

Figure 3. Effects of ERK1 and ERK2 on viral RNA synthesis in RD cells. RD cells pre-treated with siERK(s) or U0126 were infected with EV71 at the MOI of 0.4 (panel **A**) or 2 (panel **B**). Then at the indicated time points p.i., both supernatants and cell lysates were collected, viral VP1 gene among total viral genomes was quantified by relative qPCR. Data were the means of three independent experiments and error bars were denoted the SD. ** $p < 0.01$ *** $p < 0.001$, *versus* respective controls by Student's *t*-test.

3.4. Disruption of Either ERK1 or ERK2 Resulted in a Reduction of EV71 Protein

To further determine the impact of specific ERK isoform on EV71 protein, viral structural proteins VP1 and VP3/4 were detected by Western blot analysis in EV71-infected RD cells pre-transfected with siERK1,or siERK2, or both, respectively. As shown in Figure 4A, ERK1, or ERK2, or ERK1/2 expression were almost diminished by siERK1, or ERK2, or siERK1+2, which resulted in the activation inhibition of ERK1, ERK2, or ERK1/2 by about 80% to 90%. The silencing effect decreased the expression of VP1 and VP3/4 by around 70%, when compared with that of siNC control (Figure 4B).

It seemed that siERK1 had a better inhibitory effect on VP1 and VP3/4 but no significance difference compared with that of siERK2 and siERK1+2 (Figure 4B). In addition, U0126 reduced more than 99% of viral VP1 and VP3/4 expression which was significantly reduction than that of siERK(s)-treated groups. Our results suggested that both ERK1 and ERK2 might be crucial for EV71 protein production.

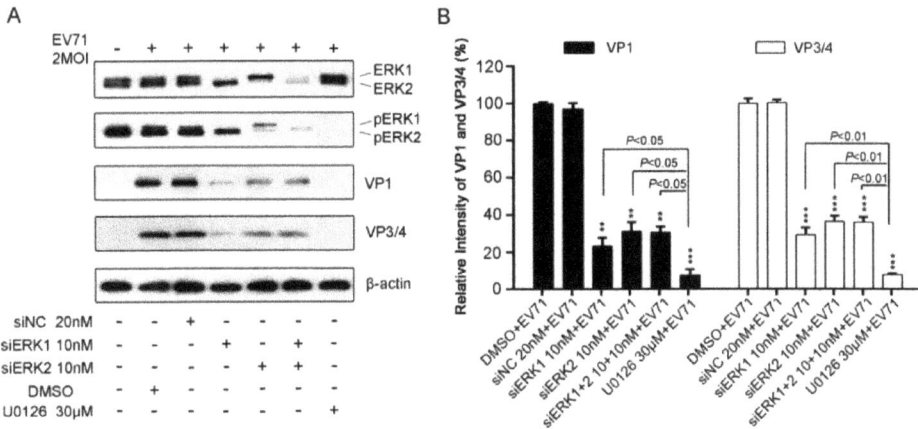

Figure 4. Impacts of ERK1 and ERK2 on protein production of EV71 in RD cells. (**A**). RD cells pre-treated with siERK(s) or U0126 were infected with 2 MOI of EV71. At 12 h p.i., total protein harvested was blotted with antibodies specific to VP1 and VP3/4. β-actin was used as the loading control. The relative intensities of VP1 (panel **B**) and VP3/4 (panel **C**) normalized to β-actin were presented by the percentage of the intensity of respective DMSO+EV71 group (100%). Data represent the average of three independent experiments and is shown as the means ± SD. ** $p < 0.01$ *** $p < 0.001$, *versus* corresponding control groups by Student's *t*-test.

3.5. Inhibiting both Early and Late Phases of ERK1/2 Activation Showed No Significant Difference from Blocking only the Late One for EV71 Replication

Since depletion of ERK1 or ERK2 both impaired virus replication severely, a time-of-drug addition assay was performed next in EV71-infected RD cells at the indicated time points to further specify the roles of the two phases of ERK1/2 activation in the viral life cycle. VP1 gene representing intracellular viral RNAs or intracellular virions was determined.

As shown in Figure 5A, disturbing activation of both ERK1 and ERK2 by treating cells with U0126 at 1 h before infection (−1), 0 h (0), 1 h(1) and 4 h (4) p.i. resulted in about 30% of suppression of intercellular viral RNAs. The similar results of intercellular virions were also obtained in the time-of-drug addition assay (Figure 5B). These data suggested that the late phase of ERK1/2 activated by EV71 might be more important for EV71 life cycle which may include viral RNA replication and/or translation.

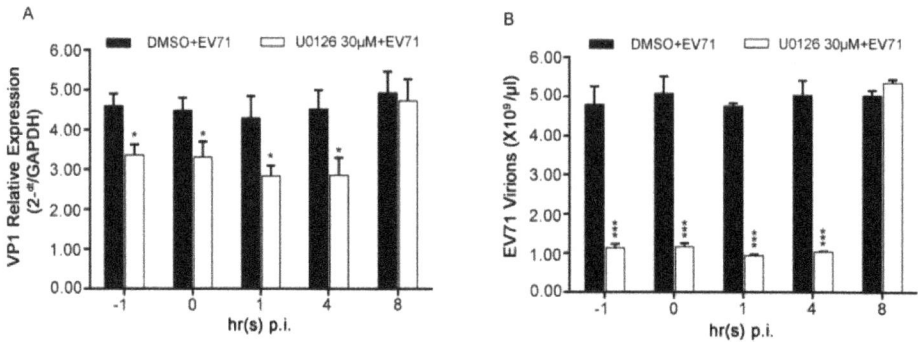

Figure 5. Effects of early ERK1/2 activation on EV71 infection in RD cells. (**A**) RD cells infected with EV71 at the MOI of 2 was treated with 30 μM of U0126 at the indicated time points p.i.. At 12 h p.i., cells were collected and viral VP1 gene among intracellular viral RNAs was quantified by relative qPCR; (**B**) The experiments were performed as described in panel A except 100 μg/mL of RNase A was used to eliminate naked viral RNAs before RNA was extracted. VP1 gene among intracellular EV71 virions was quantified by absolute qPCR. Data shown were the means $\pm$ SD ($n = 3$). * $p < 0.05$ ** $p < 0.01$ *** $p < 0.001$, *versus* respective controls by Student's *t*-test.

4. Discussion

Viruses hijack components of host's metabolic machinery for productive replication. The significance of activation of ERK pathway, as one of the specific responses to infection of varieties of viruses including EV71 [5–11,31,32], has been broadly reported. Our previous study has established the key role of MEK1 in the activation of ERK pathway induced by EV71 and proposed that MEK1-ERK1/2 signaling pathway acts as a central "hub" in EV71 replication [6]. However, it is still unclear if ERK1 and ERK2, the only known downstream kinases of MEK1 to date, play distinct role(s) in the life cycle of EV71. In fact, there are many studies focusing on distinction of ERK1 and ERK2 in several other areas rather than in virus replication. Meanwhile, the conclusions of these studies yet remain controversial. Although ERK1 and ERK2 are inclined to be thought of interchangeably [23], increasing studies have provided evidences for differential roles of ERK1 and ERK2 in cell movement [24], embryonic mice viability [33–35], and pathophysiology of central neural system *etc.* [36]. Our present work investigated distinction(s) of the individual involvement of ERK1 and ERK2 in EV71 replication.

Consistent with previous studies [6], EV71 infection caused a non-discriminative biphasic activation of ERK1/2 in RD cells. Some literatures reported that depletion of ERK1 induced higher activation of ERK2 in MEFs [34] and removal of ERK2 led to increased activation of ERK1 in NIH3T3 cells [37]. However, in our study, single silencing specific ERK isoform did not lead any significant effect on the biphasic activation of the other isoform in cells infected with or without EV71 (Figure 1), indicating that the activation of ERK1 and ERK2 might be various under different circumstances, although the mechanism is elusive.

Moreover, the present study shows both ERK1 and ERK2 are essential and not functionally redundant for EV71 replication due to the fact that individually silencing specific ERK isoform impaired EV71 propagation significantly, resulting in a marked suppression of viral progeny titer and EV71-induced CPE (Figure 2) as well as a clear reduction of viral RNA and protein synthesis (Figures 3 and 4). U0126 showed stronger inhibitory effect on virus replication than ERK siRNA(s) did, probably because U0126 was able to block the activation of ERK pathway more efficiently. Interestingly, double knockdown of ERK1/2 by siRNAs suppressed virus propagation, viral RNAs and proteins to a similar level compared with those of single knockdown groups, indicating that ERK1 and ERK2 probably work together in EV71 life cycle in RD cells. In addition, it is reasonable to hypothesize that activated ERK1 and ERK2 act as a functional ensemble rather than playing their roles separately in EV71 replication. This is the first to report that ERK1 and ERK2 are both required as key factors in EV71 efficient propagation, which is quite different from distinctive or interchangeable functions of ERK1 and ERK2 in several other areas [21–24,33,34]. Combined with our previous study on MEK1 [6], the upstream kinase of ERK1/2, these findings further verified the critical role of MEK1-ERK1/2 pathway in EV71 life cycle.

In addition, according to the time-of-drug addition assay (Figure 5), it appeared that virus successful biosynthesis process was more closely related to the late phase of ERK1/2 activation induced by EV71 than the early one. This result was consistent with and further confirmed previous studies in which the early phase of ERK1/2 activation is not involved in viral propagation directly whereas might be the consequence of the virus-receptor binding [6,11].

In conclusion, this study provided evidence that ERK1 and ERK2 are not interchangeable in the EV71 life cycle and indicated they might act their functions as a whole. The findings deepened our understanding in the roles played by MEK1-ERK1/2 signaling pathway in virus life cycle and shed light on possibilities to offer flexible choices to selectively target the one between ERK1 and ERK2 in future anti-viral strategies. Further studies will be needed to specify the effect(s) and mechanism(s) of ERK1/2 on specific EV71 replication process(es).

5. Conclusions

Overall, the present study identified the effects of specific ERK isoform on EV71 propagation as well as on the viral genomic RNAs and proteins. Our data revealed that ERK1 and ERK2 are both required and not interchangeable for EV71 life cycle.

The activations of ERK1 and ERK2 induced by EV71 were non-discriminative and independent to each another. Besides, the late phase of ERK1/2 activation under infection might be more responsible for the replication of EV71.

Acknowledgments: This work was supported by the National Natural Science Foundation of China (NSFC grant 81371816).

Author Contributions: M.Z. and Y.P. conceived and designed the experiments; M.Z. and H.D. performed the experiments; M.Z. and M.G. analyzed the data; H.Z. contributed reagents/materials/analysis tools; M.Z. wrote and Y.P. revised the manuscript, respectively.

Conflicts of Interest: The authors declare no conflict of interest.

References

1.	The Picornavirus Pages: Enterovirus A. Available online: http://www.picornaviridae.com/enterovirus/ev-a/ev-a.htm (accessed on 12 February 2015).
2.	Lin, J.Y.; Chen, T.C.; Weng, K.F.; Chang, S.C.; Chen, L.L.; Shih, S.R. Viral and host proteins involved in picornavirus life cycle. *J. Biomed. Sci.* **2009**, *16*, e103.
3.	Yi, L.; Lu, J.; Kung, H.F.; He, M.L. The virology and developments toward control of human enterovirus 71. *Crit. Rev. Microbiol.* **2011**, *37*, 313–327.
4.	Wang, S.M.; Liu, C.C. Update of enterovirus 71 infection: epidemiology, pathogenesis and vaccine. *Expert Rev. Anti Infect. Ther.* **2014**, *12*, 447–456.
5.	Tung, W.H.; Hsieh, H.L.; Lee, I.T.; Yang, C.M. Enterovirus 71 modulates a COX-2/PGE2/cAMP-dependent viral replication in human neuroblastoma cells: role of the c-Src/EGFR/p42/p44 MAPK/CREB signaling pathway. *J. Cell Biochem.* **2011**, *112*, 559–570.
6.	Wang, B.; Zhang, H.; Zhu, M.; Luo, Z.; Peng, Y. MEK1-ERKs signal cascade is required for the replication of Enterovirus 71 (EV71). *Antivir. Res.* **2012**, *93*, 110–117.
7.	Cai, Y.; Liu, Y.; Zhang, X. Suppression of coronavirus replication by inhibition of the MEK signaling pathway. *J. Virol.* **2007**, *81*, 446–456.
8.	Pleschka, S.; Wolff, T.; Ehrhardt, C.; Hobom, G.; Planz, O.; Rapp, U.R.; Ludwig, S. Influenza virus propagation is impaired by inhibition of the Raf/MEK/ERK signalling cascade. *Nat. Cell Biol.* **2001**, *3*, 301–305.
9.	Zhang, H.; Feng, H.; Luo, L.; Zhou, Q.; Luo, Z.; Peng, Y. Distinct effects of knocking down MEK1 and MEK2 on replication of herpes simplex virus type 2. *Virus Res.* **2010**, *150*, 22–27.
10.	Panteva, M.; Korkaya, H.; Jameel, S. Hepatitis viruses and the MAPK pathway: Is this a survival strategy? *Virus Res.* **2003**, *92*, 131–140.
11.	Luo, H.; Yanagawa, B.; Zhang, J.; Luo, Z.; Zhang, M.; Esfandiarei, M.; Carthy, C.; Wilson, J.E.; Yang, D.; McManus, B.M. Coxsackievirus B3 replication is reduced by inhibition of the extracellular signal-regulated kinase (ERK) signaling pathway. *J. Virol.* **2002**, *76*, 3365–3373.
12.	Murphy, L.O.; Blenis, J. MAPK signal specificity: The right place at the right time. *Trends Biochem. Sci.* **2006**, *31*, 268–275.

13. McCubrey, J.A.; Steelman, L.S.; Chappell, W.H.; Abrams, S.L.; Wong, E.W.; Chang, F.; Lehmann, B.; Terrian, D.M.; Milella, M.; Tafuri, A.; *et al.* Roles of the Raf/MEK/ERK pathway in cell growth, malignant transformation and drug resistance. *Biochim. Biophys. Acta* **2007**, *1773*, 1263–1284.

14. Zhang, W.; Liu, H.T. MAPK signal pathways in the regulation of cell proliferation in mammalian cells. *Cell Res.* **2002**, *12*, 9–18.

15. Kim, E.K.; Choi, E.J. Pathological roles of MAPK signaling pathways in human diseases. *Biochim. Biophys. Acta* **2010**, *1802*, 396–405.

16. Deschenes-Simard, X.; Kottakis, F.; Meloche, S.; Ferbeyre, G. ERKs in cancer: Friends or foes? *Cancer Res.* **2014**, *74*, 412–419.

17. Arthur, J.S.; Ley, S.C. Mitogen-activated protein kinases in innate immunity. *Nat. Rev. Immunol.* **2013**, *13*, 679–692.

18. Tanti, J.F.; Jager, J. Cellular mechanisms of insulin resistance: Role of stress-regulated serine kinases and insulin receptor substrates (IRS) serine phosphorylation. *Curr. Opin. Pharmacol.* **2009**, *9*, 753–762.

19. Chambard, J.C.; Lefloch, R.; Pouyssegur, J.; Lenormand, P. ERK implication in cell cycle regulation. *Biochim. Biophys. Acta* **2007**, *1773*, 1299–1310.

20. Boulton, T.G.; Nye, S.H.; Robbins, D.J.; Ip, N.Y.; Radziejewska, E.; Morgenbesser, S.D.; DePinho, R.A.; Panayotatos, N.; Cobb, M.H.; Yancopoulos, G.D. ERKs: A family of protein-serine/threonine kinases that are activated and tyrosine phosphorylated in response to insulin and NGF. *Cell* **1991**, *65*, 663–675.

21. Roskoski, R., Jr. ERK1/2 MAP kinases: structure, function, and regulation. *Pharmacol. Res.* **2012**, *66*, 105–143.

22. Lefloch, R.; Pouyssegur, J.; Lenormand, P. Total ERK1/2 activity regulates cell proliferation. *Cell Cycle* **2009**, *8*, 705–711.

23. Yoon, S.; Seger, R. The extracellular signal-regulated kinase: Multiple substrates regulate diverse cellular functions. *Growth Factors* **2006**, *24*, 21–44.

24. Krens, S.F.; He, S.; Lamers, G.E.; Meijer, A.H.; Bakkers, J.; Schmidt, T.; Spaink, H.P.; Snaar-Jagalska, B.E. Distinct functions for ERK1 and ERK2 in cell migration processes during zebrafish gastrulation. *Dev. Biol.* **2008**, *319*, 370–383.

25. Vantaggiato, C.; Formentini, I.; Bondanza, A.; Bonini, C.; Naldini, L.; Brambilla, R. ERK1 and ERK2 mitogen-activated protein kinases affect Ras-dependent cell signaling differentially. *J. Biol.* **2006**, *5*, e14.

26. Lloyd, A.C. Distinct functions for ERKs? *J. Biol.* **2006**, *5*, e13.

27. Cargnello, M.; Roux, P.P. Activation and function of the MAPKs and their substrates, the MAPK-activated protein kinases. *Microbiol. Mol. Biol. Rev.* **2011**, *75*, 50–83.

28. Wang, B.; Ding, L.X.; Deng, J.; Zhang, H.; Zhu, M.; Yi, T.; Liu, J.; Xu, P.; Lu, F.M.; Peng, Y.H. Replication of EV71 was suppressed by MEK1/2 inhibitor U0126. *Chin. J. Biochem. Mol. Biol.* **2010**, *26*, 538–545.

29. Dimitri, C.A.; Dowdle, W.; MacKeigan, J.P.; Blenis, J.; Murphy, L.O. Spatially separate docking sites on ERK2 regulate distinct signaling events *in vivo*. *Curr. Biol.* **2005**, *15*, 1319–1324.

30. Lu, J.; He, Y.Q.; Yi, L.N.; Zan, H.; Kung, H.F.; He, M.L. Viral kinetics of enterovirus 71 in human abdomyosarcoma cells. *World J. Gastroenterol.* **2011**, *17*, 4135–4142.

31. Rodriguez, M.E.; Brunetti, J.E.; Wachsman, M.B.; Scolaro, L.A.; Castilla, V. Raf/MEK/ERK pathway activation is required for Junin virus replication. *J. Gen. Virol.* **2014**, *95*, 799–805.

32. Pleschka, S. RNA viruses and the mitogenic Raf/MEK/ERK signal transduction cascade. *Biol. Chem.* **2008**, *389*, 1273–1282.

33. Yao, Y.; Li, W.; Wu, J.; Germann, U.A.; Su, M.S.; Kuida, K.; Boucher, D.M. Extracellular signal-regulated kinase 2 is necessary for mesoderm differentiation. *Proc. Natl. Acad. Sci. USA* **2003**, *100*, 12759–12764.

34. Pages, G.; Guerin, S.; Grall, D.; Bonino, F.; Smith, A.; Anjuere, F.; Auberger, P.; Pouyssegur, J. Defective thymocyte maturation in p44 MAP kinase (Erk 1) knockout mice. *Science* **1999**, *286*, 1374–1377.

35. Nekrasova, T.; Shive, C.; Gao, Y.; Kawamura, K.; Guardia, R.; Landreth, G.; Forsthuber, T.G. ERK1-deficient mice show normal T cell effector function and are highly susceptible to experimental autoimmune encephalomyelitis. *J. Immunol.* **2005**, *175*, 2374–2380.

36. Yu, C. Distinct roles for ERK1 and ERK2 in pathophysiology of CNS. *Front. Biol.* **2012**, *7*, 267–276.

37. Lefloch, R.; Pouyssegur, J.; Lenormand, P. Single and combined silencing of ERK1 and ERK2 reveals their positive contribution to growth signaling depending on their expression levels. *Mol. Cell Biol.* **2008**, *28*, 511–527.

Chapter 4:
Virus Replication Organelles

The Virus-Host Interplay: Biogenesis of +RNA Replication Complexes

Colleen R. Reid, Adriana M. Airo and Tom C. Hobman

Abstract: Positive-strand RNA (+RNA) viruses are an important group of human and animal pathogens that have significant global health and economic impacts. Notable members include West Nile virus, Dengue virus, Chikungunya, Severe acute respiratory syndrome (SARS) Coronavirus and enteroviruses of the *Picornaviridae* family.Unfortunately, prophylactic and therapeutic treatments against these pathogens are limited. +RNA viruses have limited coding capacity and thus rely extensively on host factors for successful infection and propagation. A common feature among these viruses is their ability to dramatically modify cellular membranes to serve as platforms for genome replication and assembly of new virions. These viral replication complexes (VRCs) serve two main functions: To increase replication efficiency by concentrating critical factors and to protect the viral genome from host anti-viral systems. This review summarizes current knowledge of critical host factors recruited to or demonstrated to be involved in the biogenesis and stabilization of +RNA virus VRCs.

Reprinted from *Viruses*. Cite as: Reid, C.R.; Airo, A.M.; Hobman, T.C. The Virus-Host Interplay: Biogenesis of +RNA Replication Complexes. *Viruses* **2015**, *7*, 4385–4413.

1. Introduction

Positive-sense RNA (+RNA) viruses including the Flaviviruses, enteroviruses of the *Picornaviridae* family, Alphaviruses, and Coronaviruses all dramatically modify cellular membranes to serve as platforms for replication and assembly of new virions. The biogenesis of these replication compartments is a complex interplay of interactions between virus and host proteins. Although considerable progress has been made in identifying host proteins that interact with virus-encoded proteins, much remains to be learned regarding the significance of these interactions. Despite morphological differences in the replication complexes formed by members of each viral family, these viruses have evolved to use common cellular pathways to complete biogenesis. Some of the shared pathways highlighted in this review include lipid metabolism, autophagy, signal transduction and proteins involved in intracellular trafficking (Table 1). Remarkably, even within the higher order of shared pathways, differences within members of specific families (such as *Flaviviridae*) exist, highlighting that the assembly and function of viral replication complexes (VRCs) varies considerably. As such, this review focuses on a broad view of host factors in which there is significant functional evidence linking them to VRCs in effort to highlight commonalities or differences and further advance the understanding of virus-host interactions.

2. *Flaviviridae*

The *Flaviviridae* family includes many significant global pathogens including Hepatitis C virus (HCV), West Nile virus (WNV), and Dengue virus (DENV). This family is comprised of four genera, with the human pathogens belonging to the genera *Flavivirus* and *Hepacivirus*. The *Hepacivirus* genus contains HCV, a prominent blood-borne human pathogen that causes chronic hepatitis and is estimated to have infected 170 million people worldwide. The *Flavivirus* genus includes DENV, WNV, Yellow Fever virus (YFV) and other viruses causing either haemorrhagic or encephalitic disease. Except for YFV and Japanese Encephalitis virus (JEV), vaccines for use in humans are not available against members of this family. Current treatment options are very limited and supportive care is often the only option. Arthropod vectors, mainly mosquitos and ticks are used by most flaviviruses to infect their hosts.

In general, virions are enveloped and contain a single copy of viral genomic RNA (~11 kilobase (kb)) encoding a single polyprotein that is cleaved by viral and host proteases into three structural and seven non-structural proteins [1]. After binding to cell surface receptors, the virions enter cells through endocytic pathways. Within the acidic environment of endosomes, the virions fuse with endosomal membrane resulting in release of the nucleocapsid into the cytoplasm. After the nucleocapsid disassembles, the viral RNA is translated into a polyprotein, which is then processed into individual viral proteins. VRCs form soon after and serve as platforms for RNA replication. Assembly of nascent virions occurs in close proximity to VRCs on the endoplasmic reticulum (ER). After budding into the ER, virions traverse the secretory pathway before release from the cell.

2.1. Genus Hepacivirus

The biogenesis of the HCV VRCs and the stabilization of these structures have been extensively studied [2,3]. Electron microscopic analysis of infected cells revealed that HCV replicates on altered ER membranes that are closely associated with lipid droplets; termed the "membranous web" [4]. The membrane-associated non-structural protein 4B (NS4B) plays a key role in the formation of this network [5], which consists of double membrane vesicles (DMVs) protruding out of ER. Of note, the DMVs are similar to ER-associated structures induced by members of *Picornaviridae* and *Coronaviridae* [6]. A plethora of host factors involved in lipid metabolism, intracellular signalling, protein folding, and vesicular trafficking are known to be important for HCV VRC activity. Due to the availability of extensive literature on the subject, they will not be discussed here. Instead, we refer readers to the following recent reviews [3,7,8].

2.2. Biogenesis of the Flavivirus Replication Complex

Of the studies investigating the membrane alterations induced by members of this genus, most have focused on DENV and WNV. Infection of mammalian cells with either the Australian attenuated strain WNV$_{KUN}$, or the highly pathogenic WNV$_{NY99}$ strain results in similar phenotypic disruptions of cellular membranes [9,10]. Early studies of cells infected with WNV or DENV revealed dramatic changes in cellular membranes and the formation of single membrane vesicular packets (VPs) and convoluted membranes (CM), which are in close association with smooth membranes and the rough-ER [9,11]. Paracrystalline arrays (PC) were also described in WNV$_{KUN}$-infected cells [9]. Infection of cells derived from the viral vector (mosquito) with DENV or WNV also led to dramatic alterations of membranes resulting in spherules associated with ER membranes [12,13]. These virus-induced structures are thought to segregate viral replication from protein translation [14]. VPs are the sites of viral replication as evidenced by the fact that they contain double-stranded RNA (dsRNA), a replication intermediate, and the viral RNA-dependent RNA polymerase, NS5 [9,11,15–17]. Two other virus-encoded non-structural proteins NS1 and NS3 also associate with these elements. VPs in DENV- and WNV-infected cells are ~85 nm in diameter indicating the conserved nature of these structures. CMs and PCs in WNV$_{KUN}$-infected cells are enriched for NS3/2b, the viral protease and do not contain dsRNA [9,15]. This suggests that CM/PCs may be the sites of viral translation and/or proteolytic processing of the viral proteins. The origins of these membranes vary between viruses. DENV-induced VP membranes contain the ER resident proteins protein disulphide isomerase and calnexin [15], whereas in cells infected with WNV$_{KUN}$, VPs that are positive for dsRNA, contain the *trans*-Golgi network protein, galactosyltransferase, possibly indicating that these structures are derived from Golgi membranes [18]. Moreover, the ER-Golgi intermediate compartment marker ERGIC53, is associated with CMs and PCs. WNV$_{NY99}$ is similar to DENV in that the VRCs colocalize with protein disulphide isomerase, suggesting these structures are ER-derived [10]. Electron tomography was utilized to further characterize the VRCs of DENV [15] (Figure 1B), WNV$_{KUN}$ [9,19], WNV$_{NY99}$ [20], and Tick-borne encephalitis (TBEV)-like virus Langat virus [21]. These "vesicles" in fact appear to be invaginations of the ER membrane with small neck-like openings (~11.2 nm for DENV) that may facilitate trafficking of molecules into and out of these replication sites. In some cases, there were connections between these vesicles within the modified-ER membrane. DENV VPs closely associated with budding sites appear as electron dense invaginations (~60 nm) and can be seen on opposing cisternae [15]. Despite there being good structural information on DENV and WNV VRCs, the exact mechanism by which these membranous organelles form, remains unclear. However, the ER localized non-structural viral proteins, WNV_{NY99} NS4B [20] and DENV-2 NS4A [22] are thought play a role in the initial membrane curvature.

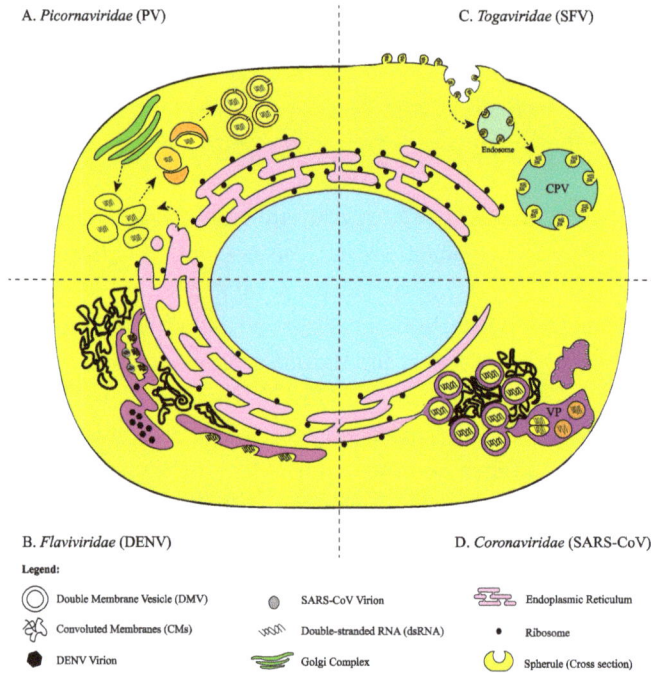

Figure 1. Biogenesis of Viral Replication Complexes (VRCs): Representative diagram of the structure and biogenesis of the VRCs for each family based on electron microscopy from the following references: Poliovirus (PV) [23] Semliki Forest virus (SFV) [24], Dengue virus (DENV) [15], Severe acute respiratory syndrome coronavirus (SARS-CoV) [25]. Diagram not to scale. (**A**) Formation of the PV VRC: Early in infection single membrane vesicles that contain dsRNA are derived from the ER and Golgi components. Progression of infection results in vesicles wrapping around each other inducing the formation of DMVs; (**B**) Formation of the DENV VRC: Spherule structures containing dsRNA form on modified rough-ER membranes surrounded by convoluted membranes. Assembly sites form on opposing cisternae where newly formed virions are stored; (**C**) Formation of SFV VRC: Spherules that contain dsRNA form at the plasma membrane. Internalization of these structures follows the endo-lysosomal pathway resulting in formation of cytopathic vacuoles (CPV); (**D**) Proposed formation of the SARS-CoV VRC: DMVs containing dsRNA are formed and their outer membranes are continuous with the ER. Convoluted membranes surround the DMVs. Formation of vesicle packets (VPs) is thought to result from DMV fusion. Newly formed virions are associated with these structures.

Following this step, other viral and host factors are likely required for the biogenesis and stabilization of the flavivirus VRC. To date, a large number of host proteins involved in flavivirus replication have been identified by proteomic and transcriptomic studies of infected cells [26–28], mapping the host cell interactome of viral proteins [29–31] and through systematic RNA interference (RNAi) screens [32,33]. Perhaps not surprisingly, common host pathways that affect flavivirus replication include those involved in lipid metabolism, signal transduction, and cell structure. While many host factors that are thought to play a role in virus replication have been identified for WNV and DENV, the corresponding functional and validation studies are comparatively limited. As such, in this review, we have focused mainly on host factors in which there are significant functional data linking them to VRCs.

2.3. Potential Role for Autophagy in Flavivirus VRC Biogenesis

Autophagy is a homeostatic process involving the formation of double membrane vesicles from the ER that fuse with lysosomes and degrade cellular material. Recently, it has been linked to VRC biogenesis for multiple viruses, including the *Picornaviridae* and *Coronaviridae* family (covered later in this review). The requirement for autophagy in flavivirus VRCs varies significantly. WNV propagation for example, is not affected by induction or repression of autophagy [34], nor is autophagy required for biogenesis of VRCs from the ER [35]. This is in contrast with DENV and JEV, which both exploit autophagy for virus propagation [36,37]. It has been proposed that DENV uses autophagy-induction to aid in release of fatty acids from lipid droplets, increase β-oxidation and ATP production [38]. Little is known about the role of autophagy in VRC biogenesis or replication of other members of the *Flavivirus* genus.

2.4. Membrane Remodelling and Lipid Metabolism

Biogenesis of VRCs requires massive expansion of ER-associated membranes and alteration of their lipid compositions. HCV and members of the *Picornaviridae* family are known to recruit phosphatidyl-inositol kinases (PI4Ks) to their VRC networks for the conversion of phosphatidylinositol (PI) to phosphatidylinositol 4-phosphate (PI4P) lipids, a process that is essential for viral replication [39,40]. PI4P lipids may serve to recruit host proteins and/or lipid components to these organelles. Interestingly, neither WNV nor DENV seem to require PI4P lipids [41,42] indicating that assembly and function of VRCs varies considerably among the *Flaviviridae* family. However, DENV infection does alter the membrane lipid composition of human cells and two host cell enzymes involved in fatty acid metabolism, fatty acid synthase (FAS) and Acetyl-CoA carboxylase 1 (ACACA) are both important for replication [42].

FAS is recruited to the VRC through interaction with the viral protease NS3, where it upregulates the formation of fatty acids from acetyl-CoA. As the length of fatty acid chains can affect membrane curvature, this process may be important for the formation of the VRC [43].

Newly synthesized fatty acids are incorporated into DENV VRCs and pharmacological inhibition of FAS by cerulenin or C75 negatively affects DENV replication in mammalian [42] as well as mosquito cells [44]. In one current model, the viral protein NS4A initially induces membrane curvature followed by the recruitment of FAS by NS2B/NS3 to the VRC resulting in local production of fatty acids and the expansion of the ER membrane in a more fluid state [42]. Similar to DENV, WNV requires FAS activity for replication [41]. WNV$_{NY99}$ infection also increases the intracellular concentration of sphingolipids and glycerophospholipids, a process that affects the make up of virion envelopes [45]. Virus assembly and release is also dependent on lipid biosynthesis, particularly ceramide. The role of glycerophospholipids such as phosphatidylcholine (PtdCho) in flavivirus replication is less clear, but there is evidence that PtdCho is part of the VRCs and is incorporated into the lipid bilayers of nascent virions [45].

The level of cholesterol also modulates the curvature and plasticity of membranes [46]; a process that is controlled by ER-localized transcription factor sensors, sterol-regulatory element binding proteins (SREBPs). When cholesterol levels are low, SREBP is released from the ER and enters the nucleus where it activates transcription of the genes for FAS, 3-hydroxy-methylglutaryl-CoA reductase (HMGCR), and/or low-density lipoprotein receptor (LDLR). HMGCR catalyzes the rate-limiting step in synthesis of mevalonate, a precursor for cholesterol biosynthesis, whereas LDLR is a cell surface receptor that binds and internalizes cholesterol-containing complexes [47,48]. It is well documented that flaviviruses modulate cholesterol levels in infected cells. During WNV$_{KUN}$ infection, total cholesterol levels rise and this is correlated with upregulation and association of HMGCR with virus-induced membrane structures [49]. This may indicate HMGCR aids VRC formation by producing cholesterol at these sites.

Interestingly, elevated cholesterol (total) levels were not observed in DENV-infected cells [50], even though LDLR transcripts, an indicator of elevated ER cholesterol, were increased. However, inhibition of HMG-CoA reduced replication of DENV replicons indicating cholesterol is important for replication. Another host factor involved in cholesterol biosynthesis, mevalonate diphosphate decarboxylase (MVD), was shown to be important for DENV replication [50]. Thus cholesterol seems to be a key lipid component of the flavivirus VRC and while total cellular cholesterol may not increase in all flavivirus infected cells, this membrane is targeted to the ER membrane where VRCs are produced. Future studies with JEV, YFV, TBEV, and St. Louis Encephalitis Virus (SLEV), are needed to determine how they may alter membrane composition in favour of VRC biogenesis.

2.5. Stabilization and Scaffolding Proteins at the Flavivirus VRC

Microfilaments, microtubules and intermediate filaments are cytoskeletal components essential for cell shape and motility as well as a myriad of other functions including intracellular trafficking, cell division and cell signalling. Identification of host proteins involved in actin polymerization and vesicular trafficking were shown to be important for DENV and WNV replication [32,42], however, comparatively little is known about how this affects VRC formation or function. Conceivably, changes to the cellular structural framework could aid in biogenesis and/or stabilization of newly formed VRCs. Reorganization of the intermediate filament component vimentin occurs after phosphorylation by calcium/calmodulin-dependent protein kinase II; an event that is necessary for productive DENV replication [51]. Moreover, knockdown of vimentin alters the distribution of the VRCs in host cells, indicating that this protein may function in scaffolding/stabilization of these structures. NS4A interactions with vimentin may be the link between intermediate filaments and VRCs [51]. Finally, Stathmin 1 (STMN1), a microtubule destabilizing protein, is another host factor linked to biogenesis of VRCs [52]. DENV infection upregulates STMN1 by reducing levels of miR-223, a microRNA that normally targets the mRNA for STMN1.

3. Picornaviridae

The *Picornaviridae* family contains many important human and animal pathogens. Prior to the development of a vaccine, poliovirus (PV) crippled hundreds of thousands of people per year, primarily children. The World Health Organization global PV eradication program started in 1988 has not yet been successful in fully eradicating the virus. Other prominent members of this family include Coxsackie virus (CV), human rhinoviruses (HRV), and the causative agent of hand-foot-and-mouth disease Enterovirus 71 (EV71). Unlike PV, effective vaccines against these pathogens have yet to be developed. Infection by CV, HRV, and EV71 cause a variety of illnesses in humans from self-limiting colds to more serious presentations of encephalitis, myocarditis, and paralysis. Children, elderly, and the immuno-compromised individuals are at highest risk for severe disease.

The majority of research has focused on members of the *Enterovirus* genus including PV, CV, and HRV, and as such our focus will be on host factors linked to formation and stabilization of their VRCs. As with all +RNA viruses, enteroviruses extensively rearrange cellular membranes to facilitate virus replication and assembly. Following entry of the virion, the 5'capped genomic RNA (~7.5 kb) is unpackaged

307

after which translation is initiated from an internal ribosome sequence. The genomic RNA encodes a single large polyprotein that is proteolytically processed into four structural proteins that form the virion, and seven non-structural proteins that function in replication and subverting the host-cellular immune system.

3.1. Biogenesis of the Enterovirus VRC

Early electron microscopy studies of PV-infected cells by Dales and Palade revealed drastic remodelling of the cell cytoplasm [53]. At 5 hours post-infection (hpi), membrane-enclosed bodies were observed in the perinuclear zone. At the peak of viral translation (2.5 hpi), nascent VRCs were not observed but rather, formed later during RNA replication [54]. Isolation of PV VRCs revealed that in addition to dsRNA, a replication intermediate, proteins encoded by the P2 genomic region, specifically 2C containing non-structural proteins, were bound to these membranes [55,56]. More recently, electron tomographic studies were used to examine the biogenesis of these structures in more detail [23]. Early in infection (~2 hpi), 100–200 nm single membrane tubular structures that are involved in RNA synthesis form followed by clustering and bending of these structures at 4 hpi. Later, DMVs, which can be larger (100–300 nm), form through a membrane wrapping process (Figure 1A). These structures evolve from *cis*-Golgi membrane and arise from positive membrane curvature or budding [23]. Despite a tremendous amount of experimental investigation, the origins of the enteroviral VRC remain controversial [57].

3.2. Membrane Remodelling during Enteroviral Infection

Earlier studies suggested that PV VRCs are derived from multiple membrane sources, including lysosome, ER, and Golgi, but do not fully resemble their parent membrane sources [58,59]. Multiple hypotheses exist for how these structures arise during PV infection, including through autophagy [58]. Expression of the PV proteins 2BC and 3A results in the formation of DMVs [59] that colocalize with lysosomal-associated membrane protein 1 (LAMP-1) and LC3-phosphatidylethanolamine conjugate (LC3-II), indicative of autophagic vesicles, early in PV infection [60]. Data consistent with the PV studies were observed with the related enteroviruses HRV-2 and HRV-14 and inhibition of autophagy decreased the amount of intracellular and extracellular virus produced [60–63]. Of note, CVB3 infection induces autophagy in a mouse model *in vivo*, indicated by increased LC3-II [61]. Despite this evidence, the role of autophagy remains controversial. Recently, it was shown that PV vesicles that stained for dsRNA did not colocalize with autophagic marker LC3 early (3 hpi) in infection [64]. However, at 5 hpi LC3 was detected by immuno-electron microscopy in association with dsRNA. In light of the seemingly discrepant data, it has been postulated that autophagy is important for late steps in infection (3 hpi) [64].

COPII-coated vesicles, which are involved in transport of cargo from the ER, have been linked to biogenesis of enterovirus VRCs. During early infection, enteroviruses disrupt anterograde transport and reroute these vesicles to sites of viral replication. PV infection or expression of protein 3A alone has been shown to block ER-Golgi protein transport [65,66]. Moreover, the movement of VRCs is dependent on microtubules leading from the ER to the microtubule organizing center in the Golgi region of the cell [67]. PV proteins 2B and its precursor 2BC colocalize with the COPII component Sec31 [68] and PV infection enhances COPII vesicle budding [69]. However, the effect is transient and is not observed late in infection. In contrast, HRV-1A and -16 have been reported to cause fragmentation of the Golgi without blocking protein secretion [70,71]. Furthermore, recent evidence citing lack of colocalization between dsRNA and the COPII component Sec31 has been interpreted to mean that PV VRC formation is not dependent on COPII [64]. One potential mechanism to account for this discrepancy is that COPII aids in the formation of an intermediary compartment from which nascent VRCs bud. However, attenuation of COPII vesicle formation did not interfere with PV infection suggesting that this budding mechanism is not absolutely required [72]. Clearly, more research is required to fully understand the role of COPII in picornavirus VRC biogenesis and function.

PV infection is sensitive to brefeldin A (BFA), a drug that inhibits activation of ADP-ribosylating factor GTPases (Arfs), which are necessary for formation of COPI vesicles [73,74]. Arfs cycle between a GDP (inactive) and a GTP (active) bound state mediated by guanine-nucleotide exchange factors (GEFs). When bound to GTP, Arfs remodel intracellular membranes to promote COPI dependent budding [75]. COPI mediates budding of vesicles from the Golgi that retrograde traffic to the ER and was identified as a host factor that is required for replication of *Drosophila* C virus, a picorna-like insect virus [72]. Reducing expression of α-COP, a COPI component, was later found to reduce PV infection [72]. Expression of PV 3A or 3CD promotes the association of Arf3 and Arf5 with membranes where viral RNA replication occurs, and this association can be blocked by BFA [76]. Two other GEFs, BIG1 and BIG2 are recruited to VRCs by expression of PV 3CD which then leads to the activation of Arf [77]. This indicates that Arf activation may induce vesicle formation and VRC biogenesis (reviewed in [78]). GBF1, yet another GEF, is also a target of BFA and is the main activator of Arf during PV infection [79]. PV 3A binds GBF1 and recruits it to virus-induced vesicles [77]. While VRCs can still form in the presence of BFA, they are unable to recruit Arf1. This may result in formation of defective VRCs or mislocalization of their contents thereby reducing viral replication and assembly [79]. Although BFA targets BIG1, BIG2, and GBF1, only GBF1 is required for CVB3 replication [80].

More recent studies examined the localization of Arf1 and GBF1 throughout CVB3 and PV infections [39]. Arf and GBF1 colocalize with viral RNA and the viral RNA polymerase, indicating their relocalization to VRCs during infection. Moreover, Arf1 knockdown negatively impacts virus replication [39]. However, since Arf1 interacts strongly with GBF1 and is found in COPI vesicles, it cannot be ruled out that recruitment of Arf1 to the VRC is a consequence of GBF1 recruitment. β-COP, a COPI component, does not colocalize with dsRNA during PV infection and thus, the COPI coat itself may not be involved budding and biogenesis of VRCs [64]. Recruitment of Arf1/GBF1 to VRCs appears to differ among enteroviruses. Expression of CVB3 3A induces the recruitment of GBF1 to membranes, whereas the homologous proteins of HRV-2 or -14 do not [81,82]. These viral proteins also function in recruitment of Phosphatidylinositol-4-OH kinase type III beta (PI4KIIIβ) to VRCs and this is discussed in further detail below.

3.3. Lipid Metabolism

Biogenesis of picornavirus VRCs also requires synthesis and trafficking of specific lipids to membrane organelles. Unlike enveloped viruses such as flaviviruses and togaviruses (also covered in this review) whose replication compartments are formed by invagination into membranes, picornaviruses induce protrusion of cellular membranes to form convoluted tubular-like structures [23,83]. Alterations in the lipid composition of these membranes are needed to allow appropriate curvature and the expansion of membranes that eventually form the VRCs. Early evidence of altered lipid metabolism came from the observation that PV increases PtdCho levels in the cells by upregulating the rate-limiting enzyme phosphocholine cytidylyltransferase [84]. PtdCho is a main component of lipid bilayers at the ER and Golgi network and increased levels of this phospholipid would enable proliferation of membranes (reviewed in [85]). This may indicate that formation of VRCs involves *de novo* lipid synthesis. The role of fatty acids was first reported when the addition of cerulenin, an inhibitor of the enzyme FAS, resulted in a block of PV replication but did not affect viral RNA translation or proteolytic processing [86]. Blocking fatty acid synthesis by inhibiting FAS also reduced proliferation of VRC membranes. Later, the same group reported that specific fatty acids are important for PV replication as evidenced by the observation that incorporation of oleic acid into cellular membranes made them incapable of supporting PV replication [87]. More recently the role of FAS for CVB3 replication has also been demonstrated. FAS upregulation was first observed during a proteomic screen of CVB3 infected cells [88]. FAS protein production is upregulated as early as one hour post CVB3 infection and does not require viral replication, suggesting that FAS gene transcription and translation may be upregulated following signalling cascades induced by CVB3 virions binding [89]. Components of the fatty acid biosynthesis pathway including SREBP and the protein product of the gene it directly regulates, *CG3523* encoding FAS are also required for picorna-like virus *Drosophila C* virus replication, suggesting that FAS may be a common host factor exploited by viruses to alter membrane lipid metabolism [72]. In addition to altering metabolism in cells, picornaviruses may increase uptake of lipids from the extracellular environment. PV infection for example, enhances import of long-chain fatty acids into cells and the viral protein 2A is involved in the initiation of this process [90]. Normally, fatty acids are trafficked and stored in lipid droplets, but in infected cells they colocalize with VRCs. Moreover, activity of long chain acyl-coenzyme A (Acyl-CoA) synthetase Acsl3, involved in the synthesis of PtdCho, was upregulated at 2 hpi, thus further supporting the notion that VRCs are formed from newly synthesized lipids.

Because Arf1 and GBF1 are recruited to VRC membranes, it was thought that Arf1 effectors might also be important for enterovirus replication. Lipid modifying enzymes including (PI4Ks) are downstream effectors of Arf. PI4KIIIβ is normally associated with the Golgi and is involved in the production of (PI4P) [91]. The PI4KIIIβ inhibitor enviroxime exhibits potent antiviral activity against enteroviruses *in vitro*, however in clinical trials, its efficacy was limited [92,93]. Other PI4KIIIβ inhibitors, including GW5074 and BF738735, also efficiently inhibit enteroviral replication *in vitro* and *in vivo* in mice, however in some mice strains inhibition of PI4KIIIβ resulted in harmful side effects thereby limiting the likely therapeutic benefit of this strategy [92,94,95]. PI4KIIIβ colocalizes with Arf1 at replication complexes during CVB3 infection, while other Arf1 effectors, such as COPI components, are lost from these sites [39]. Recruitment of PI4KIIIβ to Arf1-positive membranes can be stimulated by expression of CVB3 3A alone. During infection by PV or CVB3, PI4P levels increase 5-fold and pharmacological inhibition of PI4KIIIβ activity by PIK93, which inhibits PI4P production, also reduces viral replication. Furthermore, PI4KIIIβ interacts with CVB3 3D^{pol}, which strongly interacts with PI4P-containing membranes and thus their production may facilitate the organization and/or association of viral proteins in the VRC [39]. This kinase also interacts with PV 3A and Acyl-CoA binding domain containing 3 (ACBD3) protein [96]. Recruitment of PI4KIIIβ to the VRC seems to be conserved among enteroviruses but the interactions between this kinase and viral as well as other host proteins remain to be fully elucidated.

The production of PI4P lipids, by PI4KIIIβ at replication sites may also be important for recruitment of cholesterol. Both PI4P and cholesterol are enriched at VRCs of CVB3, PV, and HRV-2 [94]. Recently, it was suggested that cholesterol is shuttled from the endosome to the VRC where it colocalizes with CVB3 3A [97] suggesting that the virus re-routes pre-existing pools of cholesterol to VRCs. Moreover, it has also been demonstrated that uptake of extracellular cholesterol by clathrin-mediated endocytosis (CME) is essential for PV and CVB3 replication [98]. After uptake, cholesterol is targeted to recycling endosomes, which then fuse with existing VRCs. Depletion of CME components results in a trafficking of cholesterol from the plasma membrane to lipid droplets, reducing VRC formation [98]. Incorporation of cholesterol into VRCs, which imparts rigidity to membranes, may be important for their curvature and stabilization.

4. *Coronaviridae*

Members of the four genera in the family *Coronaviridae* have enveloped virions that contain very large +RNA, capped and polyadenylated RNA genomes of 26–32 kb. These viruses infect a wide range of mammals and birds causing upper respiratory, gastrointestinal, hepatic, or central nervous system diseases [99]. Members belonging to the genus *Betacoronavirus* include the important human pathogens Severe acute respiratory syndrome coronavirus (SARS-CoV) and Middle East respiratory syndrome coronavirus (MERS-CoV). Prior to the SARS outbreak in 2003, the majority of coronavirus (CoV) research focused on mouse hepatitis virus (MHV) as a model.

Following virion entry into host cells, the +RNA genome is released into the cytoplasm. Coronaviruses employ a rather complex program of gene expression. The *ORF1* encodes the replicase required for transcription of the full-length (genomic) minus-strand template and subgenomic (discontinuous transcription) minus-strand synthesis. Synthesis and processing of the genome results in production of up to 16 nonstructural proteins, of which the predicted multi-spanning membrane proteins nsp3 [100], nsp4 [101] and nsp6 [102] are believed to be involved in biogenesis and stability of the coronavirus replication/transcription complex (RTC). The study of host factors required for biogenesis and stabilization of the coronavirus RTC is a growing field of interest. The involvement of lipid rafts for virus-entry and cell-cell fusion was demonstrated for MHV [103], however, there is a lack of information on how CoVs modulate host lipid composition as previously shown for many +RNA viruses [104]. A recent kinome screen (using small interfering RNAs) has provided a glimpse of the complexity of pro-viral and anti-viral host factors involved at the SARS CoV-cell interplay, including proteins involved in lipid metabolism [105]. Future studies addressing the interplay between CoVs and lipids and the effects on viral replication would also be of considerable interest.

4.1. Coronavirus Replication/Transcription Complex (RTC)

Similar to some of the viruses described above, formation of DMVs is observed during CoV infection in mammalian cells [106]. Early in the SARS-CoV infection process, DMVs, ranging in size from 150 to 300 nm, are distributed throughout the cytoplasm [25]. The ORF1a-encoded multi-spanning transmembrane proteins, nsp3, nsp4 and nsp6 are thought to form the scaffold that facilitates DMV formation and anchors the RTC to intracellular membranes [101,107–109]. The RTC is likely formed through a complex network of interactions involving all 16 CoV non-structural proteins [108]. DsRNA can be detected in the interior of DMVs (Figure 1D) [25], suggesting that these structures serve as sites of RNA replication. Indeed this is supported by the observation that RTC activity correlates with the number of DMVs [110]. Electron tomographic analysis revealed that DMVs are not isolated vesicles but rather, an interconnected network of membranes continuous with the rough-ER [25]. Late in infection, DMVs become concentrated in the perinuclear area and CMs of 0.2 to 2 µm in diameter form in close proximity to the DMV clusters [25]. Compared to DMVs, CMs are highly enriched in SARS-CoV nonstructural proteins that include the replicase proteins [25]; leading to the notion that the active replicase complex is localized to the CM and "dead" dsRNA molecules are found in the DMVs, perhaps as a way to evade immune recognition. Electron microscopy studies revealed that similar membrane formations occur in MERS-CoV-infected cells [111]. The outer membranes of DMVs are thought to fuse together and transition into vesicle packets (VPs) late in infection. VPs are large (1–5 µm) membrane structures where virus budding occurs [25,112] (Figure 1D).

Unexpectedly, electron tomographic analyses failed to reveal connections between the interior of DMVs and the cytosol [25]. This suggests that both membranes of CoV DMVs are sealed raising the question of how the import of metabolites and export of viral RNA occurs from these structures. In contrast, small neck-like openings are clearly discernable in DMVs induced by HCV [6]. In this case, it has been speculated that replication takes place in DMVs as long as the connection to the cytosol is maintained. The question remains as to whether CoVs make use of transport molecules as a means to regulate transport of products in and out of DMVs.

Table 1. Identified cellular interacting proteins with viral replication complexes of +RNA viruses.

	Lipids and Membrane Remodeling	Cellular Trafficking and Signaling Proteins
Flaviviridae (*Flaviviruses*)	Sphingolipids (WNV$_{NY99}$ [45])	Actin polymerization (WNV [32], DENV [42])
	Glycerophospholipids (WNV$_{NY99}$ [45])	Vimentin (DENV [51])
	FAS (DENV [42], WNV$_{NY99}$ [41])	STMN1 (DENV [52])
	ACACA (DENV [42])	
	Cholesterol (DENV [50], WNV$_{KUN}$ [49])	
Picornaviridae	FAS (PV [86,87], CVB3 [88,89])	Arf (CVB3, PV) [39]
	Long chain fatty acids (PV [90])	GBF1 (CVB3, PV) [39]
	PtdCho (PV [84])	
	PI4KIIIβ (CVB3 [39], PV [96])	
	ACBD3 (CVB3, PV) [96]	
	Cholesterol (CVB3, PV) [98]	
Coronaviridae		PDI (SARS-CoV [107])
		Sec61α (SARS-CoV [110])
		EDEM1 (MHV [113])
		OS-9 (MHV [113])
Togaviridae (*Alphaviruses*)	Cholesterol and sphingomyelin lipids (SINV [114])	Vimentin (SINV [115])
		PI3K (SFV [116])
		Amphiphysins (SFV, SINV, CHIKV) [117]

Abbreviations: DENV, Dengue Virus; WNV, West Nile Virus; YFV, Yellow Fever Virus; JEV, Japanese Encephalitis Virus; PV, Poliovirus; CVB3, Coxsackievirus B3; HRV-14, Human Rhinovirus 14; EV71, Enterovirus 71; IBV, Infectious Bronchitis virus; SARS-CoV, Severe Acute Respiratory Syndrome coronavirus; MHV, Mouse Hepatitis Virus; SINV, Sindbis Virus; SFV, Semliki Forest Virus; CHIKV, Chikungunya Virus.

4.2. Potential Role of Autophagy in DMV Formation

Morphological similarities between CoV DMVs and autophagosomes and the co-localization between specific CoV replicase proteins (nsp8, nsp2, nsp3) with microtubule-associated protein Light chain 3 (LC3), a protein marker for autophagic vacuoles [118], are consistent with autophagy playing a role in DMV formation. Moreover, during MHV infection of murine cells lacking the ATG5 gene which functions in the early stages of autophagosome formation [119], no DMVs formed and virus replication was impaired [120]. Replication was restored by expression of the Atg5 protein further supporting the role of autophagy for formation of DMVs, at least in embryonic stem cells. However, another study involving MHV infection of bone marrow-derived macrophages or embryonic fibroblasts concluded that neither Atg5 nor an intact autophagic pathway is required for viral replication [121]. Other morphological studies also found no evidence for autophagy in DMV formation; specifically a lack of co-localization between the autophagy marker LC3 and the SARS-CoV replication complex [107]. Moreover, MHV replication was unaffected in autophagy-deficient cells although depletion of LC3 severely affected CoV replication [113]. Interestingly, MHV replicative structures are decorated with LC3 [106], generally regarded to as the non-functional precursor to the lipidated autophagosome marker LC3-II [122]. The involvement of autophagy and LC3 in DMV formation was further clarified when LC3 was shown to colocalize with MHV proteins nsp2/nsp3, dsRNA, and ER-associated degradation (ERAD) vesicle markers ER degradation-enhancing alpha-mannosidase-like 1 (EDEM1), Osteosarcoma amplified 9 (OS-9) in embryonic fibroblasts [113]. Down-regulation of LC3 inhibits MHV replication and virion production [113] whereas knocking out autophagy has no effect. Inhibition of MHV replication and virion production was attributed to a defect in DMV biogenesis, which negatively impacts non-structural protein production. These results suggest that LC3 and the ERAD pathway are necessary for DMV formation and the biogenesis of RTCs required for a productive infection. Finally, quantitative proteomics analysis revealed that SARS-CoV infection significantly upregulates BCL2-associated athanogene 3 (BAG3), a protein linked to regulation of the autophagy pathway [123]. Inhibition of BAG3 expression by RNA interference results in significantly reduced replication of SARS-CoV.

Unfortunately, conflicting data make it difficult to derive a definitive conclusion regarding the role of autophagy in CoV RTC formation. Some of these differences may be the result of using different cell lines and different CoVs. What is clear though is that proteins with known roles in autophagy are involved in CoV replication; however, the process of autophagy *per se* may not be functionally relevant to the formation of CoV DMVs.

4.3. The Secretory Pathway and CoV Replication

A number of studies indicate that the ER is involved in the biogenesis of the SARS-Co-V-induced reticulovesicular network (RVN), a membrane compartment involved in virus replication. Specifically, partial co-localization of CoV replicase proteins with the ER resident protein disulfide isomerase (PDI) [107] and the observation that the ER translocon subunit Sec61α redistributes to replicative structures during SARS-CoV infection support this idea [110]. In addition EDEMosome cargo proteins EDEM1 and OS-9, two proteins involved in ER quality control and ER associated degradation (ERAD), associate with CoV replicative structures [113]. However, many protein trafficking and membrane fusion proteins that function downstream of the ER in the early secretory pathway such as Sec13, syntaxin 5, GBF1, and Arf1 have not been detected at the RVN [110]. The involvement of the COPI complex was investigated using the drug BFA, which as mentioned above, blocks COPI-mediated vesicular transport at the ER-Golgi interface. When added to cells early in infection, BFA inhibits RVN formation and decreases, but does not completely abolish viral RNA synthesis [110]. Although the precise role of COPI is unknown, other positive RNA viruses discussed above such as PV and *Drosophila* C virus require the COPI-mediated vesicular transport for replication. This supports the notion of COPI as a common host factor required in viral infection.

5. *Togaviridae*

Togavirus virions are enveloped, spherical particles (50–70 nm in diameter) that contain a single-strand +RNA with a 5-cap and 3′ poly A-tail [124]. The family includes the genera *Rubivirus* and *Alphavirus*. Rubella virus (RUBV) is the sole member in the *Rubivirus* genus and is the causative agent of Rubella (also known as German Measles). The *Alphavirus* genus contains at least 30 members that are separated into New World and Old World viruses. The New World viruses include Venezuelan equine encephalitis (VEEV), Western equine encephalitis (WEEV) and Eastern equine encephalitis virus (EEEV). Old World alphaviruses evolved separately [125] and members include Semliki Forest virus (SFV), Sindbis virus (SINV), Chikungunya virus (CHIKV) and O'nyong'nyong virus. Transmission occurs mainly through mosquito vectors and human infections are often associated with fever, rash, severe joint pain (arthralgia) and stiffness that can last weeks to months in duration. Some pathogens in this group can cause much more severe illnesses including encephalitis in humans and animals.

Togavirus replication complexes originate from late endosomes and lysosomes and are morphologically similar for RUBV [126,127] and alphavirus-infected cells [128]. Alphavirus VRC biogenesis is comparatively well characterized and as such, we will focus on these structures. Most host factors that interact with alphavirus replication complexes were identified through pull down assays with alphavirus non-structural proteins. In this section we focused on host factors partners that have been speculated or clearly demonstrated to play functional roles in RC biogenesis or function. For a list of additional interacting partners that are not discussed here, readers are referred to the following articles: [129–131].

5.1. Alphavirus Replication

Shortly after alphaviruses infect host cells, small single-membrane bulb-shaped invaginations (~50 nm diameter) called "spherules" form on the external surface of the plasma membrane [132]. The spherules, which are associated with viral nonstructural proteins (nsPs) and dsRNA, each contain a neck-like opening to the cytoplasm (5–10 nm in diameter) that permits exchange of metabolites and export of nascent viral RNA [128]. The fact that dsRNA can be detected inside the spherules and the presence of partially processed non-structural proteins (P123 and nsP4) on the spherule necks [132], suggests that these structures are the sites of viral RNA synthesis [116]. Internalization of spherules by endo-lysosomal membranes gives rise to type 1 cytopathic vacuoles (CPV1), which are 600–2000 nm in diameter [24] (Figure 1C). The endosomal origin of CPVs is confirmed by the observation that these structures are often positive for both endosomal and lysosomal markers [128]. As infection progresses, the non-structural polyprotein precursors are further processed to yield individual non-structural proteins and negative-strand synthesis is inactivated. The fully processed non-structural proteins together form the mature replicase [115,133,134], which is required for efficient synthesis of positive-sense genomic and subgenomic RNA. Spherules are devoid of ribosomes and virus capsid proteins but these structures/proteins are frequently found juxtaposed to the spherule openings [128], suggesting that the site of translation is in close proximity to replication sites.

5.2. Membrane Lipids

The nsP1 protein of alphaviruses, which is required for 5′ capping of viral RNAs [135], is involved in attachment of the replication complex to membranes [136] via a highly conserved amphipathic helix [137]. When expressed in the absence of other viral proteins, nsP1 is targeted to the inner surface of the plasma membrane but this not sufficient for cytoplasmic vacuole formation [138]. NsP1 is modified by acylation; however, the significance of this process in its function has yet to be determined [139]. Cholesterol and sphingomyelin in the plasma membrane are important for alphavirus fusion [140–142] and budding [143,144]. Importance of the latter is evidenced by the observation that SINV infection of Niemann-Pick disease-A fibroblasts (NPAF), which cannot degrade sphingomyelin, results in reduced levels of genomic RNA as well as an altered ratio of subgenomic-to-genomic RNA. The authors suggest that due to the build-up of cholesterol and sphingolipids in late endosomes/lysosomes, biogenesis of replication complexes are negatively affected [114]. Interestingly, the alphavirus virions produced in NPAFs were 26 times more infectious than those produced in normal human fibroblasts; resulting in increased titers and cell death. This suggests that cellular production of less infectious virus may be a consequence of host restriction on virus replication.

5.3. Membrane Trafficking Proteins

The study of how alphaviruses enter and exit mammalian cells led to a number of fundamental discoveries about membrane trafficking. Therefore, it is somewhat surprising that comparatively little is known about how membrane trafficking components affect replication. However, it does appear that cytoskeletal elements are involved. Infection of cells with a recombinant SINV encoding GFP-tagged nsP3, followed by anti-GFP pull-downs revealed that this viral protein associates with cytoskeletal proteins, chaperones, elongation factor 1A, and heterogeneous nuclear ribonucleoproteins [115]. Others reported that nsP3 also binds actin, tubulin and myosin [145] [144], vimentin (an intermediate filament protein) [115], and the cytosolic molecular chaperone Heat shock cognate protein 70 (Hsc70) [115,146]. These interaction data suggest that the SINV RCs associate with the cytoskeleton. The concern that the highly abundant cytoskeletal proteins were mere contaminants in the nsP3-binding studies [130] was partially assuaged by imaging studies showing that nsP3 associates with vimentin patches [115]. Imaging studies also provided evidence for Hsc70-ns3p having a role in alphavirus RC formation and/or function [115,146]. Some of the many functions of Hsc70 (reviewed in [147]) are to target proteins for degradation, regulate the translocation of proteins into different cellular organelles such as ER and mitochondria, and regulate apoptosis. Hsc70 has been linked to the replication of many viruses [148] but whether or not this is largely a reflection of its general role as a chaperone has yet to be determined.

NsP3 proteins of SFV, SINV, and CHIKV have also been shown to interact in an SH3 domain-dependent manner with amphiphysin-1 and Bin1/amphiphysin-2, both of which are involved in endocytosis and membrane trafficking [117]. The re-localization of amphiphysins to alphavirus RCs promotes replication, further solidifying the role of these host proteins in formation or stabilization of the replication sites. Finally, recent data suggest a role for phosphatidylinositol 3-kinases (PI3Ks) in alphavirus RC formation. Specifically, the activity of these kinases is required for the initial internalization of spherules at the plasma membrane as well as their subsequent trafficking on microtubule and actin networks [116].

5.4. Ras GTPase-Activating Protein-Binding Proteins

The RNA-binding proteins Ras GTPase-activating protein-binding protein (G3BP)1 and G3BP2 are structural components of stress granules: large cytoplasmic ribonucleoprotein complexes that function in regulating translation. Several studies have identified interactions between G3BP proteins and alphavirus nsP2 and nsP3 [115,129,130,146]. To differentiate between host proteins associated with replication complexes from those that interact with individual nsP2 or nsP3 proteins, Varjak *et al.* used small dextran-covered magnetic beads that incorporated into CPV-1 structures in infected cells, thus permitting the isolation of membranous vesicles. This study confirmed the enrichment of G3BP1 and G3BP2 in CPV-1 vesicles in SFV infected cells [131], however, it was not possible to determine when these host proteins were recruited. Interaction of G3BPs with alphavirus nsPs as evidenced by the observation that the insect G3BP1 homolog Rasputin, was detected in nsP3-containing complexes isolated from mosquito cells infected with SINV [146]. While G3BP proteins may serve an important and conserved function in alphavirus infections, it remains unclear as to how these host proteins function in RC biogenesis. While recruitment of G3PB proteins to RCs is a common feature of alphavirus infection, this was not observed in cells infected with the flavivirus YFV, indicating that G3BP is not a host factor for RCs of all positive strand RNA viruses [130].

SINV may recruit G3BP as a means to block G3BP-dependent export of host mRNAs to the cytoplasm or that host RNAs undergoing nuclear export are sequestered, resulting in translational shutoff. Alternatively, this may reflect a specific host response to counteract infection [130]. The interaction between nsP3 and G3BP1/G3BP2 in CHIKV-infected cells occurs late in the replication cycle where the bulk of G3BP1 and G3BP2 is not associated with the viral RC but rather, is sequestered in nsP3-G3BP aggregates in the cytoplasm [149]. This may indicate that G3BPs play different roles early and late in infection. For example, late in infection interaction between nsP3 and G3BPs, which are nucleating factors for stress granules, could prevent the formation of these RNA granules. Stress granules have been implicated in the antiviral response [150] and have been reported to form late during alphavirus infection, a process that correlates with host translational shutdown [151]. These seemingly contrasting observations may be explained by the temporal dynamics of the nsp-G3BP association (see above for CHIKV). Given the role of G3BPs in stress granule formation, it seems likely that nsPs associate with G3BPs as a means to inhibit their formation; a theory that is supported by recent data showing that translational shut-off in cells infected with the alphavirus VEEV (whose replication mechanisms do not appear to involve G3BPs) is comparatively slower [146].

6. Summary

A common feature of all +RNA viruses is their ability to modulate cellular membranes aiding in the concealment of replication product intermediates from recognition by cellular immune sensors. Despite differences in membrane curvature (DMVs, InVs, *etc.*), all replication complexes are made up of viral proteins, RNA, and cellular factors. In this review, we have focused on identifying host factors that are proposed or validated to play a role in the replication complexes of human pathogens belonging to four different virus families. The cellular proteins at the replication complexes include proteins involved in lipid metabolism, intracellular trafficking, autophagy, secretory pathways, transcription, and translation. Understanding precisely how these host proteins function in virus replication may open avenues for development of novel anti-viral therapeutics.

Acknowledgments: We thank Anil Kumar for his insightful feedback and assistance in editing this review. We apologize to our colleagues whose work was not adequately cited due to space limitations. C.R.R. and A.M.A. hold graduate scholarships from the Canadian Institutes of Health Research (CIHR). T.C.H. holds a Canada Research Chair in RNA virus host interactions. Research in the Hobman laboratory is supported in part by the CIHR and the Li Ka Shing Institute of Virology.

Author Contributions: C.R.R., A.M.A., and T.C.H. wrote this review. C.R.R. provided the illustration for Figure 1.

Appendix

Table A1. Commonly used abbreviations in this review article.

Abbreviation	Full Nomenclature
ACACA	Acetyl-CoA carboxylase 1
ACBD3	Acyl-coenzyme A binding domain containing 3 protein
Arf	ADP-Ribosylation factor
ATG5	Autophagy protein 5
BAG3	Bcl-2-associated athanogene 3
BFA	Brefeldin A
ACACA	Acetyl-CoA carboxylase 1
ACBD3	Acyl-coenzyme A binding domain containing 3 protein
Arf	ADP-Ribosylation factor
ATG5	Autophagy protein 5
BIG	Brefeldin A-inhibited guanine nucleotide-exchange factor
CHIKV	Chikungunya virus
CM	Convoluted Membranes
CME	Clathrin-mediate endocytosis
CPV-1	Type 1 cytopathic vacuoles
CV	Coxsackie virus
DENV	Dengue virus
DMV	Double-Membrane Vesicles
dsRNA	Double-stranded RNA
EDEM1	ER degradation-enhancing alpha-mannosidase-like 1
EEEV	Eastern equine encephalitis virus
EV	Enterovirus
FAS	Fatty acid synthase
G3BP	Ras GTPase-activating protein (SH3 domain) binding protein
GBF1	Golgi brefeldin A resistant guanine nucleotide exchange factor 1
HCV	Hepatitis C virus
HMGCR	3-hydroxy-methyglutaryl-CoA reductase
HRV	Human Rhinovirus
Hsc70 (also known as HSPA8)	Heat shock cognate 71 kDa protein
JEV	Japanese Encephalitis virus
LAMP-1	Lysosomal-associated membrane protein 1
LC3 (Cytosolic form; LC-I)	Microtubule-associated protein 1A/1B-light chain 3
LC3-II (Lipidated form)	LC3-phosphatidylethanolamine conjugate
LDLR	Low density lipoprotein receptor
MERS-CoV	Middle East respiratory syndrome coronavirus
MHV	Mouse Hepatitis virus
MVD	Mevalonate diphosphate decarboxylase
NPAF	Niemann-Pick disease-A fibroblasts
OS-9	Osteosarcoma amplified 9, ER lectin
PC	Paracrystalline Array
PDI	Protein disulfide isomerase
PI	Phosphatidylinositol
PI4K	Phosphatidylinositol-4-OH kinase
PI4P	Phosphatidylinositol 4-phosphate
PtdCho	Phosphatidylcholine
PV	Poliovirus

Table A1. *Cont.*

RTC	Replication/Transcription Complex
RUBV	Rubella virus
RVN	Reticulovesicular Network
SARS-CoV	Severe Acute respiratory syndrome coronavirus
Sec-13, -31, -61	ER translocon proteins
SFV	Semliki Forest virus
SINV	Sindbis virus
SLEV	St. Louis Encephalitis virus
SREBP	Sterol-regulatory element binding protein
STMN1	Stathmin 1/oncoprotein 18
TBEV	Tick-borne Encephalitis virus
VEEV	Venezuelan equine encephalitis virus
VP	Vesicular Packet
VRC	Virus Replication Complex
WEEV	Western equine encephalitis virus
WNV	West Nile virus
YFV	Yellow Fever virus

Conflicts of Interest: The authors declare no conflict of interest.

References

1. Mukhopadhyay, S.; Kuhn, R.J.; Rossmann, M.G. A structural perspective of the flavivirus life cycle. *Nat. Rev. Microbiol.* **2005**, *3*, 13–22.

2. Chatel-Chaix, L.; Bartenschlager, R. Dengue virus- and hepatitis c virus-induced replication and assembly compartments: The enemy inside–caught in the web. *J. Virol.* **2014**, *88*, 5907–5911.

3. Lohmann, V. Hepatitis c virus rna replication. *Curr. Top. Microbiol. Immunol.* **2013**, *369*, 167–198.

4. Egger, D.; Wolk, B.; Gosert, R.; Bianchi, L.; Blum, H.E.; Moradpour, D.; Bienz, K. Expression of hepatitis c virus proteins induces distinct membrane alterations including a candidate viral replication complex. *J. Virol.* **2002**, *76*, 5974–5984.

5. Paul, D.; Romero-Brey, I.; Gouttenoire, J.; Stoitsova, S.; Krijnse-Locker, J.; Moradpour, D.; Bartenschlager, R. Ns4b self-interaction through conserved c-terminal elements is required for the establishment of functional hepatitis c virus replication complexes. *J. Virol.* **2011**, *85*, 6963–6976.

6. Romero-Brey, I.; Merz, A.; Chiramel, A.; Lee, J.-Y.; Chlanda, P.; Haselman, U.; Santarella-Mellwig, R.; Habermann, A.; Hoppe, S.; Kallis, S. Three-dimensional architecture and biogenesis of membrane structures associated with hepatitis c virus replication. *PLoS Pathog.* **2012**, *8*, e1003056.

7. Paul, D.; Madan, V.; Bartenschlager, R. Hepatitis c virus rna replication and assembly: Living on the fat of the land. *Cell Host Microbe* **2014**, *16*, 569–579.

8. Romero-Brey, I.; Bartenschlager, R. Membranous replication factories induced by plus-strand rna viruses. *Viruses* **2014**, *6*, 2826–2857.

9. Westaway, E.G.M.; Mackenzie, J.M.; Kenny, M.T.; Jones, M.K.; Khromykh, A.A. Ultrastructure of kunjin virus-infected cells: Colocalization of ns1 and ns3 with

double-stranded rna, and of ns2b with ns3, in virus-induced membrane structures. *J. Virol.* **1997**, *71*, 6650–6661.

10. Whiteman, M.C.; Popov, V.; Sherman, M.B.; Wen, J.; Barrett, A.D.T. Attenuated west nile virus mutant ns1130-132qqa/175a/207a exhibits virus-induced ultrastructural changes and accumulation of protein in the endoplasmic reticulum. *J. Virol.* **2015**, *89*, 1474–1478.

11. Mackenzie, J.M.; Jones, M.K.; Young, P.R. Immunolocalization of the dengue virus nonstructural glycoprotein ns1 suggests a role in viral rna replication. *Virology* **1996**, *220*, 232–240.

12. Girard, Y.A.; Popov, V.; Wen, J.; Han, V.; Higgs, S. Ultrastructural study of west nile virus pathogenesis in culex pipiens quinquefasciatus (diptera: Culicidae). *J. Med. Entomol.* **2005**, *42*, 429–444.

13. Gangodkar, S.; Jain, P.; Dixit, N.; Ghosh, K.; Basu, A. Dengue virus-induced autophagosomes and changes in endomembrane ultrastructure imaged by electron tomography and whole-mount grid-cell culture techniques. *J. Electron. Microsc.* **2010**, *59*, 503–511.

14. Uchil, P.D.; Satchidanandam, V. Architecture of the flaviviral replication complex. Protease, nuclease, and detergents reveal encasement within double-layered membrane compartments. *J. Biol. Chem.* **2003**, *278*, 24388–24398.

15. Welsch, S.; Miller, S.; Romero-Brey, I.; Merz, A.; Bleck, C.K.; Walther, P.; Fuller, S.D.; Antony, C.; Krijnse-Locker, J.; Bartenschlager, R. Composition and three-dimensional architecture of the dengue virus replication and assembly sites. *Cell Host Microbe* **2009**, *5*, 365–375.

16. Mackenzie, J.M.; Kenney, M.T.; Westaway, E.G. West nile virus strain kunjin ns5 polymerase is a phosphoprotein localized at the cytoplasmic site of viral rna synthesis. *J. Gen. Virol.* **2007**, *88*, 1163–1168.

17. Westaway, E.G.; Khromykh, A.A.; Mackenzie, J.M. Nascent flavivirus rna colocalized *in situ* with double-stranded rna in stable replication complexes. *Virology* **1999**, *258*, 108–117.

18. Mackenzie, J.M.; Jones, M.K.; Westaway, E.G. Markers for trans-golgi membranes and the intermediate compartment localize to induced membranes with distinct replication functions in flavivirus-infected cells. *J. Virol.* **1999**, *73*, 9555–9567.

19. Gillespie, L.K.; Hoenen, A.; Morgan, G.; Mackenzie, J.M. The endoplasmic reticulum provides the membrane platform for biogenesis of the flavivirus replication complex. *J. Virol.* **2010**, *84*, 10438–10447.

20. Kaufusi, P.H.; Kelley, J.F.; Yanagihara, R.; Nerurkar, V.R. Induction of endoplasmic reticulum-derived replication-competent membrane structures by west nile virus non-structural protein 4b. *PLoS ONE* **2014**, *9*, e84040.

21. Offerdahl, D.K.; Dorward, D.W.; Hansen, B.T.; Bloom, M.E. A three-dimensional comparison of tick-borne flavivirus infection in mammalian and tick cell lines. *PLoS ONE* **2012**, *7*, e47912.

22. Miller, S.; Kastner, S.; Krijnse-Locker, J.; Buhler, S.; Bartenschlager, R. The non-structural protein 4a of dengue virus is an integral membrane protein inducing membrane alterations in a 2k-regulated manner. *J. Biol. Chem.* **2007**, *282*, 8873–8882.

23. Belov, G.A.; Nair, V.; Hansen, B.T.; Hoyt, F.H.; Fischer, E.R.; Ehrenfeld, E. Complex dynamic development of poliovirus membranous replication complexes. *J. Virol.* **2012**, *86*, 302–312.

24. Grimley, P.M.; Berezesky, I.K.; Friedman, R.M. Cytoplasmic structures associated with an arbovirus infection: Loci of viral ribonucleic acid synthesis. *J. Virol.* **1968**, *2*, 1326–1338.

25. Knoops, K.; Kikkert, M.; van den Worm, S.H.; Zevenhoven-Dobbe, J.C.; van der Meer, Y.; Koster, A.J.; Mommaas, A.M.; Snijder, E.J. Sars-coronavirus replication is supported by a reticulovesicular network of modified endoplasmic reticulum. *PLoS Biol.* **2008**, *6*, e226.

26. Zhang, L.K.; Chai, F.; Li, H.Y.; Xiao, G.; Guo, L. Identification of host proteins involved in japanese encephalitis virus infection by quantitative proteomics analysis. *J. Proteome Res.* **2013**, *12*, 2666–2678.

27. Mishra, K.P.; Diwaker, D.; Ganju, L. Dengue virus infection induces upregulation of hn rnp-h and pdia3 for its multiplication in the host cell. *Virus Res.* **2012**, *163*, 573–579.

28. Campbell, C.L.; Harrison, T.; Hess, A.M.; Ebel, G.D. Microrna levels are modulated in aedes aegypti after exposure to dengue-2. *Insect Mol. Biol.* **2014**, *23*, 132–139.

29. Mairiang, D.; Zhang, H.; Sodja, A.; Murali, T.; Suriyaphol, P.; Malasit, P.; Limjindaporn, T.; Finley, R.L., Jr. Identification of new protein interactions between dengue fever virus and its hosts, human and mosquito. *PLoS ONE* **2013**, *8*, e53535.

30. Saha, S. Common host genes are activated in mouse brain by japanese encephalitis and rabies viruses. *J. Gen. Virol.* **2003**, *84*, 1729–1735.

31. Sengupta, N.; Ghosh, S.; Vasaikar, S.V.; Gomes, J.; Basu, A. Modulation of neuronal proteome profile in response to japanese encephalitis virus infection. *PLoS ONE* **2014**, *9*, e90211.

32. Krishnan, M.N.; Ng, A.; Sukumaran, B.; Gilfoy, F.D.; Uchil, P.D.; Sultana, H.; Brass, A.L.; Adametz, R.; Tsui, M.; Qian, F.; *et al.* Rna interference screen for human genes associated with west nile virus infection. *Nature* **2008**, *455*, 242–245.

33. Sessions, O.M.; Barrows, N.J.; Souza-Neto, J.A.; Robinson, T.J.; Hershey, C.L.; Rodgers, M.A.; Ramirez, J.L.; Dimopoulos, G.; Yang, P.L.; Pearson, J.L.; *et al.* Discovery of insect and human dengue virus host factors. *Nature* **2009**, *458*, 1047–1050.

34. Beatman, E.; Oyer, R.; Shives, K.D.; Hedman, K.; Brault, A.C.; Tyler, K.L.; Beckham, J.D. West nile virus growth is independent of autophagy activation. *Virology* **2012**, *433*, 262–272.

35. Vandergaast, R.; Fredericksen, B.L. West nile virus (wnv) replication is independent of autophagy in mammalian cells. *PLoS ONE* **2012**, *7*, e45800.

36. Lee, Y.-R.; Lei, H.-Y.; Liu, M.-T.; Wang, J.-R.; Chen, S.-H.; Jiang-Shieh, Y.-F.; Lin, Y.-S.; Yeh, T.-M.; Liu, C.-C.; Liu, H.-S. Autophagic machinery activated by dengue virus enhances virus replication. *Virology* **2008**, *374*, 240–248.

37. Li, J.-K.; Liang, J.-J.; Liao, C.-L.; Lin, Y.-L. Autophagy is involved in the early step of japanese encephalitis virus infection. *Microbes Infect.* **2012**, *14*, 159–168.

38. Heaton, N.S.; Randall, G. Dengue virus-induced autophagy regulates lipid metabolism. *Cell Host Microbe* **2010**, *8*, 422–432.

39. Hsu, N.Y.; Ilnytska, O.; Belov, G.; Santiana, M.; Chen, Y.H.; Takvorian, P.M.; Pau, C.; van der Schaar, H.; Kaushik-Basu, N.; Balla, T.; *et al.* Viral reorganization of the secretory pathway generates distinct organelles for rna replication. *Cell* **2010**, *141*, 799–811.

40. Berger, K.L.; Cooper, J.D.; Heaton, N.S.; Yoon, R.; Oakland, T.E.; Jordan, T.X.; Mateu, G.; Grakoui, A.; Randall, G. Roles for endocytic trafficking and phosphatidylinositol 4-kinase iii alpha in hepatitis c virus replication. *Proc. Natl. Acad. Sci. USA* **2009**, *106*, 7577–7582.

41. Martin-Acebes, M.A.; Blazquez, A.B.; Jimenez de Oya, N.; Escribano-Romero, E.; Saiz, J.C. West nile virus replication requires fatty acid synthesis but is independent on phosphatidylinositol-4-phosphate lipids. *PLoS ONE* **2011**, *6*, e24970.

42. Heaton, N.S.; Perera, R.; Berger, K.L.; Khadka, S.; Lacount, D.J.; Kuhn, R.J.; Randall, G. Dengue virus nonstructural protein 3 redistributes fatty acid synthase to sites of viral replication and increases cellular fatty acid synthesis. *Proc. Natl. Acad. Sci. USA* **2010**, *107*, 17345–17350.

43. Seddon, J.M.; Templer, R.H.; Warrender, N.A.; Huang, Z.; Cevc, G.; Marsh, D. Phosphatidylcholine-fatty acid membranes: Effects of headgroup hydration on the phase behaviour and structural parameters of the gel and inverse hexagonal (h(ii)) phases. *Biochim. Biophys. Acta* **1997**, *1327*, 131–147.

44. Perera, R.; Riley, C.; Isaac, G.; Hopf-Jannasch, A.S.; Moore, R.J.; Weitz, K.W.; Pasa-Tolic, L.; Metz, T.O.; Adamec, J.; Kuhn, R.J. Dengue virus infection perturbs lipid homeostasis in infected mosquito cells. *PLoS Pathog.* **2012**, *8*, e1002584.

45. Martín-Acebes, M.A.; Merino-Ramos, T.; Blázquez, A.-B.; Casas, J.; Escribano-Romero, E.; Sobrino, F.; Saiz, J.-C. The composition of west nile virus lipid envelope unveils a role of sphingolipid metabolism in flavivirus biogenesis. *J. Virol.* **2014**, *88*, 12041–12054.

46. Stapleford, K.A.; Miller, D.J. Role of cellular lipids in positive-sense rna virus replication complex assembly and function. *Viruses* **2010**, *2*, 1055–1068.

47. Bengoechea-Alonso, M.T.; Ericsson, J. Srebp in signal transduction: Cholesterol metabolism and beyond. *Curr. Opin. Cell Biol.* **2007**, *19*, 215–222.

48. May, P.; Bock, H.H.; Herz, J. Integration of endocytosis and signal transduction by lipoprotein receptors. *Sci. STKE* **2003**, *2003*, PE12.

49. Mackenzie, J.M.; Khromykh, A.A.; Parton, R.G. Cholesterol manipulation by west nile virus perturbs the cellular immune response. *Cell Host Microbe* **2007**, *2*, 229–239.

50. Rothwell, C.; Lebreton, A.; Young Ng, C.; Lim, J.Y.; Liu, W.; Vasudevan, S.; Labow, M.; Gu, F.; Gaither, L.A. Cholesterol biosynthesis modulation regulates dengue viral replication. *Virology* **2009**, *389*, 8–19.

51. Teo, C.S.H.; Chu, J.J.H. Cellular vimentin regulates construction of dengue virus replication complexes through interaction with ns4a protein. *J. Virol.* **2014**, *88*, 1897–1913.

52. Wu, N.; Gao, N.; Fan, D.; Wei, J.; Zhang, J.; An, J. miR-223 inhibits dengue virus replication by negatively regulating the microtubule-destabilizing protein STMN1 in EAhy926 cells. *Microbes Infect.* **2014**, *16*, 911–922.

53. Dales, S.; Eggers, H.J.; Tamm, I.; Palade, G.E. Electron microscopic study of the formation of poliovirus. *Virology* **1965**, *26*, 379–389.

54. Bienz, K.; Egger, D.; Rasser, Y.; Bossart, W. Intracellular distribution of poliovirus proteins and the induction of virus-specific cytoplasmic structures. *Virology* **1983**, *131*, 39–48.

55. Bienz, K.; Egger, D.; Troxler, M.; Pasamontes, L. Structural organization of poliovirus rna replication is mediated by viral proteins of the p2 genomic region. *J. Virol.* **1990**, *64*, 1156–1163.

56. Bienz, K.; Egger, D.; Pfister, T.; Troxler, M. Structural and functional characterization of the poliovirus replication complex. *J. Virol.* **1992**, *66*, 2740–2747.

57. Belov, G.A.; Sztul, E. Rewiring of cellular membrane homeostasis by picornaviruses. *J. Virol.* **2014**, *88*, 9478–9489.

58. Schlegel, A.; Giddings, T.H.; Ladinsky, M.S.; Kirkegaard, K. Cellular origin and ultrastructure of membranes induced during poliovirus infection. *J. Virol.* **1996**, *70*, 6576–6588.

59. Suhy, D.A.; Giddings, T.H.; Kirkegaard, K. Remodeling the endoplasmic reticulum by poliovirus infection and by individual viral proteins: An autophagy-like origin for virus-induced vesicles. *J. Virol.* **2000**, *74*, 8953–8965.

60. Jackson, W.T.; Giddings, T.H., Jr.; Taylor, M.P.; Mulinyawe, S.; Rabinovitch, M.; Kopito, R.R.; Kirkegaard, K. Subversion of cellular autophagosomal machinery by rna viruses. *PLoS Biol.* **2005**, *3*, e156.

61. Kemball, C.C.; Alirezaei, M.; Flynn, C.T.; Wood, M.R.; Harkins, S.; Kiosses, W.B.; Whitton, J.L. Coxsackievirus infection induces autophagy-like vesicles and megaphagosomes in pancreatic acinar cells in vivo. *J. Virol.* **2010**, *84*, 12110–12124.

62. Klein, K.A.; Jackson, W.T. Human rhinovirus 2 induces the autophagic pathway and replicates more efficiently in autophagic cells. *J. Virol.* **2011**, *85*, 9651–9654.

63. Robinson, S.M.; Tsueng, G.; Sin, J.; Mangale, V.; Rahawi, S.; McIntyre, L.L.; Williams, W.; Kha, N.; Cruz, C.; Hancock, B.M.; *et al.* Coxsackievirus b exits the host cell in shed microvesicles displaying autophagosomal markers. *PLoS Pathog.* **2014**, *10*, e1004045.

64. Richards, A.L.; Soares-Martins, J.A.P.; Riddell, G.T.; Jackson, W.T. Generation of unique poliovirus rna replication organelles. *MBio* **2014**, *5*, e00833–e00813.

65. Choe, S.S.; Dodd, D.A.; Kirkegaard, K. Inhibition of cellular protein secretion by picornaviral 3a proteins. *Virology* **2005**, *337*, 18–29.

66. Beske, O.; Reichelt, M.; Taylor, M.P.; Kirkegaard, K.; Andino, R. Poliovirus infection blocks ergic-to-golgi trafficking and induces microtubule-dependent disruption of the golgi complex. *J. Cell Sci.* **2007**, *120*, 3207–3218.

67. Egger, D.; Bienz, K. Intracellular location and translocation of silent and active poliovirus replication complexes. *J. Gen. Virol.* **2005**, *86*, 707–718.

68. Rust, R.C.; Landmann, L.; Gosert, R.; Tang, B.L.; Hong, W.; Hauri, H.P.; Egger, D.; Bienz, K. Cellular copii proteins are involved in production of the vesicles that form the poliovirus replication complex. *J. Virol.* **2001**, *75*, 9808–9818.

69. Trahey, M.; Oh, H.S.; Cameron, C.E.; Hay, J.C. Poliovirus infection transiently increases copii vesicle budding. *J. Virol.* **2012**, *86*, 9675–9682.

70. Quiner, C.A.; Jackson, W.T. Fragmentation of the golgi apparatus provides replication membranes for human rhinovirus 1a. *Virology* **2010**, *407*, 185–195.

71. Mousnier, A.; Swieboda, D.; Pinto, A.; Guedan, A.; Rogers, A.V.; Walton, R.; Johnston, S.L.; Solari, R. Human rhinovirus 16 causes golgi apparatus fragmentation without blocking protein secretion. *J. Virol.* **2014**, *88*, 11671–11685.

72. Cherry, S.; Kunte, A.; Wang, H.; Coyne, C.; Rawson, R.B.; Perrimon, N. Copi activity coupled with fatty acid biosynthesis is required for viral replication. *PLoS Pathog.* **2006**, *2*, e102.

73. Irurzun, A.; Perez, L.; Carrasco, L. Involvement of membrane traffic in the replication of poliovirus genomes: Effects of brefeldin a. *Virology* **1992**, *191*, 166–175.

74. Maynell, L.A.; Kirkegaard, K.; Klymkowsky, M.W. Inhibition of poliovirus rna synthesis by brefeldin a. *J. Virol.* **1992**, *66*, 1985–1994.

75. Behnia, R.; Munro, S. Organelle identity and the signposts for membrane traffic. *Nature* **2005**, *438*, 597–604.

76. Belov, G.A.; Fogg, M.H.; Ehrenfeld, E. Poliovirus proteins induce membrane association of gtpase adp-ribosylation factor. *J. Virol.* **2005**, *79*, 7207–7216.

77. Belov, G.A.; Habbersett, C.; Franco, D.; Ehrenfeld, E. Activation of cellular arf gtpases by poliovirus protein 3cd correlates with virus replication. *J. Virol.* **2007**, *81*, 9259–9267.

78. Belov, G.A.; Ehrenfeld, E. Involvement of cellular membrane traffic proteins in poliovirus replication. *Cell Cycle* **2007**, *6*, 36–38.

79. Belov, G.A.; Feng, Q.; Nikovics, K.; Jackson, C.L.; Ehrenfeld, E. A critical role of a cellular membrane traffic protein in poliovirus rna replication. *PLoS Pathog.* **2008**, *4*, e1000216.

80. Lanke, K.H.; van der Schaar, H.M.; Belov, G.A.; Feng, Q.; Duijsings, D.; Jackson, C.L.; Ehrenfeld, E.; van Kuppeveld, F.J. Gbf1, a guanine nucleotide exchange factor for arf, is crucial for coxsackievirus b3 rna replication. *J. Virol.* **2009**, *83*, 11940–11949.

81. Dorobantu, C.M.; van der Schaar, H.M.; Ford, L.A.; Strating, J.R.; Ulferts, R.; Fang, Y.; Belov, G.; van Kuppeveld, F.J. Recruitment of pi4kiiibeta to coxsackievirus b3 replication organelles is independent of acbd3, gbf1, and arf1. *J. Virol.* **2014**, *88*, 2725–2736.

82. Dorobantu, C.M.; Ford-Siltz, L.A.; Sittig, S.P.; Lanke, K.H.; Belov, G.A.; van Kuppeveld, F.J.; van der Schaar, H.M. Gbf1- and acbd3-independent recruitment of pi4kiiibeta to replication sites by rhinovirus 3a proteins. *J. Virol.* **2015**, *89*, 1913–1918.

83. Limpens, R.W.; van der Schaar, H.M.; Kumar, D.; Koster, A.J.; Snijder, E.J.; van Kuppeveld, F.J.; Barcena, M. The transformation of enterovirus replication structures: A three-dimensional study of single- and double-membrane compartments. *MBio* **2011**, *2*.

84. Vance, D.E.; Trip, E.M.; Paddon, H.B. Poliovirus increases phosphatidylcholine biosynthesis in hela cells by stimulation of the rate-limiting reaction catalyzed by ctp: Phosphocholine cytidylyltransferase. *J. Biol. Chem.* **1980**, *255*, 1064–1069.

85. Van Meer, G.; Voelker, D.R.; Feigenson, G.W. Membrane lipids: Where they are and how they behave. *Nat. Rev. Mol. Cell Biol.* **2008**, *9*, 112–124.

86. Guinea, R.; Carrasco, L. Phospholipid biosynthesis and poliovirus genome replication, two coupled phenomena. *EMBO J.* **1990**, *9*, 2011–2016.

87. Guinea, R.; Carrasco, L. Effects of fatty acids on lipid synthesis and viral rna replication in poliovirus-infected cells. *Virology* **1991**, *185*, 473–476.

88. Rassmann, A.; Henke, A.; Zobawa, M.; Carlsohn, M.; Saluz, H.-P.; Grabley, S.; Lottspeich, F.; Munder, T. Proteome alterations in human host cells infected with coxsackievirus b3. *J. Gen. Virol.* **2006**, *87*, 2631–2638.

89. Wilsky, S.; Sobotta, K.; Wiesener, N.; Pilas, J.; Althof, N.; Munder, T.; Wutzler, P.; Henke, A. Inhibition of fatty acid synthase by amentoflavone reduces coxsackievirus b3 replication. *Arch. Virol.* **2012**, *157*, 259–269.

90. Nchoutmboube, J.A.; Viktorova, E.G.; Scott, A.J.; Ford, L.A.; Pei, Z.; Watkins, P.A.; Ernst, R.K.; Belov, G.A. Increased long chain acyl-coa synthetase activity and fatty acid import is linked to membrane synthesis for development of picornavirus replication organelles. *PLoS Pathog.* **2013**, *9*, e1003401.

91. Godi, A.; Pertile, P.; Meyers, R.; Marra, P.; di Tullio, G.; Iurisci, C.; Luini, A.; Corda, D.; de Matteis, M.A. Arf mediates recruitment of ptdins-4-oh kinase-beta and stimulates synthesis of ptdins(4,5)p2 on the golgi complex. *Nat. Cell Biol.* **1999**, *1*, 280–287.

92. Arita, M.; Kojima, H.; Nagano, T.; Okabe, T.; Wakita, T.; Shimizu, H. Phosphatidylinositol 4-kinase iii beta is a target of enviroxime-like compounds for antipoliovirus activity. *J. Virol.* **2011**, *85*, 2364–2372.

93. DeLong, D.C.; Reed, S.E. Inhibition of rhinovirus replication in in organ culture by a potential antiviral drug. *J. Infect. Dis.* **1980**, *141*, 87–91.

94. Van der Schaar, H.M.; Leyssen, P.; Thibaut, H.J.; de Palma, A.; van der Linden, L.; Lanke, K.H.; Lacroix, C.; Verbeken, E.; Conrath, K.; Macleod, A.M.; *et al.* A novel, broad-spectrum inhibitor of enterovirus replication that targets host cell factor phosphatidylinositol 4-kinase iiibeta. *Antimicrob. Agents Chemother.* **2013**, *57*, 4971–4981.

95. Spickler, C.; Lippens, J.; Laberge, M.K.; Desmeules, S.; Bellavance, E.; Garneau, M.; Guo, T.; Hucke, O.; Leyssen, P.; Neyts, J.; *et al.* Phosphatidylinositol 4-kinase iii beta is essential for replication of human rhinovirus and its inhibition causes a lethal phenotype *in vivo. Antimicrob. Agents Chemother.* **2013**, *57*, 3358–3368.

96. Greninger, A.L.; Knudsen, G.M.; Betegon, M.; Burlingame, A.L.; Derisi, J.L. The 3a protein from multiple picornaviruses utilizes the golgi adaptor protein acbd3 to recruit pi4kiiibeta. *J. Virol.* **2012**, *86*, 3605–3616.

97. Albulescu, L.; Wubbolts, R.; van Kuppeveld, F.J.; Strating, J.R. Cholesterol shuttling is important for rna replication of coxsackievirus b3 and encephalomyocarditis virus. *Cell. Microbiol.* **2015**, *17*, 1144–1156.

98. Ilnytska, O.; Santiana, M.; Hsu, N.Y.; Du, W.L.; Chen, Y.H.; Viktorova, E.G.; Belov, G.; Brinker, A.; Storch, J.; Moore, C.; *et al.* Enteroviruses harness the cellular endocytic machinery to remodel the host cell cholesterol landscape for effective viral replication. *Cell Host Microbe* **2013**, *14*, 281–293.

99. Kuhn, J.H.; Li, W.; Radoshitzky, S.R.; Choe, H.; Farzan, M. Severe acute respiratory syndrome coronavirus entry as a target of antiviral therapies. *Antivir. Ther.* **2006**, *12*, 639–650.

100. Kanjanahaluethai, A.; Chen, Z.; Jukneliene, D.; Baker, S.C. Membrane topology of murine coronavirus replicase nonstructural protein 3. *Virology* **2007**, *361*, 391–401.

101. Oostra, M.; Te Lintelo, E.; Deijs, M.; Verheije, M.; Rottier, P.; de Haan, C. Localization and membrane topology of coronavirus nonstructural protein 4: Involvement of the early secretory pathway in replication. *J. Virol.* **2007**, *81*, 12323–12336.

102. Baliji, S.; Cammer, S.A.; Sobral, B.; Baker, S.C. Detection of nonstructural protein 6 in murine coronavirus-infected cells and analysis of the transmembrane topology by using bioinformatics and molecular approaches. *J. Virol.* **2009**, *83*, 6957–6962.

103. Choi, K.S.; Aizaki, H.; Lai, M.M. Murine coronavirus requires lipid rafts for virus entry and cell-cell fusion but not for virus release. *J. Virol.* **2005**, *79*, 9862–9871.

104. Heaton, N.S.; Randall, G. Multifaceted roles for lipids in viral infection. *Trends Microbiol.* **2011**, *19*, 368–375.

105. De Wilde, A.H.; Wannee, K.F.; Scholte, F.E.; Goeman, J.J.; ten Dijke, P.; Snijder, E.J.; Kikkert, M.; van Hemert, M.J. A Kinome-Wide Small Interfering RNA Screen Identifies Proviral and Antiviral Host Factors in Severe Acute Respiratory Syndrome Coronavirus Replication, Including Double-Stranded RNA-Activated Protein Kinase and Early Secretory Pathway Proteins. *J. Virol.* **2015**, *89*, 8318–8333.

106. Hagemeijer, M.C.; Rottier, P.J.; de Haan, C.A. Biogenesis and dynamics of the coronavirus replicative structures. *Viruses* **2012**, *4*, 3245–3269.

107. Snijder, E.J.; van der Meer, Y.; Zevenhoven-Dobbe, J.; Onderwater, J.J.; van der Meulen, J.; Koerten, H.K.; Mommaas, A.M. Ultrastructure and origin of membrane vesicles associated with the severe acute respiratory syndrome coronavirus replication complex. *J. Virol.* **2006**, *80*, 5927–5940.

108. Imbert, I.; Snijder, E.J.; Dimitrova, M.; Guillemot, J.-C.; Lécine, P.; Canard, B. The sars-coronavirus plnc domain of nsp3 as a replication/transcription scaffolding protein. *Virus Res.* **2008**, *133*, 136–148.

109. Angelini, M.M.; Akhlaghpour, M.; Neuman, B.W.; Buchmeier, M.J. Severe acute respiratory syndrome coronavirus nonstructural proteins 3, 4, and 6 induce double-membrane vesicles. *MBio* **2013**, *4*, e00524–e00513.

110. Knoops, K.; Swett-Tapia, C.; van den Worm, S.H.; Te Velthuis, A.J.; Koster, A.J.; Mommaas, A.M.; Snijder, E.J.; Kikkert, M. Integrity of the early secretory pathway promotes, but is not required for, severe acute respiratory syndrome coronavirus rna synthesis and virus-induced remodeling of endoplasmic reticulum membranes. *J. Virol.* **2010**, *84*, 833–846.

111. De Wilde, A.H.; Raj, V.S.; Oudshoorn, D.; Bestebroer, T.M.; van Nieuwkoop, S.; Limpens, R.W.; Posthuma, C.C.; van der Meer, Y.; Bárcena, M.; Haagmans, B.L. Mers-coronavirus replication induces severe in vitro cytopathology and is strongly inhibited by cyclosporin a or interferon-α treatment. *J. Gen. Virol.* **2013**, *94*, 1749–1760.

112. Goldsmith, C.S.; Tatti, K.M.; Ksiazek, T.G.; Rollin, P.E.; Comer, J.A.; Lee, W.W.; Rota, P.A.; Bankamp, B.; Bellini, W.J.; Zaki, S.R. Ultrastructural characterization of sars coronavirus. *Emerg. Infect. Dis.* **2004**, *10*, 320–326.

113. Reggiori, F.; Monastyrska, I.; Verheije, M.H.; Calì, T.; Ulasli, M.; Bianchi, S.; Bernasconi, R.; de Haan, C.A.; Molinari, M. Coronaviruses hijack the lc3-i-positive edemosomes, er-derived vesicles exporting short-lived erad regulators, for replication. *Cell Host Microbe* **2010**, *7*, 500–508.

114. Ng, C.G.; Coppens, I.; Govindarajan, D.; Pisciotta, J.; Shulaev, V.; Griffin, D.E. Effect of host cell lipid metabolism on alphavirus replication, virion morphogenesis, and infectivity. *Proc. Natl. Acad. Sci.* **2008**, *105*, 16326–16331.

115. Frolova, E.; Gorchakov, R.; Garmashova, N.; Atasheva, S.; Vergara, L.A.; Frolov, I. Formation of nsp3-specific protein complexes during sindbis virus replication. *J. Virol.* **2006**, *80*, 4122–4134.

116. Spuul, P.; Balistreri, G.; Kääriäinen, L.; Ahola, T. Phosphatidylinositol 3-kinase-, actin-, and microtubule-dependent transport of semliki forest virus replication complexes from the plasma membrane to modified lysosomes. *J. Virol.* **2010**, *84*, 7543–7557.

117. Neuvonen, M.; Kazlauskas, A.; Martikainen, M.; Hinkkanen, A.; Ahola, T.; Saksela, K. Sh3 domain-mediated recruitment of host cell amphiphysins by alphavirus nsp3 promotes viral rna replication. *PLoS Pathog.* **2011**, *7*, e1002383.

118. Prentice, E.; McAuliffe, J.; Lu, X.; Subbarao, K.; Denison, M.R. Identification and characterization of severe acute respiratory syndrome coronavirus replicase proteins. *J. Virol.* **2004**, *78*, 9977–9986.

119. Mizushima, N.; Noda, T.; Yoshimori, T.; Tanaka, Y.; Ishii, T.; George, M.D.; Klionsky, D.J.; Ohsumi, M.; Ohsumi, Y. A protein conjugation system essential for autophagy. *Nature* **1998**, *395*, 395–398.

120. Prentice, E.; Jerome, W.G.; Yoshimori, T.; Mizushima, N.; Denison, M.R. Coronavirus replication complex formation utilizes components of cellular autophagy. *J. Biol. Chem.* **2004**, *279*, 10136–10141.

121. Zhao, Z.; Thackray, L.B.; Miller, B.C.; Lynn, T.M.; Becker, M.M.; Ward, E.; Mizushima, N.; Denison, M.R.; Virgin, I.; Herbert, W. Coronavirus replication does not require the autophagy gene atg5. *Autophagy* **2007**, *3*, 581–585.

122. Hayat, M. *Autophagy: Cancer, Other Pathologies, Inflammation, Immunity, Infection, and Aging: Volume 3-Role in Specific Diseases*; Academic Press: Waltham, UK, 2013; Volume 3.

123. Zhang, L.; Zhang, Z.-P.; Zhang, X.-E.; Lin, F.-S.; Ge, F. Quantitative proteomics analysis reveals bag3 as a potential target to suppress severe acute respiratory syndrome coronavirus replication. *J. Virol.* **2010**, *84*, 6050–6059.

124. Westaway, E.; Brinton, M.; Gaidamovich, S.Y.; Horzinek, M.; Igarashi, A.; Kääriäinen, L.; Lvov, D.; Porterfield, J.; Russell, P.; Trent, D. Togaviridae. *Intervirology* **1985**, *24*, 125–139.

125. Garmashova, N.; Gorchakov, R.; Volkova, E.; Paessler, S.; Frolova, E.; Frolov, I. The old world and new world alphaviruses use different virus-specific proteins for induction of transcriptional shutoff. *J. Virol.* **2007**, *81*, 2472–2484.

126. Lee, J.-Y.; Marshall, J.; Bowden, D. Replication complexes associated with the morphogenesis of rubella virus. *Arch. Virol.* **1992**, *122*, 95–106.

127. Magliano, D.; Marshall, J.A.; Bowden, D.S.; Vardaxis, N.; Meanger, J.; Lee, J.-Y. Rubella virus replication complexes are virus-modified lysosomes. *Virology* **1998**, *240*, 57–63.

128. Froshauer, S.; Kartenbeck, J.; Helenius, A. Alphavirus rna replicase is located on the cytoplasmic surface of endosomes and lysosomes. *J. Cell Biol.* **1988**, *107*, 2075–2086.

129. Atasheva, S.; Gorchakov, R.; English, R.; Frolov, I.; Frolova, E. Development of sindbis viruses encoding nsp2/gfp chimeric proteins and their application for studying nsp2 functioning. *J. Virol.* **2007**, *81*, 5046–5057.

130. Cristea, I.M.; Carroll, J.-W.N.; Rout, M.P.; Rice, C.M.; Chait, B.T.; MacDonald, M.R. Tracking and elucidating alphavirus-host protein interactions. *J. Biol. Chem.* **2006**, *281*, 30269–30278.

131. Varjak, M.; Saul, S.; Arike, L.; Lulla, A.; Peil, L.; Merits, A. Magnetic fractionation and proteomic dissection of cellular organelles occupied by the late replication complexes of semliki forest virus. *J. Virol.* **2013**, *87*, 10295–10312.

132. Frolova, E.I.; Gorchakov, R.; Pereboeva, L.; Atasheva, S.; Frolov, I. Functional sindbis virus replicative complexes are formed at the plasma membrane. *J. Virol.* **2010**, *84*, 11679–11695.

133. Lemm, J.A.; Rice, C.M. Roles of nonstructural polyproteins and cleavage products in regulating sindbis virus rna replication and transcription. *J. Virol.* **1993**, *67*, 1916–1926.

134. Lemm, J.A.; Rümenapf, T.; Strauss, E.G.; Strauss, J.H.; Rice, C. Polypeptide requirements for assembly of functional sindbis virus replication complexes: A model for the temporal regulation of minus-and plus-strand rna synthesis. *EMBO J.* **1994**, *13*, 2925.

135. Leung, J.Y.-S.; Ng, M.M.-L.; Chu, J.J.H. Replication of alphaviruses: A review on the entry process of alphaviruses into cells. *Adv. Virol.* **2011**, *2011*, 249640.

136. Ahola, T.; Lampio, A.; Auvinen, P.; Kääriäinen, L. Semliki forest virus mrna capping enzyme requires association with anionic membrane phospholipids for activity. *EMBO J.* **1999**, *18*, 3164–3172.

137. Rozanov, M.N.; Koonin, E.V.; Gorbalenya, A.E. Conservation of the putative methyltransferase domain: A hallmark of the 'sindbis-like'supergroup of positive-strand rna viruses. *J. Gen. Virol.* **1992**, *73*, 2129–2134.

138. Peränen, J.; Laakkonen, P.; Hyvönen, M.; Kääriäinen, L. The alphavirus replicase protein nsp1 is membrane-associated and has affinity to endocytic organelles. *Virology* **1995**, *208*, 610–620.

139. Laakkonen, P.; Ahola, T.; Kääriäinen, L. The effects of palmitoylation on membrane association of semliki forest virus rna capping enzyme. *J. Biol. Chem.* **1996**, *271*, 28567–28571.

140. Smit, J.M.; Bittman, R.; Wilschut, J. Low-ph-dependent fusion of sindbis virus with receptor-free cholesterol-and sphingolipid-containing liposomes. *J. Virol.* **1999**, *73*, 8476–8484.

141. Kielian, M.C.; Helenius, A. Role of cholesterol in fusion of semliki forest virus with membranes. *J. Virol.* **1984**, *52*, 281–283.

142. Kielian, M.; Chatterjee, P.K.; Gibbons, D.L.; Lu, Y.E. Specific roles for lipids in virus fusion and exit examples from the alphaviruses. In *Fusion of Biological Membranes and Related Problems*; Springer: Berlin, Germany, 2002; pp. 409–455.

143. Marquardt, M.T.; Phalen, T.; Kielian, M. Cholesterol is required in the exit pathway of semliki forest virus. *J. Cell Biology* **1993**, *123*, 57–65.

144. Lu, Y.E.; Cassese, T.; Kielian, M. The cholesterol requirement for sindbis virus entry and exit and characterization of a spike protein region involved in cholesterol dependence. *J. Virol.* **1999**, *73*, 4272–4278.

145. Barton, D.J.; Sawicki, S.G.; Sawicki, D.L. Solubilization and immunoprecipitation of alphavirus replication complexes. *J. Virol.* **1991**, *65*, 1496–1506.

146. Gorchakov, R.; Garmashova, N.; Frolova, E.; Frolov, I. Different types of nsp3-containing protein complexes in sindbis virus-infected cells. *J. Virol.* **2008**, *82*, 10088–10101.

147. Liu, T.; Daniels, C.K.; Cao, S. Comprehensive review on the hsc70 functions, interactions with related molecules and involvement in clinical diseases and therapeutic potential. *Pharmacol. Ther.* **2012**, *136*, 354–374.

148. Mayer, M. Recruitment of hsp70 chaperones: A crucial part of viral survival strategies. In *Reviews of Physiology, Biochemistry and Pharmacology*; Springer: Berlin, Germany, 2005; pp. 1–46.

149. Scholte, F.E.; Tas, A.; Albulescu, I.C.; Žusinaite, E.; Merits, A.; Snijder, E.J.; van Hemert, M.J. Stress granule components g3bp1 and g3bp2 play a proviral role early in chikungunya virus replication. *J. Virol.* **2015**, *89*, 4457–4469.

150. Panas, M.D.; Varjak, M.; Lulla, A.; Eng, K.E.; Merits, A.; Hedestam, G.B.K.; McInerney, G.M. Sequestration of g3bp coupled with efficient translation inhibits stress granules in semliki forest virus infection. *Mol. Biol. Cell* **2012**, *23*, 4701–4712.

151. McInerney, G.M.; Kedersha, N.L.; Kaufman, R.J.; Anderson, P.; Liljeström, P. Importance of eif2α phosphorylation and stress granule assembly in alphavirus translation regulation. *Mol. Biol. Cell* **2005**, *16*, 3753–3763.

The Role of Electron Microscopy in Studying the Continuum of Changes in Membranous Structures during Poliovirus Infection

Evan D. Rossignol, Jie E. Yang and Esther Bullitt

Abstract: Replication of the poliovirus genome is localized to cytoplasmic replication factories that are fashioned out of a mixture of viral proteins, scavenged cellular components, and new components that are synthesized within the cell due to viral manipulation/up-regulation of protein and phospholipid synthesis. These membranous replication factories are quite complex, and include markers from multiple cytoplasmic cellular organelles. This review focuses on the role of electron microscopy in advancing our understanding of poliovirus RNA replication factories. Structural data from the literature provide the basis for interpreting a wide range of biochemical studies that have been published on virus-induced lipid biosynthesis. In combination, structural and biochemical experiments elucidate the dramatic membrane remodeling that is a hallmark of poliovirus infection. Temporal and spatial membrane modifications throughout the infection cycle are discussed. Early electron microscopy studies of morphological changes following viral infection are re-considered in light of more recent data on viral manipulation of lipid and protein biosynthesis. These data suggest the existence of distinct subcellular vesicle populations, each of which serves specialized roles in poliovirus replication processes.

Reprinted from *Viruses*. Cite as: Rossignol, E.D.; Yang, J.E.; Bullitt, E. The Role of Electron Microscopy in Studying the Continuum of Changes in Membranous Structures during Poliovirus Infection. *Viruses* **2015**, *7*, 5305–5318.

1. Introduction

The focus of this review is on changes in cellular morphology during replication of poliovirus, a positive sense (+) RNA virus of approximately 7500 nucleotides. The poliovirus genome is encapsidated in a non-enveloped icosahedral virion, 28–30 nm in diameter [1]. The virion first attaches to human cells via the receptor CD155 [2], and binding initiates an irreversible conformational change in the capsid [3,4]. This structural change exposes capsid proteins that attach the particle to the membrane in a now receptor-independent manner [5,6], and cell entry occurs via endocytosis that does not require clathrin, but is actin- and tyrosine kinase- dependent [7].

A recent review by Paul and Wimmer [8] includes details on the process of poliovirus replication after cell entry.

Poliovirus RNA replication is always membrane-associated [9] and it has been shown that pre-existing cell membranes do not support genome replication [10]. Therefore, it is not surprising that lipid synthesis is upregulated [11] in infected cells, resulting in dramatic cellular changes. Newly formed membranous structures include specialized sites of RNA replication that have been described as "replication factories" [12]. To facilitate the replication process, replication factories are comprised of cellular membranes that are remodeled and expanded with newly synthesized lipids.

Excellent reviews on RNA viruses have been written recently [13–15], including a table in the Neuman *et al.*, review enumerating membrane rearrangements from many (+) RNA viruses [15], as well as in this issue [16]. Our review focuses on the major role that electron microscopy has played in elucidating the extensive protein-lipid structures that are induced to support replication of the picornavirus poliovirus (PV) RNA. In addition, discussion is included on upregulation and changes in the lipid species that are incorporated into replication factories.

Microscopy of poliovirus-infected cells began almost 60 years ago to document the wide-ranging morphological changes that occur within infected cells. Reports beginning in 1956 detailed the first visible cellular changes, appearing by three hours post infection (hpi), using phase light microscopy at $100\times$ magnification [17]. In 1958, electron microscopy was first used to visualize details of cytopathic effects and ultrastructure changes in poliovirus-infected cells [18]. Early electron microscopy studies include morphological data that has not been discussed in detail in recent literature. These early results, and more recent work, are described below, and included in a discussion of the membrane remodeling and expansion required for replication of the poliovirus genome.

2. Poliovirus-Modulated Lipid Metabolism and Trafficking, and Their Effect on Cellular Morphology

(+) RNA viruses are dependent on cellular lipid metabolism and transport machinery for efficient genome replication. Phospholipid synthesis is significantly up-regulated [19–21] as early as 2 hpi in poliovirus infected cells [11]. Specific upregulated lipid species (Table 1) include phosphatidylcholine [19], sphingomyelin [11], and phosphatidylinositol-4-phosphate (PI4P) [22,23]. Fatty acid chain lengths of phosphatidylcholine are altered, with a longer acyl chain (C18/C18 and C18/C16) observed in infected cells as compared to mock-infected cells (C14/C16 and C16/C16) [24]. This alteration in the structural composition in poliovirus-infected cells indicates that membranes of poliovirus replication complexes are clearly distinct from pre-existing cellular membranes [25]. Lipid metabolism itself is critical for virus replication, as blocking de novo generation of phospholipids inhibits poliovirus RNA synthesis [26], indicating that poliovirus replication and virally induced lipid biosynthesis are coupled.

The mechanism by which the altered lipid metabolism induced by poliovirus infection contributes to virus replication is unclear. Various lengths of acyl chains and sizes of head groups of lipid components favor different membrane curvatures (reviewed in [27]). The spontaneous curvature caused by phosphatidylcholine (C10-14) is zero, forming flat lamellar structures [28]. Longer acyl chain lengths (C16-18) adapt phospholipids for forming more negative curvature membrane (formation of a concave surface) [29]. The longer acyl chains in infected cells [24] suggest that these newly synthesized phosphatidylcholines [11] may help induce the tubular deformation of proliferating membrane.

PI4P and cholesterol are enriched in the newly synthesized membrane structures and are critical for building poliovirus replication machinery and for efficient progeny RNA production. During infection, the phospholipid-modifying enzyme PI4KIIIβ is recruited to the replication site by poliovirus 3A (Table 1), resulting in a local elevation of its catalyzed product PI4P [23,30,31]. Since the poliovirus RNA-dependent RNA polymerase (3D) can bind directly to PI4P *in vitro*, a model has been proposed in which one of the functions of PI4P in poliovirus replication is to recruit and maintain association of soluble 3D within membranous replication factories [23,32]. This interaction could be mediated by the negatively charged head group of PI4P and the net positive charge of 3D. Additionally, PI4P incorporation has been demonstrated to increase membrane curvature by interacting with membrane-remodeling proteins via its head group [33,34] as has been shown in cells infected with hepatitis C virus (HCV) [35]. Such interactions are expected to induce membrane remodeling that facilitates efficient RNA replication.

Cholesterol is a critical determinant of membrane fluidity and flexibility. It enters the cell from the extracellular medium through clathrin-mediated endocytosis from the plasma (reviewed in [42]), and is trafficked to replication factories with the aid of poliovirus protein 3A [41] (see Table 1). Formation of such cholesterol-enriched domains within replication membranes is proposed to counter potential over-fluidity and over-bending caused by the presence of the PI4P enrichment [41]. It has been reported that cholesterol helps build active virus replication complexes consisting of lipid rafts (detergent-resistant membrane microdomains that are enriched in cholesterol and sphingolipids), in HCV, where RNA and proteins are enclosed and protected from cellular immune responses [43,44]. Observed non-uniform distribution of free cholesterol, segregated into patches, in poliovirus-infected cells [41] strongly suggests lipid rafts play a similar role in poliovirus replication.

Table 1. Poliovirus induced alterations in lipids and proteins.

Lipids/Proteins	Changes upon Infection (Superscripts Correspond to Methods Described in Next Column)	Experimental Methods
	Lipid Biosynthesis and Lipid Composition	
Phosphatidylcholine	• Synthesis increases [a] • Fatty acid tail length increases (from C14/C16 to C18/C18) [b]	a. Radioactive pulse-chase [11] b. Mass spectrometry [24]
Sphingomyelin	• Synthesis increases [a]	a. Radioactive pulse-chase [11]
PI4P	• Synthesis increases [a,b]	a. Radioactive pulse-chase [11] b. Protein lipid overlay assay [36]
Host long chain acyl-CoA synthetase	• Activity increases [a] • Long-chain acyl-CoA (phospholipid precursor) increases [a]	a. *In vitro* measurement of newly synthesized fluorescent long chain acyl-CoA [24]
PI4KIII β	• Generates PI4P [a,b]	a. Fluorescence microscopy [23] b. Protein lipid overlay assay [36]
Poliovirus protein 2A	• Activates free fatty acid import [a,b]	a. Individual protein expression in cells [24] b. Measurement of uptake of fluorescent fatty acids [24]
Poliovirus protein 3A	• Recruits PI4KIII β to replication machinery [a,b]	a. Fluorescence Microscopy [23] b. Immunoprecipitation [23]
	Lipid Transport and Lipid Redistribution	
PI4P	• Recruits OSBP to replication machinery [a,b] • May increase membrane fluidity [c,d]	a. Immunoprecipitation [37] b. Fluorescent Microscopy [37] c. Fluorescence correlation spectroscopy [38] d. Fluorescence recovery after photobleaching [38]
Cholesterol	• May counter over-fluidity caused by PI4P [a,b]	a. Fluorescence correlation spectroscopy [39] b. Fluorescence recovery after photobleaching [38]
OSBP	• Mediates PI4P/cholesterol exchange on the membrane [a,b]	a. Chemical inhibitors: Itraconazole [37], OSBP ligands [40] b. *In vitro* liposome lipid-exchange assay [37]
Poliovirus protein 2BC	• Activates PI4KIII β to generate more PI4P [a,b] • Stimulates host OSBP activity [a,b]	a. Individual protein expression in cells [40] b. Flow cytometry [40]
Poliovirus protein 3A	• Recruits clathrin-mediated cholesterol-rich endosomes to replication machinery [a,b]	a. Immunoprecipitation [41] b. Fluorescence Microscopy [41]

PI4P: phosphatidylinositol-4-phosphate, acyl-CoA: Acetyl coenzyme A; OSBP: Oxysterol-binding protein.

A question arose regarding how PI4P and cholesterol are transported and enriched at the site of virus replication. Oxysterol-binding protein (OSBP) has been demonstrated to be responsible for the transport of cholesterol and PI4P between the endoplasmic reticulum (ER) and Golgi [45], a pathway stimulated during poliovirus infection (see Table 1). Disruption of PI4KIIIβ or OSBP activity dramatically suppresses poliovirus RNA replication, possibly by obstructing the formation of viral replication complexes [40]. Together, PI4P and cholesterol enrichment, and synthesis of phospholipids with longer acyl chains, cooperate with viral proteins (discussed below) to modulate membrane remodeling for efficient RNA replication.

Membrane-associated poliovirus proteins are key participants in the membrane remodeling that is essential for viral RNA replication [46]. The role of individual proteins has been explored through gene expression or direct delivery of poliovirus proteins in cultured cells, with visualization of their cellular effects by electron microscopy. Results of these experiments are described in Table 2.

Table 2. Membrane remodeling induced by individual poliovirus proteins in the cell. Roles of individual poliovirus proteins in membrane remodeling.

Poliovirus Protein	Membrane Structures Induced upon Protein Expression/Delivery in Cells
2BC	• Vesicles 50–350 nm in diameter [47,48] • Single-membrane empty vacuoles, located in peripheral regions of the cells [49]
2C	• Vesicles 50–350 nm in diameter [47,50] • Clusters of large, clear single-membrane vesicles [50] • Membrane structures of Myelin-like swirls (sectioned cross-sectionally) [47] • Tubular sheets (visible when sectioned longitudinally) [47]
3AB	• Upon direct delivery, dilates endoplasmic reticulum (ER) and Golgi lumen [51] • Enlarged and aggregated vacuoles [51] • Horseshoe structures [51]
3A	• Swollen ER membrane [48,49] • Dilated single-membraned tubular-structures [48,49]
2BC and 3A	• Small double-membrane clustered vesicles [49] • Large clear vacuoles [49]
2C and 3A	• Swollen ER membranes enriched in cellular material with similar electron density to cytoplasm [49]

Based on the lipid biochemical experiments discussed above and the studies shown in Table 2, ER membrane structures have been proposed to be, at least partially, the origin of poliovirus-induced vesicles [30,49]. It has also been shown that an individual viral protein is capable of invaginating liposomes [51], suggesting that viral proteins directly, in the absence of cellular proteins, are sufficient for membrane remodeling. Optimal poliovirus replication is dependent on newly synthesized lipids (Table 1) and virus proteins (Table 2) that together produce efficient replication factories.

3. Viral Replication Cycle as Observed by Electron Microscopy

3.1. Overview of Cellular Changes

Poliovirus-infected cells display distinct morphological changes throughout infection. In overview, during early infection light microscopy data show a loss of chromatin from the central region of the nucleus and chromatin appears to be condensed near the nuclear membrane [17]. At 2.5 hpi, poliovirus-induced blebbing from the ER membrane is observed by electron microscopy [52], and at 3–3.5 hpi vesicles appear within the cytoplasm [52–55]. Large numbers of cytoplasmic bodies, originally called "U bodies", were first observed by electron microscopy at 4–7 hpi [18]. Vesicles become more numerous over the course of infection, until the cytoplasm is vesicle-filled, as seen by electron microscopy of thin-sectioned cells [18,53,54]. Virus-induced vesicles are generally smooth-walled, with sizes that range from 50–200 nm in diameter. Late in infection, 7.5–9 hpi, cells round up and the cytoplasm has small and large clear vacuoles [18]. In addition, the nucleus is further distorted and can be difficult to distinguish, mitochondria are swollen, and U bodies are no longer visible. No further changes are observed in the cellular ultrastructure after this time and before cell lysis [56], and it has been suggested that major alterations in cell morphology beyond 7 hpi are an indication of the disruption of all physiological systems in the cell rather than related to virus replication [18].

The morphology of poliovirus-induced vesicles has been investigated for nearly 60 years. Here we interpret these data, and discuss morphology information from these studies. While we are not able to include the original images, scaled, schematized versions are shown in Figure 1. It is important to note that while these drawings are our interpretation of the data, results from different groups show similar features and are consistent with what we describe here. A schematic of an uninfected HeLa cell, shown in Figure 1A, schematized from Schlegel *et al.*, 1996 [54], provides a point of reference for cell morphology changes that are visible in poliovirus-infected cells.

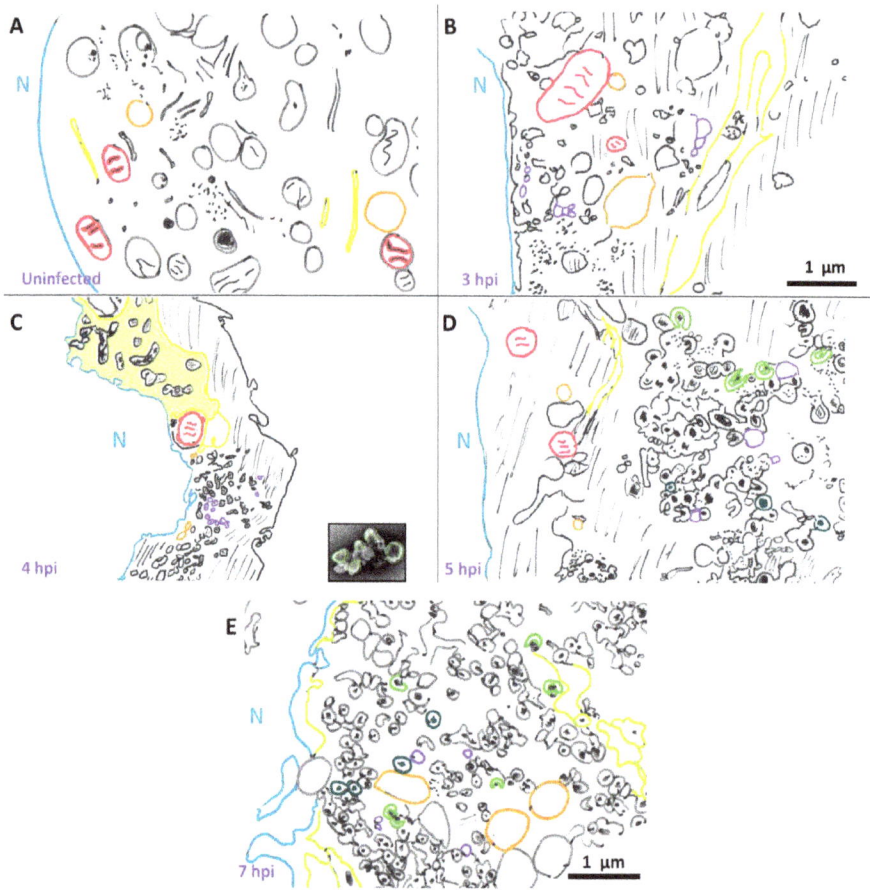

Figure 1. Changes in cellular morphology after poliovirus infection. Schematized cells from electron micrographs in Schlegel *et al.*, (**A,D**) [54], Dales *et al.*, (**B,E**) [53], and Mattern and Daniels (**C**) [57]. Representative structures are marked: nuclear membrane (blue), mitochondria (red), vacuoles (orange). Lines denote cytoplasm, and dots represent "viroplasm". Endoplasmic reticulum (yellow) becomes enlarged and sometimes called "nuclear extrusions" (as in panel **C**). Early in infection single membrane vesicles (purple) are visible, and later the appearance of U bodies/horseshoe-shaped vesicles (green) and double membrane vesicles (teal). In panel **C**, an image of a fractionated rosette (from Figure 3) is shown to scale, highlighting its similarity to structures seen within infected cells. Scale bar 1 μm.

3.2. Morphological Details of Virus-Induced Membranes throughout Infection

The first morphological changes within the infected cell are apparent in the ER at 2.5 hpi [52]. Dramatic cellular changes are observed at 3 hpi. In 1965, Dales *et al.* [53] showed the appearance of single membrane vesicular bodies in the perinuclear region of the cytoplasm in infected cells at 3 hpi, as seen in Figure 1B. More recently, Belov *et al.* [55] collected three-dimensional data and have shown that what appeared to be vesicles in two dimensions are, in fact, inter-connected branching tubular structures. In addition to the appearance of single membrane tubular structures, the endoplasmic reticulum itself has undergone a transformation by 3 hpi. The lumenal space in the endoplasmic reticulum is enlarged, forming structures described as "nuclear extrusions" by Mattern and Daniel [57], as depicted in Figure 1C. These extended, altered endoplasmic reticulum-derived structures appear adjacent to newly formed perinuclear membranous structures. It is tempting to envision that poliovirus-induced vesicles derive from such nuclear extrusions.

By 1996, Schlegel *et al.* [54] were able to achieve enhanced sample preservation using high pressure freezing and freeze substitution. Their data show similar structures to those seen in earlier studies, with additional details visible. Particularly notable are double membrane structures preserved by tannic acid. Vesicles observed by these researchers 5 hpi (Figure 1D) are substantially different than those seen at 3 hpi. While single membrane vesicles are still visible at 5 hpi, there are now multiple clusters of poliovirus-induced vesicles evident within the cytoplasm, varying in location from perinuclear to peripheral. Vesicles within these clusters appear to be predominantly round or oval, and frequently display double membrane morphology. We note that some of these vesicles show lumenal content with increased densities compared to cytoplasm.

Even more striking is that some of these double membrane morphologies are not completely closed, but instead maintain a horseshoe-like appearance with a gated lumen that is connected to the cytoplasm through a neck approximately 15 nm in diameter. These structures appear similar in shape and size to U bodies described earlier [18] and resemble those discussed above in work by Dales *et al.* [53]. The U bodies may represent a transition state from single to double membrane vesicles or *vice versa*, but this is not yet clear. As seen in Figure 1D, at 5 hpi vesicles range in size from 100–200 nm in diameter, and clusters of vesicles are on the order of 1–2 µm across.

By 7 hpi (Figure 1E, schematized from [53]), there are more double membrane vesicles, as well as fewer single membrane vesicles. The changes are not as dramatic from 5 to 7 hpi as they are from 3–5 hpi, but poliovirus capsids are visible dispersed through the vesicle clusters, occasionally appearing as crystals, and frequently within the lumen of double-membrane vesicles.

Electron tomography data from Belov *et al.*, 2012 [55] revealed that the three-dimensional morphology of poliovirus vesicles is even more complicated than what had been visualized in two dimensions. At 3 hpi, vesicles in the cytoplasm consist of irregularly shaped, branching clusters of single membrane vesicles. The diameters of individual vesicles range from approximately 25–300 nm. These tubular vesicles appear tightly clustered, with their morphologies influenced by neighboring vesicles. At 4 hpi, individual vesicles appear larger in volume and some have associated viruses on their cytoplasmic side. At 7 hpi, most of the vesicular structures are composed of double membrane vesicles, with tightly apposed membranes. These more oval structures appear to be better separated, and less convoluted, with larger interior regions than the membranous structures visible at 3 and 4 hpi. As mentioned above, some of these double membrane structures appear to have gated lumenal areas, with a roughly round central lumen of diameter ~250 nm and a gate opening of ~20 nm. These three-dimensional data may correspond to similar horseshoe shaped structures observed earlier [18,54].

4. Cell Fractionation for Investigation of Poliovirus Replication Factories

Fractionated samples may represent isolation of clusters of poliovirus vesicles. Membrane-associated cellular and viral processes such as virus replication and protein translation are well separated by subcellular fractionation, as shown in Figure 2. The work of Caliguiri and Tamm [58] showed that sucrose density fractions arising from infected cells display significant increases in lipids and that the fraction containing twice the phospholipid content, as compared to the corresponding mock-infected sample, comprises vesicles with a smooth membrane (Fraction 2 in Figure 2) and contains newly synthesized phospholipids. The relative amount of smooth membrane increases during the course of virus infection, as compared to rough membrane [59]. Both Caliguiri and Mosser [60] and Butterworth et al. [61] provided evidence that the most abundant protein in this smooth membrane fraction is a 37 kDa polypeptide "X", which has since been renamed poliovirus protein 2C (see [62] for a table of poliovirus protein nomenclature). Poliovirus protein 2C is associated with membrane remodeling (see Table 2), and is known to play a critical role in virus replication, which occurs on smooth membrane. A fraction containing rough membrane (membrane decorated with ribosomes; Fraction 5 in Figure 2), contains the majority of infectious virus. These data indicate that fractionation successfully separates RNA replication factories from virions as evidenced by a discrete distribution of replicative intermediates. This ability to resolve, by buoyancy, vesicles associated with infectious virus from those associated with RNA production, indicates that vesicles associated with replication are distinct from those associated with protein translation or virion packaging.

Within Fraction 2 (Figure 2), replication complexes active in replication have been named "rosettes" and characterized by the work of Bienz et al., 1990 [63]. As seen in Figure 3, rosettes appear as multivesicular aggregates, the overall configuration appearing as an oval shape approximately 1 μm by 500 nm, with an unknown height. These structures were called rosettes because they are composed of large (200–400 nm diameter) vesicles ("petals") surrounding a more compact central structure [54,63–65]. While rosette structure is not necessary for RNA replication [21], the overall organization and morphology between isolated samples and clusters seen in sections prepared from virus-infected cells appears to be conserved, as seen in a to-scale schematic of Figure 3 included in panel D of Figure 1.

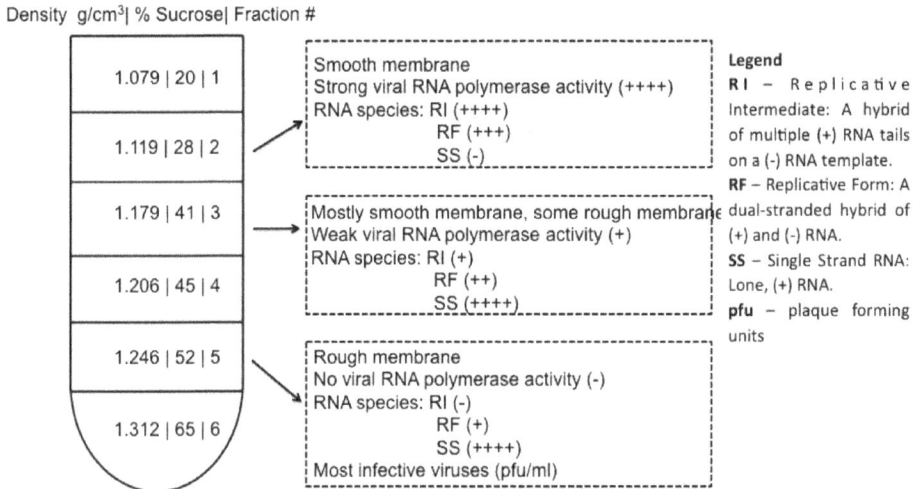

Density g/cm³| % Sucrose| Fraction #

Density g/cm³ | % Sucrose | Fraction #	Properties	Legend
1.079 | 20 | 1	Smooth membrane Strong viral RNA polymerase activity (++++) RNA species: RI (++++) RF (+++) SS (-)	**Legend** **RI** – Replicative Intermediate: A hybrid of multiple (+) RNA tails on a (-) RNA template.
1.119 | 28 | 2		
1.179 | 41 | 3	Mostly smooth membrane, some rough membrane Weak viral RNA polymerase activity (+) RNA species: RI (+) RF (++) SS (++++)	**RF** – Replicative Form: A dual-stranded hybrid of (+) and (-) RNA. **SS** – Single Strand RNA: Lone, (+) RNA.
1.206 | 45 | 4		
1.246 | 52 | 5	Rough membrane No viral RNA polymerase activity (-) RNA species: RI (-) RF (+) SS (++++) Most infective viruses (pfu/ml)	**pfu** – plaque forming units
1.312 | 65 | 6		

Figure 2. Subcellular fractionation of functionally distinct vesicles associated with replication processes. In work by Caliguiri and Tamm, Dounce homogenates of HeLa cells were layered in a discontinuous sucrose gradient, centrifuged at $86,000 \times g$. Upon fractionation of the gradient, membrane bands with the indicated densities were collected. Properties of fractions 2, 3, and 5 are shown. Vesicle content was examined by electron microscopy [58]; RNA polymerase activity of the isolated fraction was measured by tritiated ATP incorporation; the abundance of RNA species (RI, RF, SS) was determined by gel electrophoresis; and viral titer was measured by plaque assay [58,66]. Note: data from Caliguiri and Tamm [58,66,67].

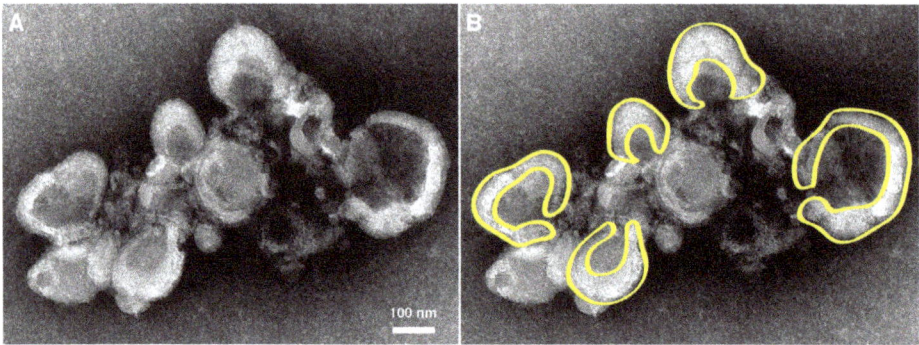

Figure 3. Morphology of rosette isolated from poliovirus-infected cells. (**A**) U body-like densities surround a dense central region of smaller vesicles. Cell fractionated material was prepared and imaged by Evan Rossignol using methods from Schlegel *et al.*, 1996 [54]. Sample from 4 hpi is negatively stained and imaged by electron microscopy; (**B**) Cartoon highlights the horseshoe-shaped (U body-like) configurations of individual vesicles of the cluster. Magnification bar, 100 nm.

5. Conclusions

During poliovirus infection there is a continuum of membranous structures that are formed in concert with assembly of replication factories and in support of RNA replication, as shown schematized in Figure 4. Virus-induced membranous structures begin to appear at 2.5 hpi [52] (Figure 4A), and dominate the cytoplasmic space by 7 hpi [18,53]. These structures serve multiple roles in virus-infected cells, including as a site for RNA replication [68] (Figure 4B,C), for RNA packaging (Figure 4B) [69,70], and for virus maturation [71–75] (Figure 4D,E). It is not clear whether membranous structures transition between roles or are independently formed. Indeed, if there are independent membrane formation paths for different viral roles, this could explain the observed presence in replication factories of markers from different cellular origins (ER, Golgi, *etc.*).

On a cellular scale, RNA synthesis is compartmentalized in infected cells. Negative-sense RNA is synthesized in a perinuclear region of ER, and (+) RNA in a more peripheral location [76]. Despite the fact that packaging of RNA is coupled to its replication [77], differing buoyant densities of fractionated vesicles indicate that packaging of RNA into virions occurs in a population of vesicles separate from those active in replication [64]. This suggests spatial compartmentalization of the viral processes, where (+) RNA migrates from the membranous site of RNA synthesis to membranous packaging sites, as shown in Figure 4B. Thus active replication involves only a fraction of the available membranous structures.

Early in infection poliovirus morphology and other (+) RNA viruses all appear to produce single membrane tubular structures [55,78]. The progression of the structures to double membrane vesicles is also shared. Evidence shows that RNA replication occurs within vesicle lumenal regions for multiple (+) RNA viruses (e.g., [79–82]). These structures may serve to sequester from host cell defenses the double-stranded RNA intermediate during RNA synthesis. In the case of a prototypical arterivirus, equine arteritis virus, the double membrane vesicles formed during infection label strongly for double-stranded RNA. These labeled vesicles display the same dark phenotype seen in a population of poliovirus vesicles. Analysis of the phosphorous content using electron spectroscopy reveals these vesicles in arterivirus infection contain the phosphorous content equivalent to roughly a dozen copies of the viral genome [81]. Although the method by which RNA products can exit the lumen of a double membrane vesicle is unclear for some viruses, poliovirus-infected cells display a possible intermediate. The U-bodies or horseshoe vesicles commonly observed late in poliovirus infection [18,53,54] display a gated lumenal morphology that is consistent with membrane fusion between the outer and inner vesicles, a feat possibly mediated in part by 3AB [51] (see Figure 4C).

Autophagosome-like vesicles are large double membrane structures, about twice the size of U bodies. They play a critical role in maturation of viruses, through acidification of the vesicle, which promotes the final step in virus maturation and viral transmission [46,71,72,83,84]. Better understanding of the pathways of vesicle transitions can help elucidate the infection cycle and the compartmentalization of function.

The complex cellular changes that occur throughout infection require specialized phospholipid biosynthesis that evolves over the time of infection [11,24,59]. In our model each vesicle population serves specialized roles (and perhaps some shared roles) that support the poliovirus replication processes. Different membranous species can co-exist temporally, providing a complex environment for efficient coordination of all aspects of poliovirus replication.

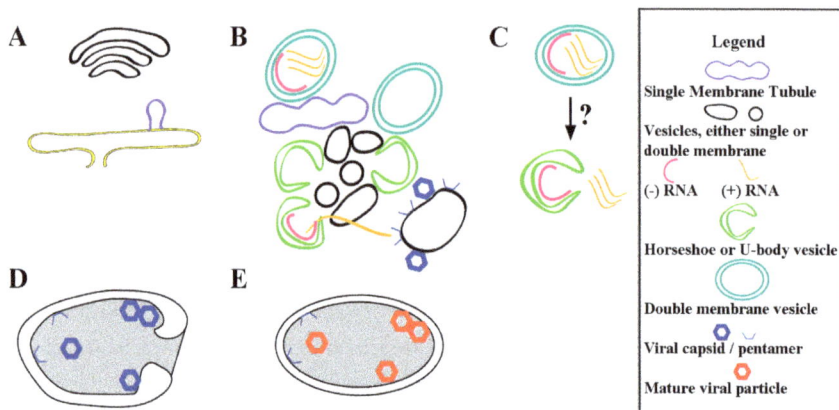

Figure 4. Virus-induced membrane remodeling and expansion for poliovirus replication. (**A**) As early as 2.5 hpi, lipid buds off the ER [52], producing structures comprised of mostly virally induced lipids [11]; (**B**) New and pre-existing cellular vesicles are dynamically remodeled [55] and invaginated from tubules into vesicles, and U-bodies, using a unique lipid composition and specific virus proteins (listed in Tables 1 and 2). Only a fraction of the membranous structures are involved in active RNA replication within lumenal spaces. Throughout infection (+) RNA exits the replication factory, destined for translation or packaging. For virus assembly, (+) RNA interacts with membrane-associated capsid protein pentamers on membranes that are distinct from replication factory membranes [70]; (**C**) Possible mechanism by which closed DMVs transitions to U-body with gated lumen. These structures are visible at timepoints beginning 4 hpi and are believed to continue active replication; (**D**) Autophagosome-like vesicle formation engulfs immature virus particles. These vesicles have twice the diameter of replication vesicles [71], and we posit that these membranes are distinct from those in RNA replication factories, which are still present in the cell; (**E**) Acidification of the autophagosome-like vesicles produces mature virus particles [72].

Acknowledgments: This work was supported by NIH grant no. GM102474 to E.B.

Author Contributions: All authors participated in writing and editing this manuscript.

Conflicts of Interest: The authors declare no conflict of interest.

References

1.   Filman, D.J.; Syed, R.; Chow, M.; Macadam, A.J.; Minor, P.D.; Hogle, J.M. Structural factors that control conformational transitions and serotype specificity in type 3 poliovirus. *EMBO J.* **1989**, *8*, 1567–1579.

2. Mendelsohn, C.L.; Wimmer, E.; Racaniello, V.R. Cellular receptor for poliovirus: Molecular cloning, nucleotide sequence, and expression of a new member of the immunoglobulin superfamily. *Cell* **1989**, *56*, 855–865.

3. Joklik, W.K.; Darnell, J.E., Jr. The adsorption and early fate of purified poliovirus in HeLa cells. *Virology* **1961**, *13*, 439–447.

4. De Sena, J.; Mandel, B. Studies on the *in vitro* uncoating of poliovirus. II. Characteristics of the membrane-modified particle. *Virology* **1977**, *78*, 554–566.

5. Fricks, C.E.; Hogle, J.M. Cell-induced conformational change in poliovirus: Externalization of the amino terminus of VP1 is responsible for liposome binding. *J. Virol.* **1990**, *64*, 1934–1945.

6. Tuthill, T.J.; Bubeck, D.; Rowlands, D.J.; Hogle, J.M. Characterization of early steps in the poliovirus infection process: Receptor-decorated liposomes induce conversion of the virus to membrane-anchored entry-intermediate particles. *J. Virol.* **2006**, *80*, 172–180.

7. Brandenburg, B.; Lee, L.Y.; Lakadamyali, M.; Rust, M.J.; Zhuang, X.; Hogle, J.M. Imaging Poliovirus Entry in Live Cells. *PLoS Biol.* **2007**, *5*, e183.

8. Paul, A.V.; Wimmer, E. Initiation of protein-primed picornavirus RNA synthesis. *Virus Res.* **2015**, *206*, 12–26.

9. Koch, F.; Koch, G. *The Molecular Biology of Poliovirus*; Springer: Vienna, Austria, 1985; p. 209.

10. Egger, D.; Teterina, N.; Ehrenfeld, E.; Bienz, K. Formation of the poliovirus replication complex requires coupled viral translation, vesicle production, and viral RNA synthesis. *J. Virol.* **2000**, *74*, 6570–6580.

11. Cornatzer, W.E.; Fischer, R.G. Effect of poliomyelitis virus on phospholipid metabolism of HeLa cell. *JAMA* **1961**, *178*, 912–914.

12. Cook, P.R. The organization of replication and transcription. *Science* **1999**, *284*, 1790–1795.

13. Den Boon, J.A.; Diaz, A.; Ahlquist, P. Cytoplasmic viral replication complexes. *Cell Host Microbe* **2010**, *8*, 77–85.

14. Romero-Brey, I.; Bartenschlager, R. Membranous replication factories induced by plus-strand RNA viruses. *Viruses* **2014**, *6*, 2826–2857.

15. Neuman, B.W.; Angelini, M.M.; Buchmeier, M.J. Does form meet function in the coronavirus replicative organelle? *Trends Microbiol.* **2014**, *22*, 642–647.

16. Linden, L.; Wolthers, K.C.; van Kuppeveld, F.J.M. Replication and Inhibitors of Enteroviruses and Parechoviruses. *Viruses* **2015**, *7*, 4529–4562.

17. Howes, D.W.; Melnick, J.L.; Reissig, M. Sequence of morphological changes in epithelial cell cultures infected with poliovirus. *J. Exp. Med.* **1956**, *104*, 289–304.

18. Kallman, F.; Williams, R.C.; Dulbecco, R.; Vogt, M. Fine Structure of Changes Produced in Cultured Cells Sampled at Specified Intervals During a Single Growth Cycle of Polio Virus. *J. Biophys. Biochem. Cytol.* **1958**, *4*, 301–308.

19. Vance, D.E.; Trip, E.M.; Paddon, H.B. Poliovirus increases phosphatidylcholine biosynthesis in HeLa cells by stimulation of the rate-limiting reaction catalyzed by CTP: Phosphocholine cytidylyltransferase. *J. Biol. Chem.* **1980**, *255*, 1064–1069.

20. Guinea, R.; Carrasco, L. Effects of fatty acids on lipid synthesis and viral RNA replication in poliovirus-infected cells. *Virology* **1991**, *185*, 473–476.

21. Fogg, M.H.; Teterina, N.L.; Ehrenfeld, E. Membrane requirements for uridylylation of the poliovirus VPg protein and viral RNA synthesis *in vitro*. *J. Virol.* **2003**, *77*, 11408–11416.

22. Penman, S.; Summers, D. Effects on host cell metabolism following synchronous infection with poliovirus. *Virology* **1965**, *27*, 614–620.

23. Hsu, N.Y.; Ilnytska, O.; Belov, G.; Santiana, M.; Chen, Y.H.; Takvorian, P.M.; Pau, C.; van der Schaar, H.; Kaushik-Basu, N.; Balla, T.; *et al.* Viral reorganization of the secretory pathway generates distinct organelles for RNA replication. *Cell* **2010**, *141*, 799–811.

24. Nchoutmboube, J.A.; Viktorova, E.G.; Scott, A.J.; Ford, L.A.; Pei, Z.; Watkins, P.A.; Ernst, R.K.; Belov, G.A. Increased Long Chain acyl-Coa Synthetase Activity and Fatt383y Acid Import Is Linked to Membrane Synthesis for Development of Picornavirus Replication Organelles. *PLoS Pathog.* **2013**, *9*, e1003401.

25. Mosser, A.G.; Caliguiri, L.A.; Tamm, I. Incorporation of lipid precursors into cytoplasmic membranes of poliovirus-infected HeLa cells. *Virology* **1972**, *47*, 39–47.

26. Guinea, R.; Carrasco, L. Phospholipid biosynthesis and poliovirus genome replication, two coupled phenomena. *EMBO J.* **1990**, *9*, 2011–2016.

27. McMahon, H.T.; Gallop, J.L. Membrane curvature and mechanisms of dynamic cell membrane remodelling. *Nature* **2005**, *438*, 590–596.

28. Lewis, R.N.A.H.; Mannock, D.A.; McElhaney, R.N.; Turner, D.C.; Gruner, S.M. Effect of fatty acyl chain length and structure on the lamellar gel to liquid-crystalline and lamellar to reversed hexagonal phase transitions of aqueous phosphatidylethanolamine dispersions. *Biochemistry* **1989**, *28*, 541–548.

29. Szule, J.A.; Fuller, N.L.; Peter Rand, R. The Effects of Acyl Chain Length and Saturation of Diacylglycerols and Phosphatidylcholines on Membrane Monolayer Curvature. *Biophys. J.* **2015**, *83*, 977–984.

30. Belov, G.A.; Altan-Bonnet, N.; Kovtunovych, G.; Jackson, C.L.; Lippincott-Schwartz, J.; Ehrenfeld, E. Hijacking components of the cellular secretory pathway for replication of poliovirus RNA. *J. Virol.* **2007**, *81*, 558–567.

31. Arita, M.; Kojima, H.; Nagano, T.; Okabe, T.; Wakita, T.; Shimizu, H. Oxysterol-binding protein family I is the target of minor enviroxime-like compounds. *J. Virol.* **2013**, *87*, 4252–4260.

32. Richards, A.L.; Soares-Martins, J.A.P.; Riddell, G.T.; Jackson, W.T. Generation of unique poliovirus RNA replication organelles. *MBio* **2014**, *5*.

33. Ishiyama, N.; Hill, C.M.; Bates, I.R.; Harauz, G. The formation of helical tubular vesicles by binary monolayers containing a nickel-chelating lipid and phosphoinositides in the presence of basic polypeptides. *Chem. Phys. Lipids* **2002**, *114*, 103–111.

34. Lenoir, M.; Grzybek, M.; Majkowski, M.; Rajesh, S.; Kaur, J.; Whittaker, S.B.-M.; Coskun, Ü.; Overduin, M. Structural Basis of Dynamic Membrane Recognition by trans-Golgi Network Specific FAPP Proteins. *J. Mol. Biol.* **2015**, *427*, 966–981.

35. Khan, I.; Katikaneni, D.S.; Han, Q.; Sanchez-Felipe, L.; Hanada, K.; Ambrose, R.L.; Mackenzie, J.M.; Konan, K.V. Modulation of Hepatitis C Virus Genome Replication by Glycosphingolipids and Four-Phosphate Adaptor Protein 2. *J. Virol.* **2014**, *88*, 12276–12295.

36. Dowler, S.; Kular, G.; Alessi, D.R. Protein Lipid Overlay Assay. *Sci. Signal.* **2002**, *2002*, pl6.

37. Strating, J.R.P.M.; van der Linden, L.; Albulescu, L.; Bigay, J.; Arita, M.; Delang, L.; Leyssen, P.; van der Schaar, H.M.; Lanke, K.H.W.; Thibaut, H.J.; *et al.* Itraconazole Inhibits Enterovirus Replication by Targeting the Oxysterol-Binding Protein. *Cell Rep.* **2015**, *10*, 600–615.

38. Adkins, E.M.; Samuvel, D.J.; Fog, J.U.; Eriksen, J.; Jayanthi, L.D.; Vaegter, C.B.; Ramamoorthy, S.; Gether, U. Membrane Mobility and Microdomain Association of the Dopamine Transporter Studied with Fluorescence Correlation Spectroscopy and Fluorescence Recovery after Photobleaching. *Biochemistry* **2007**, *46*, 10484–10497.

39. Owen, D.M.; Williamson, D.; Rentero, C.; Gaus, K. Quantitative Microscopy: Protein Dynamics and Membrane Organisation. *Traffic* **2009**, *10*, 962–971.

40. Arita, M. Phosphatidylinositol-4 kinase III beta and oxysterol-binding protein accumulate unesterified cholesterol on poliovirus-induced membrane structure. *Microbiol. Immunol.* **2014**, *58*, 239–256.

41. Ilnytska, O.; Santiana, M.; Hsu, N.Y.; Du, W.L.; Chen, Y.H.; Viktorova, E.G.; Belov, G.; Brinker, A.; Storch, J.; Moore, C.; Dixon, J.L.; Altan-Bonnet, N. Enteroviruses harness the cellular endocytic machinery to remodel the host cell cholesterol landscape for effective viral replication. *Cell Host Microbe* **2013**, *14*, 281–293.

42. Chang, T.-Y.; Chang, C.C.Y.; Ohgami, N.; Yamauchi, Y. Cholesterol Sensing, Trafficking, and Esterification. *Annu. Rev. Cell Dev. Biol.* **2006**, *22*, 129–157.

43. Aizaki, H.; Lee, K.-J.; Sung, V.M.-H.; Ishiko, H.; Lai, M.M.C. Characterization of the hepatitis C virus RNA replication complex associated with lipid rafts. *Virology* **2004**, *324*, 450–461.

44. Paul, D.; Hoppe, S.; Saher, G.; Krijnse-Locker, J.; Bartenschlager, R. Morphological and Biochemical Characterization of the Membranous Hepatitis C Virus Replication Compartment. *J. Virol.* **2013**, *87*, 10612–10627.

45. Mesmin, B.; Bigay, J.; Moser von Filseck, J.; Lacas-Gervais, S.; Drin, G.; Antonny, B. A four-step cycle driven by PI(4)P hydrolysis directs sterol/PI(4)P exchange by the ER-Golgi tether OSBP. *Cell* **2013**, *155*, 830–843.

46. Jackson, W.T. Poliovirus-induced changes in cellular membranes throughout infection. *Curr. Opin. Virol.* **2014**, *9*, 67–73.

47. Cho, M.W.; Teterina, N.; Egger, D.; Bienz, K.; Ehrenfeld, E. Membrane rearrangement and vesicle induction by recombinant poliovirus 2C and 2BC in human cells. *Virology* **1994**, *202*, 129–145.

48. Doedens, J.R.; Giddings, T.H.; Kirkegaard, K. Inhibition of endoplasmic reticulum-to-Golgi traffic by poliovirus protein 3A: Genetic and ultrastructural analysis. *J. Virol.* **1997**, *71*, 9054–9064.

49. Suhy, D.A.; Giddings, T.H.; Kirkegaard, K. Remodeling the endoplasmic reticulum by poliovirus infection and by individual viral proteins: An autophagy-like origin for virus-induced vesicles. *J. Virol.* **2000**, *74*, 8953–8965.

50. Teterina, N.L.; Gorbalenya, A.E.; Egger, D.; Bienz, K.; Ehrenfeld, E. Poliovirus 2C protein determinants of membrane binding and rearrangements in mammalian cells. *J. Virol.* **1997**, *71*, 8962–8972.

51. Wang, J.; Ptacek, J.B.; Kirkegaard, K.; Bullitt, E. Double-membraned liposomes sculpted by poliovirus 3AB protein. *J. Biol. Chem.* **2013**, *288*, 27287–27298.

52. Bienz, K.; Egger, D.; Pasamontes, L. Association of polioviral proteins of the P2 genomic region with the viral replication complex and virus-induced membrane synthesis as visualized by electron microscopic immunocytochemistry and autoradiography. *Virology* **1987**, *160*, 220–226.

53. Dales, S.; Eggers, H.J.; Tamm, I. Electron Microscopic Study of the Formation of Poliovirus. *Virology* **1965**, *26*, 379–389.

54. Schlegel, A.; Giddings, T.H.; Ladinsky, M.S.; Kirkegaard, K. Cellular origin and ultrastructure of membranes induced during poliovirus infection. *J. Virol.* **1996**, *70*, 6576–6588.

55. Belov, G.A.; Nair, V.; Hansen, B.T.; Hoyt, F.H.; Fischer, E.R.; Ehrenfeld, E. Complex Dynamic Development of Poliovirus Membranous Replication Complexes. *J. Virol.* **2012**, *86*, 302–312.

56. Bienz, K.; Egger, D.; Wolff, D.A. Virus replication, cytopathology, and lysosomal enzyme response of mitotic and interphase Hep-2 cells infected with poliovirus. *J. Virol.* **1973**, *11*, 565–574.

57. Mattern, C.F.T.; Daniel, W.A. Replication of poliovirus in HeLa cells: Electron microscopic observations. *Virology* **1965**, *26*, 646–663.

58. Caliguiri, L.A.; Tamm, I. The role of cytoplasmic membranes in poliovirus biosynthesis. *Virology* **1970**, *42*, 100–111.

59. Mosser, A.G.; Caliguiri, L.A.; Scheid, A.S.; Tamm, I. Chemical and enzymatic characteristics of cytoplasmic membranes of poliovirus-infected HeLa cells. *Virology* **1972**, *47*, 30–38.

60. Caliguiri, L.A.; Mosser, A.G. Proteins associated with the poliovirus RNA replication complex. *Virology* **1971**, *46*, 375–386.

61. Butterworth, B.E.; Shimshick, E.J.; Yin, F.H. Association of the polioviral RNA polymerase complex with phospholipid membranes. *J. Virol.* **1976**, *19*, 457–466.

62. Pallansch, M.A.; Kew, O.M.; Semler, B.L.; Omilianowski, D.R.; Anderson, C.W.; Wimmer, E.; Rueckert, R.R. Protein processing map of poliovirus. *J. Virol.* **1984**, *49*, 873–880.

63. Bienz, K.; Egger, D.; Troxler, M.; Pasamontes, L. Structural organization of poliovirus RNA replication is mediated by viral proteins of the P2 genomic region. *J. Virol.* **1990**, *64*, 1156–1163.

64. Bienz, K.; Egger, D.; Pfister, T.; Troxler, M. Structural and functional characterization of the poliovirus replication complex. *J. Virol.* **1992**, *66*, 2740–2747.

65. Egger, D.; Pasamontes, L.; Bolten, R.; Boyko, V.; Bienz, K. Reversible dissociation of the poliovirus replication complex: Functions and interactions of its components in viral RNA synthesis. *J. Virol.* **1996**, *70*, 8675–8683.

66. Caliguiri, L.A.; Tamm, I. Characterization of poliovirus-specific structures associated with cytoplasmic membranes. *Virology* **1970**, *42*, 112–122.

67. Caliguiri, L.A.; Tamm, I. Membranous structures associated with translation and transcription of poliovirus RNA. *Science* **1969**, *166*, 885–886.

68. Bienz, K.; Egger, D.; Rasser, Y.; Bossart, W. Kinetics and location of poliovirus macromolecular synthesis in correlation to virus-induced cytopathology. *Virology* **1980**, *100*, 390–399.

69. Horne, R.W.; Nagington, J. Electron microscope studies of the development and structure of poliomyelitis virus. *J. Mol. Biol.* **1959**, *1*, 333.

70. Pfister, T.; Pasamontes, L.; Troxler, M.; Egger, D.; Bienz, K. Immunocytochemical localization of capsid-related particles in subcellular fractions of poliovirus-infected cells. *Virology* **1992**, *188*, 676–684.

71. Chen, Y.-H.; Du, W.; Hagemeijer, M.C.; Takvorian, P.M.; Pau, C.; Cali, A.; Brantner, C.A.; Stempinski, E.S.; Connelly, P.S.; Ma, H.-C.; *et al.* Phosphatidylserine Vesicles Enable Efficient En Bloc Transmission of Enteroviruses. *Cell* **2015**, *160*, 619–630.

72. Richards, A.L.; Jackson, W.T. Intracellular Vesicle Acidification Promotes Maturation of Infectious Poliovirus Particles. *PLoS Pathog.* **2012**, *8*, e1003046.

73. Taylor, M.P.; Burgon, T.B.; Kirkegaard, K.; Jackson, W.T. Role of microtubules in extracellular release of poliovirus. *J. Virol.* **2009**, *83*, 6599–6609.

74. Jackson, W.T.; Giddings, T.H.; Taylor, M.P.; Mulinyawe, S.; Rabinovitch, M.; Kopito, R.R.; Kirkegaard, K. Subversion of cellular autophagosomal machinery by RNA viruses. *PLoS Biol.* **2005**, *3*, e156.

75. Bird, S.W.; Maynard, N.D.; Covert, M.W.; Kirkegaard, K. Nonlytic viral spread enhanced by autophagy components. *Proc. Natl. Acad. Sci. USA* **2014**, *111*, 13081–13086.

76. Egger, D.; Bienz, K. Intracellular location and translocation of silent and active poliovirus replication complexes. *J. Gen. Virol.* **2005**, *86*, 707–718.

77. Nugent, C.I.; Johnson, K.L.; Sarnow, P.; Kirkegaard, K. Functional coupling between replication and packaging of poliovirus replicon RNA. *J. Virol.* **1999**, *73*, 427–435.

78. Limpens, R.W.A.L.; van der Schaar, H.M.; Kumar, D.; Koster, A.J.; Snijder, E.J.; van Kuppeveld, F.J.M.; Bárcena, M. The transformation of enterovirus replication structures: A three-dimensional study of single- and double-membrane compartments. *MBio* **2011**, *2*.

79. Gosert, R.; Kanjanahaluethai, A.; Egger, D.; Bienz, K.; Baker, S.C. RNA replication of mouse hepatitis virus takes place at double-membrane vesicles. *J. Virol.* **2002**, *76*, 3697–3708.

80. Knoops, K.; Kikkert, M.; van den Worm, S.H.E.; Zevenhoven-Dobbe, J.C.; van der Meer, Y.; Koster, A.J.; Mommaas, A.M.; Snijder, E.J. SARS-coronavirus replication is supported by a reticulovesicular network of modified endoplasmic reticulum. *PLoS Biol.* **2008**, *6*, 1957–1974.

81. Knoops, K.; Barcena, M.; Limpens, R.W.A.L.; Koster, A.J.; Mommaas, A.M.; Snijder, E.J. Ultrastructural Characterization of Arterivirus Replication Structures: Reshaping the Endoplasmic Reticulum to Accommodate Viral RNA Synthesis. *J. Virol.* **2012**, *86*, 2474–2487.

82. Romero-Brey, I.; Merz, A.; Chiramel, A.; Lee, J.-Y.; Chlanda, P.; Haselman, U.; Santarella-Mellwig, R.; Habermann, A.; Hoppe, S.; Kallis, S.; *et al.* Three-dimensional architecture and biogenesis of membrane structures associated with hepatitis C virus replication. *PLoS Pathog.* **2012**, *8*, e1003056.

83. Taylor, M.P.; Kirkegaard, K. Modification of Cellular Autophagy Protein LC3 by Poliovirus. *J. Virol.* **2007**, *81*, 12543–12553.

84. Robinson, S.M.; Tsueng, G.; Sin, J.; Mangale, V.; Rahawi, S.; McIntyre, L.L.; Williams, W.; Kha, N.; Cruz, C.; Hancock, B.M.; *et al.* Coxsackievirus B Exits the Host Cell in Shed Microvesicles Displaying Autophagosomal Markers. *PLoS Pathog.* **2014**, *10*, e1004045.

Amino Terminal Region of Dengue Virus NS4A Cytosolic Domain Binds to Highly Curved Liposomes

Yu-Fu Hung, Melanie Schwarten, Silke Hoffmann, Dieter Willbold,
Ella H. Sklan and Bernd W. Koenig

Abstract: Dengue virus (DENV) is an important human pathogen causing millions of disease cases and thousands of deaths worldwide. Non-structural protein 4A (NS4A) is a vital component of the viral replication complex (RC) and plays a major role in the formation of host cell membrane-derived structures that provide a scaffold for replication. The N-terminal cytoplasmic region of NS4A(1–48) is known to preferentially interact with highly curved membranes. Here, we provide experimental evidence for the stable binding of NS4A(1–48) to small liposomes using a liposome floatation assay and identify the lipid binding sequence by NMR spectroscopy. Mutations L6E;M10E were previously shown to inhibit DENV replication and to interfere with the binding of NS4A(1–48) to small liposomes. Our results provide new details on the interaction of the N-terminal region of NS4A with membranes and will prompt studies of the functional relevance of the curvature sensitive membrane anchor at the N-terminus of NS4A.

Reprinted from *Viruses*. Cite as: Hung, Y.-F.; Schwarten, M.; Hoffmann, S.; Willbold, D.; Sklan, E.H.; Koenig, B.W. Amino Terminal Region of Dengue Virus NS4A Cytosolic Domain Binds to Highly Curved Liposomes. *Viruses* **2015**, *7*, 4119–4130.

1. Introduction

Dengue virus (DENV), the causative agent of dengue fever, is a positive strand RNA, enveloped virus belonging to the *Flaviviridae* family. The viral RNA is translated into a single polyprotein which is processed by cellular and viral proteases into three structural proteins (capsid, premembrane, and envelope) and the seven non-structural (NS) proteins NS1, NS2A, NS2B, NS3, NS4A, NS4B, and NS5 [1]. The NS proteins are not found in the mature virion but are crucial for viral replication. Synthesis of viral RNA takes place in replication complexes (RCs) that contain essential NS proteins, viral RNA and host cell factors [2]. Upon DENV infection, a complex and continuous network of ER membrane-derived vesicular structures and convoluted membranes is formed. These structures contain the viral replication complexes and the sites of virion assembly [3–5].

NS4A is small integral membrane protein containing four predicted transmembrane segments (pTMSs) [1]. Although pTMS4, often referred to as the 2k

351

fragment, is not part of the mature NS4A, it serves as a signal peptide for the ER localization of NS4B and is cleaved from the mature NS4A [6]. Experimental data verify that pTMS1 and pTMS3 span the membrane while pTMS2 is embedded in the luminal leaflet of the ER membrane and does not span it [1].

NS4A is crucial for the formation of the virus-induced membrane structures. Expression of NS4A lacking the 2k fragment alone is sufficient to induce membrane alterations that resemble the DENV-induced highly curved membranes that harbor the RCs [1]. Clearly, to induce these structures NS4A will have to closely interact with host membranes. However, the mechanism by which NS4A induces the curved morphology of these newly formed membranes is still unknown. Insertion of amphipathic helices into one leaflet of a membrane bilayer, as well as oligomerization of membrane proteins are among the mechanisms known to participate in the induction of membrane curvature [7]. Molecular dynamics (MD) simulations suggest that pTMS2 of NS4A could support membrane undulations upon stable association with the membrane [8]. Curved vesicular structures might also be induced via homooligomerization of NS4A [9,10]. While we have previously shown that the NS4A N-terminal cytoplasmic region is implicated in its oligomerization [10], a recent study demonstrated that pTMS1 is the major determinant in this process [9]. Introduction of two point mutations at the N-terminal of NS4A (L6E and M10E) reduced both the amphipathic character of this region and NS4A homooligomerization and abolished viral replication [10].

NS4A is an essential component of the viral RC [1]. Direct interaction of NS4A with the cytoskeletal protein vimentin was reported to be necessary for correct localization of the RC at the perinuclear site [11]. The vimentin binding site was found to be located at the N-terminal 50 residues of NS4A [11]. NS4A was also reported to bind NS4B, another component of the RC, via pTMS1 [9]. Mutational analysis suggests a functional relevance of this interaction for viral replication [9]. It was speculated that the interaction between NS4A and NS4B in concert with NS4A oligomerization and NS4B dimerization may play a role in the spatial and temporal regulation of distinct molecular complexes involved in the viral infection cycle [9].

Our previous circular dichroism (CD) data demonstrated that NS4A(1–48) interacts with highly curved small unilamellar liposomes under α-helix formation, while mutated NS4A(1–48, L6E;M10E) does not [12]. Surface plasmon resonance data indicated a seven fold-greater association of wild type NS4A(1–48) with immobilized liposomes compared to the mutant [12]. The structure of NS4A(1–48) in presence of membrane mimicking SDS micelles was characterized by NMR [12]. To further extend these results we used liposome flotation for direct proof of NS4A(1–48) binding to free liposomes. The exact location of lipid binding sites in the amino acid sequence of NS4A(1–48) was addressed by NMR spectroscopy. Our findings provide a basis for specific structure-function studies that will enhance our understanding of the role of NS4A and might provide future targets for anti-viral intervention.

2. Materials and Methods

2.1. Peptide Production

The peptide NS4A(1–48) corresponds to amino acid residues 1–48 from the *N*-terminal of NS4A of dengue virus serotype 2 (NCBI Protein database accession number: NP739588). A mutant peptide containing the mutations L6E and M10E was designated NS4A(1–48, L6E;M10E). The two NS4A peptides were recombinantly produced in *E. coli* BL21 cells and enzymatically cleaved from the affinity tag as described earlier [13]. Uniform isotope labeling with ^{15}N or ^{13}C, ^{15}N was achieved by expression in M9 medium containing ^{15}N ammonium chloride and ^{13}C glucose (Eurisotop, Saarbrücken, Germany) as the sole source of nitrogen and carbon, respectively. Unlabeled peptides were expressed in LB media.

2.2. Fluorescence Labeling

Alexa Fluor 488 succinimidyl ester (NHS ester) was purchased from Life Technologies, Darmstadt, Germany. The dye was dissolved in anhydrous DMSO at a concentration of 3 mM immediately prior to the labeling reaction. For the reaction 300 μL from a 100 μM NS4A(1–48) or NS4A(1–48, L6E;M10E) stock in sample buffer (50 mM sodium phosphate, pH 6.8, 150 mM NaCl) were combined with 100 μL of 0.4 M NaHCO$_3$ and the pH was adjusted to 8.3. This 400 μL peptide solution was supplemented with 100 μL of the Alexa Fluor 488 NHS ester in DMSO resulting in an approximately ten-fold excess of dye-over-peptide. The labeling reaction was wrapped in aluminum foil and incubated on a rocking platform shaker at 4 °C for 16 h. Labeled protein and free dye were separated on a Superdex 75 10/300 GL column (GE Healthcare, Freiburg, Germany) operated on an ÄKTApurifier system (GE Healthcare).

2.3. Liposome Preparation

The lipid 1-palmitoyl-2-oleoyl-*sn*-glycero-3-phosphocholine (POPC) in chloroform solution was purchased from Avanti Polar Lipids (Alabaster, AL, USA). Small unilamellar lipid vesicles (SUVs) were prepared from chloroform-free POPC dispersions (20 mg·mL^{-1}) in sample buffer as described earlier [12]. SUVs were obtained by sequential extrusion through 50 nm (15 times) and 30 nm (15 times) Nuclepore polycarbonate membranes (GE Healthcare) with nominal pore diameter of either 50 or 30 nm, followed by sonication with a 3 mm microtip of a Branson 250 sonifier (15 cycles of sonication, 20 s each, interrupted by cooling for 2 min after each cycle). Sonicated SUVs were centrifuged for 10 min at 16,100× g and 10 °C in a refrigerated Eppendorf 5415 R tabletop centrifuge to remove any titanium abrasion of the microtip from the sample. The hydrodynamic radius of each liposome preparation was determined by dynamic light scattering (DLS) using a Dyna Pro instrument (Protein Solutions, Lakewood, NJ, USA) equipped with a 3 mm path length 45 µL quartz cell. Liposome solutions (20 mg of POPC per mL) were diluted 100-fold with buffer directly after extrusion or sonication and measured immediately. Data were analyzed with Dynamics V6 software distributed with the instrument. Experimental data were fitted to the model of Rayleigh spheres.

2.4. Liposome Floatation Assay

Equal volumes of 80 µM suspensions of Alexa Fluor 488-labeled peptides or free dye and sonicated POPC SUV (20 mg· mL^{-1}) in sample buffer were combined and mixed at room temperature for 5 min. 100 µL of each of the three resulting samples were thoroughly mixed with 100 µL of a 70% (w/v) sucrose solution to obtain homogeneous solutions containing 20 µM of either Alexa Fluor 488-labeled peptide or free Alexa Fluor 488 dye, POPC liposomes (5 mg· mL^{-1}) and 35% (w/v) sucrose in sample buffer. Sucrose solution (400 µL of a 70% (w/v) solution in sample buffer) was transferred to the bottom of a Polyallomer centrifuge tube (11 mm × 34 mm; Beckman Coulter) followed by a second layer formed by the 200 µL of 35% (w/v) sucrose solution containing one of the labeled NS4A peptides or the free dye and POPC liposomes. Finally, each sample was carefully overlaid with two cushions of decreasing sucrose concentration, i.e., 1.2 mL of 20% (w/v) followed by 200 µL of 10% (w/v) sucrose in sample buffer (cf. scheme in Figure 1, left). Samples were centrifuged for 14 h at 259,000× g and 4 °C in a Beckman Coulter Optima Max-XP ultracentrifuge using a TLS 55 swinging bucket rotor. Fluorescence images were taken in front of a Mini Transilluminator (Bio-Rad, Munich, Germany) prior to and immediately after centrifugation.

2.5. Nuclear Magnetic Resonance (NMR) Spectroscopy

NMR experiments were conducted at 30 °C on Bruker Avance III HD NMR and Varian VNMRS instruments, equipped with cryogenic Z-axis pulse-field-gradient (PFG) triple resonance probes operating at proton frequencies of 700 and 900 MHz, respectively. Samples for resonance assignment contained 300 µM [U-^{15}N, ^{13}C]-labeled NS4A(1–48) in sample buffer (50 mM sodium phosphate, pH 6.8, 150 mM NaCl) as used for the liposome flotation experiments but supplemented with 10% (v/v) deuterium oxide and 0.03% (w/v) NaN$_3$ (referred to as NMR buffer). Assignment of protein backbone resonances was accomplished using a combined set of heteronuclear multidimensional NMR experiments: 2D (^{1}H–^{15}N)-HSQC [14,15], 2D (^{1}H–^{13}C)-HSQC [16], 3D HNCA [17], 3D BT-HNCO [18], and 3D HNcaCO [19]. ^{1}H and ^{13}C chemical shifts were referenced directly to internal 4,4-dimethyl-4-silapentane-1-sulfonic acid (DSS) at 0 ppm and ^{15}N chemical shifts were referenced indirectly to DSS using the absolute ratio of the ^{15}N and ^{1}H zero point frequencies [20]. NMR data were processed using NMRPipe, v.8.1 [21] and evaluated with CcpNmr v.2.4 [22].

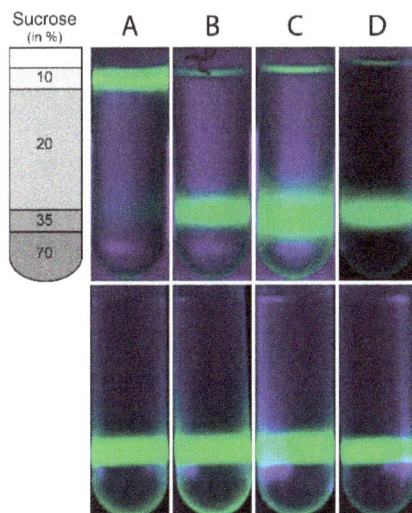

Figure 1. Liposome floatation assay of wild type and mutant NS4A(1–48). Alexa-488-labeled NS4A(1–48) (**A**); Alexa-488-labeled NS4A(1–48, L6E;M10E) (**B**); or free Alexa Fluor 488 dye (**C**) were mixed with sonicated POPC liposomes and loaded with the 35% (*w/v*) sucrose layer of a sucrose step gradient schematically shown on the left; Alexa-488-labeled NS4A(1–48) without liposomes was loaded with the 35% (*w/v*) sucrose layer in lane (**D**). Note: The narrow green line at the top of tubes (**B**), (**C**) and (**D**) results from reflection of fluorescence light at the air buffer interface rather than the presence of dye. Fluorescence images of the four tubes recorded prior to ultracentrifugation are shown in the lower row.

The interaction of NS4A(1–48) and NS4A(1–48, L6E;M10E) with sonicated POPC liposomes was studied based on a series of 2D (^{1}H–^{15}N) Heteronuclear Single Quantum Coherence (HSQC) spectra recorded with 40 µM peptide in NMR buffer but with different amounts of liposomes (0; 2.5; 5, or 10 mg of POPC per mL). Spectra were acquired with 150 complex data points in the ^{15}N time domain, up to 128 scans per *t*1 increment and a recycle delay of 1.5 s. Data were processed and analyzed using NMRPipe.

3. Results

3.1. Wild Type NS4A(1–48) Binds to Highly Curved POPC SUVs

Binding of fluorescently labeled proteins to liposomes can be visualized using a simple floatation assay [23]. POPC bilayers have a mass density very close to $1 \, g \cdot cm^{-3}$ [24] at room temperature. Thus, POPC liposomes will migrate to the top layer containing the lowest sucrose concentration in a step gradient of decreasing sucrose concentrations upon centrifugation. In contrast, a small protein of ~5 kDa molecular weight is expected to have a mass density well above $1.41 \, g \cdot cm^{-3}$ [25] and thus will accumulate in a sucrose rich layer of high mass density. Alexa-488-labeled NS4A(1–48) apparently binds to small sonicated POPC liposomes and migrates with the liposomes to the top of the gradient (Figure 1A). In contrast, the mutated peptide NS4A(1–48, L6E;M10E) remains in the high density layer with 35% (w/v) sucrose even after 14 h of centrifugation (Figure 1B), indicating that this peptide does not bind to liposomes. As a negative control, the Alexa-488 free dye was loaded to the 35% (w/v) sucrose layer with POPC liposomes. The free dye remained in the high-density region after centrifugation (Figure 1C) indicating a lack of association with the POPC SUVs. The fluorescent band of the free dye has a larger vertical extension than that of the labeled NS4A(1–48, L6E;M10E) after 14 h of centrifugation (Figure 1B,C). This probably reflects the larger diffusion coefficient of the low molecular weight dye molecule ($643.4 \, g \cdot mol^{-1}$). In a second control experiment Alexa-488-labeled NS4A(1-48) was loaded to the 35% (w/v) sucrose layer without adding POPC liposomes. As expected, the peptide did not significantly migrate during the 14 h of centrifugation and remained almost completely in the 35% (w/v) sucrose layer (Figure 1D). The floatation results confirm the binding of the wild type peptide to small liposomes.

3.2. NMR Identifies Regions of NS4A(1–48) Associated with POPC Liposomes

NMR spectroscopy was used to identify the regions of NS4A(1–48) associated with POPC liposomes. Almost complete assignment of the expected ^1HN, ^{15}N, $^{13}C\alpha$ and $^{13}C'$ backbone resonances was accomplished for both NS4A(1–48) and NS4A(1–48, L6E;M10E) in NMR buffer. No assignments were obtained for S1 and N42. In addition to P14 there are four residues which only lack amide (^1HN, ^{15}N) assignments. Assignments have been deposited at the Biological Magnetic Resonance Data Bank (BMRB) under accession number 25586 for NS4A(1–48) and accession number 25676 for NS4A(1–48, L6E;M10E).

Amide ^1HN, ^{15}N cross peaks in HSQC spectra of NS4A(1–48) were used to monitor peptide interaction with sonicated POPC liposomes in an amino acid residue resolved manner. The hydrodynamic radius of the POPC liposomes was ~26 nm based on DLS measurements. Backbone cross peaks for most of the 48 amino acid residues of NS4A(1–48) were identified in buffer without liposomes except for S1, L2, P14, M17, H32, N42 and H43. The observed cross peaks characterize the free peptide conformation. Addition of increasing amounts of sonicated POPC liposomes at constant peptide concentration caused gradual peak intensity reductions in a peptide region specific manner (Figure 2A). Interestingly, peak positions did not significantly change, except for R12, which showed a small shift. Such a behavior is typical for slow or intermediate exchange of the peptide between the free and liposome-bound state. Strongest peak intensity reduction is observed in the N-terminal region extending up to K20 (Figure 2A). Some peaks in this region completely disappear already at 2.5 mg· mL^{-1} POPC (N5, I7, E9, G11, K20) while the others are reduced to less than 40%. At the highest liposome concentration studied (10 mg· mL^{-1}) only two cross peaks from this peptide region remain visible and show low intensity. It is likely, that a number of amino acid residues of the N-terminal region bind directly to the liposome. Binding will change the chemical environment and strongly reduce the rotational correlation time of the amino acid residues in direct contact with lipids. NMR signals of bound residues are likely broadened beyond detection. Exchange dynamics may differ somewhat among the amino acid residues in this region, explaining the variable intensity reduction of the free state cross peaks.

Cross peaks of the central region from A21 through L31 of NS4A(1–48) show rather uniform intensity reduction upon gradual liposome addition (Figure 2A). All free state peaks remain visible and retain about 10% of their original intensity even in the presence of 10 mg· mL^{-1} POPC. Different scenarios might contribute to the reduction of the free state peak intensities. Peptides that are anchored with their N-terminal region in the liposome might retain their free state conformation in the central domain, albeit with a reduced overall rotational correlation time and thus lower peak intensities. In addition, amino acid residues of the central region of some NS4A peptides might bind directly to the liposome leading to the disappearance of the corresponding NMR signals. The uniform peak intensity reduction pattern in the central region may suggest a concerted binding of this amino acid stretch, e.g., as one secondary structure element, to the liposome.

Finally, amino acid residues in the C-terminal region of NS4A(1–48) from T33 to L48 show the smallest reduction in the free state peak intensities upon liposome addition (Figure 2A). Perhaps, these small intensity reductions might be entirely caused by anchoring of the peptides via amino acid residues in the N-terminal and perhaps the central regions.

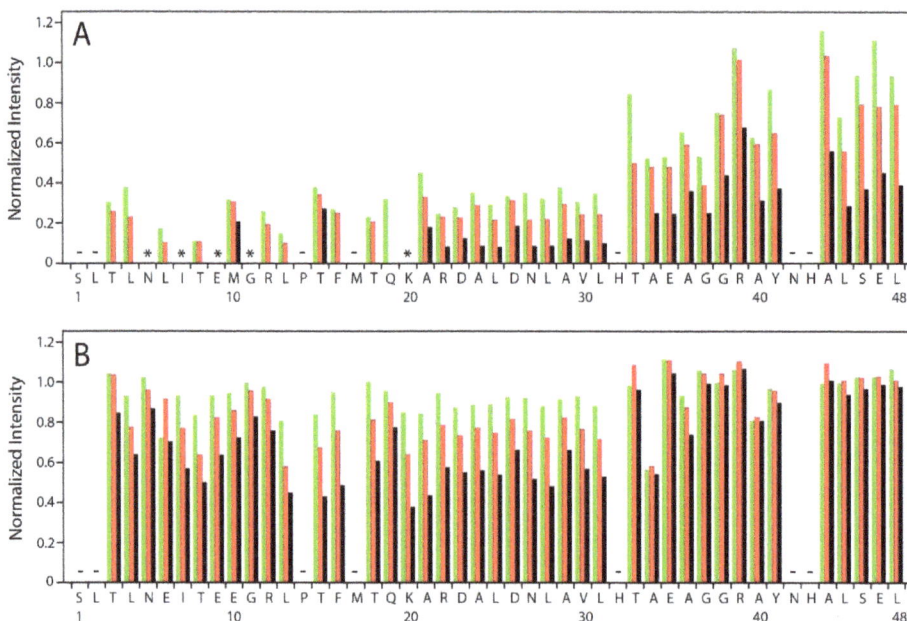

Figure 2. Intensity of backbone amide cross peaks in HSQC spectra of NS4A(1–48) (**A**) and NS4A(1–48, L6E;M10E) (**B**) recorded at various lipid concentrations. Peak intensities measured at 2.5 (green), 5 (red) and 10 mg· mL^{-1} POPC (black) in the sample were normalized to the intensity of the same signal observed in lipid-free buffer and are shown as a function of the amino acid sequence of the studied peptide. Cross peaks not observed in the lipid-free sample are indicated by minus signs. Cross peaks that are present in buffer but completely disappear after addition of 2.5 mg· mL^{-1} POPC are indicated by asterisk.

3.3. NMR Data Suggest Diminished Interaction of NS4A(1–48, L6E;M10E) with POPC Liposomes

Inspection of HSQC spectra of mutant NS4A(1–48, L6E;M10E) in buffer and with increasing amounts of sonicated POPC liposomes revealed no changes in cross peak positions and only a minor influence of lipid addition on cross peak intensities of backbone amide signals. All cross peaks observed in buffer remain visible in presence of liposomes and retain at least 40% of their original intensity upon addition of 10 mg· mL^{-1} of POPC liposomes (Figure 2B). Almost no peak reduction is observed for the C-terminal region of NS4A(1–48, L6E;M10E). Peak intensity reductions in the N-terminal and central regions are quite moderate, in particular at low liposome concentration (2.5 and 5 mg· mL^{-1} POPC). Even at 10 mg· mL^{-1} POPC liposomes cross peak intensities remain between 40% and 80% (N-terminal region) or between 40% and 60% (central region) with respect to intensities measured in buffer.

358

Apparently some interaction of the mutant peptide with liposomes is retained at the highest lipid concentration studied. Amino acid residues in the central and perhaps in the N-terminal part of the mutant peptide are likely to make transient contact with the liposome. However, the observed peak intensity reductions are much weaker in case of NS4A(1–48, L6E;M10E) than for wild type.

4. Discussion

The L6E and M10E mutations that disrupt the amphipathic character of the NS4A *N*-terminal abolish viral replication, indicating that the 48 N-terminal residues of NS4A play a crucial role in replication. Furthermore, these mutations had a similar effect when inserted as single mutations [10]. We have previously shown that NS4A(1–48) interacts preferentially with highly curved liposomes. CD spectroscopy demonstrated that these two point mutations severely compromise this interaction [12]. The interaction of wild type NS4A(1–48) with highly curved membranes has been demonstrated for three different lipid compositions, *i.e.*, pure POPC, a POPC/DOPS mixture at a molar ratio of 4:1, and a blend of synthetic lipids resembling the composition of membranes in the ER (ER lipid mix), but no dependence on lipid composition was detected [12]. Here, we confirm these results using a liposome floatation assay and NMR. Initial floatation experiments were conducted with pure POPC and with the ER lipid mix. Again, no influence of lipid composition on NS4A(1–48) binding was observed. The detailed analysis presented in the current manuscript was conducted with single component POPC SUVs. Liposome floatation experiments (Figure 1) clearly show that wild type NS4A(1–48) binds to highly curved POPC liposomes, this interaction was not observed with the mutant peptide. The interaction between NS4A(1–48) and liposomes was further characterized using NMR. These experiments indicate that the main lipid binding sites of the peptide are located at the N-terminal 20 amino acid residues of NS4A(1–48). Residues A21 through L31 may also be involved in liposome binding while the remaining C-terminal residues are only weakly affected by liposome binding and do not seem to play a direct role in this process. The backbone resonance assignment of NS4A(1–48) and NS4A(1–48, L6E;M10E) reported in this manuscript contains information on the secondary structure of the two peptides in lipid-free buffer. Analysis with the TALOS-N software [26] clearly shows the lack of secondary structure for both peptides in buffer. The NMR data on NS4A(1–48) recorded in presence of liposomes do not allow any straightforward conclusion on the structure of the liposome-bound peptide. However, interaction of NS4A(1–48) with sodium dodecyl sulfate (SDS) micelles induces the formation of two amphipathic helices (AH) encompassing residues N5 to E9 (AH1) and T15 to L31 (AH2) of NS4A(1–48) [12]. CD spectra of NS4A(1–48) recorded in presence of either SDS micelles or small POPC liposomes are very similar [12]. Therefore, it is conceivable that the two amphipathic helices AH1 and AH2 are also formed in the liposome-bound peptide. The N-terminal region of NS4A(1–48) that forms AH1, an interhelical linker and the N-terminal half of AH2 in SDS micelles seem to be crucial for peptide binding to liposomes. Interestingly, this region also contains the two mutations L6E and M10E, which abolished liposome binding in the floatation experiment (Figure 1). Moreover, mutagenesis of other residues in this region including P14A [10], R12A and K20A [9] was also shown to reduce or abolish DENV replication.

The central part of NS4A(1–48) encompassing residues A21 to L31 shows less pronounced NMR signal intensity reductions upon titration with POPC liposomes than its N-terminal 20 amino acid residues (Figure 2A). A comparatively weak NMR signal intensity reduction is observed for both the N-terminal and central regions of NS4A(1–48, L6E;M10E) (Figure 2B) indicating some residual interaction with liposomes. However, the liposome floatation assay clearly shows that this interaction of the mutant peptide with POPC liposomes is too weak for stable anchoring of the peptide at the membrane. The amino acid sequence of A21 through L31 is identical in both peptides. We conclude that this amino acid stretch A21 to L31 is not sufficient for stable membrane anchoring of NS4A(1–48).

DENV NS4A apparently contains two separate membrane anchors. The membrane spanning helices pTMS1 and pTMS3 stably integrate the protein into the membrane. The N-terminal region of the cytosolic domain specifically binds to the convex surface of highly curved membranes [12] and may serve as a second membrane anchor. Therefore, one hypothesis might be that NS4A can bridge two adjacent membranes or connect separate patches of the same membrane that come into close proximity due to membrane convolution. We speculate that membrane bridging by NS4A might play a crucial role in stabilizing the complex morphology of DENV-induced ER-derived membrane structures, which include stacks of convoluted membranes (CM), double membrane vesicles and tubes [5]. The vesicles were described as invaginations of the ER, which are connected to the cytosol via pore-like openings [5]. NS4A may play different roles in the reorganization of these ER-derived membranes. Asymmetric insertion of pTMS2 into the luminal leaflet of the inner membrane of the vesicles as well as oligomerization of NS4A may induce concave membrane curvature required for vesicle formation [7]. Binding of the N-terminal region of NS4A to the saddle-shaped neck region connecting the vesicle and the pore may further stabilize the vesicular structures.

All positive-strand RNA viruses form their replication complexes on modified host membranes. However, the source of the membranes and the nature of the modifications vary (for review see [2]). In general, the role of these modifications is twofold; to provide a scaffold concentrating and correctly positioning the viral and host factors for efficient viral replication and to protect the replicating virus from detection by the host immune system. The mechanisms driving the formation of these structures are still incompletely understood. Convoluted membranes are a form of membrane modification induced by several positive-strand RNA viruses including, in addition to DENV, severe acute respiratory syndrome coronavirus (SARS-CoV) and Kunjin virus (KUNV) [5,27,28], for example. In KUNV convoluted membranes are thought to be the site of polyprotein processing [27]. While in DENV the role of these structures is still unclear, they are thought to be a depot for factors required for replication [5].

In summary, the liposome floatation data provide direct proof for specific binding of NS4A(1–48) to highly curved free liposomes. The main lipid binding sites in NS4A(1–48) are located within the N-terminal 20 amino acid residues. The exact role of this specific interaction in the viral life cycle is still under investigation. Nevertheless, this important structural information may assist in further understanding of the role of NS4A and the mechanism by which it induces the membrane alterations underlying the viral RC formation.

Acknowledgments: We would like to thank Matthias Stoldt for excellent NMR support. D.W. acknowledges funding from the Sonderforschungsbereich SFB974.

Author Contributions: Bernd W. Koenig, Silke Hoffmann, Yu-Fu Hung and Melanie Schwarten conceived and designed the study; Yu-Fu Hung and Melanie Schwarten performed the experiments and analyzed the data, Ella H. Sklan, Silke Hoffmann and Dieter Willbold discussed the results and revised the manuscript; Bernd W. Koenig wrote the manuscript.

Conflicts of Interest: The authors declare no conflict of interest.

References

1. Miller, S.; Kastner, S.; Krijnse-Locker, J.; Buhler, S.; Bartenschlager, R. The non-structural protein 4A of dengue virus is an integral membrane protein inducing membrane alterations in a 2K-regulated manner. *J. Biol. Chem.* **2007**, *282*, 8873–8882.
2. Miller, S.; Krijnse-Locker, J. Modification of intracellular membrane structures for virus replication. *Nat. Rev. Microbiol.* **2008**, *6*, 363–374.
3. Salonen, A.; Ahola, T.; Kaariainen, L. Viral RNA replication in association with cellular membranes. *Curr. Top. Mirobiol. Immunol.* **2005**, *285*, 139–173.
4. Mackenzie, J. Wrapping things up about virus RNA replication. *Traffic* **2005**, *6*, 967–977.
5. Welsch, S.; Miller, S.; Romero-Brey, I.; Merz, A.; Bleck, C.K.E.; Walther, P.; Fuller, S.D.; Antony, C.; Krijnse-Locker, J.; Bartenschlager, R. Composition and three-dimensional architecture of the dengue virus replication and assembly sites. *Cell Host Microbe* **2009**, *5*, 365–375.
6. Lin, C.; Amberg, S.M.; Chambers, T.J.; Rice, C.M. Cleavage at a novel site in the NS4A region by the yellow fever virus NS2B-3 proteinase is a prerequisite for processing at the downstream 4A/4B signalase site. *J. Virol.* **1993**, *67*, 2327–2335.
7. McMahon, H.T.; Gallop, J.L. Membrane curvature and mechanisms of dynamic cell membrane remodelling. *Nature* **2005**, *438*, 590–596.

8. Lin, M.H.; Hsu, H.J.; Bartenschlager, R.; Fischer, W.B. Membrane undulation induced by NS4A of Dengue virus: A molecular dynamics simulation study. *J. Biomol. Struct. Dyn.* **2013**, *32*, 1552–1562.

9. Lee, C.M.; Xie, X.; Zou, J.; Li, S.H.; Lee, M.Y.; Dong, H.; Qin, C.F.; Kang, C.; Shi, P.Y. Determinants of dengue virus NS4A protein oligomerization. *J. Virol.* **2015**, *89*, 6171–6183.

10. Stern, O.; Hung, Y.F.; Valdau, O.; Yaffe, Y.; Harris, E.; Hoffmann, S.; Willbold, D.; Sklan, E.H. An N-Terminal amphipathic helix in dengue virus nonstructural protein 4A mediates oligomerization and is essential for replication. *J. Virol.* **2013**, *87*, 4080–4085.

11. Teo, C.S.; Chu, J.J. Cellular vimentin regulates construction of dengue virus replication complexes through interaction with NS4A protein. *J. Virol.* **2014**, *88*, 1897–1913.

12. Hung, Y.F.; Schwarten, M.; Schunke, S.; Thiagarajan-Rosenkranz, P.; Hoffmann, S.; Sklan, E.H.; Willbold, D.; Koenig, B.W. Dengue virus NS4A cytoplasmic domain binding to liposomes is sensitive to membrane curvature. *Biochim. Biophys. Acta* **2015**, *1848*, 1119–1126.

13. Hung, Y.F.; Valdau, O.; Schunke, S.; Stern, O.; Koenig, B.W.; Willbold, D.; Hoffmann, S. Recombinant production of the amino terminal cytoplasmic region of dengue virus non-structural protein 4A for structural studies. *PLoS ONE* **2014**, *9*, e86482.

14. Bodenhausen, G.; Ruben, D.J. Natural abundance nitrogen-15 NMR by enhanced heteronuclear spectroscopy. *Chem. Phys. Lett.* **1980**, *69*, 185–189.

15. Grzesiek, S.; Bax, A. Amino acid type determination in the sequential assignment procedure of uniformly 13C/15N-enriched proteins. *J. Biomol. NMR* **1993**, *3*, 185–204.

16. Kay, L.E.; Keifer, P.; Saarinen, T. Pure absorption gradient enhanced heteronuclear single quantum correlation spectroscopy with improved sensitivity. *J. Am. Chem. Soc.* **1992**, *114*, 10663–10665.

17. Ikura, M.; Kay, L.E.; Bax, A. A novel approach for sequential assignment of 1H, 13C, and 15N spectra of proteins: Heteronuclear triple-resonance three-dimensional NMR spectroscopy. Application to calmodulin. *Biochemistry* **1990**, *29*, 4659–4667.

18. Solyom, Z.; Schwarten, M.; Geist, L.; Konrat, R.; Willbold, D.; Brutscher, B. BEST-TROSY experiments for time-efficient sequential resonance assignment of large disordered proteins. *J. Biomol. NMR* **2013**, *55*, 311–321.

19. Yamazaki, T.; Lee, W.; Arrowsmith, C.H.; Muhandiram, D.R.; Kay, L.E. A suite of triple resonance NMR experiments for the backbone assignment of ^{15}N, ^{13}C, ^{2}H labeled proteins with high sensitivity. *J. Am. Chem. Soc.* **1994**, *116*, 11655–11666.

20. Wishart, D.S.; Bigam, C.G.; Holm, A.; Hodges, R.S.; Sykes, B.D. ^{1}H, ^{13}C and ^{15}N random coil NMR chemical shifts of the common amino acids. I. Investigations of nearest-neighbor effects. *J. Biomol. NMR* **1995**, *5*, 67–81.

21. Delaglio, F.; Grzesiek, S.; Vuister, G.W.; Zhu, G.; Pfeifer, J.; Bax, A. NMRPipe: A multidimensional spectral processing system based on UNIX pipes. *J. Biomol. NMR* **1995**, *6*, 277–293.

22. Vranken, W.F.; Boucher, W.; Stevens, T.J.; Fogh, R.H.; Pajon, A.; Llinas, M.; Ulrich, E.L.; Markley, J.L.; Ionides, J.; Laue, E.D. The CCPN data model for NMR spectroscopy: Development of a software pipeline. *Proteins* **2005**, *59*, 687–696.
23. Bigay, J.; Casella, J.F.; Drin, G.; Mesmin, B.; Antonny, B. ArfGAP1 responds to membrane curvature through the folding of a lipid packing sensor motif. *EMBO J.* **2005**, *24*, 2244–2253.
24. Koenig, B.W.; Gawrisch, K. Specific volumes of unsaturated phosphatidylcholines in the liquid crystalline lamellar phase. *Biochim. Biophys. Acta* **2005**, *1715*, 65–70.
25. Fischer, H.; Polikarpov, I.; Craievich, A.F. Average protein density is a molecular-weight-dependent function. *Protein Sci.* **2004**, *13*, 2825–2828.
26. Shen, Y.; Bax, A. Protein backbone and sidechain torsion angles predicted from NMR chemical shifts using artificial neural networks. *J. Biomol. NMR* **2013**, *56*, 227–241.
27. Westaway, E.G.; Khromykh, A.A.; Kenney, M.T.; Mackenzie, J.M.; Jones, M.K. Proteins C and NS4B of the flavivirus Kunjin translocate independently into the nucleus. *Virology* **1997**, *234*, 31–41.
28. Knoops, K.; Kikkert, M.; Worm, S.H.; Zevenhoven-Dobbe, J.C.; van der Meer, Y.; Koster, A.J.; Mommaas, A.M.; Snijder, E.J. SARS-coronavirus replication is supported by a reticulovesicular network of modified endoplasmic reticulum. *PLoS Biol.* **2008**, *6*, e226.

Chapter 5:
Antiviral Drugs

Anti-HBV Drugs: Progress, Unmet Needs, and New Hope

Lei Kang, Jiaqian Pan, Jiaofen Wu, Jiali Hu, Qian Sun and Jing Tang

Abstract: Approximately 240 million people worldwide are chronically infected with hepatitis B virus (HBV), which represents a significant challenge to public health. The current goal in treating chronic HBV infection is to block progression of HBV-related liver injury and inflammation to end-stage liver diseases, including cirrhosis and hepatocellular carcinoma, because we are unable to eliminate chronic HBV infection. Available therapies for chronic HBV infection mainly include nucleos/tide analogues (NAs), non-NAs, and immunomodulatory agents. However, none of them is able to clear chronic HBV infection. Thus, a new generation of anti-HBV drugs is urgently needed. Progress has been made in the development and testing of new therapeutics against chronic HBV infection. This review aims to summarize the state of the art in new HBV drug research and development and to forecast research and development trends and directions in the near future.

Reprinted from *Viruses*. Cite as: Kang, L.; Pan, J.; Wu, J.; Hu, J.; Sun, Q.; Tang, J. Anti-HBV Drugs: Progress, Unmet Needs, and New Hope. *Viruses* **2015**, *7*, 4960–4977.

1. Introduction

Hepatitis B virus (HBV) is a hepatotropic DNA virus that primarily infects hepatocytes and causes liver disease [1,2]. It is estimated that approximately 240 million people are chronically infected with HBV worldwide [3]. In a national survey in China, an HBV endemic country, the hepatitis B surface antigen (HBsAg) positive rate in the general population was reported to be about 10% [4]. Chronic HBV infection is a leading cause of chronic hepatitis and advanced-stage liver diseases, including cirrhosis and hepatocellular carcinoma (HCC). The pathogenesis of HBV infection remains poorly understood, and current HBV treatments are unsatisfactory. New efforts are being directed to develop new and more effective anti-HBV therapeutics [5–9]. This review reports the state of the art in research and development of new drugs against HBV.

2. HBV

2.1. Morphology and Viral Proteins

HBV infection produces a virion (also called Dane particle) that is 42 nm in size and subviral structures consisting of 22-nm spherical and filamentous particles [10]. The virion is composed of an outer shell and inner core. The virion shell (envelope) consists of HBsAg and phospholipids that are drawn from infected cells. The core consists of a nucleocapsid that contains 240 copies of HBV core protein and a partially double-stranded HBV genome. The HBsAg subviral particle does not contain the viral genome and is not infectious [11]. The HBV genome is 3.2 kb long and consists of four partially-overlapping open reading frames (ORFs) that encode pre-S/S, pre-core/core, Pol, and X proteins. The pre-S/S ORF encodes 163- and 226-amino acid pre-S and S (major envelope protein) peptides, respectively [12]. The pre-core/core ORF encodes HBeAg and HBcAg as well as polymerase protein. The X ORF encodes a 154-amino acid polypeptide (HBx).

2.2. Life Cycle

The HBV genome is partially double-stranded DNA. The viral P protein contains a RNA reverse transcriptase (RT) domain, which is required to reverse transcribe pregenomic RNA (pgRNA) to viral minus strand DNA. The HBV lifecycle involves many steps and has yet to be fully elucidated. The HBV virion binds to hepatocytes through cellular receptors, one of which has been identified as sodium taurocholate co-transporting polypeptide (NTCP) [13]. The virion, once it enters the cell, is decoated, and the released capsid is delivered to the nucleus to release the viral genome. The partially double-stranded DNA is converted to covalently closed circular DNA (cccDNA) to serve as a transcription template. Four RNA transcripts including 3.5-, 2.4-, 2.1-, and 0.7-kb viral RNAs are synthesized and migrate into the cytoplasm for translation of viral proteins. The pgRNA interacts with P and core proteins to form a nucleocapsid where the viral minus stranded DNA is reverse transcribed from the pgRNA. Plus stranded DNA is synthesized using the minus stranded DNA as a template to become a mature nucleocapsid. The mature capsid can be assembled into virions for the production of infectious viral particles or transported back to the nucleus to replenish the cccDNA pool (intracellular pathway), which is required for establishing and maintaining HBV infection. HBsAg synthesis occurs in the rough endoplasmic reticulum and is then transported to the Golgi apparatus for the assembly of virion and subviral particles [14]. Gaggar *et al.* [15] schematically described the key steps of the HBV lifecycle.

3. Current State of Anti-HBV Drugs

3.1. Immunoregulators

Immuno-regulator drugs [16] have already been used in treating severe pneumonia, immunodeficiency, and chronic hepatitis B. Many hepatitis B carriers, who are currently free of clinical hepatitis manifestations, can experience flare-ups of liver injury later. Immunoregulatory drugs can improve the patients' immune response, especially the specific immunity to HBV. Immunoregulators may help the immune cells to recognize and destroy HBV-infected cells, resulting in clearance of HBV in those destructed cells.

3.1.1. Interferon (IFN)

IFN [17] is a secretory glycoprotein and functions as an antiviral, anti-proliferation, and immune regulatory cytokine. Production of IFN is triggered when the host cells responsd to various stimuli. IFN is categorized as α-(white blood cells), β-(fibroblasts), and γ-(lymphocytes) based on the producing cells. Currently, the major type of IFN used to treat chronic HBV infection is the IFN that binds to specific receptors on the cell surface to trigger a series of signal transduction events, resulting in the production of antiviral protein (AVP), which degrades viral mRNA and inhibits viral replication. It also simultaneously strengthens the activity of natural killer (NK) cells, macrophages, and T lymphocytes, demonstrating that IFN regulates the immune response in addition to the direct antiviral ability.

The half-life of IFN can be extended through chemical modification. At present, polyethylene glycol (PEG) is the most widely used agent to form multiple copies of IFN molecules that are more slowly degraded once injected. The pegIFN acquires a longer half-life up to 40 hours and remains effective in inhibiting HBV replication for 168 h. Therefore, pegIFN only needs to be injected once a week, which makes compliance with the treatment schedule more convenient and easier for patients.

However, IFN treatment is expensive, and the adverse effects of both short- and long-acting IFNs are severe, including fever, hair loss, and reduction of white blood cells. In addition, the antiviral efficacy in patients with normal or mildly-elevated alanine aminotransferase (ALT) is poor.

3.1.2. Thymosin-α1

Thymosin-α1 is one of the immunoregulators, and its main functions are to promote differentiation of T cells to a mature stage and enhance the response to antigens and other excitants. The action of boosting the host immune system helps the host mount a defense against chronic HBV infection, and thymosin-α1 also exhibits a modest antiviral effect. It can be combined with IFN to treat chronic hepatitis B. The guidelines of the Asian Pacific Association for the Study of the Liver (APASL) [18] reviewed recent clinical studies of thymosin-α1 and indicated that fixed therapeutic duration and minimal side effects are major advantages. More well-designed large-scale studies to confirm its efficacy and combination therapy with pegIFN or nucleoside analogues are direction for future development.

3.1.3. Cytokines

Cytokines are synthetized and secreted by a multiple types of immune cells (such as monocytes, macrophages, T cells, B cells, and NK cells) and non-immune cells (such as endothelial cells, epidermal cells, and fibroblasts) upon stimulation. Cytokines are characterized by a small molecular size and extensive biological functions [19], with some showing strong antiviral activity. For instance, interleukin (IL)-12 can induce TO cells into T1 cells, leading to IFN production.

3.2. Nucleos/Tide Analogues (NAs)

3.2.1. Lamivudine

Lamivudine, a pyrimidine nucleoside drug, was the first NA approved for treating chronic HBV infection. Lamivudine is a reverse transcriptase (RT) inhibitor and inhibits HBV replication by inhibiting RT activity to lower the HBV DNA level. The drug is orally administered with fast absorption and high bioavailability. It can effectively decrease the serum HBV DNA level in treated patients.

Lamivudine needs a long-term course to maintain inhibition. HBV replication returns back quickly once lamivudine is withdrawn. Another challenge to administration of lamivudine is a high frequency of resistance mutations even after a short course treatment [20,21]. The common drug resistant mutations include tyrosine-methionine-aspartate-aspartate mutations (known as the YMDD mutant).

3.2.2. Telbivudine

Telbivudine is a specific, selective and oral drug used for treating chronic hepatitis B. It shows unique advantages in inhibition of the synthesis of both strands of HBV DNA. In clinical practice, telbivudine is the only drug that can be safely used for treating pregnant female patients. The antiviral efficacy of telbivudine is stronger than that of lamivudine. The number of patients with undetectable HBV DNA after telbivudine therapy is significantly higher than that after treatment with lamivudine [22,23].

3.2.3. Entecavir

Entecavir is a guanine nucleoside drug that offers potent selective inhibition of HBV polymerase [18]. It becomes an active triphosphate form after phosphorylation in the cells. ETV inhibits all three functions of viral polymerase: (1) initiation of HBV polymerase; (2) synthesis of minus strand DNA from the pregenomic RNA template (RT step); and (3) synthesis of HBV DNA chain. The inhibitive efficacy of Entecavir is 300 times greater than that of other NAs, such as lamivudine or/and adefovir, and the resistance mutation frequency is very low [3,18,24]. It shares certain cross tolerance with lamivudine. The ETV regime should be fulfilled before cessation; otherwise, rapid deterioration may occur.

3.2.4. Adefovir

Adefovir [25] is a purine nucleoside prodrug. Phosphorylation of adefovir is required for inhibition, and it occurs to form the active metabolite adefovir diphosphate once it enters the cell. Adefovir diphosphate can substitute for normal substrates of dATP (adenosine). Once incorporated into the viral DNA chain, it stops the elongation of viral DNA synthesis. It is suggested that adefovir can also induce production of α-IFN, increase NK cell activity, and stimulate the hosts' immune response. Adefovir shows stronger efficacy in inhibiting HBV replication, but a lower resistance rate as an add-on therapy for LAM resistance than switching to ADV only [26,27]. A larger dose (30 mg or higher per day) can cause renal toxicity, but the standard dose at 10 mg/d does not affect renal function seriously [18]. Close monitoring of renal function is required when adefovir is in use.

3.2.5. Tenofovir

Tenofovir is an acyclic adenine nucleotide analogue that inhibits both HBV and HIV replication at the RT step [3]. Tenofovir prodrug is converted to tenofovir diphosphate via catalysis by a cellular kinase. Tenofovir is recommended because of its potent inhibition, high barrier to drug resistance, and the overall balance between benefits and risks.

A summary of the structures the anti-HBV drugs are shown in Table 1, and the information is cited as provided in the manufacturers' instructions.

3.3. Traditional Chinese Medicine

3.3.1. Sedum sarmentosum Granules

Sedum sarmentosum granules are derived from either the fresh or dried whole plant of sarmentosum, which belongs to the Crassulaceae family. Investigation of effective pharmacological components has shown that the liver protective function is mainly based on constituents of megastigmane glycosides [28]. Experiments using an animal liver injury model have shown that Sedum sarmentosum granules can reduce serum ALT and aspartate aminotransferase (AST) levels in animals with acute liver injury [29]. In clinical practice, it is mainly used for treating acute or chronic hepatitis, especially patients whose cereal third transaminase level is increased [30]. In addition, Sedum sarmentosum granules can up-regulate cellular immunity even in a state of inhibition, which is crucial for facilitating recovery from liver injury.

At present, the mechanism of action and toxicology of this traditional Chinese medicine remain unknown, and therefore, reasonable dosages can only be empirically determined based on patients' conditions.

3.3.2. Oxymatrine

Oxymatrine, an alkaloid extracted from the herb *Sophora alopecuraides* L, exhibits an anti-HBV effect in both HBV transgenic mice [36,37] and patients with chronic HBV infection [38], as shown in Table 1. Chen *et al.* [36] and Lu *et al.* [37] independently found that oxymatrine can suppress the levels of HBsAg and HBcAg in liver and HBV DNA in serum of transgenic mice. Recently, a clinical study [39] found that the combination of oxymatrine with lamivudine can prevent the development of lamivudine resistance in chronic HBV-infected patients. However, the antiviral mechanisms or targets of oxymatrine against HBV remain unknown. Xu *et al.* [40] suggested that oxymatrine may interfere with the packaging process of pgRNA into the nucleocapsid or suppress the activity of viral DNA polymerase. Wang *et al.* [41] found that oxymatrine may interfere with the reverse transcription process from pgRNA to DNA by destabilizing heat stress cognate 70 (Hsc70) mRNA.

Although oxymatrine was mentioned in the 2005 Chinese Guideline for Prevention and Treatment of Chronic HBV Infection [42], further longer-term, multicenter, and randomized, controlled clinical trials with large numbers of cases are needed to validate antiviral efficacy.

Table 1. Summary of the structures of the anti-HBV drugs.

	Structures	Active Form Structures	EC$_{50}$	Therapeutic Target
Lamivudine (EPIVIR-HBV [31])			Varies from 0.01 μM (2.3 ng/mL) to 5.6 μM (1.3 μg/mL) depending upon the duration of exposure of cells to lamivudine, the cell model system, and the protocol used	Inhibition of the RNA- and DNA-dependent polymerase activities of HBV reverse transcriptase
Telbivudine (Sebivo [32])			HBV first strand (EC$_{50}$ = 0.4–1.3 μM) and second strand (EC$_{50}$ = 0.12–0.24 μM)	Inhibition of HBV DNA polymerase (reverse transcriptase) by competing with the natural substrate, thymidine 5′-triphosphate

Table 1. *Cont.*

Structures	Active Form Structures	EC$_{50}$	Therapeutic Target
Entecavir (Baraclude [33])		0.004 μM in human HepG2 cells transfected with wild-type HBV, 0.026 μM (range, 0.010–0.059 μM) against lamivudine resistance HBV (rtL180M and rtM204V)	Inhibition of (1) priming of the HBV polymerase; (2) reverse transcription of the negative strand DNA from the pregenomic messenger RNA; and (3) synthesis of the positive strand HBV DNA
Adefovir dipivoxil (SigmaPharm Laboratories, LLC [34])		Ranges from 0.2 to 2.5 μM in HBV-transfected human hepatoma cell lines	Inhibition of HBV DNA polymerase (reverse transcriptase) by competing with the natural substrate deoxyadenosine triphosphate and by causing DNA chain termination after its incorporation into viral DNA

374

Table 1. *Cont.*

Structures	Active Form Structures	EC$_{50}$	Therapeutic Target
Tenofovir (VIREAD) [35]		Ranges from 0.04 µM to 8.5 µM	Inhibition of the activity of HBV reverse transcriptase by competing with the natural substrate deoxyadenosine 5'-triphosphate and, after incorporation into DNA, by DNA chain termination
Oxymatrine		Unknown	Unknown

4. New Anti-HBV Drugs under Development and Evaluation

4.1. New Drugs that Target the Viral Components

4.1.1. MCC-478

MCC-478, an adefovir derivative, has been in clinical Phase I trials for safety and efficacy assessment [43,44]. It acts similarly to other NAs and can inhibit HBV replication by inhibiting P protein packaging reaction. MCC-478 can be effective against both wild-type HBV and lamivudine-resistant mutants [45,46].

4.1.2. cccDNA

Currently, the first-line antiviral therapy is NA-based treatment, which functions to suppress HBV replication. Although new cccDNA amplification is restrained, the hepatocytes remain infected due to persistent cccDNA in the nucleus, which escapes via the error-prone viral polymerase [47] and drug-resistance mutants [48]. Therefore, sustained elimination of cccDNA from infected hepatocytes represents a major challenge, and one possible solution is immunological therapy, such as cytokine-mediated or immune-associated receptor-mediated cccDNA degradation.

A very recent study by Lucifora *et al.* [49] demonstrated that high-dose IFN-α can induce cccDNA eradication in HBV-infected hepatocytes. Although the approved IFN-α therapy is effective in some patients, a relatively low response rate [50], contraindications, less convenient parenteral administration [51], and certain serious adverse effects [52] all limit its clinical application. Therefore, the better therapeutic option is the alternative receptor-mediated cccDNA degradation. Through the use of specific antibodies, lymphotoxin (LT) β receptor (LTβR) activation was also shown to induce cccDNA eradication in HBV-infected hepatocytes, without causing any detectable hepatocytotoxicity. With respect to the underlying mechanism, LTβR activation can up-regulate the expression of nuclear APOBEC3 (A3) deaminases and subsequently induce deamination and A purinic/A pyrimidinic (AP) site formation in HBV cccDNA, resulting in its degradation, without affecting host genomic DNA. The A3 family members of A3A and A3B, which are located in the nucleus [53], play an essential role in the eradication of foreign DNA [54,55]. They might be targeted to cccDNA by their interaction with the HBV core protein, suggesting a selective mechanism for distinguishing HBV cccDNA from host genomic DNA. Additionally, IFN-α was found to induce a similar effect, which indicates that, through use of LTβR agonists or adoptive T cell therapy [56], receptor-mediated cccDNA degradation, if confirmed in clinical trials, could lead to clearance of chronic HBV infection from the liver.

In addition, Ahmed *et al.* [57] summarized current therapeutic strategies against cccDNA production. In addition to INFs and LTβR agonists, factors (methylation and acetylation) affecting the process of cccDNA transcription and translation and DNA cleavage enzymes (zinc-finger protein nucleases and transcription-activator-like effectors), which interrupt the structure and/or functions of cccDNA, are some of the other possible approaches.

4.1.3. HBsAg Gene

The major component of HBsAg is the small S protein with 226 amino acids. HBsAg elicits production of neutralizing antibodies. However, HBsAg is overwhelmingly produced and stably maintained in chronically HBV-infected patients, which contributes to the suppression of an HBV-specific immune response. In addition to the immune response, HBsAg is required for assembling viral particles. A group of researchers [58] investigated the cellular gene expression profile in cells containing transfected HBsAg gene via microarray analysis. It was found that among 1152 gene analyzed, 30 were significantly upregulated, including tumor necrosis factor (TNF)-related apoptosis-inducing ligand (TRAIL) receptor, cell division cycle protein (CDC23), and FKBP-associated protein (FAP48). Moreover, 29 genes were significantly down-regulated, including TNF receptor-associated factor (TRAF), TNF receptor-associated protein (AD022), and TNF receptor-associated factor 2 (TRAF2). These genes impact cell growth, apoptosis, signal transduction, immune regulation, and tumorigenesis. Both up- and down-regulation of specific proteins is involved in the process of apoptosis, suggesting that HBsAg is probably involved in the regulation of apoptosis. Furthermore, the HBsAg gene is thought to be involved in HCC development. Taken together, these findings suggest that HBsAg is a potential target for HCC gene therapy.

4.1.4. Chinese Herbal Medicines

Helioxanthin (HE-145) is an arylnaphthalene ligand isolated from *Taiwania Cryptomerioides*. HE-145 and its analogues 5-4-2 [59], 8-1 [60], 32 [61], and 15 [62] have been reported to exhibit potent anti-HBV activity *in vitro*. They not only suppress the expression of HBV RNAs and proteins, but also suppress viral DNA replication of both wild-type and lamivudine-resistant mutants. The mode of action involves decreasing the DNA binding activity of hepatocyte nuclear extracts to specific *cis*-elements in the HBV core promoter, whereas the ectopic expression of the *cis*-elements relieves such suppression. Thus, HE-145 suppresses HBV gene expression and replication by selectively modulating the host transcriptional machinery [63].

HE-145 analogue 8-1 [60] reduces activity at all HBV promoters by post-transcriptionally decreasing the expression of critical transcription factors in

HBV-producing cells, which diminishes their binding to the precore/core promoter enhancer II region. Thus, it blocks viral gene expression to negatively impact viral DNA replication.

Pang *et al.* [64] found that the ethanol extract from *Ampelopsis sinica* root (EASR) effectively suppresses the levels of HBsAg, HBeAg, and extracellular HBV DNA *in vitro* by selectively inhibiting the activities of several HBV promoters and the p53-associated signaling pathway.

4.2. New Drugs that Target Cellular Factors

4.2.1. HBV Receptors

HBV is enveloped with viral envelope proteins. HBV envelope proteins not only protect the virus but also make it infectious as they are required for viral entry, a first step for initiating HBV infection. The viral entry is mediated by specific interactions between viral envelope proteins and receptors on hepatocytes. Studies have shown that the myristoylated preS1 domain of HBV L-protein plays a pivotal role in viral infectivity by mediating attachment to a hepatocyte-specific receptor [65,66]. Recently, one of the assumed HBV receptors was identified as sodium taurocholate cotransporting polypeptide (NTCP) [13]. NTCP-mediated HBV and HDV entry has been independently confirmed by other groups [67,68].

NTCP represents a new target for the development of therapeutics that can block HBV entry. Myrcludex-B, a synthetic lipopeptide derived from the HBV preS1 domain sequence, has been shown to specifically bind to NTCP. It can efficiently block *de novo* HBV infection and prevent intrahepatic viral spreading both *in vitro* and *in vivo* [67,69]. In addition, NTCP also functions as a hepatic bile acid transporter that mediates the uptake of most sodium-dependent bile salts into hepatocytes. Yan *et al.* [68] found that two uptake functions of NTCP seem to be mutually exclusive, implying that the regulation of uptake of bile acids or their derivatives could impact HBV entry, a potentially new strategy for the development of novel antiviral drugs.

NTCP expression is subjected to cellular regulation. It was suggested that retinoic acid receptor (RAR) regulates the promoter activity of the human NTCP (hNTCP) gene [70]. Tsukuda *et al.* [70] demonstrated that a RAR-selective antagonist Ro41-5253 decreases cellular susceptibility to HBV infection by inhibiting hNTCP promoter activity. Furthermore, IL-6 was found to regulate NTCP expression, and the effect of IL-6 on HBV entry was also noted. Bouezzedine *et al.* [71] found that NTCP mRNA expression is reduced by 98%, along with an 80% decrease in NTCP-mediated taurocholate uptake and 90% inhibition of HBV entry, upon pretreatment of HepaRG cells with IL-6. Such findings require further validation in more stringent infection systems.

Other cellular factors that are involved in the HBV lifecycle include the Toll-like receptors (TLRs). TLRs are known to mediate the innate immune response to infection. TLRs recognize pathogen-associated molecular patterns and respond by activating a series of antiviral mechanisms. The TLR-ligand interaction results in the production and release of antiviral molecules such as IFNs, pro-inflammatory cytokines, and chemokines. However, HBV can disrupt TLR expression and hinder intracellular signaling cascades as a strategy for evading the innate immune response to chronic HBV infection. An emerging new treatment strategy involves combining antiviral treatment with adjuvant therapy using a TLR agonist to restore the innate immune response. TLR ligands that can activate the TLR-mediated innate immune response may represent promising adjuvant drug candidates. Isogawa *et al.* [72] found that all ligands, except the one for TLR2, inhibit HBV replication in a IFN-α- and -β-dependent manner after ligands specific for TLR 2-5, 7, and 9 are individually administered to HBV transgenic mice, suggesting that ligand-TLR interaction can elicit an effective immune response to inhibit HBV replication.

GS-9620, an orally-administered agonist of TLR7, was investigated for its safety, tolerability, pharmacokinetics, and pharmacodynamics in healthy volunteers and hepatitis C virus (HCV)- and HBV-infected patients. Three phase I clinical trials were completed [73–75]. GS9620 was well-absorbed and tolerated at oral doses of 0.3, 1, 2, 4, 6, 8, or 12 mg per day. The tested dosages were finally adjusted to 0.3, 1, 2, and 4 mg because chemokines/cytokines and IFN-stimulated genes (ISGs) can be induced at doses $\geqslant$2 mg. The most common adverse events were flu-like symptoms and headache. In healthy volunteers, minimal adverse events were similar to symptoms associated with an increased serum IFN-α level. In the majority of HBV- and HCV-infected patients, adverse events varied from mild to moderate in severity and the serum IFN-α levels became detectable in 16.7% (8/48) and 12% of HCV and HBV patients, respectively. A transient dose-dependent ISG15 induction was observed, peaking within 48 hours and followed by a decrease to baseline within 7 days. However, there were no significant changes in HCV RNA in HCV-infected patients, and no significant reductions of the HBsAg or HBV DNA level in HBV-infected patients either.

Recently, Zhang *et al.* [76] described a subgroup of the TLR family and reported that TLR3 and TLR2/4-mediated innate immune responses can control HBV infection. Kapoor *et al.* [77] summarized the roles of TLR7 and TLR9 in chronic HBV infection.

4.2.2. Novel Target: La Protein Inhibitor (HBSC11)

La protein [78] is a phosphoprotein with a molecular weight of 47 kDa. It was initially thought to be a self-antigen produced in patients with systemic lupus erythematosus (SLE) and primary Sjogren's syndrome (pSS). Now, it is known that human La protein is a multifunctional RNA-binding protein that is also involved in

HBV RNA metabolism. A previous study showed that La protein exhibits a protective effect against HBV, and the protein kinase CK2 (tyrosine kinase II) enables La protein to be phosphorylated at serine 366, which activates the La protein functions [79]. Recently, *in vitro* experiments have also shown that La protein is the HBV RNA transcription factor. It transfers the HBsAg-specific cytotoxic T cells (CLTs) to the liver of HBV-infected transgenic mice and degrades HBV RNA, which leads to the disappearance of mouse La protein [80]. Based on the findings described above La protein is involved in HBV replication. Researchers have utilized virtual screening techniques to filter the La protein binding sites through multi-level molecular docking and target screening process. HBSC11, a novel inhibitor that targets human La protein was shown to have an anti-HBV effect, via the use of the Specs database and laboratory chemical database. This *in vitro* validation shows that HBSC11 could potently inhibit the transcription and expression of La protein [81].

4.2.3. Transforming Growth Factor-β (TGF-β)

The TGF-β superfamily consists of a group of bioactive polypeptides with related structures and similar functions in regulating cell growth, differentiation, migration, death, and extracellular matrix (ECM) production. Thus, TGF-β can regulate the growth and differentiation of endothelial cells, inflammatory cell chemotaxis, fibroblast proliferation, carcinogenesis, and ECM synthesis and degradation. Smad3, a key protein in the Smads signaling pathway, plays a positive role in the regulation of the TGF-β1 pathway, which is highly involved in organ fibrosis. In a clinical study [82], researchers detected an abnormally higher level of TGF-β1 in liver cancer tissue via immunohistochemistry. Furthermore, TGF-β levels were elevated in the liver cancer tissues, regardless of whether they represented primary or metastatic cancer. In addition, many studies have suggested that TGF-β1 is involved in the pathogenic process of hepatitis and liver fibrosis.

In an investigation of TGF-β1 expression in chronic HBV infection, Peng *et al.* [83] found that the serum TGF-β1 level increased gradually with the progressive severity of liver damage in 89 cases with mild, moderate, and severe chronic HBV infection. TGF-β1 can promote ECM synthesis and deposition, which is a pivotal factor for inducing liver fibrosis. Thus, it plays a key role in chronic HBV infection-induced liver inflammation and fibrosis. At present, many studies have also examined the molecular structure of TGF-β1 to define the relationship between TGF-β1 and HBV from a perspective of gene polymorphisms [84]. However, many of the results are inconsistent and unsatisfactory because of ethnic and geographical differences and case selection bias. The only consistent finding is that these gene polymorphism sites are mainly located in the promoter and control regions.

4.2.4. MicroRNAs

MicroRNAs (miRNA) are small single-stranded RNAs with a final length of 20–23 bases. miRNAs are generally transcribed from non-coding regions of cellular genes. In 1993, a miRNA was first detected in *Caenorhabditis elegans*, and then in humans, plants, and other organisms. miRNAs are non-coding RNAs, but with regulatory functions at the mRNA and protein translational levels. miRNA regulation contributes to the control of physiological processes such as cell growth, differentiation, and apoptosis, lipid metabolism, and hormone secretion. In recent years, many studies have shown that a variety of human tumors are associated with aberrant expression of miRNAs. For instance, miRNA-221 is upregulated in pancreatic cancer. miRNA normally participates in the maintenance of cell homeostasis by regulating the target mRNA and its translation. Studies have indicated that miRNA expression is altered during cancer development. Abnormal expression of multiple miRNAs was detected in HCC cells. Furthermore, miRNA-199a-3p and miRNA-210 can effectively reduce the expression of HBsAg in HBV infection, indicating that miRNAs can not only regulate tumorigenesis but also mediate the interaction between the virus and the host [85]. However, no significant differences in miRNA expression were found between cirrhosis and HCC patients, suggesting that the abnormal expression of miRNAs already occurred in the early phase of the process. Changes in miRNA expression are presumed to be an initiating factor. However, the underlying mechanisms have yet to be clearly elucidated. There are two opinions: the most popular opinion is that miRNAs can degrade target mRNA molecules by complementary binding to the 3′-end of the untranslated region (UTR) of the target mRNA. The other opinion is that miRNAs inhibit translation of the target mRNA to reduce the protein level of the targeted gene. Due to the fact that miRNAs are small molecules, lack immunogenicity, and exhibit diverse regulatory functions at the mRNA and protein translational levels, miRNAs directly degrade specific mRNAs. miRNAs can be used as a molecular tool to target HBV RNA to inhibit the HBV lifecycle. However, a challenge is that a single miRNA can have multiple targets, and this multi-specificity for target genes may limit the clinical application of a given miRNA.

4.3. Immune Checkpoints

Immune checkpoints refer to a homeostatic function of the immune system and are responsible for the balance of co-stimulatory and co-inhibitory signals [86]. Under normal physiological conditions, immune checkpoints play an essential role in maintaining self-tolerance, whereas upon pathogen infection, they function to regulate the amplitude and duration of immune responses [87]. Tumor cells and viruses can take advantage of these immune checkpoint pathways and exploit them for immune evasion. The two major immune checkpoint targets, cytotoxic

T-lymphocyte antigen 4 (CTLA-4) and programmed cell-death protein 1 (PD1), both are negative immunomodulatory molecules and can inhibit T cell-mediated immune responses. Blocking of these two molecules is thought to prompt the immune system to regain strength to destroy the tumor cells. Ipilimumab (Yervoy; Bristol-Myers Squibb; Middlesex, UK), nivolumab (Opdivo; Bristol-Myers Squibb/Ono Pharmaceuticals; Middlesex, UK), and pembrolizumab (Keytruda; Merck & Co; Hertfordshire, UK.) are three immune checkpoint inhibitors currently approved for the treatment of malignant melanoma [86]. They are a CTLA-4- and PD1-specific monoclonal antibody and an anti-PD1 therapy, respectively. Expanding indications for other cancers and combination therapy [88] are the future directions for development of these drugs. In addition, a number of other immune checkpoint inhibitors are currently in the development pipeline [86].

However, the concept of blocking immune checkpoint inhibitors for HBV therapy is still in its infancy [89]. One important aspect is that blocking of immune checkpoint inhibitors should be a targeted therapy in order to avoid autoimmune-like side effects, as treatments targeting the T cell immunoglobulin-3 (TIM-3) pathway [90] and LTβR-mediated cccDNA degradation [49] do. Another important aspect is that the structural information and experimental data for the blocking agents should be understood as much as possible. The prospective from small molecules [91,92] and state-of-the-art technologies [93,94] can help to understand and expand the relevant knowledge. Therefore, based on oncological application (blocking agents) and the developing knowledge of immune checkpoint inhibitors, it is hoped that the application of certain blocking agents for treating chronic viral disease will not be too far in the future [89,95].

5. Guidelines for Currently Approved Medications

In 2015, the World Health Organization (WHO) published guidelines for the prevention of HBV infection and care and treatment of persons with chronic HBV infection [3]. The WHO guidelines recommend that the NAs tenofovir and entecavir, which have a high genetic barrier to drug resistance, should be the first-line treatments in all patients over 12 years of age, and entecavir should be used for children aged 2–11 years. Tenofovir is recommended as the second-line treatment in the same pediatric group. Other NAs (such as lamivudine, adefovir, and telbivudine) with a low barrier to drug resistance are explicitly not recommended. In addition, IFN is not considered a treatment option due to resource-limited settings and contraindications.

Before the WHO guidelines were published, the European Association for the Study of the Liver (EASL) [24] and the Asian Pacific Association for the Study of the Liver (APASL) [18] separately released their guidelines for the treatment of chronic HBV infection.

Lamivudine, entecavir, telbivudine, adefovir, tenofovir, and pegIFN have been approved in Europe for HBV treatment. The EASL guidelines of 2012 provided two treatment strategies, including finite-duration treatment with pegIFN and a long-term treatment with NAs. pegIFN, if chosen, should be used with caution due to the contraindications, and its combination with NAs is not recommended. The NA used for finite-duration treatment should be the most potent agent with the highest barrier to drug resistance. Tenofovir and entecavir are the first-line monotherapies. The evidence shows that the vast majority patients with monotherapy for ⩾3 years maintain a full virological response [96,97]. Solutions to antiviral treatment failure are also recommended in the WHO guidelines. In case of primary non-response or partial response, a drug switch to tenofovir or entecavir from the initial drug is recommended. Once virological breakthrough is detected and incomplete patient compliance is excluded as a possible cause, a new therapeutic regimen should be adopted as early as possible after monitoring HBV DNA loads and identifying the pattern of resistance mutations.

The 2012 APASL guidelines state that HBV does not directly lead to cytopathic consequences and chronic HBV infection is a dynamic process of interaction among the virus, hepatocytes, and the host immune system. Therefore, chronic HBV infection therapeutics include immunomodulatory agents and antivirals used with NAs. The immunomodulatory agents include conventional IFN-α, pegIFN, and thymosin α1. The listed NAs are lamivudine, telbivudine, adefovir, tenofovir, and entecavir. Entecavir or tenofovir is the first-line drug, whereas the others are considered as second-line drugs. Notably, cost is still one of the most important factors for drug selection aside from the drug efficacy in the Asia-Pacific region. This restraint can be only addressed by developing more effective, but less expensive, new drugs.

6. Conclusions

In summary, the number of patients chronically infected with HBV continues to grow, and chronic HBV infection can lead to cirrhosis and HCC, causing an unbearable burden to patients and society. Current antiviral therapies can potently inhibit HBV replication and improve liver pathology but are rarely able to clear chronic HBV infection. Both new antiviral strategies and drugs are urgently needed. Several new and encouraging drug candidates are under development, but much research is still needed before they can be applied clinically.

Acknowledgments: This study was supported by the Natural Science Foundation of China (No. 81470852), the Science and Technology Commission of Shanghai Science and Technology support project (No. 13431900503), the Medical and Technology Across project of Shanghai Jiao Tong University (No. YG2012MS02), and the Young Talents Plan of the Shanghai Health System (No. XYQ2013091).

Author Contributions: Lei Kang and Jiaqian Pan contributed equally to the manuscript in collecting the data and describing the structures. Jiaofen Wu, Jiali Hu and Qian Sun revised the manuscript and added some current references. Jing Tang designed the whole manuscript and completed the revisions.

Conflicts of Interest: The authors declare no conflict of interest.

References

1. Grimm, D.; Thimme, R.; Blum, H.E. HBV life cycle and novel drug targets. *Hepatol. Int.* **2011**, *5*, 644–653.
2. Stein, L.L.; Loomba, R. Drug targets in hepatitis B virus infection. *Infect. Disord. Drug Targets* **2009**, *9*, 105–116.
3. WHO. Guidelines for the Prevention, Care and Treatment of Persons with Chronic Hepatitis B Infection. Available online: http://www.who.int/hiv/pub/hepatitis/hepatitis-b-guidelines/en/ (accessed on 8 July 2015).
4. Lu, F.M.; Zhuang, H. Management of hepatitis B in China. *Chin. Med. J.* **2009**, *122*, 3–4.
5. Bhattacharya, D.; Thio, C.L. Review of hepatitis B therapeutics. *Clin. Infect. Dis.* **2010**, *51*, 1201–1208.
6. Lozano, R.; Naghavi, M.; Foreman, K.; Lim, S.; Shibuya, K.; Aboyans, V.; Abraham, J.; Adair, T.; Aggarwal, R.; Ahn, S.Y.; *et al.* Global and regional mortality from 235 causes of death for 20 age groups in 1990 and 2010: A systematic analysis for the global burden of disease study 2010. *Lancet* **2012**, *380*, 2095–2128.
7. Chen, A.; Panjaworayan, T.T.N.; Brown, C.M. Prospects for inhibiting the post-transcriptional regulation of gene expression in hepatitis B virus. *World J. Gastroenterol.* **2014**, *20*, 7993–8004.
8. Levrero, M.; Pollicino, T.; Petersen, J.; Belloni, L.; Raimondo, G.; Dandri, M. Control of cccDNA function in hepatitis B virus infection. *J. Hepatol.* **2009**, *51*, 581–592.
9. Bharadwaj, M.; Roy, G.; Dutta, K.; Misbah, M.; Husain, M.; Hussain, S. Tackling hepatitis B virus-associated hepatocellular carcinoma—The future is now. *Cancer Metastasis Rev.* **2013**, *32*, 229–268.
10. Singer, G.A.; Zielsdorf, S.; Fleetwood, V.A.; Alvey, N.; Cohen, E.; Eswaran, S.; Shah, N.; Chan, E.Y.; Hertl, M.; Fayek, S.A. Limited hepatitis b immunoglobulin with potent nucleos(t)ide analogue is a cost-effective prophylaxis against hepatitis b virus after liver transplantation. *Transplant. Proc.* **2015**, *47*, 478–484.
11. Seeger, C.; Mason, W.S. Molecular biology of hepatitis b virus infection. *Virology* **2015**, *479–480*, 672–686.
12. Zhao, Z.M.; Jin, Y.; Gan, Y.; Zhu, Y.; Chen, T.Y.; Wang, J.B.; Sun, Y.; Cao, Z.G.; Qian, G.S.; Tu, H. Novel approach to identifying the hepatitis B virus pre-s deletions associated with hepatocellular carcinoma. *World J. Gastroenterol.* **2014**, *20*, 13573–13581.
13. Yan, H.; Zhong, G.; Xu, G.; He, W.; Jing, Z.; Gao, Z.; Huang, Y.; Qi, Y.; Peng, B.; Wang, H.; *et al.* Sodium taurocholate cotransporting polypeptide is a functional receptor for human hepatitis B and d virus. *eLife* **2012**, *1*, e00049.

14. Chua, P.K.; Wang, R.Y.; Lin, M.H.; Masuda, T.; Suk, F.M.; Shih, C. Reduced secretion of virions and hepatitis B virus (HBV) surface antigen of a naturally occurring HBV variant correlates with the accumulation of the small s envelope protein in the endoplasmic reticulum and golgi apparatus. *J. Virol.* **2005**, *79*, 13483–13496.

15. Gaggar, A.; Coeshott, C.; Apelian, D.; Rodell, T.; Armstrong, B.R.; Shen, G.; Subramanian, G.M.; McHutchison, J.G. Safety, tolerability and immunogenicity of gs-4774, a hepatitis B virus-specific therapeutic vaccine, in healthy subjects: A randomized study. *Vaccine* **2014**, *32*, 4925–4931.

16. Jiang, W. Blockade of b7-h1 enhances dendritic cell-mediated T cell response and antiviral immunity in HBV transgenic mice. *Vaccine* **2012**, *30*, 758–766.

17. Tian, Y.; Chen, W.L.; Ou, J.H. Effects of interferon-alpha/beta on HBV replication determined by viral load. *PLoS Pathog.* **2011**, *7*, e1002159.

18. Liaw, Y.F.; Kao, J.H.; Piratvisuth, T.; Chien, R.N. Asian-pacific consensus statement on the management of chronic hepatitis B: A 2012 update. *Hepatol. Int.* **2012**, *6*, 531–561.

19. Xiang, X.G.; Xie, Q. IL-35: A potential therapeutic target for controlling hepatitis B virus infection. *J. Dig. Dis.* **2015**, *16*, 1–6.

20. Lai, C.L.; Dienstag, J.; Schiff, E.; Leung, N.W.; Atkins, M.; Hunt, C.; Brown, N.; Woessner, M.; Boehme, R.; Condreay, L. Prevalence and clinical correlates of YMDD variants during lamivudine therapy for patients with chronic hepatitis B. *Clin. Infect. Dis.* **2003**, *36*, 687–696.

21. Yao, G.B.; Zhu, M.; Cui, Z.Y.; Wang, B.E.; Yao, J.L.; Zeng, M.D. A 7-year study of lamivudine therapy for hepatitis B virus E antigen-positive chronic hepatitis B patients in china. *J. Dig. Dis.* **2009**, *10*, 131–137.

22. Hou, J.; Yin, Y.K.; Xu, D.; Tan, D.; Niu, J.; Zhou, X.; Wang, Y.; Zhu, L.; He, Y.; Ren, H.; *et al.* Telbivudine *versus* lamivudine in chinese patients with chronic hepatitis B: Results at 1 year of a randomized, double-blind trial. *Hepatology* **2008**, *47*, 447–454.

23. Liaw, Y.F.; Gane, E.; Leung, N.; Zeuzem, S.; Wang, Y.; Lai, C.L.; Heathcote, E.J.; Manns, M.; Bzowej, N.; Niu, J.; *et al.* 2-year globe trial results: Telbivudine is superior to lamivudine in patients with chronic hepatitis B. *Gastroenterology* **2009**, *136*, 486–495.

24. Papatheodoridis, G.; Buti, M.; Cornberg, M.; Janssen, H.; Mutimer, D. Easl clinical practice guidelines: Management of chronic hepatitis B virus infection. *J. Hepatol.* **2012**, *57*, 167–185.

25. Zhang, Q.; Han, T.; Nie, C.Y.; Ha, F.S.; Liu, L.; Liu, H. Tenofovir rescue regimen following prior suboptimal response to entecavir and adefovir combination therapy in chronic hepatitis B patients exposed to multiple treatment failures. *J. Med. Virol.* **2015**, *87*, 1013–1021.

26. Yatsuji, H.; Suzuki, F.; Sezaki, H.; Akuta, N.; Suzuki, Y.; Kawamura, Y.; Hosaka, T.; Kobayashi, M.; Saitoh, S.; Arase, Y.; *et al.* Low risk of adefovir resistance in lamivudine-resistant chronic hepatitis B patients treated with adefovir plus lamivudine combination therapy: Two-year follow-up. *J. Hepatol.* **2008**, *48*, 923–931.

27. Lee, J.M.; Park, J.Y.; Kim do, Y.; Nguyen, T.; Hong, S.P.; Kim, S.O.; Chon, C.Y.; Han, K.H.; Ahn, S.H. Long-term adefovir dipivoxil monotherapy for up to 5 years in lamivudine-resistant chronic hepatitis B. *Antivir. Ther.* **2010**, *15*, 235–241.

28. Ninomiya, K.; Morikawa, T.; Zhang, Y.; Nakamura, S.; Matsuda, H.; Muraoka, O.; Yoshikawa, M. Bioactive constituents from chinese natural medicines. Xxiii. Absolute structures of new megastigmane glycosides, sedumosides a(4), a(5), a(6), H, and I, and hepatoprotective megastigmanes from sedum sarmentosum. *Chem. Pharm. Bull.* **2007**, *55*, 1185–1191.

29. Lian, L.H.; Jin, X.; Wu, Y.L.; Cai, X.F.; Lee, J.J.; Nan, J.X. Hepatoprotective effects of sedum sarmentosum on d-galactosamine/lipopolysaccharide-induced murine fulminant hepatic failure. *J. Pharmacol. Sci.* **2010**, *114*, 147–157.

30. He, A.; Wang, M.; Hao, H.; Zhang, D.; Lee, K.H. Hepatoprotective triterpenes from sedum sarmentosum. *Phytochemistry* **1998**, *49*, 2607–2610.

31. Epivir-hbv Prescribing Information. GlaxoSmithKline. Available online: http://www.accessdata.fda.gov/ drugsatfda_docs/label/2013/021003s015, 021004s015lbl.pdf.20-Dec-2013 (accessed on 16 August 2015).

32. Sebivo prescribing information. Novartis Pharmaceuticals UK Ltd. Available online: http://www.medicines.org.uk/emc/medicine/19740. 02-Feb-2015 (accessed on 16 August 2015).

33. Baraclude Prescribing Information. Bristol-Myers Squibb Pharmaceutical Limited. Available online: http://www.medicines.org.uk/emc/medicine/18377 (accessed on 16 August 2015).

34. AdefovirDipivoxil Prescribing Information. SigmaPharm Laboratories, LLC. Available online: http://www.accessdata.fda.gov/drugsatfda_docs/label/2013/202051Orig1s000lbl.pdf (accessed on 16 August 2015).

35. Viread Prescribing Information. Gilead Sciences, Inc. Available online: http://www.accessdata.fda.gov/drugsatfda_ docs/label/2015/021356s049, 022577s007lbl.pdf (accessed on 16 August 2015).

36. Chen, X.S.; Wang, G.J.; Cai, X.; Yu, H.Y.; Hu, Y.P. Inhibition of hepatitis b virus by oxymatrine *in vivo*. *World J. Gastroenterol.* **2001**, *7*, 49–52.

37. Lu, L.G.; Zeng, M.D.; Mao, Y.M.; Fang, J.Y.; Song, Y.L.; Shen, Z.H.; Cao, A.P. Inhibitory effect of oxymatrine on serum hepatitis B virus DNA in HBV transgenic mice. *World J. Gastroenterol.* **2004**, *10*, 1176–1179.

38. Lu, L.G.; Zeng, M.D.; Mao, Y.M.; Li, J.Q.; Wan, M.B.; Li, C.Z.; Chen, C.W.; Fu, Q.C.; Wang, J.Y.; She, W.M.; *et al.* Oxymatrine therapy for chronic hepatitis B: A randomized double-blind and placebo-controlled multi-center trial. *World J. Gastroenterol.* **2003**, *9*, 2480–2483.

39. Wang, Y.P.; Zhao, W.; Xue, R.; Zhou, Z.X.; Liu, F.; Han, Y.X.; Ren, G.; Peng, Z.G.; Cen, S.; Chen, H.S.; *et al.* Oxymatrine inhibits hepatitis b infection with an advantage of overcoming drug-resistance. *Antivir. Res.* **2011**, *89*, 227–231.

40. Xu, W.S.; Zhao, K.K.; Miao, X.H.; Ni, W.; Cai, X.; Zhang, R.Q.; Wang, J.X. Effect of oxymatrine on the replication cycle of hepatitis B virus *in vitro*. *World J. Gastroenterol.* **2010**, *16*, 2028–2037.

41. Wang, Y.P.; Liu, F.; He, H.W.; Han, Y.X.; Peng, Z.G.; Li, B.W.; You, X.F.; Song, D.Q.; Li, Z.R.; Yu, L.Y.; *et al.* Heat stress cognate 70 host protein as a potential drug target against drug resistance in hepatitis B virus. *Antimicrob. Agents Chemother.* **2010**, *54*, 2070–2077.

42. Chinese Society of Hepatology and Chinese Society of Infectious Diseases, Chinese Medical Association. The guidelines of prevention and treatment for chronic hepatitis B. *Zhonghua Gan Zang Bing Za Zhi* **2005**, *13*, 881–891. (In Chinese).

43. Soon, D.K.; Lowe, S.L.; Teng, C.H.; Yeo, K.P.; McGill, J.; Wise, S.D. Safety and efficacy of alamifovir in patients with chronic hepatitis b virus infection. *J. Hepatol.* **2004**, *41*, 852–858.

44. Chan, C.; Abu-Raddad, E.; Golor, G.; Watanabe, H.; Sasaki, A.; Yeo, K.P.; Soon, D.; Sinha, V.P.; Flanagan, S.D.; He, M.M.; *et al.* Clinical pharmacokinetics of alamifovir and its metabolites. *Antimicrob. Agents Chemother.* **2005**, *49*, 1813–1822.

45. Ono-Nita, S.K.; Kato, N.; Shiratori, Y.; Carrilho, F.J.; Omata, M. Novel nucleoside analogue mcc-478 (ly582563) is effective against wild-type or lamivudine-resistant hepatitis B virus. *Antimicrob. Agents Chemother.* **2002**, *46*, 2602–2605.

46. Kamiya, N.; Kubota, A.; Iwase, Y.; Sekiya, K.; Ubasawa, M.; Yuasa, S. Antiviral activities of mcc-478, a novel and specific inhibitor of hepatitis B virus. *Antimicrob. Agents Chemother.* **2002**, *46*, 2872–2877.

47. Wu, T.T.; Coates, L.; Aldrich, C.E.; Summers, J.; Mason, W.S. In hepatocytes infected with duck hepatitis B virus, the template for viral RNA synthesis is amplified by an intracellular pathway. *Virology* **1990**, *175*, 255–261.

48. Locarnini, S.A.; Yuen, L. Molecular genesis of drug-resistant and vaccine-escape HBV mutants. *Antivir. Ther.* **2010**, *15*, 451–461.

49. Lucifora, J.; Xia, Y.; Reisinger, F.; Zhang, K.; Stadler, D.; Cheng, X.; Sprinzl, M.F.; Koppensteiner, H.; Makowska, Z.; Volz, T.; *et al.* Specific and nonhepatotoxic degradation of nuclear hepatitis B virus cccdna. *Science* **2014**, *343*, 1221–1228.

50. Thomas, H.C.; Karayiannis, P.; Brook, G. Treatment of hepatitis b virus infection with interferon. Factors predicting response to interferon. *J. Hepatol.* **1991**, *1*, S4–S7.

51. Zhang, F.; Wang, G. A review of non-nucleoside anti-hepatitis B virus agents. *Eur. J. Med. Chem.* **2014**, *75*, 267–281.

52. Fattovich, G.; Giustina, G.; Favarato, S.; Ruol, A. A survey of adverse events in 11,241 patients with chronic viral hepatitis treated with Alfa interferon. *J. Hepatol.* **1996**, *24*, 38–47.

53. Muckenfuss, H.; Hamdorf, M.; Held, U.; Perkovic, M.; Lower, J.; Cichutek, K.; Flory, E.; Schumann, G.G.; Munk, C. Apobec3 proteins inhibit human line-1 retrotransposition. *J. Biol. Chem.* **2006**, *281*, 22161–22172.

54. Stenglein, M.D.; Burns, M.B.; Li, M.; Lengyel, J.; Harris, R.S. Apobec3 proteins mediate the clearance of foreign DNA from human cells. *Nat. Struct. Mol. Biol.* **2010**, *17*, 222–229.

55. Carpenter, M.A.; Li, M.; Rathore, A.; Lackey, L.; Law, E.K.; Land, A.M.; Leonard, B.; Shandilya, S.M.; Bohn, M.F.; Schiffer, C.A.; *et al.* Methylcytosine and normal cytosine deamination by the foreign DNA restriction enzyme apobec3a. *J. Biol. Chem.* **2012**, *287*, 34801–34808.

56. Krebs, K.; Böttinger, N.; Huang, L.R.; Chmielewski, M.; Arzberger, S.; Gasteiger, G.; Jäger, C.; Schmitt, E.; Bohne, F.; Aichler, M.; *et al.* T cells expressing a chimeric antigen receptor that binds hepatitis b virus envelope proteins control virus replication in mice. *Gastroenterology* **2013**, *145*, 456–465.

57. Ahmed, M.; Wang, F.; Levin, A.; Le, C.; Eltayebi, Y.; Houghton, M.; Tyrrell, L.; Barakat, K. Targeting the achilles heel of the hepatitis b virus: A review of current treatments against covalently closed circular DNA. *Drug Discov. Today* **2015**, *20*, 548–561.

58. Jagya, N.; Varma, S.P.; Thakral, D.; Joshi, P.; Durgapal, H.; Panda, S.K. Rna-seq based transcriptome analysis of hepatitis e virus (HEV) and hepatitis b virus (HBV) replicon transfected huh-7 cells. *PLoS ONE* **2014**, *9*, e87835.

59. Cheng, Y.C.; Ying, C.X.; Leung, C.H.; Li, Y. New targets and inhibitors of HBV replication to combat drug resistance. *J. Clin. Virol.* **2005**, *34*, S147–S150.

60. Ying, C.; Li, Y.; Leung, C.H.; Robek, M.D.; Cheng, Y.C. Unique antiviral mechanism discovered in anti-hepatitis B virus research with a natural product analogue. *Proc. Natl. Acad. Sci. USA* **2007**, *104*, 8526–8531.

61. Janmanchi, D.; Tseng, Y.P.; Wang, K.C.; Huang, R.L.; Lin, C.H.; Yeh, S.F. Synthesis and the biological evaluation of arylnaphthalene lignans as anti-hepatitis B virus agents. *Bioorg. Med. Chem.* **2010**, *18*, 1213–1226.

62. Janmanchi, D.; Lin, C.H.; Hsieh, J.Y.; Tseng, Y.P.; Chen, T.A.; Jhuang, H.J.; Yeh, S.F. Synthesis and biological evaluation of helioxanthin analogues. *Bioorg. Med. Chem.* **2013**, *21*, 2163–2176.

63. Tseng, Y.P.; Kuo, Y.H.; Hu, C.P.; Jeng, K.S.; Janmanchi, D.; Lin, C.H.; Chou, C.K.; Yeh, S.F. The role of helioxanthin in inhibiting human hepatitis B viral replication and gene expression by interfering with the host transcriptional machinery of viral promoters. *Antivir. Res.* **2008**, *77*, 206–214.

64. Pang, R.; Tao, J.Y.; Zhang, S.L.; Chen, K.L.; Zhao, L.; Yue, X.; Wang, Y.F.; Ye, P.; Zhu, Y.; Wu, J.G. Ethanol extract from ampelopsis sinica root exerts anti-hepatitis B virus activity via inhibition of p53 pathway *in vitro. Evid. Based Complement. Altern. Med.* **2011**, *2011*, e939205.

65. Le Seyec, J.; Chouteau, P.; Cannie, I.; Guguen-Guillouzo, C.; Gripon, P. Infection process of the hepatitis b virus depends on the presence of a defined sequence in the pre-s1 domain. *J. Virol.* **1999**, *73*, 2052–2057.

66. Blanchet, M.; Sureau, C. Infectivity determinants of the hepatitis b virus pre-s domain are confined to the *N*-terminal 75 amino acid residues. *J. Virol.* **2007**, *81*, 5841–5849.

67. Volz, T.; Allweiss, L.; ḾBarek, M.B.; Warlich, M.; Lohse, A.W.; Pollok, J.M.; Alexandrov, A.; Urban, S.; Petersen, J.; Lütgehetmann, M.; *et al.* The entry inhibitor myrcludex-b efficiently blocks intrahepatic virus spreading in humanized mice previously infected with hepatitis b virus. *J. Hepatol.* **2013**, *58*, 861–867.

68. Yan, H.; Peng, B.; Liu, Y.; Xu, G.; He, W.; Ren, B.; Jing, Z.; Sui, J.; Li, W. Viral entry of hepatitis b and d viruses and bile salts transportation share common molecular determinants on sodium taurocholate cotransporting polypeptide. *J. Virol.* **2014**, *88*, 3273–3284.

69. Slijepcevic, D.; Kaufman, C.; Wichers, C.G.; Gilglioni, E.H.; Lempp, F.A.; Duijst, S.; de Waart, D.R.; Elferink, R.P.; Mier, W.; Stieger, B.; *et al.* Impaired uptake of conjugated bile acids and hepatitis b virus pres1-binding in na(+) -taurocholate cotransporting polypeptide knockout mice. *Hepatology* **2015**, *62*, 207–219.

70. Tsukuda, S.; Watashi, K.; Iwamoto, M.; Suzuki, R.; Aizaki, H.; Okada, M.; Sugiyama, M.; Kojima, S.; Tanaka, Y.; Mizokami, M.; *et al.* Dysregulation of retinoic acid receptor diminishes hepatocyte permissiveness to hepatitis b virus infection through modulation of sodium taurocholate cotransporting polypeptide (NTCP) expression. *J. Biol. Chem.* **2015**, *290*, 5673–5684.

71. Bouezzedine, F.; Fardel, O.; Gripon, P. Interleukin 6 inhibits HBV entry through NTCP down regulation. *Virology* **2015**, *481*, 34–42.

72. Isogawa, M.; Robek, M.D.; Furuichi, Y.; Chisari, F.V. Toll-like receptor signaling inhibits hepatitis b virus replication *in vivo*. *J. Virol.* **2005**, *79*, 7269–7272.

73. Lopatin, U.; Wolfgang, G.; Tumas, D.; Frey, C.R.; Ohmstede, C.; Hesselgesser, J.; Kearney, B.; Moorehead, L.; Subramanian, G.M.; McHutchison, J.G. Safety, pharmacokinetics and pharmacodynamics of gs-9620, an oral toll-like receptor 7 agonist. *Antivir. Ther.* **2013**, *18*, 409–418.

74. Lawitz, E.; Gruener, D.; Marbury, T.; Hill, J.; Webster, L.; Hassman, D.; Nguyen, A.H.; Pflanz, S.; Mogalian, E.; Gaggar, A.; *et al.* Safety, pharmacokinetics and pharmacodynamics of the oral toll-like receptor 7 agonist gs-9620 in treatment-naive patients with chronic hepatitis C. *Antivir. Ther.* **2014**.

75. Gane, E.J.; Lim, Y.S.; Gordon, S.C.; Visvanathan, K.; Sicard, E.; Fedorak, R.N.; Roberts, S.; Massetto, B.; Ye, Z.; Pflanz, S.; *et al.* The oral toll-like receptor-7 agonist gs-9620 in patients with chronic hepatitis b virus infection. *J. Hepatol.* **2015**, *63*, 320–328.

76. Zhang, E.; Lu, M. Toll-like receptor (tlr)-mediated innate immune responses in the control of hepatitis b virus (HBV) infection. *Med. Microbiol. Immunol.* **2015**, *204*, 11–20.

77. Kapoor, R.; Kottilil, S. Strategies to eliminate hbv infection. *Future Virol.* **2014**, *9*, 565–585.

78. Tang, J.; Zhang, Z.H.; Huang, M.; Heise, T.; Zhang, J.; Liu, G.L. Phosphorylation of human la protein at ser 366 by casein kinase II contributes to hepatitis B virus replication and expression *in vitro*. *J. Viral Hepat.* **2013**, *20*, 24–33.

79. Tang, J.; Zhang, Z.H.; Liu, G.L. A systematic analysis of the predicted human la protein targets identified a hepatitis b virus infection signature. *J. Viral Hepat.* **2013**, *20*, 12–23.

80. Heise, T.; Guidotti, L.G.; Chisari, F.V. Characterization of nuclear rnases that cleave hepatitis B virus RNA near the la protein binding site. *J. Virol.* **2001**, *75*, 6874–6883.

81. Tang, J.; Huang, Z.M.; Chen, Y.Y.; Zhang, Z.H.; Liu, G.L.; Zhang, J. A novel inhibitor of human la protein with anti-HBV activity discovered by structure-based virtual screening and *in vitro* evaluation. *PLoS ONE* **2012**, *7*, e36363.

82. Yu, X.; Guo, R.; Ming, D.; Deng, Y.; Su, M.; Lin, C.; Li, J.; Lin, Z.; Su, Z. The tgf-beta1/il-31 pathway is up-regulated in patients with acute-on-chronic hepatitis b liver failure and is associated with disease severity and survival. *Clin. Vaccine Immunol.* **2015**, *22*, 484–492.

83. Qi, P.; Chen, Y.M.; Wang, H.; Fang, M.; Ji, Q.; Zhao, Y.P.; Sun, X.J.; Liu, Y.; Gao, C.F. −509c > t polymorphism in the tgf-beta1 gene promoter, impact on the hepatocellular carcinoma risk in Chinese patients with chronic hepatitis b virus infection. *Cancer Immunol. Immunother.* **2009**, *58*, 1433–1440.

84. Wu, Y.; Zhao, J.; He, M. Correlation between tgf-beta1 gene 29 t > c single nucleotide polymorphism and clinicopathological characteristics of osteosarcoma. *Tumour Biol.* **2015**, *36*, 5149–5156.

85. Yin, W.; Zhao, Y.; Ji, Y.J.; Tong, L.P.; Liu, Y.; He, S.X.; Wang, A.Q. Serum/plasma micrornas as biomarkers for HBV-related hepatocellular carcinoma in China. *Biomed. Res. Int.* **2015**, *2015*.

86. Webster, R.M. The immune checkpoint inhibitors: Where are we now? *Nat. Rev. Drug Discov.* **2014**, *13*, 883–884.

87. Pardoll, D.M. The blockade of immune checkpoints in cancer immunotherapy. *Nat. Rev. Cancer* **2012**, *12*, 252–264.

88. Wolchok, J.D.; Kluger, H.; Callahan, M.K.; Postow, M.A.; Rizvi, N.A.; Lesokhin, A.M.; Segal, N.H.; Ariyan, C.E.; Gordon, R.A.; Reed, K.; *et al.* Nivolumab plus ipilimumab in advanced melanoma. *N. Engl. J. Med.* **2013**, *369*, 122–133.

89. Barakat, K. Immune checkpoints: The search for a single antiviral-anticancer magic bullet. *J. Pharma Care Health Syst.* **2015**, *2*, e125.

90. Gao, X.; Zhu, Y.; Li, G.; Huang, H.; Zhang, G.; Wang, F.; Sun, J.; Yang, Q.; Zhang, X.; Lu, B. Tim-3 expression characterizes regulatory t cells in tumor tissues and is associated with lung cancer progression. *PLoS ONE* **2012**, *7*, e30676.

91. Barakat, K. Do we need small molecule inhibitors for the immune checkpoints? *J. Pharma Care Health Syst.* **2014**, *1*.

92. Viricel, C.; Ahmed, M.; Barakat, K. Human pd-1 binds differently to its human ligands: A comprehensive modeling study. *J. Mol. Graph. Model.* **2015**, *57*, 131–142.

93. Barakat, K. Computer-aided drug design. *J. Pharma Care Health Syst.* **2014**, *1*.

94. Barakat, K.H.; Jordheim, L.P.; Perez-Pineiro, R.; Wishart, D.; Dumontet, C.; Tuszynski, J.A. Virtual screening and biological evaluation of inhibitors targeting the xpa-ercc1 interaction. *PLoS ONE* **2012**, *7*, e51329.

95. Couzin-Frankel, J. Breakthrough of the year 2013. Cancer immunotherapy. *Science* **2013**, *342*, 1432–1433.

96. Chang, T.T.; Lai, C.L.; Kew Yoon, S.; Lee, S.S.; Coelho, H.S.; Carrilho, F.J.; Poordad, F.; Halota, W.; Horsmans, Y.; Tsai, N.; *et al.* Entecavir treatment for up to 5 years in patients with hepatitis b e antigen-positive chronic hepatitis B. *Hepatology* **2010**, *51*, 422–430.

97. Heathcote, E.J.; Marcellin, P.; Buti, M.; Gane, E.; De Man, R.A.; Krastev, Z.; Germanidis, G.; Lee, S.S.; Flisiak, R.; Kaita, K.; *et al.* Three-year efficacy and safety of tenofovir disoproxil fumarate treatment for chronic hepatitis B. *Gastroenterology* **2011**, *140*, 132–143.

Replication and Inhibitors of Enteroviruses and Parechoviruses

Lonneke van der Linden, Katja C. Wolthers and Frank J.M. van Kuppeveld

Abstract: The *Enterovirus* (EV) and *Parechovirus* genera of the picornavirus family include many important human pathogens, including poliovirus, rhinovirus, EV-A71, EV-D68, and human parechoviruses (HPeV). They cause a wide variety of diseases, ranging from a simple common cold to life-threatening diseases such as encephalitis and myocarditis. At the moment, no antiviral therapy is available against these viruses and it is not feasible to develop vaccines against all EVs and HPeVs due to the great number of serotypes. Therefore, a lot of effort is being invested in the development of antiviral drugs. Both viral proteins and host proteins essential for virus replication can be used as targets for virus inhibitors. As such, a good understanding of the complex process of virus replication is pivotal in the design of antiviral strategies goes hand in hand with a good understanding of the complex process of virus replication. In this review, we will give an overview of the current state of knowledge of EV and HPeV replication and how this can be inhibited by small-molecule inhibitors.

Reprinted from *Viruses*. Cite as: van der Linden, L.; Wolthers, K.C.; van Kuppeveld, F.J.M. Replication and Inhibitors of Enteroviruses and Parechoviruses. *Viruses* **2015**, *7*, 4529–4562.

1. Enterovirus and Parechovirus Associated Diseases

1.1. Picornaviridae

The virus family *Picornaviridae* is one of the largest virus families (Figure 1), classified into 29 genera (Figure 1) [1,2]. This review focuses on human pathogens belonging to the genera *Enterovirus* and *Parechovirus*. We will discuss the replication strategies of these viruses in the light of antiviral therapy.

1.2. Enteroviruses

The genus *Enterovirus* (EV) of the picornavirus family contains many important human pathogens, which are among the most common infections in mankind. Overall, it was estimated that around 10–15 million symptomatic EV infections occur annually in the United States alone [3]. This figure excludes the very prevalent rhinovirus infections. The EVs have been classified into 12 species, including EVs A-D, RV A-C, and five EV species that only infect animals (EV-E to EV-J)

391

(Figure 1) [1,2]. These viruses include coxsackie A and B viruses (CVA and CVB, respectively), echoviruses, polioviruses (PVs), numbered EVs, and rhinoviruses (RV).

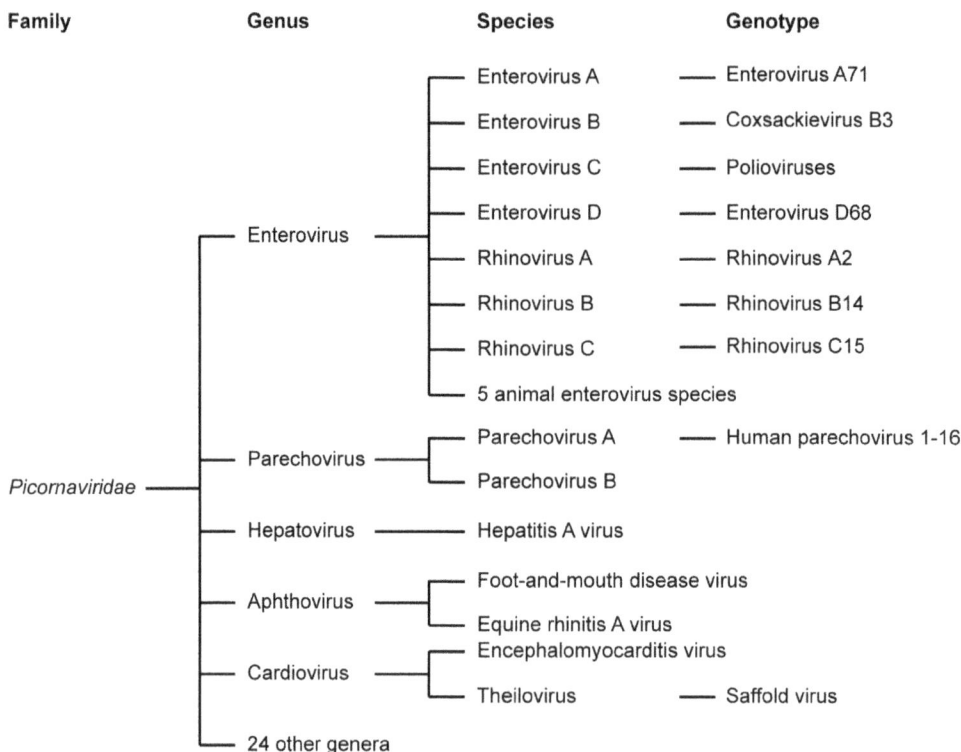

Family	Genus	Species	Genotype
		Enterovirus A	Enterovirus A71
		Enterovirus B	Coxsackievirus B3
		Enterovirus C	Polioviruses
	Enterovirus	Enterovirus D	Enterovirus D68
		Rhinovirus A	Rhinovirus A2
		Rhinovirus B	Rhinovirus B14
		Rhinovirus C	Rhinovirus C15
		5 animal enterovirus species	
Picornaviridae	Parechovirus	Parechovirus A	Human parechovirus 1-16
		Parechovirus B	
	Hepatovirus	Hepatitis A virus	
	Aphthovirus	Foot-and-mouth disease virus	
		Equine rhinitis A virus	
	Cardiovirus	Encephalomyocarditis virus	
		Theilovirus	Saffold virus
	24 other genera		

Figure 1. Classification of the virus family *Picornaviridae*. The clinically most important genera are depicted. For a selection of these, the species and some examples of genotypes/serotypes are given.

EVs are transmitted via the fecal-oral route or via respiratory transmission, depending on the type. EVs have two primary replication sites, the gastrointestinal tract and the respiratory tract, from where the virus can spread to the target organs via the blood circulation. This can result in severe, potentially fatal diseases.

PV is the prototype EV. As in many EV infections, PV infections are mostly clinically mild [4]. However, PV infections can progress to non-paralytic aseptic meningitis (in 1%–2% of cases) or to poliomyelitis, a form of flaccid paralysis (in 0.1%–1% of cases) [4]. Due to intense vaccination programs and surveillance, PV has nearly become extinct, but nevertheless, the virus remains endemic in three countries (Afghanistan, Nigeria and Pakistan) and sporadic PV outbreaks occur.

Coxsackie A and B viruses, echoviruses, and the numbered EVs are associated with a great variety of manifestations, varying from mild respiratory and gastrointestinal infections, herpangina, and hand-foot-and-mouth disease (HFMD), to more severe disease like pleurodynia, hepatitis, myopericarditis, pancreatitis, meningitis, encephalitis, paralysis, and neonatal sepsis leading to mortality [5]. EVs are the most important cause for viral meningitis, accounting for 85%–95% of all cases for which an etiological agent was identified [6].

The genotype/serotype EV-A71 is an emerging pathogen that has caused several outbreaks since the late 1990s. EV-A71 epidemics have been reported worldwide, but mostly in the Asia-Pacific region [7]. HFMD is the most common manifestation of EV-A71 and affects mostly children and infants. However, EV-A71 infections may also result in severe diseases such as pulmonary edema and neurological complications, which may be fatal.

EV-D68 has recently drawn attention because of an outbreak in the United States and to a smaller extent in the rest of the world (e.g., [8–11]. These EV-D68 infected patients presented with severe respiratory illness. Furthermore, the virus was frequently detected in patients with AFP, suggesting the virus may in rare cases be neurotropic [8,12].

RVs can infect both the upper and the lower respiratory tract and are the major cause of the common cold. Though on the less severe side of the spectrum of the diseases caused by EVs, the common cold results in major costs by, among other things, loss of working days, amounting in the United States to $40 billion annually [13]. In addition to the common cold, RVs can cause severe lower respiratory tract infections, such as pneumonia and bronchiolitis [5]. Moreover, RV infections are a serious threat to patients with asthma, chronic obstructive pulmonary disorder (COPD), or cystic fibrosis in whom respiratory tract infections with RVs can lead to exacerbations [14–24]. RVs are subdivided into the species A, B and C. RV-C has been discovered only recently with the help of molecular diagnostic techniques. Initial studies suggested that RV-C is associated with more severe lower respiratory disease than the other species, but later reports suggest that RV-A may be equally pathogenic [25].

1.3. Parechoviruses

When the HPeVs were first identified they were initially classified in the *Enterovirus* genus as echovirus 22 and 23 on the basis of cell-culture characteristics. However, phylogenetic analysis showed these viruses to be genetically distinct from any other picornavirus genus [26,27] and these strains were reclassified in a new genus named *Parechovirus* [28]. Currently, the species Parechovirus *A* contains 16 HPeV types [1,2]. HPeV prevalence has been underestimated, but current data show that HPeVs are at least as prevalent as EVs [29] and that HPeV is a major pathogen in young children [30]. The most commonly circulating HPeV is HPeV-1, which mainly causes mild gastrointestinal and respiratory disease although sometimes in young children more severe disease can be observed [31]. HPeVs are the second most important cause of viral sepsis-like illness and meningitis in infants [32–34]. The majority of these cases are caused by HPeV-3 [33], which is the most pathogenic HPeV type. It is associated with paralysis, neonatal sepsis-like illness and sudden death in infected infants [33,35–43]. The HPeVs have received very little attention from the scientific community in the past, but continuing reports of HPeV circulation all over the world are increasing awareness of the significance of this virus group.

1.4. The Need for Antiviral Drugs against Enteroviruses and Parechoviruses

Although EVs and HPeVs represent a major medical threat, the tools available to fight these viruses are limited. Vaccines are only available against PV: An inactivated PV vaccine (IPV) and an attenuated oral PV vaccine (OPV). Worldwide vaccination schemes started in 1988 have managed to reduce the incidence of PV enormously, but not completely [44]. Sporadic outbreaks may be caused by wildtype virus strains or by vaccine-derived poliovirus (VDPV), which originates from the attenuated OPV. Therefore, as long as OPV is used, the risk of epidemics with VDPV remains. In addition, significant advances have been made towards the development of a vaccine against EV-A71, leading to three inactivated whole virus vaccines that have completed Phase III clinical trials and have applied for approval from the Chinese Food and Drug Administration [45–47]. Although vaccine development for specific pathogens such as PV and EV-A71 is possible, developing vaccines against all members of the EV and Parechovirus genera is not feasible due to the sheer quantity of serotypes (more than 250 EVs, and 16 HPeVs) [2].

To date, no antiviral drugs have been approved for the treatment of picornavirus infections, so treatment is currently limited to supportive care. The only option available for treatment is the administration of pooled immunoglobulin from multiple blood donors (intravenous immunoglobulin, IVIG), but success of this treatment modality depends on the presence of specific neutralizing antibodies in the preparation [48–51]. Hence, there is a strong need for broad-range antiviral drugs to combat EV and HPeV infections. In addition to treating infected patients, antiviral drugs might aid in the eradication of PV by helping to contain VDPV outbreaks and by treatment of immunodeficient patients who are chronically shedding PV. In addition, antiviral drugs may be useful to contain posteradication outbreaks of PV [52].

For the development of antiviral drugs, fundamental knowledge is required about the replication cycle of EVs and HPeVs. Several aspects of virus replication will be summarized in the next section, where the focus will be on the replication of EVs since these viruses have been studied much more intensively and in more detail.

2. Enterovirus Replication Cycle

2.1. Enterovirus Virions and Viral Genome Organization

Picornaviruses are small positive-strand RNA viruses. The genome is encapsidated by an icosahedral capsid, forming a virion of around 30 nm in size without an envelope.

The viral genome contains a single open reading frame with highly structured untranslated regions (UTR) at the 5′- and 3′-end and a 3′-poly(A) tail (Figure 2A). The viral genome is uncapped and instead the 5′-end is covalently coupled to the viral protein 3B, in this context usually termed VPg (viral protein genome-linked). The 5′-UTR contains an internal ribosomal entry site (IRES) which mediates cap-independent translation. Overall, the organization of the open reading frame is similar in all picornaviruses, but there are some differences between genera. In the case of EVs, the open reading frame encodes a polyprotein that contains structural proteins (VP1-4) in the P1 region and the nonstructural proteins (2A–2C and 3A–3D) in the P2 and P3 regions (Figure 2B).

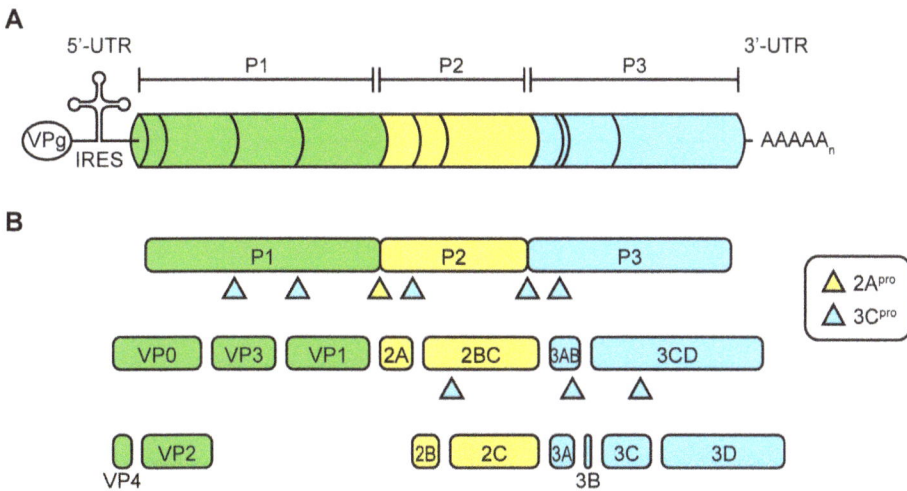

Figure 2. Enterovirus genome. (**A**) Depicted is a schematic representation of the enterovirus genome on scale. The enterovirus genome encodes a single polyprotein divided into a P1, P2, and P3 area. At the 5′- and 3′-end the genome contains untranslated regions (UTR), which are highly structured. The 5′-UTR contains an internal ribosomal entry site for cap-independent translation. At the 5′-end, the RNA genome is covalently bound to the viral protein VPg which is used as a primer during RNA replication; (**B**) The polyprotein is processed into the viral proteins and some stable precursors by the viral proteases $2A^{pro}$ and $3C^{pro}$ (and its precursors).

2.2. Protein Translation and Processing

The EV replication cycle, depicted in Figure 3, is initiated by binding to a receptor. The receptor used differs per virus [53]. For many EVs, the receptor binds at a depression in the capsid called the canyon, which surrounds the fivefold axis of symmetry. Subsequently, the virus is internalized and the viral RNA is released into the cytoplasm. The single polyprotein that is produced, is proteolytically processed by the viral proteases $2A^{pro}$ and $3C^{pro}$ to release the structural and nonstructural viral proteins and some stable precursors (Figure 2B).

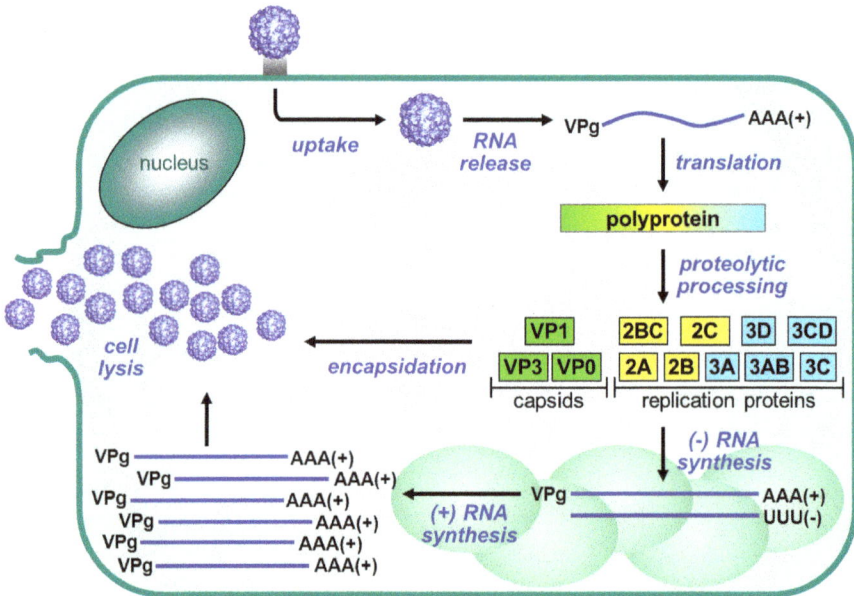

Figure 3. Enterovirus replication cycle. The Enterovirus replication cycle is initiated by binding of the virus to the receptor and internalization into the cell. Subsequently, the viral RNA genome is released from the virion and translated into a single polyprotein which is then processed by the viral proteases to release the viral proteins. Next, the nonstructural proteins mediate the replication of the RNA genome via a negative-stranded intermediate. This takes place on replication organelles that are formed as a result of a rearrangement of cellular membranes. Newly synthesized positive-stranded RNA molecules can then either enter another round of translation and replication (not depicted) or they can be packaged into the viral capsid proteins to form new infectious virus particles which are released upon cell lysis and through several non-lytic mechanisms.

Apart from processing of the viral protein, the viral proteases cleave cellular targets, which serves to optimize the environment for viral proliferation. Cleavage of eIF4G and poly(A)-binding protein (PABP) by 2A^{pro} and 3C^{pro} results in a blockage of translation of cellular proteins, a so-called host shut-off [54–56]. In addition, viral proteases cleave several other cellular factors in order to support virus reproduction and/or suppress innate antiviral responses [57–62].

2.3. Enteroviral RNA Replication

Once liberated from the polyprotein, the nonstructural proteins mediate the replication of the viral genome. RNA genome replication is initiated by uridylylation of the protein primer VPg by the viral RNA-dependent RNA polymerase 3D^{pol} using a secondary RNA structure in the viral genome called cis-acting replication element (Cre) as a template [63–66]. VPgpUpU is then elongated by 3D^{pol} to produce a negative-stranded intermediate which in turn is used as a template for synthesis of positive-stranded RNA molecules. Positive-stranded RNA molecules can then either enter another round of translation and replication or they can be packaged into capsids to produce infectious virus particles.

2.4. The Role of Viral Proteins and Host Factors in Membrane Rearrangements

Typical for positive-strand RNA viruses, replication of the viral RNA takes place on cellular membranes which are drastically reorganized during virus infection [67,68]. In EV-infected cells, both single- and double-membrane structures are observed (Figure 4) [69,70]. Electron tomography studies with PV and CVB3 have revealed that early in virus infection (when RNA replication is already maximal) single-membrane tubular structures are predominant, whereas in later stages these structures appear to flatten, curve, and fuse to form double-membrane vesicles (DMV) [69,70]. These DMVs can then be wrapped by multiple additional cisternae and form multilamellar structures.

The exact origin of the membranes of these organelles is yet unclear. Evidence has been presented that the membranes originated from the early secretory pathway while other data suggested they were derived from the autophagy pathway. The results from the electron tomography studies suggest that there may be some truth in both theories, with the early secretory pathway acting as a source for the single membrane tubules and the autophagy pathway being involved in DMV formation.

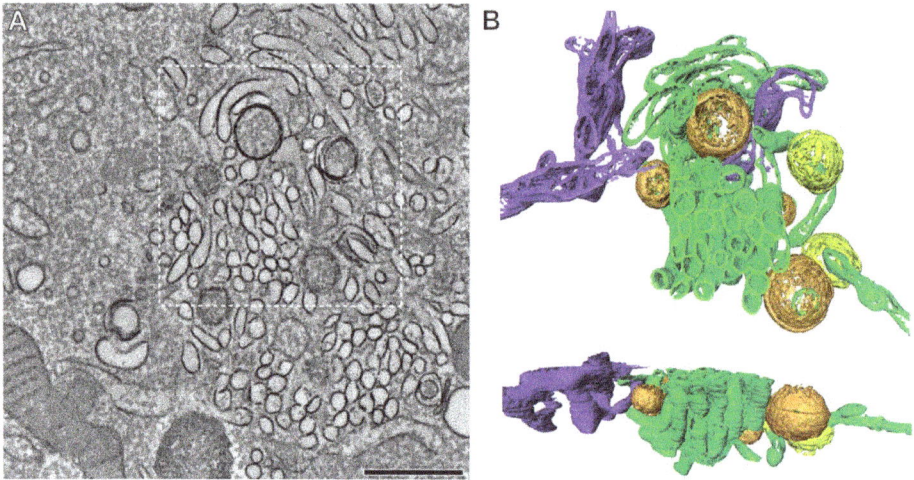

Figure 4. Extensive membrane rearrangements in Enterovirus-infected cells. An electron tomographic slice through a serial tomogram, bar = 500 nm (**A**); and top and side views of the surface-rendered model of the boxed area (**B**) show the presence of single-membrane tubules (**green**), open (**orange**) and closed (**yellow**) double-membrane vesicles in a cell infected with coxsackievirus B3 at 5 h post infection. The ER is depicted in blue. Reprinted from Limpens *et al.* [70], mBio 2011 with permission from the authors, © 2011 by the American Society for Microbiology.

Important observations that support a role for the early secretory pathway in the membrane rearrangements are that Brefeldin A (BFA), a well-known inhibitor of ER-to-Golgi transport, completely abolishes EV replication [71–74] and that several proteins from the secretory pathway are essential for virus replication and can be detected on replication organelles. One of these is Golgi-specific BFA-resistance factor 1 (GBF1), which is a target of BFA. In uninfected cells, GBF1 stimulates GTP exchange of the GTPase ADP-ribosylation factor 1 (Arf1), which is localized on the Golgi complex and the ER-Golgi intermediate compartment. Upon activation, Arf1-GTP becomes membrane-bound and mediates the recruitment of effector proteins such as the COP-I coat complex, thereby inducing the formation of secretory vesicles. Arf1 is thus a key regulator of protein and lipid transport within the early secretory pathway. Upon infection, the viral protein 3A recruits GBF1 and indirectly Arf1 to replication organelles (*i.e.*, virus-induced vesicles plus associated replication complexes) through a direct interaction with GBF1 (Figure 5) [71,75,76]. Through a yet unknown mechanism, this leads to the loss of COP-I from membranes, resulting in a disturbance of the secretory pathway and blockage of protein secretion [75,77–79]. This impairs the expression of MHC class I on the cell surface and cytokine secretion [80,81], implying that the virus-induced membrane rearrangement not only serves to facilitate viral RNA replication but also to suppress infection-limiting host immune responses.

Phosphatidylinositol-4-kinase III beta (PI4KIIIβ), another Golgi-localized protein, is also an essential host factor for EV replication [77]. PI4KIIIβ is a kinase that synthesizes phosphatidylinositol-4-phosphates (PI4P). As a precursor for PI(4,5)P$_2$, PI4P lipids are part of PI3K and phospholipase C signaling pathways, but PI4P lipids also recruit proteins with a PI4P-binding pleckstrin homology (PH) domain to membranes, thereby regulating membrane biogenesis, lipid homeostasis, and vesicle-mediated trafficking at the Golgi complex [82–84]. During infection, PI4KIIIβ is recruited to replication sites by 3A and consequently local levels of PI4P lipids increase (Figure 5) [77]. The mechanism behind the recruitment of PI4KIIIβ by 3A remains to be determined but was shown to be independent of GBF1, Arf1, and ACBD3 [85,86]. *In vitro*, PI4P lipids specifically bound 3D^{pol}, suggesting that they may serve to recruit the viral polymerase to replication complexes [77], but firm proof for this idea is lacking. Alternatively, or additionally, the function of PI4P lipids in virus replication may be to recruit PH domain-containing proteins for example with membrane-modifying properties.

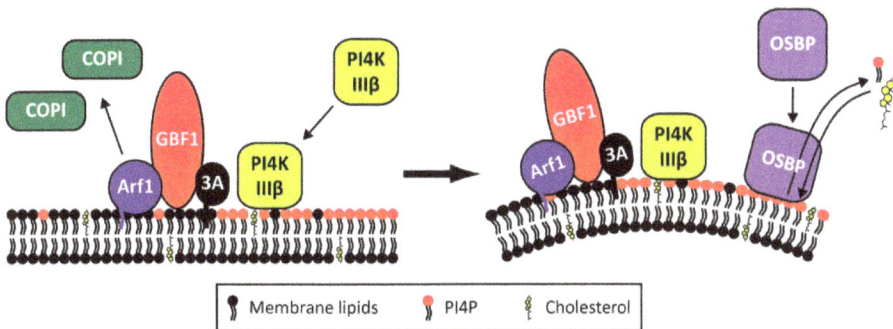

Figure 5. The proposed role of Golgi-localized host factors in Enterovirus replication. Upon infection, the viral 3A protein recruits GBF1 and indirectly Arf1 to the replication organelles. As a result, COP-I is lost from the membranes. At the same time, PI4KIIIβ is recruited by 3A through a GBF1/Arf-independent mechanism, resulting in an increase in PI4P lipids. OSBP then binds to the PI4P lipids and mediates a PI4P/cholesterol counterflow between the membranes of the replication organelles and the ER.

One such protein appears to be oxysterol-binding protein (OSBP), a PI4P-binding protein that transports PI4P lipids produced by PI4KIIIβ from the Golgi complex to the ER, in exchange for cholesterol which is transported in the opposite direction [87]. Recent studies by others and ourselves have revealed that OSBP binds to PI4P-enriched replication organelle membranes and mediates a PI4P/cholesterol counterflow between these membranes and the ER (Figure 5) [88–90]. As a result, the cholesterol content of membranes of the replication organelles is increased. In addition to this mechanism, increased uptake of cholesterol and a role of endosomal cholesterol have been suggested to contribute to the accumulation of cholesterol in the membranes of the replication [90–92]. The cholesterol content of membranes determines the membrane fluidity and formation of lipid microdomains and therefore the virus-induced accumulation of cholesterol may serve to induce the membrane deformations required to generate replication organelles. All in all, it appears that the regulation of PI4P and cholesterol levels is very important to support replication.

DMVs are reminiscent of autophagosomes with respect to their appearance and formation, which originated the idea that the autophagic pathway is involved in the formation of replication organelles. The recent observation that DMVs occur mostly in later stages of infection suggests that this pathway is mostly important in the advanced stages of membrane rearrangements [69,70]. Inhibition of the autophagy pathway impairs viral replication, but only to a modest extent [93,94]. A recent publication has provided evidence that vesicular acidification promotes maturation of PV particles (*i.e.*, VP0 cleavage, see next section), implicating a role for autophagy and DMVs in the last step of the replication cycle [95]. Furthermore, it has been suggested that the DMVs might be involved in non-lytic release of virus particles by fusion with the plasma membrane, challenging the dogma that EVs egress only through cell lysis. Together, these data suggest that the early secretory pathway and the autophagy pathway have a distinct, but important function during EV replication.

Genetic and biochemical evidence suggests that the viral proteins 2BC and 3A are involved in the formation of replication organelles during infection [96–98]. These proteins have hydrophobic domains and extensively interact with cellular membranes. 3A is probably important in membrane reorganization through its (direct and indirect) interactions with GBF1, Arf1, and PI4KIIIβ. 2B is a viroporin that enhances the permeability of ER and Golgi membranes [99–103]. Overexpression of 2B leads to disturbed ion homeostasis, impaired transport of proteins through the Golgi complex, and increased targeting of endocytic vesicles to the Golgi complex [78,101,104,105]. How and if these activities are involved in the formation of replication organelles is unknown. 2C has been postulated to contribute to the membrane remodeling by insertion of its hydrophobic domains in the membranes, as well as through its interaction with reticulon proteins [106]. These latter proteins are membrane-shaping proteins that induce and stabilize positive membrane curvature, and may be involved in the formation of the positively curved membranes that are essential for the morphogenesis of the viral replication organelles.

As has become clear from this brief overview, membrane remodeling involves many viral and host proteins and lipids and is a very complicated process that is not completely understood.

2.5. Morphogenesis and Virus Release

Once synthesized, viral RNA of positive polarity is encapsidated by capsid proteins to form new virions. This process is coupled to active replication as only newly synthesized genomes are encapsidated [107,108]. This is not guided by an RNA encapsidation signal or RNA-protein interactions, but rather by an interaction between 2C, which is located in the replication complex, and the capsid protein VP3 [109].

The assembly of new virions (Figure 6) is initiated by the release of the P1 capsid precursor from the polyprotein. This is subsequently folded by the chaperone protein Hsp90 and processed by 3CD[pro] to release VP0 (the precursor of VP4 and VP2), VP1, and VP3 [110,111]. In a spontaneous process, these capsid proteins assemble to form a protomer. Five protomers together then form a pentamer which in turn assemble to form a provirion. Several, but not all, EVs require the presence of glutathione for the formation and/or stability of the pentamers [112–115]. The last step is a maturation of the virion by RNA-induced cleavage of VP0 into VP2 and VP4, yielding an infectious virus particle. This process has been suggested to be enhanced by the acidic environment in autophagosome-like DMVs [95].

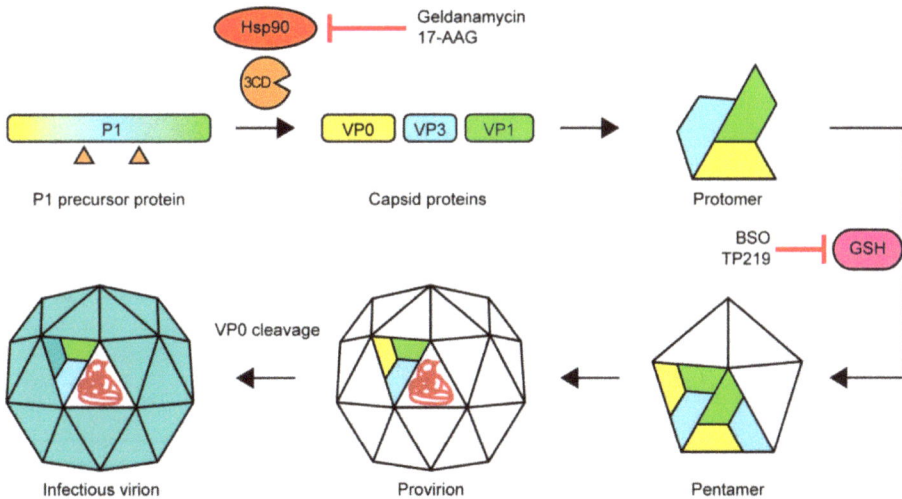

Figure 6. Morphogenesis of enteroviruses and targets for assembly inhibitors. Hsp90 ensures the proper folding of the P1 precursor protein enabling the cleavage by 3CD[pro] into capsid proteins VP0, VP3, and VP1 which then form protomers. For part of the EVs, glutathione (GSH) is required either for the transition of protomers into pentamers or for the stabilization of pentamers. Twelve pentamers plus the viral genome (in red) combine to form a provirion, followed by a maturation step in which the VP0 protein is cleaved into VP4 and VP2. Treatment with Hsp90 inhibitors or glutathione depletors results in impaired morphogenesis.

The dogma has always been that newly formed infectious particles are released by lysis of the host cell, but recent reports have suggested additional methods of egress that do not require cell lysis, such as non-lytic release through DMVs and release of phosphatidylserine lipid-enriched vesicles packed with virions [116,117]. This mechanism is reminiscent of the release of hepatitis A virus, another picornavirus, which was recently shown to be released as membrane-wrapped virus particles in membrane structures resembling exosomes [118].

3. Antiviral Compounds

Antiviral therapy can have a variety of mechanisms of action and each step of the virus replication cycle can be targeted. Virus replication can be impaired by targeting either viral proteins or host factors that are required for virus replication. Decades of searching for compounds with antipicornaviral activity has yielded a range of compounds that inhibit EV replication, although none of them have reached approval for clinical use. The current status of research on and development of small-molecule antiviral compounds that target the replication of EVs is summarized here below.

3.1. Enterovirus Inhibitors

Capsid binders. The three compounds that are currently under clinical evaluation are pleconaril (Viropharma, Exton, USA, licensed to Schering-Plough in 2003), BTA798 (Biota Pharmaceuticals, Alpharetta, USA), and pocapavir (V-073, ViroDefense Inc., Rockville, USA)). All three compounds bind the capsid in the canyon, a depression in the capsid which is responsible for receptor binding. By doing this they impair the first steps in the viral replication cycle, *i.e.*, receptor attachment and/or viral uncoating. Pleconaril displays broad-spectrum activity against most, but not all EVs and had some beneficial effects in clinical trials [119–124], but was rejected by the FDA in 2002 for the treatment of the common cold due to safety issues [125]. In a later clinical phase II trial on the prevention of asthma exacerbations and common cold symptoms in asthma patients challenged with RV, pleconaril treatment had no significant effect [126] while the results of a trial to test the efficacy of pleconaril in infants with enteroviral sepsis syndrome have not been made available yet [126,127].

The compound BTA798 is currently under investigation for the treatment of RV infections in asthmatic patients. In March 2012, a clinical phase IIb trial testing the efficacy of BTA798 in asthma patients with naturally acquired RV infections was completed [128]. Biota reported significant reductions in symptoms upon treatment with BTA798 and is currently planning another phase IIb trial in RV-infected patients with moderate-to-severe asthma [129].

Pocapavir is under clinical investigation to be used in the eradication of PV [130], but has also been shown to have activity against other EVs [131]. Recently, a case study was reported in which pocapavir was used to treat an infant with severe enteroviral sepsis [132]. Pocapavir appeared to be tolerated well, but a clear antiviral effect was absent.

One of the biggest issues with capsid binders is that EVs readily develop resistance [133]. The quick emergence of resistance may be one of the reasons that the performance of this class of compounds in clinical trials is disappointing [134]. In addition, one of the problems with capsid binders is that they do not inhibit all EVs [135–137] and we have recently reported a naturally occurring pleconaril-resistant E11 strain [50,138]. Because of these resistance issues it is questionable whether capsid inhibitors will make it to the market.

Protease inhibitors. Viral proteases have shown to be useful targets for antiviral therapy in treatments against human immunodeficiency virus (HIV) and hepatitis C virus. The only EV protease inhibitors that have made it to clinical studies are the 3C^{pro} inhibitors rupintrivir (also known as AG7088), and AG7404 (also known as compound 1). These protease inhibitors have been designed to mimic the protease substrate. The compound rupintrivir was developed as a potent inhibitor of RV 3C^{pro}, but further studies revealed that rupintrivir also inhibits replication of other EVs [139–143]. Rupintrivir is an irreversible peptidomimetic with an α,β-unsaturated ester. Because of limited activity in clinical trials with natural RV infections [144], the clinical development was halted. AG7404 was developed as an analog of rupintrivir with improved oral bioavailability compared to rupintrivir [145]. AG7404 displays antiviral activity *in vitro* and is safe and well-tolerated *in vivo*, but clinical development was discontinued [144].

Polymerase inhibitors. Evidently, inhibition of the viral polymerase 3D^{pol} is detrimental for virus replication and is therefore an attractive strategy for antiviral approaches. There are two classes of polymerase inhibitors. The nucleoside analogs are compounds that have structural similarity to endogenous nucleosides. Incorporation of these analogs by 3D^{pol} results in chain termination and/or incorporation of incorrect nucleosides leading to mutations. One such compound is ribavirin, a synthetic purin analog. Treatment with ribavirin leads to lethal mutagenesis [146,147]. The other class of polymerase inhibitors encompasses the non-nucleoside inhibitors. These inhibitors can have a variety of mechanisms of action. For example, the polymerase inhibitor amiloride increases the error rate of 3D^{pol} and competes with incoming nucleoside triphosphates and Mg^{2+} [148]. In addition, we have just reported the discovery of a compound (GPC-N114) which interferes with productive binding of the template-primer to 3D^{pol} by binding at the template-binding site [149]. However, for most compounds the mechanism of action has been poorly characterized (gliotoxin, DTrip-22, aurintricarboxilic acid, BPR-3P0128) [150–154]. Up to now, no 3D^{pol} inhibitors have been tested in clinical trials.

2C-targeting compounds. Another group of inhibitors is that of compounds that select for mutations in 2C. The functions of the nonstructural protein 2C are not fully understood, but the protein has been implicated in RNA replication [106,155–160], RNA binding [161–163], the induction of membrane rearrangements [97,98,103,164], encapsidation [109,165,166], and uncoating [167]. On the basis of several conserved motifs, 2C was also predicted to be a helicase [168], although the RNA unwinding activity has not yet been experimentally confirmed. 2C has a nucleoside triphosphate-binding motif and displays ATPase activity [169,170]. A variety of antiviral compounds have been shown to target 2C (guanidine hydrochloride, fluoxetine, HBB, MRL-1237, and TBZE-029) [171–176], but their mechanisms of action have not been resolved yet. This is in a great part due to the inability to obtain functional recombinant 2C and the lack of a crystal structure. None of the 2C inhibitors have progressed to *in vivo* studies in humans.

PI4KIIIβ inhibitors. In the late 1970s, the compound enviroxime was reported to have potent antiviral activity against EVs [177], but it took until 2012 to identify PI4KIIIβ as the target of enviroxime following the recognition of PI4KIIIβ as an essential host factor [77,178]. Aided by cross-resistance of virus variants with mutations in 3A, also GW5074, T-00127-HEV1, and compound 1 were identified as PI4KIIIβ inhibitors [178–183]. Enviroxime was tested in several clinical studies in which it was able to reduce symptoms, but overall the *in vivo* efficacy was disappointing [184–189]. This in combination with gastrointestinal side-effects and a poor pharmacokinetic profile caused the clinical development to be halted [188].

For a number of other PI4KIIIβ inhibitors little if any toxicity in cell lines as well as in primary human fibroblasts was reported [178,183,190,191]. One of these inhibitors (*i.e.*, compound 1) also showed no obvious adverse effects upon testing in a CVB4-induced pancreatitis mouse model, while it resulted in a strong inhibitory effect on *in vivo* CVB4 replication and CVB4-induced pathology [183]. However, in another study it was described that chemical inhibition of PI4KIIIβ by a small molecule identified by Boehringer Ingelheim as well as by T-00127-HEV1 was lethal to mice [192]. Moreover, PI4KIIIβ inhibitors developed by Novartis were found to interfere with lymphocyte proliferation and chemokine secretion in a mixed leukocyte reaction [190]. These latter results have raised many concerns about the potential application of PI4KIIIβ inhibitors to treat EV infections.

OSBP inhibitors. The 3A-mutant viruses that were resistant against PI4KIIIβ inhibitors, also displayed a resistant phenotype against a set of antiviral compounds that had no effect on PI4KIIIβ activity. This included the compounds itraconazole, OSW-1, AN-12-H5 and T-00127-HEV2, and 25-hydroxycholesterol. We and others revealed that the target of these compounds was OSBP [88,193,194]. Studying the mechanism-of-action of these compounds has been significantly instrumental in discovering that OSBP is required for virus replication. Itraconazole has already been approved by the FDA in 1992 as an antifungal drug and therefore already much is known about the behavior of this compound *in vivo*. In recent years, itraconazole has also gained interest as an anticancer reagent, and is currently being tested in clinical trials [195]. The wealth of information gathered while testing itraconazole for these other applications should facilitate testing the antiviral effect of itraconazole in mouse models.

Assembly inhibitors. Assembly of new virus particles is a stepwise process which can therefore be interrupted at multiple steps (Figure 6). The use of inhibitors has greatly contributed to our knowledge of the assembly process. The chaperone Hsp90 interacts with the P1 capsid precursor and ensures a correct folding that enables recognition and/or cleavage by 3CD^{pro} [110,196]. Inhibition of Hsp90 with the compounds geldanamycin or its analog 17-AAG (17-allyamino-17-demethoxygeldanamycin) still allows the interaction between Hsp90 and P1, but processing of P1 is impaired. As a result, morphogenesis is impaired. The role of glutathione in the assembly of a subgroup of EVs has come to light by the study of the glutathione synthesis inhibitor BSO (buthionine sulfoximine) and the glutathione scavenger TP219 [112–115]. BSO has been tested in clinical trials for testing its ability to reverse the resistance to chemotherapy in cancer treatments caused by high glutathione levels. These showed that BSO treatments were tolerated relatively well [197–199], but to our knowledge, no attempts or plans have been made to test the antiviral activity of BSO or other glutathione-depleting compounds *in vivo*.

3.2. Parechovirus Replication and Inhibitors

Despite the clinical significance of HPeVs, these viruses have not been the topic of many in-depth molecular studies and very little work has been done to identify compounds with antiviral activity against this virus group. The fact that HPeVs and EVs are members of the same virus family, suggests that HPeVs may benefit from the ample work being done on EVs. However, work done so far, has only highlighted the differences between the two virus genera.

The capsid binder pleconaril has been described not to possess any antiviral activity against HPeVs [200,201]. Though the HPeV capsid structure is similar to that of EVs in that it is icosahedral and the capsid proteins have the typical β-barrel structure, the HPeV capsid has a shallower canyon than in EVs [202]. Moreover, the canyon does not mediate receptor binding, at least not for the HPeVs that utilize integrins as a receptor [202]. Thus, in this light it is not surprising that pleconaril does not possess antiviral activity against HPeV. Since all of the capsid binders developed against EVs target the canyon, these are not expected to display activity against HPeVs.

Besides capsid inhibitors, 3C^{pro} inhibitors were suggested to be the most promising candidates for anti-HPeV therapy [201]. However, we showed recently that rupintrivir, the best-established EV 3C inhibitor, and SG85, a recently developed analogue [203], do not inhibit the proteolytical activity of HPeV 3C^{pro} [204]. The reason for this is unknown, since the structure of HPeV 3C^{pro} has not been resolved yet. An explanation may lie in the difference between the cleavage sites recognized by the EV and HPeV proteases, since these are peptidomimetics designed to resemble the cleavage site recognized by 3C^{pro}. Rupintrivir is a tripeptide-derived molecule with a Michael acceptor moiety [205]. Its structure is based on the leucine-phenylalanine-glutamine peptide sequence at the P3-P2-P1 position observed in EV 3C^{pro} cleavage sites [205,206]. However, none of the HPeV 3C^{pro} cleavage sites observed contain a leucine at P3 [207]. On the other hand, SG85 contains a t-butyl ether of serine at P3, and a serine has been observed at this position in HPeV cleavage sites [203,207]. Though the t-butyl ether modification may be responsible for the inactivity of SG85 against HPeV 3C^{pro}, this suggests that differences in cleavage sites alone cannot completely explain the disparity between the susceptibilities of the EV and HPeV 3C proteases to rupintrivir and SG85. Detailed analysis of cleavage sites recognized by HPeV 3C^{pro} and especially crystal structures are required to understand the insusceptibility of HPeV 3C^{pro} to rupintrivir and SG85 and for the design of HPeV 3C^{pro} inhibitors.

In contrast to 2A of EVs, the 2A protein of HPeV does not have proteolytic activity and the P1/P2 cleavage is carried out by 3C^{pro} [208]. Hence, HPeV 2A does not induce a host shut-off through cleavage of translation factors [209]. Instead, 2A has been described to bind the 3′-end of the HPeV genome, suggesting a role in HPeV RNA replication [210]. Accordingly, EV and HPeV 2A have only very limited amino acid identity [211]. HPeV 2A will therefore not benefit from antiviral studies targeting EV 2A^{pro}, though these have been rare.

On the other hand, for both genera the 2C protein is predicted to be a helicase and has been shown to have ATPase activity [212]. 2C represents a promising target for EVs, but the EV 2C inhibitors GuHCl and HBB have no effect on HPeV 2C [213]. Since these compounds have the broadest spectrum of anti-enteroviral activity of the 2C inhibitors available [171,175,176,213,214], it is to be expected that the others do not inhibit HPeV 2C either. For fluoxetine and TBZE-029, for example, the susceptibility seems to require the presence of the so-called AGSINA motif, which is not present in HPeV 2C [176].

The replication strategies employed by HPeV are mostly a black box, but it is clear that they are different from those used by EVs. A first clue came from the partial resistance of HPeV to the GBF1 inhibitor BFA, which completely blocks EV replication [72]. In addition, COP-I is dispersed in cells infected with HPeV while in EV-infected cells, COP-I is found on replication organelles [72]. Accordingly, the amino acid identity between HPeV 3A and EV 3A is very low [211] and the replication organelles formed upon infection have a different morphology and appear to have a different origin than those in EV-infected cells [72,215]. It is not known whether HPeV uses PI4KIIIβ like EVs, or any other PI4K for replication. No PI4K inhibitors have been tested with HPeV thus far, but doing so will help to elucidate this question. Our study on itraconazole as an EV and cardiovirus inhibitor, revealed that this compound has no effect on HPeV replication, suggestive of an independence of OSBP or at the least a different usage of OSBP [88].

Despite that multiple compounds have been identified as EV 3D^{pol} inhibitors, none of these have been tested with HPeV. However, it seems reasonable to assume that the nucleoside analogues such as ribavirine will also display activity against HPeV since these have a very broad spectrum of activity as a result of a very conserved nucleotide-binding site [216]. Our preliminary results suggest that GPC-N114 has no effect on HPeV replication (unpublished results).

Other than that the maturation cleavage of VP0 into VP4 and VP2 does not occur in HPeVs, not much is known about assembly of HPeV virions. It has not been studied whether HPeV requires the chaperone Hsp90 and/or glutathione for morphogenesis.

Overall, it is clear that there are large gaps in our knowledge of HPeV biology and that the tools against HPeVs are severely limited.

3.3. Issues in Antiviral Drug Development

Several issues complicate the development of antiviral drugs against EVs and HPeVs. First of all, broad-spectrum antiviral activity is desired, to be able to treat all EV or HPeV infections with the same antiviral drug to eliminate the need for typing. In addition, it would be undesirable to have to develop dozens of antiviral drugs to be able to treat all EV/HPeV infections. Secondly, the possible emergence of drug resistance could become a major problem when drugs are introduced into the clinic. The mutation frequency for RNA polymerases is relatively high. As a result, a virus population can be described as a quasispecies. Drug-treatment will select for virus mutants with a selective advantage, allowing the outgrowth of resistant virus mutants and rendering the antiviral therapy ineffective. Although this is considered less of a problem when treating acute infections, the class of capsid binders has shown a quick emergence of resistant variants, making this class of inhibitors less suitable for treating picornaviral infections [133]. Therefore, compounds to which the virus cannot (easily) develop drug resistance are highly needed.

One strategy believed to help with both of the above-named issues, is to develop antiviral drugs that target host factors. Many essential host factors are common to all viruses within a genus, which allows targeting an entire genus without the problem of variable drug pockets in the viral proteins. Part of the host factors may even be shared by multiple genera. One illustrative example of this is the OSBP inhibitor itraconazole, for which the spectrum of antiviral activity encompasses all tested viruses of both the EV and even the cardiovirus genus [88]. In addition, it is highly unlikely that a host factor will become resistant to the drug in response to therapy. A finding that supports this hypothesis, is the inability to obtain resistant PV when passaging virus in the presence of geldanamycin, an inhibitor of the host chaperone protein Hsp90 which is required for virus assembly [110]. However, there are also examples where the virus has become resistant to host protein-targeting compounds. For example, PV could become resistant to the GBF1-inhibitor BFA by acquiring mutations in 2C and 3A [217]. Furthermore, we found that single point mutations in CVB3 3A can render CVB3 independent of PI4KIIIβ/PI4P and resistant to OSBP inhibitors, which shows that RNA viruses can even become resistant to inhibitors of essential host factors and use a bypass mechanism for replication independent of (high levels of) a critical lipid [178].

A potential downside of targeting host factors is that the chances of adverse effects of drug treatment are higher when inhibiting cellular targets than when targeting viral proteins. However, one should realize that in general most drugs have host targets. In fact, this is the case in all diseases in which no pathogen is involved, demonstrating that targeting host proteins is not necessarily problematic. For example, inhibitors of the hepatitis C virus host factor cyclophin A, a protein that makes up 0.1%–0.4% of the total cellular protein content [218], generally have a good safety profile and any adverse effects observed seem to be associated with the specific inhibitor rather than the target [219].

Overall, targeting host factors may prove a good, though not perfect, strategy, notwithstanding the usefulness of direct-acting antiviral compounds. Ultimately, the aim would be to develop combination therapy consisting of multiple antiviral drugs with different resistance profiles.

4. Future Perspectives in Antiviral Research

In the last decades, many researchers have focused on identifying inhibitors of EV replication, leading to the discovery of numerous of EV inhibitors. However, since none of these have reached the market, a lot of work still needs to be done.

A category of inhibitors that remains largely unexplored is that of the allosteric 3C^{pro} inhibitors. Most of the protease inhibitors developed so far—such as rupintrivir, compound 1, and the SG compounds—are peptidomimetic active site inhibitors which have been designed based on the cleavage site recognized by the targeted protease. Interest has increased in allosteric site inhibitors, since the occurrence of resistance mutations in the active site of competitive compounds often gives cross-resistance to other competitive inhibitors [220,221]. Last year, Wang et al. [222,223] reported the first inhibitors of EV71 3C^{pro} with a binding site outside the substrate binding site, illustrating the potential of small molecules as allosteric inhibitors of EV proteases. Further endeavors are expected to yield more allosteric inhibitors, thereby opening up a whole new class of 3C^{pro} inhibitors.

Another area where advances can be made, are the 2C inhibitors. The fact that many compounds have been found to target 2C implies that 2C is easily targeted, possibly as a result of an easily accessible compound-binding pocket. Elucidation of the structure and function of 2C has proven very difficult, but will greatly benefit the development and characterization of 2C inhibitors. For now, the most promising 2C inhibitor appears to be fluoxetine. Its modest potency in combination with the relatively low plasma concentrations achieved *in vivo* and reported side effects could possibly obstruct the use for antiviral therapy *in vivo* [176]. However, it was recently pointed out that fluoxetine reaches considerably higher concentrations in the brain than in plasma, opening up the possibility to use fluoxetine for treatment of neurological diseases caused by EV-B and EV-D viruses, such as EV-D68 [224].

While considerable progress has been made for the EVs over the past decades, research into HPeV replication and anti-parechoviral therapy is still in early stages. Clearly, anti-parechoviral studies would be greatly facilitated by a greater knowledge of HPeV replication. For example, HPeV 2A may be used as a target for antiviral therapy, once its role in the RNA replication has been deciphered. Currently we are trying to gain more insight into the replication strategies employed by HPeVs by utilizing inhibitors that target EV host factors. This provides an easy and quick method to determine whether these viruses require PI4KIIIβ, OSBP, Hsp90, and glutathione for efficient replication. The results will provide a starting point from where more in-depth replication studies can be done and may provide targets for antiviral therapy. In addition, a crystal structure of HPeV 3C^{pro} would be of great benefit for the design of inhibitors of this enzyme.

For both groups of picornaviruses, the repurposing of drugs, *i.e.*, the use of drugs already on the market for a different indication, may be a way to obtain antiviral drugs that can quickly be put to practice, because these drugs have been well-characterized and tested *in vivo*. Both the antifungal itraconazole and the antidepressant fluoxetine were found when screening FDA-approved drug libraries [88,176], and additional screening will probably yield more hits.

In the coming years, advances made in the fields of EV/HPeV replication mechanisms and antiviral therapy will develop synergistically, yielding a better understanding of replication mechanisms, thus providing novel targets for antiviral therapy and the characterization of compounds with antiviral activity, giving us more insight into aspects of virus replication. Though the development of antiviral therapy represents a major challenge, progress made in recent years raises hope that this is indeed an achievable goal.

Acknowledgments: The authors thank Jeroen Strating for providing Figure 5. This work was supported by grants from the European Union FP7: Marie Curie IAPP "AIROPico" (grant agreement number 612308) (K.C.W.), Marie Curie Initial Training Network "EUVIRNA" (grant agreement number 264286) (F.J.M.K.) and Large Scale Collaborative Project "SILVER" (grant agreement number 260644) (F.J.M.K.), from the Netherlands Organisation for Scientific Research (NWO): ALW-820.02.018 and VICI-91812628 (F.J.M.K.) and from Crucell (L.L.). The funders had no role in study design, data collection and analysis, decision to publish, or preparation of the manuscript.

Author Contributions: L.L., K.C.W. and F.J.M.K wrote the paper.

Conflicts of Interest: The authors declare no conflict of interest.

Bibliography

1. Adams, M.J.; King, A.M.; Carstens, E.B. Ratification vote on taxonomic proposals to the International Committee on Taxonomy of Viruses (2013). *Arch. Virol.* **2013**, *158*, 2023–2030.

2. Knowles, N.J.; Hovi, T.; Hyypiä, T.; King, A.M.; Lindberg, A.M.; Pallansch, M.A.; Palmenberg, A.C.; Simmonds, P.; Skern, T.; Stanway, G.; *et al.* Picornaviridae. In *Virus Taxonomy: Classification and Nomenclature of Viruses: Ninth Report of the International Committee on Taxonomy of Viruses*; King, A.M.Q., Adams, M.J., Carstens, E.B., Lefkowitz, E.J., Eds.; Elsevier: San Diego, CA, USA, 2012; pp. 855–880.

3. Strikas, R.A.; Anderson, L.J.; Parker, R.A. Temporal and geographic patterns of isolates of nonpolio enterovirus in the United States, 1970–1983. *J. Infect. Dis.* **1986**, *153*, 346–351.

4. Centers for Disease Control and Prevention. Poliomyelitis. In *Epidemiology and Prevention of Vaccine-Preventable Diseases*; Atkinson, W., Hamborsky, J., Wolfe, S., Eds.; Public Health Foundation: Washington DC, 2012; Volume 12, pp. 249–261.

5. Tapparel, C.; Siegrist, F.; Petty, T.J.; Kaiser, L. Picornavirus and enterovirus diversity with associated human diseases. *Infect. Genet. Evol.* **2013**, *14*, 282–293.

6. Rotbart, H.A. Viral meningitis. *Semin. Neurol.* **2000**, *20*, 277–292.

7. Ooi, M.H.; Wong, S.C.; Lewthwaite, P.; Cardosa, M.J.; Solomon, T. Clinical features, diagnosis, and management of enterovirus 71. *Lancet Neurol.* **2010**, *9*, 1097–1105.

8. Greninger, A.L.; Naccache, S.N.; Messacar, K.; Clayton, A.; Yu, G.; Somasekar, S.; Federman, S.; Stryke, D.; Anderson, C.; Yagi, S.; *et al.* A novel outbreak enterovirus D68 strain associated with acute flaccid myelitis cases in the USA (2012–14): A retrospective cohort study. *Lancet. Infect. Dis.* **2015**, *15*, 671–682.

9. Meijer, A.; Benschop, K.S.; Donker, G.A.; van der Avoort, H.G. Continued seasonal circulation of enterovirus D68 in the Netherlands, 2011–2014. *Euro Surveill.* **2014**, *19*, art 1.

10. Midgley, C.M.; Jackson, M.A.; Selvarangan, R.; Turabelidze, G.; Obringer, E.; Johnson, D.; Giles, B.L.; Patel, A.; Echols, F.; Oberste, M.S.; *et al.* Severe respiratory illness associated with enterovirus D68—Missouri and Illinois, 2014. *MMWR. Morb. Mortal. Wkly. Rep.* **2014**, *63*, 798–799.

11. Bragstad, K.; Jakobsen, K.; Rojahn, A.E.; Skram, M.K.; Vainio, K.; Holberg-Petersen, M.; Hungnes, O.; Dudman, S.G.; Kran, A.-M.B. High frequency of enterovirus D68 in children hospitalised with respiratory illness in Norway, Autumn 2014. *Influenza Other Respi. Viruses* **2015**, *9*, 59–63.

12. Messacar, K.; Schreiner, T.L.; Maloney, J.A.; Wallace, A.; Ludke, J.; Oberste, M.S.; Nix, W.A.; Robinson, C.C.; Glodé, M.P.; Abzug, M.J.; *et al*. A cluster of acute flaccid paralysis and cranial nerve dysfunction temporally associated with an outbreak of enterovirus D68 in children in Colorado, USA. *Lancet* **2015**, *385*, 1662–1671.

13. Fendrick, A.M.; Monto, A.S.; Nightengale, B.; Sarnes, M. The economic burden of non-influenza-related viral respiratory tract infection in the United States. *Arch. Intern. Med.* **2003**, *163*, 487–494.

14. Dimopoulos, G.; Lerikou, M.; Tsiodras, S.; Chranioti, A.; Perros, E.; Anagnostopoulou, U.; Armaganidis, A.; Karakitsos, P. Viral epidemiology of acute exacerbations of chronic obstructive pulmonary disease. *Pulm. Pharmacol. Ther.* **2012**, *25*, 12–18.

15. Gern, J.E. The ABCs of rhinoviruses, wheezing, and asthma. *J. Virol.* **2010**, *84*, 7418–7426.

16. Kherad, O.; Kaiser, L.; Bridevaux, P.-O.; Sarasin, F.; Thomas, Y.; Janssens, J.-P.; Rutschmann, O.T. Upper-respiratory viral infection, biomarkers, and COPD exacerbations. *Chest* **2010**, *138*, 896–904.

17. Mallia, P.; Message, S.D.; Kebadze, T.; Parker, H.L.; Kon, O.M.; Johnston, S.L. An experimental model of rhinovirus induced chronic obstructive pulmonary disease exacerbations: A pilot study. *Respir. Res.* **2006**, *7*, 116.

18. McManus, T.E.; Marley, A.-M.; Baxter, N.; Christie, S.N.; O'Neill, H.J.; Elborn, J.S.; Coyle, P.V.; Kidney, J.C. Respiratory viral infection in exacerbations of COPD. *Respir. Med.* **2008**, *102*, 1575–1580.

19. Papadopoulos, N.G.; Christodoulou, I.; Rohde, G.; Agache, I.; Almqvist, C.; Bruno, A.; Bonini, S.; Bont, L.; Bossios, A.; Bousquet, J.; *et al*. Viruses and bacteria in acute asthma exacerbations—A GA[2] LEN-DARE systematic review. *Allergy* **2011**, *66*, 458–468.

20. Seemungal, T.A.; Harper-Owen, R.; Bhowmik, A.; Jeffries, D.J.; Wedzicha, J.A. Detection of rhinovirus in induced sputum at exacerbation of chronic obstructive pulmonary disease. *Eur. Respir. J.* **2000**, *16*, 677–683.

21. Burns, J.L.; Emerson, J.; Kuypers, J.; Campbell, A.P.; Gibson, R.L.; McNamara, S.; Worrell, K.; Englund, J.A. Respiratory viruses in children with cystic fibrosis: viral detection and clinical findings. *Influenza Other Respi. Viruses* **2012**, *6*, 218–223.

22. De Almeida, M.B.; Zerbinati, R.M.; Tateno, A.F.; Oliveira, C.M.; Romão, R.M.; Rodrigues, J.C.; Pannuti, C.S.; da Silva Filho, L.V.F. Rhinovirus C and respiratory exacerbations in children with cystic fibrosis. *Emerg. Infect. Dis.* **2010**, *16*, 996–999.

23. Kieninger, E.; Singer, F.; Tapparel, C.; Alves, M.P.; Latzin, P.; Tan, H.-L.; Bossley, C.; Casaulta, C.; Bush, A.; Davies, J.C.; *et al*. High rhinovirus burden in lower airways of children with cystic fibrosis. *Chest* **2013**, *143*, 782–790.

24. Wat, D.; Gelder, C.; Hibbitts, S.; Cafferty, F.; Bowler, I.; Pierrepoint, M.; Evans, R.; Doull, I. The role of respiratory viruses in cystic fibrosis. *J. Cyst. Fibros.* **2008**, *7*, 320–328.

25. Hayden, F.G.; Turner, R.B. Rhinovirus genetics and virulence: looking for needles in a haystack. *Am. J. Respir. Crit. Care Med.* **2012**, *186*, 818–820.

26. Joki-Korpela, P.; Hyypiä, T. Parechoviruses, a novel group of human picornaviruses. *Ann. Med.* **2001**, *33*, 466–471.

27. Stanway, G.; Joki-Korpela, P.; Hyypiä, T. Human parechoviruses—Biology and clinical significance. *Rev. Med. Virol.* **2000**, *10*, 57–69.

28. Stanway, G.; Brown, F.; Christian, P.; Hovi, T.; Hyypiä, T.; King, A.M.Q.; Knowles, N.J.; Lemon, S.M.; Minor, P.D.; Pallansch, M.A.; *et al.* Family Picornaviridae. In *Virus Taxonomy. Eighth Report of the International Committee on Taxonomy of Viruses*; Fauquet, C.M., Mayo, M.A., Maniloff, J., Desselberger, U., Ball, L.A., Eds.; Elsevier/Academic Press: London, UK, 2005; pp. 757–778.

29. Benschop, K.; Thomas, X.; Serpenti, C.; Molenkamp, R.; Wolthers, K. High prevalence of human Parechovirus (HPeV) genotypes in the Amsterdam region and identification of specific HPeV variants by direct genotyping of stool samples. *J. Clin. Microbiol.* **2008**, *46*, 3965–3970.

30. Khatami, A.; McMullan, B.J.; Webber, M.; Stewart, P.; Francis, S.; Timmers, K.J.; Rodas, E.; Druce, J.; Mehta, B.; Sloggett, N.A.; *et al.* Sepsis-like disease in infants due to human parechovirus type 3 during an outbreak in Australia. *Clin. Infect. Dis.* **2015**, *60*, 228–236.

31. Wildenbeest, J.G.; Wolthers, K.C.; Straver, B.; Pajkrt, D. Successful IVIG treatment of human parechovirus-associated dilated cardiomyopathy in an infant. *Pediatrics* **2013**, *132*, e243–e247.

32. Benschop, K.S.; Schinkel, J.; Minnaar, R.P.; Pajkrt, D.; Spanjerberg, L.; Kraakman, H.C.; Berkhout, B.; Zaaijer, H.L.; Beld, M.G.H.M.; Wolthers, K.C. Human parechovirus infections in Dutch children and the association between serotype and disease severity. *Clin. Infect. Dis.* **2006**, *42*, 204–210.

33. Harvala, H.; Robertson, I.; Chieochansin, T.; McWilliam Leitch, E.C.; Templeton, K.; Simmonds, P. Specific association of human parechovirus type 3 with sepsis and fever in young infants, as identified by direct typing of cerebrospinal fluid samples. *J. Infect. Dis.* **2009**, *199*, 1753–1760.

34. Wolthers, K.C.; Benschop, K.S.M.; Schinkel, J.; Molenkamp, R.; Bergevoet, R.M.; Spijkerman, I.J.B.; Kraakman, H.C.; Pajkrt, D. Human parechoviruses as an important viral cause of sepsislike illness and meningitis in young children. *Clin. Infect. Dis.* **2008**, *47*, 358–363.

35. Boivin, G.; Abed, Y.; Boucher, F.D. Human parechovirus 3 and neonatal infections. *Emerg. Infect. Dis.* **2005**, *11*, 103–105.

36. Renaud, C.; Kuypers, J.; Ficken, E.; Cent, A.; Corey, L.; Englund, J.A. Introduction of a novel parechovirus RT-PCR clinical test in a regional medical center. *J. Clin. Virol.* **2011**, *51*, 50–53.

37. Sainato, R.; Flanagan, R.; Mahlen, S.; Fairchok, M.; Braun, L. Severe human parechovirus sepsis beyond the neonatal period. *J. Clin. Virol.* **2011**, *51*, 73–74.

38. Schuffenecker, I.; Javouhey, E.; Gillet, Y.; Kugener, B.; Billaud, G.; Floret, D.; Lina, B.; Morfin, F. Human parechovirus infections, Lyon, France, 2008-10: evidence for severe cases. *J. Clin. Virol.* **2012**, *54*, 337–341.

39. Selvarangan, R.; Nzabi, M.; Selvaraju, S.B.; Ketter, P.; Carpenter, C.; Harrison, C.J. Human parechovirus 3 causing sepsis-like illness in children from midwestern United States. *Pediatr. Infect. Dis. J.* **2011**, *30*, 238–242.

40. Verboon-Maciolek, M.A.; Groenendaal, F.; Hahn, C.D.; Hellmann, J.; van Loon, A.M.; Boivin, G.; de Vries, L.S. Human parechovirus causes encephalitis with white matter injury in neonates. *Ann. Neurol.* **2008**, *64*, 266–273.

41. Walters, B.; Peñaranda, S.; Nix, W.A.; Oberste, M.S.; Todd, K.M.; Katz, B.Z.; Zheng, X. Detection of human parechovirus (HPeV)-3 in spinal fluid specimens from pediatric patients in the Chicago area. *J. Clin. Virol.* **2011**, *52*, 187–191.

42. Xiang, Z.; Gonzalez, R.; Xie, Z.; Xiao, Y.; Liu, J.; Chen, L.; Liu, C.; Zhang, J.; Ren, L.; Vernet, G.; *et al.* Human rhinovirus C infections mirror those of human rhinovirus A in children with community-acquired pneumonia. *J. Clin. Virol.* **2010**, *49*, 94–99.

43. Yuzurihara, S.S.; Ao, K.; Hara, T.; Tanaka, F.; Mori, M.; Kikuchi, N.; Kai, S.; Yokota, S. Human parechovirus-3 infection in nine neonates and infants presenting symptoms of hemophagocytic lymphohistiocytosis. *J. Infect. Chemother.* **2013**, *19*, 144–148.

44. Porter, K.A.; Diop, O.M.; Burns, C.C.; Tangermann, R.H.; Wassilak, S.G.F. Tracking progress toward polio eradication—Worldwide, 2013–2014. *MMWR. Morb. Mortal. Wkly. Rep.* **2015**, *64*, 415–420.

45. Bejing Vigoo Biological Co., Ltd. A clinical trial to assess the efficacy and safety of an inactivated vaccine (vero cell) against EV71 in Chinese children aged 6–35 months. Available online: https://clinicaltrials.gov/ct2/show/NCT01508247 (accessed on 27 March 2015).

46. Longding Liu, Chinese Academy of Medical Sciences. A protected study of inactivated EV71 vaccine (human diploid cell, KMB-17) in Chinese infants and children. Available online: https://clinicaltrials.gov/ct2/show/NCT01569581 (accessed on 10 March 2015).

47. Sinovac Biotech Co., Ltd. An Efficacy Trial in inactivated enterovirus type 71 (EV71) vaccine. Available online: https://clinicaltrials.gov/ct2/show/NCT01507857 (accessed on 27 March 2015).

48. Abzug, M.J.; Keyserling, H.L.; Lee, M.L.; Levin, M.J.; Rotbart, H.A. Neonatal enterovirus infection: virology, serology, and effects of intravenous immune globulin. *Clin. Infect. Dis.* **1995**, *20*, 1201–1206.

49. Nagington, J. Echovirus 11 infection and prophylactic antiserum. *Lancet* **1982**, *1*, 446.

50. Wildenbeest, J.G.; van den Broek, P.J.; Benschop, K.S.M.; Koen, G.; Wierenga, P.C.; Vossen, A.C.; Kuijpers, T.W.; Wolthers, K.C. Pleconaril revisited: clinical course of chronic enteroviral meningoencephalitis after treatment correlates with in vitro susceptibility. *Antivir. Ther.* **2012**, *17*, 459–466.

51. Yen, M.-H.; Huang, Y.-C.; Chen, M.-C.; Liu, C.-C.; Chiu, N.-C.; Lien, R.; Chang, L.-Y.; Chiu, C.-H.; Tsao, K.-C.; Lin, T.-Y. Effect of intravenous immunoglobulin for neonates with severe enteroviral infections with emphasis on the timing of administration. *J. Clin. Virol.* **2015**, *64*, 92–96.

52. Enserink, M. Polio endgame. Wanted: Drug for a disappearing disease. *Science* **2004**, *303*.

53. Tuthill, T.J.; Groppelli, E.; Hogle, J.M.; Rowlands, D.J. Picornaviruses. *Curr. Top. Microbiol. Immunol.* **2010**, *343*, 43–89.

54. Kräusslich, H.G.; Nicklin, M.J.; Toyoda, H.; Etchison, D.; Wimmer, E. Poliovirus proteinase 2A induces cleavage of eucaryotic initiation factor 4F polypeptide p220. *J. Virol.* **1987**, *61*, 2711–2718.

55. Lloyd, R.E.; Grubman, M.J.; Ehrenfeld, E. Relationship of p220 cleavage during picornavirus infection to 2A proteinase sequencing. *J. Virol.* **1988**, *62*, 4216–4223.

56. Joachims, M.; Van Breugel, P.C.; Lloyd, R.E. Cleavage of poly(A)-binding protein by enterovirus proteases concurrent with inhibition of translation in vitro. *J. Virol.* **1999**, *73*, 718–727.

57. Badorff, C.; Lee, G.H.; Lamphear, B.J.; Martone, M.E.; Campbell, K.P.; Rhoads, R.E.; Knowlton, K.U. Enteroviral protease 2A cleaves dystrophin: Evidence of cytoskeletal disruption in an acquired cardiomyopathy. *Nat. Med.* **1999**, *5*, 320–326.

58. Barral, P.M.; Morrison, J.M.; Drahos, J.; Gupta, P.; Sarkar, D.; Fisher, P.B.; Racaniello, V.R. MDA-5 is cleaved in poliovirus-infected cells. *J. Virol.* **2007**, *81*, 3677–3684.

59. Castelló, A.; Alvarez, E.; Carrasco, L. The multifaceted poliovirus 2A protease: Regulation of gene expression by picornavirus proteases. *J. Biomed. Biotechnol.* **2011**, *2011*.

60. Lei, X.; Xiao, X.; Xue, Q.; Jin, Q.; He, B.; Wang, J. Cleavage of interferon regulatory factor 7 by enterovirus 71 3C suppresses cellular responses. *J. Virol.* **2013**, *87*, 1690–1698.

61. Mukherjee, A.; Morosky, S.A.; Delorme-Axford, E.; Dybdahl-Sissoko, N.; Oberste, M.S.; Wang, T.; Coyne, C.B. The coxsackievirus B 3C protease cleaves MAVS and TRIF to attenuate host type I interferon and apoptotic signaling. *PLoS Pathog.* **2011**, *7*, e1001311.

62. Wang, B.; Xi, X.; Lei, X.; Zhang, X.; Cui, S.; Wang, J.; Jin, Q.; Zhao, Z. Enterovirus 71 protease 2Apro targets MAVS to inhibit anti-viral type I interferon responses. *PLoS Pathog.* **2013**, *9*, e1003231.

63. Gerber, K.; Wimmer, E.; Paul, A.V. Biochemical and genetic studies of the initiation of human rhinovirus 2 RNA replication: Identification of a *cis*-replicating element in the coding sequence of 2A(pro). *J. Virol.* **2001**, *75*, 10979–10990.

64. Paul, A.V.; Rieder, E.; Kim, D.W.; van Boom, J.H.; Wimmer, E. Identification of an RNA hairpin in poliovirus RNA that serves as the primary template in the in vitro uridylylation of VPg. *J. Virol.* **2000**, *74*, 10359–10370.

65. Rieder, E.; Paul, A.V.; Kim, D.W.; van Boom, J.H.; Wimmer, E. Genetic and biochemical studies of poliovirus *cis*-acting replication element *cre* in relation to VPg uridylylation. *J. Virol.* **2000**, *74*, 10371–10380.

66. Yang, Y.; Rijnbrand, R.; McKnight, K.L.; Wimmer, E.; Paul, A.; Martin, A.; Lemon, S.M. Sequence requirements for viral RNA replication and VPg uridylylation directed by the internal *cis*-acting replication element (*cre*) of human rhinovirus type 14. *J. Virol.* **2002**, *76*, 7485–7494.

67. Paul, D.; Bartenschlager, R. Architecture and biogenesis of plus-strand RNA virus replication factories. *World J. Virol.* **2013**, *2*, 32–48.

68. Belov, G.A.; van Kuppeveld, F.J. (+)RNA viruses rewire cellular pathways to build replication organelles. *Curr. Opin. Virol.* **2012**, *2*, 740–747.

69. Belov, G.A.; Nair, V.; Hansen, B.T.; Hoyt, F.H.; Fischer, E.R.; Ehrenfeld, E. Complex dynamic development of poliovirus membranous replication complexes. *J. Virol.* **2012**, *86*, 302–312.

70. Limpens, R.W.; van der Schaar, H.M.; Kumar, D.; Koster, A.J.; Snijder, E.J.; van Kuppeveld, F.J.; Bárcena, M. The transformation of enterovirus replication structures: A three-dimensional study of single- and double-membrane compartments. *MBio* **2011**.

71. Belov, G.A.; Feng, Q.; Nikovics, K.; Jackson, C.L.; Ehrenfeld, E. A critical role of a cellular membrane traffic protein in poliovirus RNA replication. *PLoS Pathog.* **2008**, *4*, e1000216.

72. Gazina, E.V.; Mackenzie, J.M.; Gorrell, R.J.; Anderson, D.A. Differential requirements for COPI coats in formation of replication complexes among three genera of Picornaviridae. *J. Virol.* **2002**, *76*, 11113–11122.

73. Irurzun, A.; Perez, L.; Carrasco, L. Involvement of membrane traffic in the replication of poliovirus genomes: effects of brefeldin A. *Virology* **1992**, *191*, 166–175.

74. Lanke, K.H.; van der Schaar, H.M.; Belov, G.A.; Feng, Q.; Duijsings, D.; Jackson, C.L.; Ehrenfeld, E.; van Kuppeveld, F.J. GBF1, a guanine nucleotide exchange factor for Arf, is crucial for coxsackievirus B3 RNA replication. *J. Virol.* **2009**, *83*, 11940–11949.

75. Wessels, E.; Duijsings, D.; Niu, T.-K.; Neumann, S.; Oorschot, V.M.; de Lange, F.; Lanke, K.H.; Klumperman, J.; Henke, A.; Jackson, C.L.; *et al.* A viral protein that blocks Arf1-mediated COP-I assembly by inhibiting the guanine nucleotide exchange factor GBF1. *Dev. Cell* **2006**, *11*, 191–201.

76. Wessels, E.; Duijsings, D.; Lanke, K.H.; van Dooren, S.H.J.; Jackson, C.L.; Melchers, W.J.; van Kuppeveld, F.J. Effects of picornavirus 3A Proteins on Protein Transport and GBF1-dependent COP-I recruitment. *J. Virol.* **2006**, *80*, 11852–11860.

77. Hsu, N.-Y.; Ilnytska, O.; Belov, G.; Santiana, M.; Chen, Y.-H.; Takvorian, P.M.; Pau, C.; van der Schaar, H.; Kaushik-Basu, N.; Balla, T.; *et al.* Viral reorganization of the secretory pathway generates distinct organelles for RNA replication. *Cell* **2010**, *141*, 799–811.

78. Doedens, J.R.; Kirkegaard, K. Inhibition of cellular protein secretion by poliovirus proteins 2B and 3A. *EMBO J.* **1995**, *14*, 894–907.

79. Wessels, E.; Duijsings, D.; Notebaart, R.A.; Melchers, W.J.; van Kuppeveld, F.J. A proline-rich region in the coxsackievirus 3A protein is required for the protein to inhibit endoplasmic reticulum-to-golgi transport. *J. Virol.* **2005**, *79*, 5163–5173.

80. Deitz, S.B.; Dodd, D.A.; Cooper, S.; Parham, P.; Kirkegaard, K. MHC I-dependent antigen presentation is inhibited by poliovirus protein 3A. *Proc. Natl. Acad. Sci. USA* **2000**, *97*, 13790–13795.

81. Dodd, D.A.; Giddings, T.H.; Kirkegaard, K. Poliovirus 3A protein limits interleukin-6 (IL-6), IL-8, and beta interferon secretion during viral infection. *J. Virol.* **2001**, *75*, 8158–8165.

82. D'Angelo, G.; Vicinanza, M.; Di Campli, A.; De Matteis, M.A. The multiple roles of PtdIns(4)*P*—Not just the precursor of PtdIns(4,5)P_2. *J. Cell Sci.* **2008**, *121*, 1955–1963.

83. Graham, T.R.; Burd, C.G. Coordination of Golgi functions by phosphatidylinositol 4-kinases. *Trends Cell Biol.* **2011**, *21*, 113–121.

84. Santiago-Tirado, F.H.; Bretscher, A. Membrane-trafficking sorting hubs: cooperation between PI4P and small GTPases at the trans-Golgi network. *Trends Cell Biol.* **2011**, *21*, 515–525.

85. Dorobantu, C.M.; van der Schaar, H.M.; Ford, L.A.; Strating, J.R.; Ulferts, R.; Fang, Y.; Belov, G.; van Kuppeveld, F.J. Recruitment of PI4KIIIβ to coxsackievirus B3 replication organelles is independent of ACBD3, GBF1, and Arf1. *J. Virol.* **2014**, *88*, 2725–2736.

86. Téoulé, F.; Brisac, C.; Pelletier, I.; Vidalain, P.-O.; Jégouic, S.; Mirabelli, C.; Bessaud, M.; Combelas, N.; Autret, A.; Tangy, F.; *et al.* The Golgi protein ACBD3, an interactor for poliovirus protein 3A, modulates poliovirus replication. *J. Virol.* **2013**, *87*, 11031–11046.

87. Mesmin, B.; Bigay, J.; Moser von Filseck, J.; Lacas-Gervais, S.; Drin, G.; Antonny, B. A four-step cycle driven by PI(4)*P* hydrolysis directs sterol/PI(4)*P* exchange by the ER-Golgi tether OSBP. *Cell* **2013**, *155*, 830–843.

88. Strating, J.R.; van der Linden, L.; Albulescu, L.; Bigay, J.; Arita, M.; Delang, L.; Leyssen, P.; van der Schaar, H.M.; Lanke, K.H.; Thibaut, H.J.; *et al.* Itraconazole Inhibits Enterovirus Replication by Targeting the Oxysterol-Binding Protein. *Cell Rep.* **2015**, *10*, 600–615.

89. Arita, M. Phosphatidylinositol-4 kinase III beta and oxysterol-binding protein accumulate unesterified cholesterol on poliovirus-induced membrane structure. *Microbiol. Immunol.* **2014**, *58*, 239–256.

90. Roulin, P.S.; Lötzerich, M.; Torta, F.; Tanner, L.B.; van Kuppeveld, F.J.; Wenk, M.R.; Greber, U.F. Rhinovirus uses a phosphatidylinositol 4-phosphate/cholesterol counter-current for the formation of replication compartments at the ER-Golgi interface. *Cell Host Microbe* **2014**, *16*, 677–690.

91. Albulescu, L.; Wubbolts, R.; van Kuppeveld, F.J.; Strating, J.R. Cholesterol shuttling is important for RNA replication of coxsackievirus B3 and encephalomyocarditis virus. *Cell. Microbiol.* **2015**, *17*, 1144–1156.

92. Ilnytska, O.; Santiana, M.; Hsu, N.-Y.; Du, W.-L.; Chen, Y.-H.; Viktorova, E.G.; Belov, G.; Brinker, A.; Storch, J.; Moore, C.; *et al.* Enteroviruses harness the cellular endocytic machinery to remodel the host cell cholesterol landscape for effective viral replication. *Cell Host Microbe* **2013**, *14*, 281–293.

93. Wong, J.; Zhang, J.; Si, X.; Gao, G.; Mao, I.; McManus, B.M.; Luo, H. Autophagosome supports coxsackievirus B3 replication in host cells. *J. Virol.* **2008**, *82*, 9143–9153.

94. Jackson, W.T.; Giddings, T.H.; Taylor, M.P.; Mulinyawe, S.; Rabinovitch, M.; Kopito, R.R.; Kirkegaard, K. Subversion of cellular autophagosomal machinery by RNA viruses. *PLoS Biol.* **2005**, *3*, e156.

95. Richards, A.L.; Jackson, W.T. Intracellular vesicle acidification promotes maturation of infectious poliovirus particles. *PLoS Pathog.* **2012**, *8*, e1003046.

96. Barco, A.; Carrasco, L. A human virus protein, poliovirus protein 2BC, induces membrane proliferation and blocks the exocytic pathway in the yeast Saccharomyces cerevisiae. *EMBO J.* **1995**, *14*, 3349–3364.

97. Cho, M.W.; Teterina, N.; Egger, D.; Bienz, K.; Ehrenfeld, E. Membrane rearrangement and vesicle induction by recombinant poliovirus 2C and 2BC in human cells. *Virology* **1994**, *202*, 129–145.

98. Suhy, D.A.; Giddings, T.H.; Kirkegaard, K. Remodeling the endoplasmic reticulum by poliovirus infection and by individual viral proteins: An autophagy-like origin for virus-induced vesicles. *J. Virol.* **2000**, *74*, 8953–8965.

99. Agirre, A.; Barco, A.; Carrasco, L.; Nieva, J.L. Viroporin-mediated membrane permeabilization. Pore formation by nonstructural poliovirus 2B protein. *J. Biol. Chem.* **2002**, *277*, 40434–40441.

100. De Jong, A.S.; Wessels, E.; Dijkman, H.B.; Galama, J.M.; Melchers, W.J.; Willems, P.H.; van Kuppeveld, F.J. Determinants for membrane association and permeabilization of the coxsackievirus 2B protein and the identification of the Golgi complex as the target organelle. *J. Biol. Chem.* **2003**, *278*, 1012–1021.

101. De Jong, A.S.; Visch, H.-J.; de Mattia, F.; van Dommelen, M.M.; Swarts, H.G.; Luyten, T.; Callewaert, G.; Melchers, W.J.; Willems, P.H.; van Kuppeveld, F.J. The Coxsackievirus 2B Protein Increases Efflux of Ions from the Endoplasmic Reticulum and Golgi, thereby Inhibiting Protein Trafficking through the Golgi. *J. Biol. Chem.* **2006**, *281*, 14144–14150.

102. Van Kuppeveld, F.J.; Hoenderop, J.G.; Smeets, R.L.; Willems, P.H.; Dijkman, H.B.; Galama, J.M.; Melchers, W.J. Coxsackievirus protein 2B modifies endoplasmic reticulum membrane and plasma membrane permeability and facilitates virus release. *EMBO J.* **1997**, *16*, 3519–3532.

103. Aldabe, R.; Carrasco, L. Induction of membrane proliferation by poliovirus proteins 2C and 2BC. *Biochem. Biophys. Res. Commun.* **1995**, *206*, 64–76.

104. Van Kuppeveld, F.J.; Melchers, W.J.; Kirkegaard, K.; Doedens, J.R. Structure-function analysis of coxsackie B3 virus protein 2B. *Virology* **1997**, *227*, 111–118.

105. Cornell, C.T.; Kiosses, W.B.; Harkins, S.; Whitton, J.L. Coxsackievirus B3 proteins directionally complement each other to downregulate surface major histocompatibility complex class I. *J. Virol.* **2007**, *81*, 6785–6797.

106. Tang, W.-F.; Yang, S.-Y.; Wu, B.-W.; Jheng, J.-R.; Chen, Y.-L.; Shih, C.-H.; Lin, K.-H.; Lai, H.-C.; Tang, P.; Horng, J.-T. Reticulon 3 binds the 2C protein of enterovirus 71 and is required for viral replication. *J. Biol. Chem.* **2007**, *282*, 5888–5898.

107. Molla, A.; Paul, A.V.; Wimmer, E. Cell-free, de novo synthesis of poliovirus. *Science* **1991**, *254*, 1647–1651.

108. Nugent, C.I.; Johnson, K.L.; Sarnow, P.; Kirkegaard, K. Functional coupling between replication and packaging of poliovirus replicon RNA. *J. Virol.* **1999**, *73*, 427–435.

109. Liu, Y.; Wang, C.; Mueller, S.; Paul, A.V.; Wimmer, E.; Jiang, P. Direct interaction between two viral proteins, the nonstructural protein 2C and the capsid protein VP3, is required for enterovirus morphogenesis. *PLoS Pathog.* **2010**, *6*, e1001066.

110. Geller, R.; Vignuzzi, M.; Andino, R.; Frydman, J. Evolutionary constraints on chaperone-mediated folding provide an antiviral approach refractory to development of drug resistance. *Genes Dev.* **2007**, *21*, 195–205.

111. Ypma-Wong, M.F.; Dewalt, P.G.; Johnson, V.H.; Lamb, J.G.; Semler, B.L. Protein 3CD is the major poliovirus proteinase responsible for cleavage of the P1 capsid precursor. *Virology* **1988**, *166*, 265–270.

112. Mikami, T.; Satoh, N.; Hatayama, I.; Nakane, A. Buthionine sulfoximine inhibits cytopathic effect and apoptosis induced by infection with human echovirus 9. *Arch. Virol.* **2004**, *149*, 1117–1128.

113. Smith, A.D.; Dawson, H. Glutathione is required for efficient production of infectious picornavirus virions. *Virology* **2006**, *353*, 258–267.

114. Thibaut, H.J.; van der Linden, L.; Jiang, P.; Thys, B.; Canela, M.-D.; Aguado, L.; Rombaut, B.; Wimmer, E.; Paul, A.; Pérez-Pérez, M.-J.; *et al.* Binding of glutathione to enterovirus capsids is essential for virion morphogenesis. *PLoS Pathog.* **2014**, *10*, e1004039.

115. Ma, H.-C.; Liu, Y.; Wang, C.; Strauss, M.; Rehage, N.; Chen, Y.-H.; Altan-Bonnet, N.; Hogle, J.; Wimmer, E.; Mueller, S.; *et al.* An interaction between glutathione and the capsid is required for the morphogenesis of C-cluster enteroviruses. *PLoS Pathog.* **2014**, *10*, e1004052.

116. Chen, Y.-H.; Du, W.; Hagemeijer, M.C.; Takvorian, P.M.; Pau, C.; Cali, A.; Brantner, C.A.; Stempinski, E.S.; Connelly, P.S.; Ma, H.-C.; *et al.* Phosphatidylserine Vesicles Enable Efficient En Bloc Transmission of Enteroviruses. *Cell* **2015**, *160*, 619–630.

117. Robinson, S.M.; Tsueng, G.; Sin, J.; Mangale, V.; Rahawi, S.; McIntyre, L.L.; Williams, W.; Kha, N.; Cruz, C.; Hancock, B.M.; *et al.* Coxsackievirus B exits the host cell in shed microvesicles displaying autophagosomal markers. *PLoS Pathog.* **2014**, *10*, e1004045.

118. Feng, Z.; Hensley, L.; McKnight, K.L.; Hu, F.; Madden, V.; Ping, L.; Jeong, S.-H.; Walker, C.; Lanford, R.E.; Lemon, S.M. A pathogenic picornavirus acquires an envelope by hijacking cellular membranes. *Nature* **2013**, *496*, 367–371.

119. Abzug, M.J.; Cloud, G.; Bradley, J.; Sánchez, P.J.; Romero, J.; Powell, D.; Lepow, M.; Mani, C.; Capparelli, E.V.; Blount, S.; *et al.* Double blind placebo-controlled trial of pleconaril in infants with enterovirus meningitis. *Pediatr. Infect. Dis. J.* **2003**, *22*, 335–341.

120. Desmond, R.A.; Accortt, N.A.; Talley, L.; Villano, S.A.; Soong, S.-J.; Whitley, R.J. Enteroviral meningitis: Natural history and outcome of pleconaril therapy. *Antimicrob. Agents Chemother.* **2006**, *50*, 2409–2414.

121. Hayden, F.G.; Coats, T.; Kim, K.; Hassman, H.A.; Blatter, M.M.; Zhang, B.; Liu, S. Oral pleconaril treatment of picornavirus-associated viral respiratory illness in adults: Efficacy and tolerability in phase II clinical trials. *Antivir. Ther.* **2002**, *7*, 53–65.

122. Hayden, F.G.; Herrington, D.T.; Coats, T.L.; Kim, K.; Cooper, E.C.; Villano, S.A.; Liu, S.; Hudson, S.; Pevear, D.C.; Collett, M.; *et al.* Efficacy and safety of oral pleconaril for treatment of colds due to picornaviruses in adults: Results of 2 double-blind, randomized, placebo-controlled trials. *Clin. Infect. Dis.* **2003**, *36*, 1523–1532.

123. Rotbart, H.A.; Webster, A.D. Treatment of potentially life-threatening enterovirus infections with pleconaril. *Clin. Infect. Dis.* **2001**, *32*, 228–235.

124. Schiff, G.M.; Sherwood, J.R. Clinical activity of pleconaril in an experimentally induced coxsackievirus A21 respiratory infection. *J. Infect. Dis.* **2000**, *181*, 20–26.

125. Senior, K. FDA panel rejects common cold treatment. *Lancet. Infect. Dis.* **2002**, *2*, 264.

126. Merck Sharp & Dohme Corp. Effects of Pleconaril Nasal Spray on Common Cold Symptoms and Asthma Exacerbations Following Rhinovirus Exposure. Available online: https://clinicaltrials.gov/ct2/show/NCT00394914 (accessed on 23 June 2015).

127. Pleconaril Enteroviral Sepsis Syndrome. Available online: https://clinicaltrials.gov/ct2/show/NCT00031512 (accessed on 28 February 2013).

128. Biota Pharmaceuticals HRV Phase IIb Study Achieves Primary Endpoint. Available online: http://www.biota.com.au/uploaded/154/1021819_20hrvphaseiibstudyachieve.pdf (accessed on 28 March 2015).

129. Biota Pharmaceuticals Biota Commences Dosing in Vapendavir SPIRITUS Phase 2b Trial. Available online: http://investors.biotapharma.com/releasedetail.cfm?releaseid=899451 (accessed on 28 March 2015).

130. Oberste, M.S.; Moore, D.; Anderson, B.; Pallansch, M.A.; Pevear, D.C.; Collett, M.S. *In vitro* antiviral activity of V-073 against polioviruses. *Antimicrob. Agents Chemother.* **2009**, *53*, 4501–4503.

131. Buontempo, P.J.; Cox, S.; Wright-Minogue, J.; DeMartino, J.L.; Skelton, A.M.; Ferrari, E.; Albin, R.; Rozhon, E.J.; Girijavallabhan, V.; Modlin, J.F.; *et al.* SCH 48973: A potent, broad-spectrum, antienterovirus compound. *Antimicrob. Agents Chemother.* **1997**, *41*, 1220–1225.

132. Torres-Torres, S.; Myers, A.L.; Klatte, J.M.; Rhoden, E.E.; Oberste, M.S.; Collett, M.S.; McCulloh, R.J. First use of investigational antiviral drug pocapavir (v-073) for treating neonatal enteroviral sepsis. *Pediatr. Infect. Dis. J.* **2015**, *34*, 52–54.

133. De Palma, A.M.; Vliegen, I.; De Clercq, E.; Neyts, J. Selective inhibitors of picornavirus replication. *Med. Res. Rev.* **2008**, *28*, 823–884.

134. Pevear, D.C.; Hayden, F.G.; Demenczuk, T.M.; Barone, L.R.; McKinlay, M.A.; Collett, M.S. Relationship of pleconaril susceptibility and clinical outcomes in treatment of common colds caused by rhinoviruses. *Antimicrob. Agents Chemother.* **2005**, *49*, 4492–4499.

135. Kaiser, L.; Crump, C.E.; Hayden, F.G. In vitro activity of pleconaril and AG7088 against selected serotypes and clinical isolates of human rhinoviruses. *Antivir. Res.* **2000**, *47*, 215–220.

136. Ledford, R.M.; Patel, N.R.; Demenczuk, T.M.; Watanyar, A.; Herbertz, T.; Collett, M.S.; Pevear, D.C. VP1 sequencing of all human rhinovirus serotypes: Insights into genus phylogeny and susceptibility to antiviral capsid-binding compounds. *J. Virol.* **2004**, *78*, 3663–3674.

137. Pevear, D.C.; Tull, T.M.; Seipel, M.E.; Groarke, J.M. Activity of pleconaril against enteroviruses. *Antimicrob. Agents Chemother.* **1999**, *43*, 2109–2115.

138. Benschop, K.S.; Wildenbeest, J.G.; Koen, G.; Minnaar, R.P.; van Hemert, F.J.; Westerhuis, B.M.; Pajkrt, D.; van den Broek, P.J.; Vossen, A.C.; Wolthers, K.C. Genetic and antigenic structural characterization for resistance of echovirus 11 to pleconaril in an immunocompromised patient. *J. Gen. Virol.* **2015**, *96*, 571–579.

139. Dragovich, P.S.; Prins, T.J.; Zhou, R.; Webber, S.E.; Marakovits, J.T.; Fuhrman, S.A.; Patick, A.K.; Matthews, D.A.; Lee, C.A.; Ford, C.E.; *et al.* Structure-based design, synthesis, and biological evaluation of irreversible human rhinovirus 3C protease inhibitors. 4. Incorporation of P1 lactam moieties as L-glutamine replacements. *J. Med. Chem.* **1999**, *42*, 1213–1224.

140. De Palma, A.M.; Pürstinger, G.; Wimmer, E.; Patick, A.K.; Andries, K.; Rombaut, B.; De Clercq, E.; Neyts, J. Potential use of antiviral agents in polio eradication. *Emerg. Infect. Dis.* **2008**, *14*, 545–551.

141. Lee, J.-C.; Shih, S.-R.; Chang, T.-Y.; Tseng, H.-Y.; Shih, Y.-F.; Yen, K.-J.; Chen, W.-C.; Shie, J.-J.; Fang, J.-M.; Liang, P.-H.; *et al.* A mammalian cell-based reverse two-hybrid system for functional analysis of 3C viral protease of human enterovirus 71. *Anal. Biochem.* **2008**, *375*, 115–123.

142. Patick, A.K.; Binford, S.L.; Brothers, M.A.; Jackson, R.L.; Ford, C.E.; Diem, M.D.; Maldonado, F.; Dragovich, P.S.; Zhou, R.; Prins, T.J.; *et al.* In vitro antiviral activity of AG7088, a potent inhibitor of human rhinovirus 3C protease. *Antimicrob. Agents Chemother.* **1999**, *43*, 2444–2450.

143. Tsai, M.-T.; Cheng, Y.-H.; Liu, Y.-N.; Liao, N.-C.; Lu, W.-W.; Kung, S.-H. Real-time monitoring of human enterovirus (HEV)-infected cells and anti-HEV 3C protease potency by fluorescence resonance energy transfer. *Antimicrob. Agents Chemother.* **2009**, *53*, 748–755.

144. Patick, A.K.; Brothers, M.A.; Maldonado, F.; Binford, S.; Maldonado, O.; Fuhrman, S.; Petersen, A.; Smith, G.J.; Zalman, L.S.; Burns-Naas, L.A.; *et al.* In vitro antiviral activity and single-dose pharmacokinetics in humans of a novel, orally bioavailable inhibitor of human rhinovirus 3C protease. *Antimicrob. Agents Chemother.* **2005**, *49*, 2267–2275.

145. Dragovich, P.S.; Prins, T.J.; Zhou, R.; Johnson, T.O.; Hua, Y.; Luu, H.T.; Sakata, S.K.; Brown, E.L.; Maldonado, F.C.; Tuntland, T.; *et al.* Structure-based design, synthesis, and biological evaluation of irreversible human rhinovirus 3C protease inhibitors. 8. Pharmacological optimization of orally bioavailable 2-pyridone-containing peptidomimetics. *J. Med. Chem.* **2003**, *46*, 4572–4585.

146. Crotty, S.; Maag, D.; Arnold, J.J.; Zhong, W.; Lau, J.Y.; Hong, Z.; Andino, R.; Cameron, C.E. The broad-spectrum antiviral ribonucleoside ribavirin is an RNA virus mutagen. *Nat. Med.* **2000**, *6*, 1375–1379.

147. Crotty, S.; Cameron, C.E.; Andino, R. RNA virus error catastrophe: direct molecular test by using ribavirin. *Proc. Natl. Acad. Sci. USA* **2001**, *98*, 6895–6900.

148. Gazina, E.V.; Smidansky, E.D.; Holien, J.K.; Harrison, D.N.; Cromer, B.A.; Arnold, J.J.; Parker, W.W.; Cameron, C.E.; Petrou, S. Amiloride is a competitive inhibitor of coxsackievirus B3 RNA polymerase. *J. Virol.* **2011**, *85*, 10364–10374.

149. Van der Linden, L.; Vives-Adrián, L.; Selisko, B.; Ferrer-Orta, C.; Liu, X.; Lanke, K.; Ulferts, R.; De Palma, A.M.; Tanchis, F.; Goris, N.; *et al.* The RNA template channel of the RNA-dependent RNA polymerase as a target for development of antiviral therapy of multiple genera within a virus family. *PLoS Pathog.* **2015**, *11*, e1004733.

150. Hung, H.-C.; Chen, T.-C.; Fang, M.-Y.; Yen, K.-J.; Shih, S.-R.; Hsu, J.T.-A.; Tseng, C.-P. Inhibition of enterovirus 71 replication and the viral 3D polymerase by aurintricarboxylic acid. *J. Antimicrob. Chemother.* **2010**, *65*, 676–683.

151. Miller, P.A.; Milstrey, K.P.; Trown, P.W. Specific inhibition of viral ribonucleic acid replication by Gliotoxin. *Science* **1968**, *159*, 431–432.

152. Rodriguez, P.L.; Carrasco, L. Gliotoxin: inhibitor of poliovirus RNA synthesis that blocks the viral RNA polymerase 3Dpol. *J. Virol.* **1992**, *66*, 1971–1976.

153. Velu, A.B.; Chen, G.-W.; Hsieh, P.-T.; Horng, J.-T.; Hsu, J.T.-A.; Hsieh, H.-P.; Chen, T.-C.; Weng, K.-F.; Shih, S.-R. BPR-3P0128 inhibits RNA-dependent RNA polymerase elongation and VPg uridylylation activities of Enterovirus 71. *Antivir. Res.* **2014**, *112*, 18–25.

154. Chen, T.-C.; Chang, H.-Y.; Lin, P.-F.; Chern, J.-H.; Hsu, J.T.-A.; Chang, C.-Y.; Shih, S.-R. Novel antiviral agent DTriP-22 targets RNA-dependent RNA polymerase of enterovirus 71. *Antimicrob. Agents Chemother.* **2009**, *53*, 2740–2747.

155. Barton, D.J.; Flanegan, J.B. Synchronous replication of poliovirus RNA: initiation of negative-strand RNA synthesis requires the guanidine-inhibited activity of protein 2C. *J. Virol.* **1997**, *71*, 8482–8489.

156. Li, J.P.; Baltimore, D. Isolation of poliovirus 2C mutants defective in viral RNA synthesis. *J. Virol.* **1988**, *62*, 4016–4021.

157. Pfister, T.; Jones, K.W.; Wimmer, E. A cysteine-rich motif in poliovirus protein 2C(ATPase) is involved in RNA replication and binds zinc in vitro. *J. Virol.* **2000**, *74*, 334–343.

158. Teterina, N.L.; Kean, K.M.; Gorbalenya, A.E.; Agol, V.I.; Girard, M. Analysis of the functional significance of amino acid residues in the putative NTP-binding pattern of the poliovirus 2C protein. *J. Gen. Virol.* **1992**, *73 (Pt 8)*, 1977–1986.

159. Teterina, N.L.; Levenson, E.; Rinaudo, M.S.; Egger, D.; Bienz, K.; Gorbalenya, A.E.; Ehrenfeld, E. Evidence for functional protein interactions required for poliovirus RNA replication. *J. Virol.* **2006**, *80*, 5327–5337.

160. Tolskaya, E.A.; Romanova, L.I.; Kolesnikova, M.S.; Gmyl, A.P.; Gorbalenya, A.E.; Agol, V.I. Genetic studies on the poliovirus 2C protein, an NTPase. A plausible mechanism of guanidine effect on the 2C function and evidence for the importance of 2C oligomerization. *J. Mol. Biol.* **1994**, *236*, 1310–1323.

161. Banerjee, R.; Echeverri, A.; Dasgupta, A. Poliovirus-encoded 2C polypeptide specifically binds to the 3'-terminal sequences of viral negative-strand RNA. *J. Virol.* **1997**, *71*, 9570–9578.

162. Banerjee, R.; Tsai, W.; Kim, W.; Dasgupta, A. Interaction of poliovirus-encoded 2C/2BC polypeptides with the 3' terminus negative-strand cloverleaf requires an intact stem-loop B. *Virology* **2001**, *280*, 41–51.

163. Banerjee, R.; Dasgupta, A. Interaction of picornavirus 2C polypeptide with the viral negative-strand RNA. *J. Gen. Virol.* **2001**, *82*, 2621–2627.

164. Teterina, N.L.; Gorbalenya, A.E.; Egger, D.; Bienz, K.; Ehrenfeld, E. Poliovirus 2C protein determinants of membrane binding and rearrangements in mammalian cells. *J. Virol.* **1997**, *71*, 8962–8972.

165. Vance, L.M.; Moscufo, N.; Chow, M.; Heinz, B.A. Poliovirus 2C region functions during encapsidation of viral RNA. *J. Virol.* **1997**, *71*, 8759–8765.

166. Verlinden, Y.; Cuconati, A.; Wimmer, E.; Rombaut, B. The antiviral compound 5-(3,4-dichlorophenyl) methylhydantoin inhibits the post-synthetic cleavages and the assembly of poliovirus in a cell-free system. *Antivir. Res.* **2000**, *48*, 61–69.

167. Li, J.P.; Baltimore, D. An intragenic revertant of a poliovirus 2C mutant has an uncoating defect. *J. Virol.* **1990**, *64*, 1102–1107.

168. Gorbalenya, A.E.; Koonin, E.V. Viral proteins containing the purine NTP-binding sequence pattern. *Nucleic Acids Res.* **1989**, *17*, 8413–8440.

169. Mirzayan, C.; Wimmer, E. Biochemical studies on poliovirus polypeptide 2C: Evidence for ATPase activity. *Virology* **1994**, *199*, 176–187.

170. Rodríguez, P.L.; Carrasco, L. Poliovirus protein 2C has ATPase and GTPase activities. *J. Biol. Chem.* **1993**, *268*, 8105–8110.

171. De Palma, A.M.; Heggermont, W.; Lanke, K.; Coutard, B.; Bergmann, M.; Monforte, A.-M.; Canard, B.; De Clercq, E.; Chimirri, A.; Pürstinger, G.; et al. The thiazolobenzimidazole TBZE-029 inhibits enterovirus replication by targeting a short region immediately downstream from motif C in the nonstructural protein 2C. *J. Virol.* **2008**, *82*, 4720–4730.

172. Hadaschik, D.; Klein, M.; Zimmermann, H.; Eggers, H.J.; Nelsen-Salz, B. Dependence of echovirus 9 on the enterovirus RNA replication inhibitor 2-(α-Hydroxybenzyl)-benzimidazole maps to nonstructural protein 2C. *J. Virol.* **1999**, *73*, 10536–10539.

173. Pincus, S.E.; Diamond, D.C.; Emini, E.A.; Wimmer, E. Guanidine-selected mutants of poliovirus: Mapping of point mutations to polypeptide 2C. *J. Virol.* **1986**, *57*, 638–646.

174. Sadeghipour, S.; Bek, E.J.; McMinn, P.C. Selection and characterisation of guanidine-resistant mutants of human enterovirus 71. *Virus Res.* **2012**, *169*, 72–79.

175. Shimizu, H.; Agoh, M.; Agoh, Y.; Yoshida, H.; Yoshii, K.; Yoneyama, T.; Hagiwara, A.; Miyamura, T. Mutations in the 2C region of poliovirus responsible for altered sensitivity to benzimidazole derivatives. *J. Virol.* **2000**, *74*, 4146–4154.

176. Ulferts, R.; van der Linden, L.; Thibaut, H.J.; Lanke, K.H.; Leyssen, P.; Coutard, B.; De Palma, A.M.; Canard, B.; Neyts, J.; van Kuppeveld, F.J. Selective serotonin reuptake inhibitor fluoxetine inhibits replication of human enteroviruses B and D by targeting viral protein 2C. *Antimicrob. Agents Chemother.* **2013**, *57*, 1952–1956.

177. Wikel, J.H.; Paget, C.J.; DeLong, D.C.; Nelson, J.D.; Wu, C.Y.; Paschal, J.W.; Dinner, A.; Templeton, R.J.; Chaney, M.O.; Jones, N.D.; *et al.* Synthesis of syn and anti isomers of 6-[[(hydroxyimino)phenyl]methyl]-1-[(1-methylethyl)sulfonyl]-1H-benzimidazol-2-amine. Inhibitors of rhinovirus multiplication. *J. Med. Chem.* **1980**, *23*, 368–372.

178. Van der Schaar, H.M.; van der Linden, L.; Lanke, K.H.; Strating, J.R.; Pürstinger, G.; de Vries, E.; de Haan, C.A.M.; Neyts, J.; van Kuppeveld, F.J. Coxsackievirus mutants that can bypass host factor PI4KIIIβ and the need for high levels of PI4P lipids for replication. *Cell Res.* **2012**, *22*, 1576–1592.

179. Brown-Augsburger, P.; Vance, L.M.; Malcolm, S.K.; Hsiung, H.; Smith, D.P.; Heinz, B.A. Evidence that enviroxime targets multiple components of the rhinovirus 14 replication complex. *Arch. Virol.* **1999**, *144*, 1569–1585.

180. Heinz, B.A.; Vance, L.M. The antiviral compound enviroxime targets the 3A coding region of rhinovirus and poliovirus. *J. Virol.* **1995**, *69*, 4189–4197.

181. Heinz, B.A.; Vance, L.M. Sequence determinants of 3A-mediated resistance to enviroxime in rhinoviruses and enteroviruses. *J. Virol.* **1996**, *70*, 4854–4857.

182. Arita, M.; Kojima, H.; Nagano, T.; Okabe, T.; Wakita, T.; Shimizu, H. Phosphatidylinositol 4-kinase III β is a target of enviroxime-like compounds for antipoliovirus activity. *J. Virol.* **2011**, *85*, 2364–2372.

183. Van der Schaar, H.M.; Leyssen, P.; Thibaut, H.J.; de Palma, A.; van der Linden, L.; Lanke, K.H.; Lacroix, C.; Verbeken, E.; Conrath, K.; Macleod, A.M.; *et al.* A novel, broad-spectrum inhibitor of enterovirus replication that targets host cell factor phosphatidylinositol 4-kinase IIIβ. *Antimicrob. Agents Chemother.* **2013**, *57*, 4971–4981.

184. Hayden, F.G.; Gwaltney, J.M. Prophylactic activity of intranasal enviroxime against experimentally induced rhinovirus type 39 infection. *Antimicrob. Agents Chemother.* **1982**, *21*, 892–897.

185. Higgins, P.G.; Barrow, G.I.; al-Nakib, W.; Tyrrell, D.A.; DeLong, D.C.; Lenox-Smith, I. Failure to demonstrate synergy between interferon-alpha and a synthetic antiviral, enviroxime, in rhinovirus infections in volunteers. *Antivir. Res.* **1988**, *10*, 141–149.

186. Levandowski, R.A.; Pachucki, C.T.; Rubenis, M.; Jackson, G.G. Topical enviroxime against rhinovirus infection. *Antimicrob. Agents Chemother.* **1982**, *22*, 1004–1007.

187. Miller, F.D.; Monto, A.S.; DeLong, D.C.; Exelby, A.; Bryan, E.R.; Srivastava, S. Controlled trial of enviroxime against natural rhinovirus infections in a community. *Antimicrob. Agents Chemother.* **1985**, *27*, 102–106.

188. Phillpotts, R.J.; Jones, R.W.; Delong, D.C.; Reed, S.E.; Wallace, J.; Tyrrell, D.A. The activity of enviroxime against rhinovirus infection in man. *Lancet* **1981**, *1*, 1342–1344.

189. Phillpotts, R.J.; Wallace, J.; Tyrrell, D.A.; Tagart, V.B. Therapeutic activity of enviroxime against rhinovirus infection in volunteers. *Antimicrob. Agents Chemother.* **1983**, *23*, 671–675.

190. Lamarche, M.J.; Borawski, J.; Bose, A.; Capacci-Daniel, C.; Colvin, R.; Dennehy, M.; Ding, J.; Dobler, M.; Drumm, J.; Gaither, L.A.; *et al.* Anti-hepatitis C virus activity and toxicity of type III phosphatidylinositol-4-kinase beta inhibitors. *Antimicrob. Agents Chemother.* **2012**, *56*, 5149–5156.

191. Arita, M.; Wakita, T.; Shimizu, H. Characterization of pharmacologically active compounds that inhibit poliovirus and enterovirus 71 infectivity. *J. Gen. Virol.* **2008**, *89*, 2518–2530.

192. Spickler, C.; Lippens, J.; Laberge, M.-K.; Desmeules, S.; Bellavance, É.; Garneau, M.; Guo, T.; Hucke, O.; Leyssen, P.; Neyts, J.; *et al.* Phosphatidylinositol 4-kinase III beta is essential for replication of human rhinovirus and its inhibition causes a lethal phenotype in vivo. *Antimicrob. Agents Chemother.* **2013**, *57*, 3358–3368.

193. Arita, M.; Kojima, H.; Nagano, T.; Okabe, T.; Wakita, T.; Shimizu, H. Oxysterol-binding protein family I is the target of minor enviroxime-like compounds. *J. Virol.* **2013**, *87*, 4252–4260.

194. Albulescu, L.; Strating, J.R.; Thibaut, H.J.; van der Linden, L.; Shair, M.D.; Neyts, J.; van Kuppeveld, F.J. Broad-range inhibition of enterovirus replication by OSW-1, a natural compound targeting OSBP. *Antivir. Res.* **2015**, *117*, 110–114.

195. Shim, J.S.; Liu, J.O. Recent advances in drug repositioning for the discovery of new anticancer drugs. *Int. J. Biol. Sci.* **2014**, *10*, 654–663.

196. Tsou, Y.-L.; Lin, Y.-W.; Chang, H.-W.; Lin, H.-Y.; Shao, H.-Y.; Yu, S.-L.; Liu, C.-C.; Chitra, E.; Sia, C.; Chow, Y.-H. Heat shock protein 90: Role in enterovirus 71 entry and assembly and potential target for therapy. *PLoS ONE* **2013**, *8*, e77133.

197. Bailey, H.H.; Mulcahy, R.T.; Tutsch, K.D.; Arzoomanian, R.Z.; Alberti, D.; Tombes, M.B.; Wilding, G.; Pomplun, M.; Spriggs, D.R. Phase I clinical trial of intravenous L-buthionine sulfoximine and melphalan: an attempt at modulation of glutathione. *J. Clin. Oncol.* **1994**, *12*, 194–205.

198. Bailey, H.H.; Ripple, G.; Tutsch, K.D.; Arzoomanian, R.Z.; Alberti, D.; Feierabend, C.; Mahvi, D.; Schink, J.; Pomplun, M.; Mulcahy, R.T.; *et al.* Phase I study of continuous-infusion L-S,R-buthionine sulfoximine with intravenous melphalan. *J. Natl. Cancer Inst.* **1997**, *89*, 1789–1796.

199. O'Dwyer, P.J.; Hamilton, T.C.; LaCreta, F.P.; Gallo, J.M.; Kilpatrick, D.; Halbherr, T.; Brennan, J.; Bookman, M.A.; Hoffman, J.; Young, R.C.; *et al.* Phase I trial of buthionine sulfoximine in combination with melphalan in patients with cancer. *J. Clin. Oncol.* **1996**, *14*, 249–256.

200. Van de Ven, A.A.; Douma, J.W.; Rademaker, C.; van Loon, A.M.; Wensing, A.M.; Boelens, J.-J.; Sanders, E.A.M.; van Montfrans, J.M. Pleconaril-resistant chronic parechovirus-associated enteropathy in agammaglobulinaemia. *Antivir. Ther.* **2011**, *16*, 611–614.

201. Wildenbeest, J.G.; Harvala, H.; Pajkrt, D.; Wolthers, K.C. The need for treatment against human parechoviruses: How, why and when? *Expert Rev. Anti. Infect. Ther.* **2010**, *8*, 1417–1429.

202. Seitsonen, J.; Susi, P.; Heikkilä, O.; Sinkovits, R.S.; Laurinmäki, P.; Hyypiä, T.; Butcher, S.J. Interaction of αVβ3 and αVβ6 integrins with human parechovirus 1. *J. Virol.* **2010**, *84*, 8509–8519.

203. Tan, J.; George, S.; Kusov, Y.; Perbandt, M.; Anemüller, S.; Mesters, J.R.; Norder, H.; Coutard, B.; Lacroix, C.; Leyssen, P.; *et al.* 3C protease of enterovirus 68: Structure-based design of Michael acceptor inhibitors and their broad-spectrum antiviral effects against picornaviruses. *J. Virol.* **2013**, *87*, 4339–4351.

204. Van der Linden, L.; Ulferts, R.; Nabuurs, S.B.; Kusov, Y.; Liu, H.; George, S.; Lacroix, C.; Goris, N.; Lefebvre, D.; Lanke, K.H.; *et al.* Application of a cell-based protease assay for testing inhibitors of picornavirus 3C proteases. *Antivir. Res.* **2014**, *103*, 17–24.

205. Matthews, D.A.; Dragovich, P.S.; Webber, S.E.; Fuhrman, S.A.; Patick, A.K.; Zalman, L.S.; Hendrickson, T.F.; Love, R.A.; Prins, T.J.; Marakovits, J.T.; *et al.* Structure-assisted design of mechanism-based irreversible inhibitors of human rhinovirus 3C protease with potent antiviral activity against multiple rhinovirus serotypes. *Proc. Natl. Acad. Sci. USA* **1999**, *96*, 11000–11007.

206. Blom, N.; Hansen, J.; Blaas, D.; Brunak, S. Cleavage site analysis in picornaviral polyproteins: Discovering cellular targets by neural networks. *Protein Sci.* **1996**, *5*, 2203–2216.

207. Williams, C.H.; Panayiotou, M.; Girling, G.D.; Peard, C.I.; Oikarinen, S.; Hyöty, H.; Stanway, G. Evolution and conservation in human parechovirus genomes. *J. Gen. Virol.* **2009**, *90*, 1702–1712.

208. Schultheiss, T.; Emerson, S.U.; Purcell, R.H.; Gauss-Müller, V. Polyprotein processing in echovirus 22: A first assessment. *Biochem. Biophys. Res. Commun.* **1995**, *217*, 1120–1127.

209. Coller, B.A.; Chapman, N.M.; Beck, M.A.; Pallansch, M.A.; Gauntt, C.J.; Tracy, S.M. Echovirus 22 is an atypical enterovirus. *J. Virol.* **1990**, *64*, 2692–2701.

210. Samuilova, O.; Krogerus, C.; Pöyry, T.; Hyypiä, T. Specific interaction between human parechovirus nonstructural 2A protein and viral RNA. *J. Biol. Chem.* **2004**, *279*, 37822–37831.

211. Hyypiä, T.; Horsnell, C.; Maaronen, M.; Khan, M.; Kalkkinen, N.; Auvinen, P.; Kinnunen, L.; Stanway, G. A distinct picornavirus group identified by sequence analysis. *Proc. Natl. Acad. Sci. USA* **1992**, *89*, 8847–8851.

212. Samuilova, O.; Krogerus, C.; Fabrichniy, I.; Hyypiä, T. ATP hydrolysis and AMP kinase activities of nonstructural protein 2C of human parechovirus 1. *J. Virol.* **2006**, *80*, 1053–1058.

213. Tamm, I.; Eggers, H.J. Differences in the selective virus inhibitory action of 2-(α-hydroxybenzyl)-benzimidazole and guanidine HCl. *Virology* **1962**, *18*, 439–447.

214. Eggers, H.J.; Tamm, I. Spectrum and characteristics of the virus inhibitory action of 2-(α-hydroxybenzyl)-benzimidazole. *J. Exp. Med.* **1961**, *113*, 657–682.

215. Krogerus, C.; Egger, D.; Samuilova, O.; Hyypiä, T.; Bienz, K. Replication complex of human parechovirus 1. *J. Virol.* **2003**, *77*, 8512–8523.

216. Snell, N.J. Ribavirin–current status of a broad spectrum antiviral agent. *Expert Opin. Pharmacother.* **2001**, *2*, 1317–1324.

217. Crotty, S.; Saleh, M.-C.; Gitlin, L.; Beske, O.; Andino, R. The poliovirus replication machinery can escape inhibition by an antiviral drug that targets a host cell protein. *J. Virol.* **2004**, *78*, 3378–3386.

218. Koletsky, A.J.; Harding, M.W.; Handschumacher, R.E. Cyclophilin: distribution and variant properties in normal and neoplastic tissues. *J. Immunol.* **1986**, *137*, 1054–1059.

219. Lin, K.; Gallay, P. Curing a viral infection by targeting the host: the example of cyclophilin inhibitors. *Antivir. Res.* **2013**, *99*, 68–77.

220. Chang, M.W.; Giffin, M.J.; Muller, R.; Savage, J.; Lin, Y.C.; Hong, S.; Jin, W.; Whitby, L.R.; Elder, J.H.; Boger, D.L.; *et al.* Identification of broad-based HIV-1 protease inhibitors from combinatorial libraries. *Biochem. J.* **2010**, *429*, 527–532.

221. Yang, H.; Nkeze, J.; Zhao, R.Y. Effects of HIV-1 protease on cellular functions and their potential applications in antiretroviral therapy. *Cell Biosci.* **2012**, *2*, 32.

222. Wang, J.; Su, H.; Zhang, T.; Du, J.; Cui, S.; Yang, F.; Jin, Q. Inhibition of Enterovirus 71 replication by 7-hydroxyflavone and diisopropyl-flavon7-yl Phosphate. *PLoS ONE* **2014**, *9*, e92565.

223. Wang, J.; Zhang, T.; Du, J.; Cui, S.; Yang, F.; Jin, Q. Anti-enterovirus 71 effects of chrysin and its phosphate ester. *PLoS ONE* **2014**, *9*, e89668.

224. Tyler, K.L. Rationale for the evaluation of fluoxetine in the treatment of enterovirus D68-associated acute flaccid myelitis. *JAMA Neurol.* **2015**, *72*, 493–494.

MDPI AG

Klybeckstrasse 64

4057 Basel, Switzerland

Tel. +41 61 683 77 34

Fax +41 61 302 89 18

http://www.mdpi.com/

Viruses Editorial Office

E-mail: viruses@mdpi.com

http://www.mdpi.com/journal/viruses